IMAGE ANALYSIS, CLASSIFICATION AND CHANGE DETECTION IN REMOTE SENSING

With Algorithms for ENVI/IDL and Python

THIRD EDITION

IMAGE ANALYSIS, CLASSIFICATION AND CHANGE DETECTION IN REMOTE SENSING

With Algorithms for ENVI/IDL and Python

THIRD EDITION

Morton J. Canty

CRC Press
Taylor & Francis Group
Boca Raton London New York

CRC Press is an imprint of the
Taylor & Francis Group, an **informa** business

CRC Press
Taylor & Francis Group
6000 Broken Sound Parkway NW, Suite 300
Boca Raton, FL 33487-2742

Printed on acid-free paper
Version Date: 20140206

International Standard Book Number-13: 978-1-4665-7037-5 (Hardback)

Library of Congress Cataloging-in-Publication Data

Canty, Morton John.
Image analysis, classification and change detection in remote sensing : with algorithms for for ENVI/IDL and Python Morton J. Canty. -- Third edition.
pages cm
Includes bibliographical references and index.
ISBN 978-1-4665-7037-5 (hardback)
1. Remote sensing--Mathematics. 2. Image analysis--Mathematics. I. Title.

G70.4.C36 2014
621.36'70285--dc23 2013045745

Visit the Taylor & Francis Web site at
http://www.taylorandfrancis.com

and the CRC Press Web site at
http://www.crcpress.com

Contents

Preface to the First Edition

This textbook had its beginnings as a set of notes to accompany seminars and lectures conducted at the Geographical Institute of Bonn University and at its associated Center for Remote Sensing of Land Cover. Lecture notes typically continue to be refined and polished over the years until the question inevitably poses itself: "Why not have them published?" The answer of course is "By all means, if they contribute something new and useful."

So what is "new and useful" here? This is a book about remote sensing image analysis with a distinctly mathematical-algorithmic-computer-oriented flavor, intended for graduate-level teaching and with, to borrow from the remote sensing jargon, a rather restricted FOV. It does not attempt to match the wider *fields of view* of existing texts on the subject, such as Schowengerdt (1997), Richards and Jia (2006), Jensen (2005) and others. However, the topics that are covered are dealt with in considerable depth, and I believe that this coverage fills an important gap. Many aspects of the analysis of remote sensing data are quite technical and tend to be intimidating to students with moderate mathematical backgrounds. At the same time, one often witnesses a desire on the part of students to apply advanced methods and to modify them to fit their particular research problems. Fulfilling the latter wish, in particular, requires more than superficial understanding of the material.

The focus of the book is on pixel-oriented analysis of visual/infrared Earth observation satellite imagery. Among the topics that get the most attention are the discrete wavelet transform, image fusion, supervised classification with neural networks, clustering algorithms and statistical change detection methods. The first two chapters introduce the mathematical and statistical tools necessary in order to follow later developments. Chapters 3 and 4 deal with spatial/spectral transformations, convolutions and filtering of multispectral image arrays. Chapter 5 treats image enhancement and some of the preprocessing steps that precede classification and change detection. Chapters 6 and 7 are concerned, respectively, with supervised and unsupervised land cover classification. The last chapter is about change detection with heavy emphasis on the use of canonical correlation analysis. Each of the 8 chapters concludes with exercises, some of which are small programming projects, intended to illustrate or justify the foregoing development. Solutions to the exercises are included in a separate booklet. Appendix A provides some additional mathematical/statistical background and Appendix B develops two efficient training algorithms for neural networks. Finally, Appendix C describes the installation and use of the many computer programs introduced in the course of the book.

I've made considerable effort to maintain a consistent, clear mathematical style throughout. Although the developments in the text are admittedly uncompromising, there is nothing that, given a little perseverance, cannot be followed by a reader who has grasped the elementary matrix algebra and statistical concepts explained in the first two chapters. If the student has ambitions to write his or her own image analysis programs, then he or she must be prepared to "get the maths right" beforehand. There are, heaven knows, enough pitfalls to worry about thereafter.

All of the illustrations and applications in the text are programmed in RSI's ENVI/IDL. The software is available for download at the publisher's website:

`http://www.crcpress.com/e_products/downloads/default.asp`

Given the plethora of image analysis and geographic information system (GIS) software systems on the market or available under open source license, one might think that the choice of computer environment would have been difficult. It wasn't. IDL is an extremely powerful, array- and graphics-oriented, universal programming language with a versatile interface (ENVI) for importing and analyzing remote sensing data — a peerless combination for my purposes. Extending the ENVI interface in IDL in order to implement new methods and algorithms of arbitrary sophistication is both easy and fun.

So, apart from some exposure to elementary calculus (and the aforesaid perseverance), the only other prerequisites for the book are a little familiarity with the ENVI environment and the basic knowledge of IDL imparted by such excellent introductions as Fanning (2000) or Gumley (2002). For everyday problems with IDL at any level from "newbie" on upward, help and solace are available at the newsgroup

`comp.lang.idl-pvwave`

frequented by some of the friendliest and most competent gurus on the net.

I would like to express my thanks to Rudolf Avenhaus and Allan Nielsen for their many comments and suggestions for improvement of the manuscript and to CRC Press for competent assistance in its preparation. Part of the software documented in the text was developed within the Global Monitoring for Security and Stability (GMOSS) network of excellence funded by the European Commission.

Morton Canty

Preface to the Second Edition

Shortly after the manuscript for the first edition of this book went to the publisher, ENVI 4.3 appeared along with, among other new features, a support vector machine classifier. Although my decision not to include the SVM in the original text was a conscious one (I balked at the thought of writing my own IDL implementation), this event did point to a rather glaring omission in a book purporting to be partly about land use/land cover classification. So, almost immediately, I began to dream of a Revised Second Edition and to pester CRC Press for a contract. This was happily forthcoming and the present edition now has a fairly long section on supervised classification with support vector machines.

The SVM is just one example of so-called kernel methods for nonlinear data analysis, and I decided to make kernelization one of the themes of the revised text. The treatment begins with a dual formulation for ridge regression in Chapter 2 and continues through kernel principal components analysis in Chapters 3 and 4, support vector machines in Chapter 6, kernel K-means clustering in Chapter 8 and nonlinear change detection in Chapter 9. Other new topics include entropy and mutual information (Chapter 1), adaptive boosting (Chapter 7) and image segmentation (Chapter 8). In order to accommodate the extended material on supervised classification, discussion is now spread over the two Chapters 6 and 7. The exercises at the end of each chapter have been extended and re-worked and, as for the first edition, a solutions manual is provided.

I have written several additional IDL extensions to ENVI to accompany the new themes, which are available, together with updated versions of previous programs, for download on the Internet. In order to accelerate some of the more computationally intensive routines for users with access to CUDA (parallel processing on NVIDIA graphics processors), code is included which can make use of the IDL bindings to CUDA provided by Tech-X Corporation in their GPULib product:

`http://gpulib.txcorp.com`

Notwithstanding the revisions, the present edition remains a monograph on pixel-oriented analysis of intermediate-resolution remote sensing imagery with emphasis on the development and programming of statistically motivated, data-driven algorithms. Important topics such as object-based feature analysis (for high-resolution imagery), or the physics of the radiation/surface interaction (for example, in connection with hyperspectral sensing) are only touched upon briefly, and the huge field of radar remote sensing is left out completely. Nevertheless, I hope that the in-depth focus on the topics covered will continue to be of use both to practitioners as well as to teachers.

I would like to express my appreciation to Peter Reinartz and the German Aerospace Center for permission to use the traffic scene images in Chapter 9 and to NASA's Land Processes Distributed Active Archive Center for free and uncomplicated access to archived ASTER imagery. Thanks also go to Peter Messmer and Michael Galloy, Tech-X Corp., for their prompt responses to my many cries for help with GPULib. I am especially grateful to my colleagues Harry Vereecken and Allan Nielsen, the former for generously providing me with the environment and resources needed to complete this book, the latter for the continuing inspiration of our friendship and long-time collaboration.

Morton Canty

Preface to the Third Edition

A main incentive for me to write a third edition of this book stemmed from my increasing enthusiasm for the Python programming language. I began to see the advantage of illustrating the many image processing algorithms covered in earlier editions of the text not only in the powerful and convenient, but not inexpensive, ENVI/IDL world, but also on a widely available open source platform. Python, together with the Numpy and Scipy packages, can hold its own with any commercial array processing software system. Furthermore, the Geospatial Data Abstraction Library (GDAL) and its Python wrappers allow for great versatility and convenience in reading, writing and manipulating different image formats. This was enough to get me going on a revised textbook, one which I hope will have appeal beyond the ENVI/IDL community.

Another incentive for a new edition was hinted at in the preface to the previous edition, namely the lack of any discussion of the vast and increasingly important field of radar remote sensing. Obviously this would be a topic for (at least) a whole new book, so I have included material only on a very special aspect of particular interest to me, namely multivariate statistical classification and change detection algorithms applied to polarimetric synthetic aperture radar (polSAR) data. Up until recently, not many researchers or practitioners have had access to this kind of data. However with the advent of several spaceborne polarimetric SAR instruments such as the Japanese ALOS, the Canadian Radarsat-2, the German TerraSAR-X and the Italian COSMO-SkyMed missions, the situation has greatly improved. Chapters 5, 7 and 9 now include treatments of speckle filtering, image co-registration, supervised classification and multivariate change detection with multi-look polSAR data.

The software associated with the present edition includes, along with the ENVI/IDL extensions, Python scripts for all of the main processing, classification and change detection algorithms. In addition, many examples discussed in the text are illustrated with Python scripts as well as in IDL. The Appendices C and D separately document the installation and use of the ENVI/IDL and Python code. For readers who wish to use the Eclipse/Pydev development environment (something which I highly recommend), the Python scripts are provided in the form of a Pydev project.

What is missing in the Python world, of course, is the slick GUI provided by ENVI. I have made no attempt to mimic an ENVI graphical environment in Python, and the scripts provided content themselves with reading imagery from, and writing results to, the file system. A rudimentary command line script for RGB displays of multispectral band combinations in different histogram enhancement modes is included.

For an excellent introduction to scientific computing in Python see Langtangen (2009). The book by Westra (2013) provides valuable tips on geospatial development in Python, including GDAL programming. The definitive reference on IDL is now certainly Galloy (2011), an absolute must for anyone who uses the language professionally.

With version 5.0, a new ENVI graphics environment and associated API has appeared which has a very different look and feel to the old "ENVI Classic" environment, as it is officially referred to. Fortunately the classic environment is still available and, for reasons of compatibility with previous versions, the IDL programming examples in the text use the classic interface and its associated syntax. Most of the ENVI/IDL extensions as documented in Appendix C are provided both for the new as well as for the classic GUI/API.

I would like to express my appreciation to the German Aerospace Center for permission to use images from the TerraSAR-X platform and to Henning Skriver, DTU Space Denmark, for allowing me to use his EMISAR polarimetric data. My special thanks go to Allan Nielsen and Frank Thonfeld for acquainting me with SAR imagery analysis and to Rudolf Avenhaus for his many helpful suggestions in matters statistical.

Morton Canty

List of Figures

Program Listings

1

Images, Arrays, and Matrices

There are many Earth observation satellite-based sensors, both active and passive, currently in orbit or planned for the near future. Representative of these, we describe briefly the multispectral ASTER system (Abrams et al., 1999) and the TerraSAR-X synthetic aperture radar satellite (Pitz and Miller, 2010). See Jensen (2005), Richards and Jia (2006) and Mather and Koch (2010) for overviews of remote sensing satellite platforms.

The Advanced Spaceborne Thermal Emission and Reflectance Radiometer (ASTER) instrument was launched in December 1999 on the Terra spacecraft. It is being used to obtain detailed maps of land surface temperature, reflectance and elevation and consists of sensors to measure reflected solar radiance and thermal emission in three spectral intervals:

- VNIR: Visible and near-infrared bands 1, 2, 3N, and 3B, in the spectral region between 0.52 and 0.86 μm (four arrays of charge-coupled detectors (CCDs) in pushbroom scanning mode).

- SWIR: Short wavelength infrared bands 4 to 9 in the region between 1.60 and 2.43 μm (six cooled PtSi-Si Schottky barrier arrays, pushbroom scanning).

- TIR: Thermal infrared bands 10 to 14 covering a spectral range from 8.13 to 11.65 μm (cooled HgCdTe detector arrays, whiskbroom scanning).

The altitude of the spacecraft is 705 km. The across- and in-track *ground sample distances* (GSDs), i.e., the detector widths projected through the system optics onto the Earth's surface, are 15 m (VNIR), 30 m (SWIR) and 90 m (TIR).* The telescope associated with the 3B sensors is back-looking at an angle of 27.6^o to provide, together with the 3N sensors, along-track stereo image pairs. In addition, the VNIR camera can be rotated from straight down (nadir) to $\pm$ 24° across-track. The SWIR and TIR instrument mirrors can be pointed to $\pm$ 8.5^o across-track. Like most platforms in this ground resolution category, the orbit is near polar, sun-synchronous. Quantization levels are 8 bits for VNIR and SWIR and 12 bits for TIR. The sensor systems have an average duty cycle of 8% per orbit (about 650 scenes per day, each 60×60

*At the time of writing, the SWIR sensor is no longer functioning although both the VNIR and TIR systems are still producing good data.

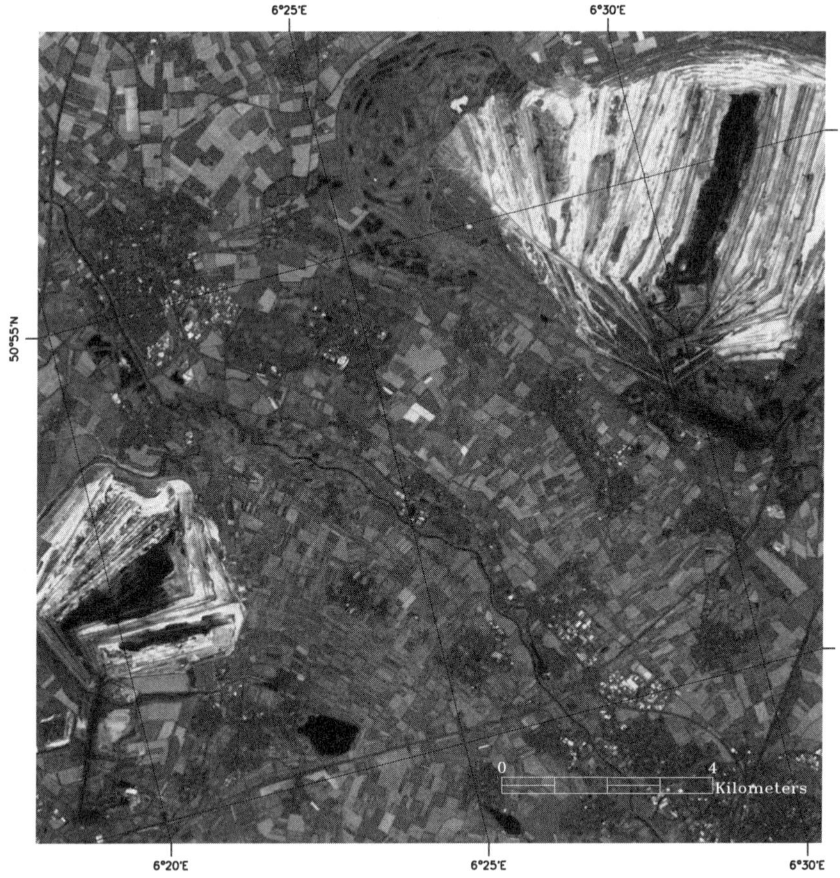

FIGURE 1.1
ASTER color composite image (1000 × 1000 pixels) of VNIR bands 1 (blue), 2 (green), and 3N (red) over the town of Jülich in Germany, acquired on May 1, 2007. The bright areas are open cast coal mines. **(See color insert.)**

km^2 in area) with revisit times between 4 and 16 days. Figure 1.1 shows a spatial/spectral subset of an ASTER scene. The image is a UTM (Universal Transverse Mercator) projection oriented along the satellite path (rotated approximately 16.4^o from north) and orthorectified using a digital terrain model generated from the stereo bands.

Unlike passive multi- and hyperspectral imaging sensors, which measure reflected solar energy or the Earth's thermal radiation, synthetic aperture radar (SAR) airborne and satellite platforms supply their own microwave radiation source, allowing observations which are independent of time of day or cloud cover. The radar antenna on the TerraSAR-X satellite, launched in

FIGURE 1.2
A 5000×5000-pixel spatial subset of the HH polarimetric band of a TerraSAR-X quad polarimetric image acquired over the Rhine River, Germany, in so-called Stripmap mode. The data are slant-range, single-look, complex. The gray-scale values correspond to the magnitudes of the complex pixel values.

June, 2007, emits and receives X-band radar (9.65 GHz) in both horizontal and vertical polarizations to provide surface imaging with a geometric resolution from about 18 m (scanSAR mode, 10 km×150 km swath) down to 1 m (high-resolution Spotlight mode, 10 km×5 km swath). It flies in a sun-synchronous, near-polar orbit at an altitude of 514 km with a revisit time for points on the equator of 11 days. Figure 1.2 shows a TerraSAR-X HH polarimetric band (horizontally polarized radiation emitted and detected) acquired over the Rhine River, Germany, in April, 2010. The data are at the "single-look, slant-range complex" processing level, and are not map-projected.

1.1 Multispectral satellite images

A multispectral image such as that shown in Figure 1.1 may be represented as a three-dimensional array of gray-scale values or pixel intensities

$$g_k(i,j), \quad 1 \le i \le c,\ 1 \le j \le r,\ 1 \le k \le N,$$

where c is the number of pixel columns (also called *samples*) and r is the number of pixel rows (or *lines*). The index k denotes the spectral band, of which there are N in all. For data at an early processing stage a pixel may be stored as a *digital number* (DN), often in a single byte, so that $0 \le g_k \le 255$. This is the case for the ASTER VNIR and SWIR bands at processing level L1A (unprocessed reconstructed instrument data), whereas the L1A TIR data are quantized to 12 bits (as unsigned integers) and thus stored as digital numbers from 0 to $2^{12} - 1 = 4095$. Processed image data may of course be stored in byte, integer or floating point format and can have negative or even complex values.

The gray-scale values in the various bands encode measurements of the radiance $L_{\Delta\lambda}(x,y)$ in wavelength interval $\Delta\lambda$ due to sunlight reflected from some point (x,y) on the Earth's surface, or due to thermal emission from that surface, and focused by the instrument's optical system along the array of sensors. Ignoring all absorption and scattering effects of the intervening atmosphere, the at-sensor radiance available for measurement from reflected sunlight from a horizontal, *Lambertian* surface, i.e., a surface which scatters reflected radiation uniformly in all directions, is given by

$$L_{\Delta\lambda}(x,y) = E_{\Delta\lambda} \cdot \cos\theta_z \cdot R_{\Delta\lambda}(x,y)/\pi. \tag{1.1}$$

The units are $[\mathrm{W}/(\mathrm{m}^2 \cdot \mathrm{sr} \cdot \mu\mathrm{m})]$, $E_{\Delta\lambda}$ is the average spectral solar irradiance in the spectral band $\Delta\lambda$, θ_z is the solar zenith angle, $R_{\Delta\lambda}(x,y)$ is the surface reflectance at coordinates (x,y), a number between 0 and 1, and π accounts for the upper hemisphere of solid angle. The conversion between DN and at-sensor radiance is determined by the sensor calibration as measured (and maintained) by the satellite image provider. For example, for ASTER VNIR and SWIR L1A data,

$$L_{\Delta\lambda}(x,y) = A \cdot DN/G + D.$$

The quantities A (linear coefficient), G (gain), and D (offset) are tabulated for each of the detectors in the arrays and included with each acquisition. Atmospheric scattering and absorption models may be used to deduce at-surface radiance, surface temperature and emissivity or surface reflectance from the observed radiance at the sensor. Reflectance and emissivity are directly related to the physical properties of the surface being imaged. See

Schowengerdt (1997) for a thorough discussion of atmospheric effects and their correction.

Various conventions are used for storing the image array $g_k(i, j)$ in computer memory or other storage media. In *band interleaved by pixel* (BIP) format, for example, a two-channel, 3×3 pixel image would be stored as

$$\begin{array}{cccccc} g_1(1,1) & g_2(1,1) & g_1(2,1) & g_2(2,1) & g_1(3,1) & g_2(3,1) \\ g_1(1,2) & g_2(1,2) & g_1(2,2) & g_2(2,2) & g_1(3,2) & g_2(3,2) \\ g_1(1,3) & g_2(1,3) & g_1(2,3) & g_2(2,3) & g_1(3,3) & g_2(3,3), \end{array}$$

whereas in *band interleaved by line* (BIL) it would be stored as

$$\begin{array}{cccccc} g_1(1,1) & g_1(2,1) & g_1(3,1) & g_2(1,1) & g_2(2,1) & g_2(3,1) \\ g_1(1,2) & g_1(2,2) & g_1(3,2) & g_2(1,2) & g_2(2,2) & g_2(3,2) \\ g_1(1,3) & g_1(2,3) & g_1(3,3) & g_2(1,3) & g_2(2,3) & g_2(3,3), \end{array}$$

and in *band sequential* (BSQ) format as

$$\begin{array}{ccc} g_1(1,1) & g_1(2,1) & g_1(3,1) \\ g_1(1,2) & g_1(2,2) & g_1(3,2) \\ g_1(1,3) & g_1(2,3) & g_1(3,3) \\ g_2(1,1) & g_2(2,1) & g_2(3,1) \\ g_2(1,2) & g_2(2,2) & g_2(3,2) \\ g_2(1,3) & g_2(2,3) & g_2(3,3). \end{array}$$

In the computer language IDL, so-called *column major indexing* is used for arrays (Gumley, 2002) and the elements in an array are numbered from zero. This means that, if a gray-scale image g is assigned to an IDL array variable `G`, then the intensity value $g(i, j)$ is addressed as `G[i-1,j-1]`. An N-band multispectral image is stored in BIP format as an $N \times c \times r$ array in IDL, in BIL format as a $c \times N \times r$ and in BSQ format as a $c \times r \times N$ array. So, for example, in BIP format the value $g_k(i, j)$ is stored at `G[k-1,i-1,j-1]`.

In numerical Python (Python with the Numpy package) arrays are also numbered from zero, but the indexing is *row major*. The array shapes are thus reversed relative to IDL. For example, a BSQ image has shape $N \times r \times c$.

Auxiliary information, such as image acquisition parameters and georeferencing, is sometimes included with the image data on the same file, and the format may or may not make use of compression algorithms. Examples are the GeoTIFF* file format used, for instance, by Space Imaging Inc. for distributing Carterra(c) imagery and which includes lossless compression, the HDF-EOS (Hierarchical Data Format-Earth Observing System) files in which ASTER images are distributed, and the PCIDSK format employed by PCI

*GeoTIFF is an open source specification and refers to TIFF files which have geographic (or cartographic) data embedded as tags within the file. The geographic data can be used to position the image in the correct location and geometry on the screen of a geographic information display.

Listing 1.1: Reading and displaying a multispectral image in ENVI/IDL.

```
PRO EX1_1

   envi_select, title='Choose␣multispectral␣image', $
             fid=fid, dims=dims,pos=pos
   IF (fid EQ -1) THEN BEGIN
      PRINT, 'cancelled'
      RETURN
   ENDIF

   envi_file_query, fid, fname=fname

; image dimensions
   cols = dims[2]-dims[1]+1
   rows = dims[4]-dims[3]+1
   bands = n_elements(pos)

; BSQ array
   mage = fltarr(cols,rows,bands)

   FOR i=0,bands-1 DO image[*,*,i] = $
     envi_get_data(fid=fid,dims=dims,pos=pos[i])

; display first band
   window, 11, xsize=cols, ysize=rows, title=fname
   tvscl, image[*,*,0]

END
```

Geomatics(c) with its image processing software, in which auxiliary information is in plain ASCII and the image data are not compressed. ENVI uses a simple "flat binary" file structure with an additional ASCII header file.

The IDL program shown in Listing 1.1 makes use of the ENVI procedures `ENVI_SELECT`, `ENVI_FILE_QUERY` and the ENVI function `ENVI_GET_DATA()` to read a multispectral image into an IDL array with BSQ interleave format and then display the first spectral band with the IDL procedures `WINDOW` and `TVSCL`. This script uses what is now referred to as the "ENVI Classic" interface to IDL, and in order to execute it, an ENVI Classic (or simply ENVI for pre 5.0 versions) interactive session should be up and running. With ENVI Version 5, a new graphical user interface and associated object-oriented application programming interface (API) was introduced. An example of its use is given in Listing 1.10 at the end of this chapter.

Listing 1.2 is a simple and fairly self-explanatory Python script which does the same thing with the aid of GDAL (the Geospatial Data Abstraction Library) and the `matplotlib.pyplot` package. It also uses the `select_infile()`,

Listing 1.2: Reading and displaying a multispectral image in Python.

```
#!/usr/bin/env python
#  Name:     ex1_1.py
IMPORT auxil.auxil as auxil
FROM numpy IMPORT *
FROM osgeo IMPORT gdal
FROM osgeo.gdalconst IMPORT GA_ReadOnly
IMPORT matplotlib.pyplot as plt

DEF main():

    gdal.AllRegister()
    infile = auxil.select_infile()
    IF infile:
        inDataset = gdal.Open(infile,GA_ReadOnly)
        cols = inDataset.RasterXSize
        rows = inDataset.RasterYSize
        bands = inDataset.RasterCount
    ELSE:
        RETURN

#  spectral and spatial subsets
    pos =  auxil.select_pos(bands)
    x0,y0,rows,cols = auxil.select_dims([0,0,rows,cols])

#  BSQ array
    image = zeros((LEN(pos),rows,cols))
    k = 0
    FOR b IN pos:
        band = inDataset.GetRasterBand(b)
        image[k,:,:]=band.ReadAsArray(x0,y0,cols,rows)\
                                    .astype(FLOAT)
        k += 1
    inDataset = None

#  display first band
    band0 = image[0,:,:]
    mn = amin(band0)
    mx = amax(band0)
    plt.imshow((band0-mn)/(mx-mn), cmap='gray' )
    plt.show()

IF __name__ == '__main__':
    main()
```

`select_pos()` and `select_dims()` functions defined in the `auxil.py` module which accompanies this book. The script is run from the command prompt in Windows or from a console window on Unix-like systems with the command `python ex1_1.py`. In the latter case, the "Shebang" `#!` in the first line allows it to be run simply by typing the filename, assuming the path to the `env` utility is `/usr/BIN`.

1.2 Synthetic aperture radar images

Synthetic aperture radar (SAR) systems differ significantly from optical/infrared sensor-based platforms. Richards (2009) and Oliver and Quegan (2004) provide thorough introductions to SAR remote sensing, SAR image statistics, image analysis and interpretation.

The power received by a radar transmitting/receiving antenna reflected from a distributed (as opposed to point) target a distance D from the antenna is given by (Richards, 2009)

$$P_R = \frac{P_T G_T G_R \lambda^2 \sigma^o \Delta_a \Delta_r}{(4\pi)^3 D^4}\ [W], \tag{1.2}$$

where P_T is the transmitted power $[W \cdot m^{-2}]$, λ is the operating wave-

Listing 1.3: Reading and displaying a polarimetric SAR image in IDL.

```
PRO ex1_2
COMPILE_OPT IDL2

   fname = dialog_pickfile(filter='*.xml', $
            title='Select␣COSAR␣image')
   IF fname EQ '' THEN RETURN
   tdx_openfile, /no_envi, filename=fname, $
                              saInfo=saInfo
   data = tdx_readspatial(band =[0],$
                          xs=1000, $
                          xe=5999, $
                          ys=1000,$
                          ye=5999,$
                          saInfo=saInfo)
   tdx_closefile, saInfo
   window, 11, xsize=5000, ysize=5000
   tvscl, abs(data)

END
```

Listing 1.4: Reading and displaying a polarimetric SAR image in Python.

```
#!/usr/bin/env python
#  Name:      ex1_2.py
IMPORT auxil.auxil as auxil
FROM numpy IMPORT *
FROM osgeo IMPORT gdal
FROM osgeo.gdalconst IMPORT GA_ReadOnly
IMPORT matplotlib.pyplot as plt

DEF main():
    gdal.AllRegister()
    infile = auxil.select_infile(filt='*.xml')
    IF infile:
        inDataset = gdal.Open(infile,GA_ReadOnly)
        cols = inDataset.RasterXSize
        rows = inDataset.RasterYSize
        bands = inDataset.RasterCount
    ELSE:
        RETURN
    pos =  auxil.select_pos(bands)
    x0,y0,rows,cols=auxil.select_dims([0,0,rows,cols])
#  BSQ array
    image = zeros((LEN(pos),rows,cols),dtype=complex64)
    k = 0
    FOR b IN pos:
        band = inDataset.GetRasterBand(b)
        image[k,:,:]=band.ReadAsArray(x0,y0,cols,rows)\
                                               .astype(COMPLEX)
        k += 1
    inDataset = None
#  display magnitude in linear 2% stretch
    band0 = ABS(image[0,:,:])
    band0 = auxil.lin2pcstr(band0)
    mn = amin(band0)
    mx = amax(band0)
    plt.imshow((band0-mn)/(mx-mn), cmap='gray')
    plt.show()

IF __name__ == '__main__':
    main()
```

length $[m]$, $G_T(G_R)$ is the transmitting (receiving) antenna gain, $\Delta_a(\Delta_r)$ is the azimuth (ground range) resolution $[m]$ and σ^o is the unitless scattering coefficient (referred to as the *radar cross section*) of the target surface. The scattering coefficient is related to the (bio)physical properties of the surface being irradiated, notably its water content.

In later chapters we will be concerned primarily with fully and partially polarimetric SAR data. A full, or *quad*, polarimetric SAR measures a 2×2 *scattering matrix* $\boldsymbol{S}$ at each resolution cell on the ground. The scattering matrix relates the incident and the backscattered electric fields $\boldsymbol{E}^i$ and $\boldsymbol{E}^b$ according to

$$\boldsymbol{E}^b = \boldsymbol{S}\boldsymbol{E}^i \quad \text{or} \quad \begin{pmatrix} E_h^b \\ E_v^b \end{pmatrix} = \begin{pmatrix} s_{hh} & s_{hv} \\ s_{vh} & s_{vv} \end{pmatrix} \begin{pmatrix} E_h^i \\ E_v^i \end{pmatrix}. \tag{1.3}$$

Here $E_h^{i(b)}$ and $E_v^{i(b)}$ denote the horizontal and vertical components of the incident (backscattered) oscillating electric fields directly at the target. These can be deduced from the transmitted and received radar signals via the so-called *far-field* approximations; see Richards (2009). If both horizontally and vertically polarized radar pulses are emitted and discriminated, then they determine, from Equation (1.3), the four complex scattering matrix elements.

Complex numbers provide a convenient representation for the amplitude E of an electric field:

$$E = |E|cos(\omega t + \phi) = \text{Re}\left(|E|e^{\mathbf{i}(\omega t+\phi)}\right),$$

where $\omega = 2\pi f$ and ϕ are the angular frequency and phase of the radiation and Re denotes "real part."* It is usually convenient to work exclusively with complex amplitudes $|E|e^{\mathbf{i}(\omega t+\phi)}$, bearing in mind that only the real part is physically significant. When the oscillating electric fields are described by complex numbers in this way, the scattering matrix elements are also complex. A full polarimetric, or *quad polarimetric* SAR image then consists of four complex bands s_{hh}, s_{hv}, s_{vh} and s_{vv}, one for each pixel-wise determination of an element of the scattering matrix. So-called *reciprocity* (Richards, 2009), which normally applies to natural targets, implies that $s_{hv} = s_{vh}$. The squared amplitudes of the scattering coefficients, i.e., $|s_{hh}|^2$, $|s_{hv}|^2$, etc., constitute the radar cross sections for each polarization combination. These in turn replace σ^o in Equation (1.2) and determine the received power in each polarization channel.

Listing 1.3 gives a basic IDL script for reading polarimetric TerraSAR-X imagery in COSAR format, an annotated binary matrix format with each sample (pixel) held in 32 bytes (16 bytes real, 16 bytes imaginary) stored with the most significant byte first (big endian format). The script makes use of the freely available TerraSAR-X/TanDEM-X SSC/CoSSC reader for IDL and ENVI (Exelis/DLR, 2012). After opening the data file, a 500×500 pixel subset of the first (HH) band is read into the IDL variable `data`. Then the magnitudes of the complex pixel values are displayed in an IDL window as in Figure 1.2.

Listing 1.4 shows a similar script in Python which makes use of GDAL and the `matplotlib` library as well as the `auxil` package to access and plot the data.

*A brief introduction to complex numbers is given in Appendix A.

1.3 Algebra of vectors and matrices

It is very convenient, in fact essential, to use a vector representation for multispectral or SAR image pixels. In this book, a pixel will be represented in the form

$$\boldsymbol{g}(i,j) = \begin{pmatrix} g_1(i,j) \\ \vdots \\ g_N(i,j) \end{pmatrix}, \tag{1.4}$$

which is understood to be a *column vector* of spectral intensities or *gray-scale values* at the image position (i,j). It can be thought of as a point in N-dimensional Euclidean space, commonly referred to as *input space* or *feature space*. In the case of SAR images, the vector components may be complex scattering amplitudes.

Since we will be making extensive use of the vector notation of Equation (1.4), some of the basic properties of vectors, and of matrices which generalize them, will be reviewed here. One can illustrate — and the reader can easily verify — many of these properties for 2-component vectors and 2×2 matrices. A list of frequently used mathematical symbols is given at the end of the book.

1.3.1 Elementary properties

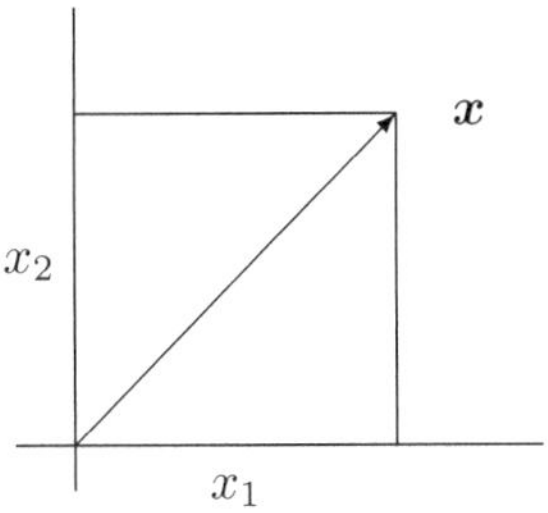

FIGURE 1.3
A vector with two components.

The *transpose* of the two-component column vector

$$\boldsymbol{x} = \begin{pmatrix} x_1 \\ x_2 \end{pmatrix}, \tag{1.5}$$

shown in Figure 1.3, is the *row vector*

$$\boldsymbol{x}^\top = (x_1, x_2). \tag{1.6}$$

The *sum* of two column vectors is given by

$$\boldsymbol{x}+\boldsymbol{y} = \begin{pmatrix} x_1 \\ x_2 \end{pmatrix} + \begin{pmatrix} y_1 \\ y_2 \end{pmatrix} = \begin{pmatrix} x_1+y_1 \\ x_2+y_2 \end{pmatrix}, \tag{1.7}$$

and their *inner product* by

$$\boldsymbol{x}^\top \boldsymbol{y} = (x_1, x_2)\begin{pmatrix} y_1 \\ y_2 \end{pmatrix} = x_1 y_1 + x_2 y_2. \tag{1.8}$$

The *length* or *Euclidean norm* of the vector $\boldsymbol{x}$ is

$$\|\boldsymbol{x}\| = \sqrt{x_1^2 + x_2^2} = \sqrt{\boldsymbol{x}^\top \boldsymbol{x}}\ . \tag{1.9}$$

The programming language IDL is especially good at manipulating array structures like vectors and matrices. In the following dialog, a column vector is assigned to the variable `X` and then printed together with its transposed form:

```
IDL> X=[[1],[2]]
IDL> PRINT,X
       1
       2
IDL> PRINT,transpose(X)
       1       2
```

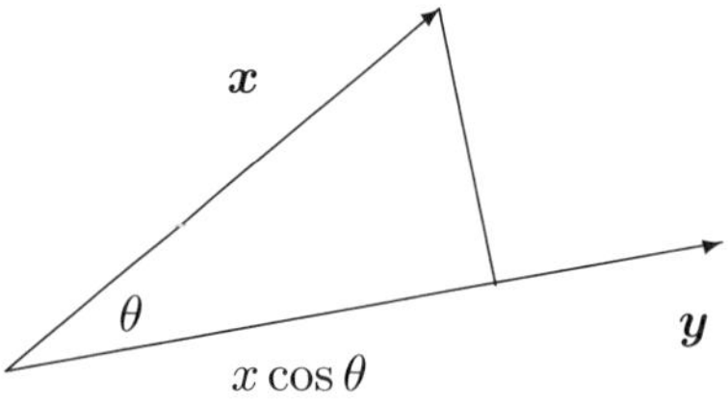

FIGURE 1.4
Illustrating the inner product.

The inner product of $\boldsymbol{x}$ and $\boldsymbol{y}$ can be expressed in terms of the vector lengths and the angle θ between the two vectors:

$$\boldsymbol{x}^\top \boldsymbol{y} = \|\boldsymbol{x}\|\|\boldsymbol{y}\|\cos\theta; \tag{1.10}$$

see Figure 1.4 and Exercise 1. If $\theta = 90^o$, the vectors are said to be *orthogonal*, in which case $\boldsymbol{x}^\top \boldsymbol{y} = 0$.

Any vector can be expressed in terms of orthogonal *unit vectors*, e.g.,

$$\boldsymbol{x} = \begin{pmatrix} x_1 \\ x_2 \end{pmatrix} = x_1 \begin{pmatrix} 1 \\ 0 \end{pmatrix} + x_2 \begin{pmatrix} 0 \\ 1 \end{pmatrix} = x_1 \boldsymbol{i} + x_2 \boldsymbol{j} \ , \tag{1.11}$$

where the symbols $\boldsymbol{i}$ and $\boldsymbol{j}$ denote vectors of unit length along the x and y directions, respectively.

A 2×2 *matrix* is written in the form

$$\boldsymbol{A} = \begin{pmatrix} a_{11} & a_{12} \\ a_{21} & a_{22} \end{pmatrix}. \tag{1.12}$$

The first index of a matrix element indicates its row, the second its column. While numerical Python indexes two-dimensional array and matrix objects in this way, the indices of the elements in Equation (1.12) *do not* correspond to the IDL convention. The first index of a 2-dimensional IDL array denotes the column, the second the row. Thus, IDL may be said to be array-oriented but not matrix-oriented. This can cause confusion and must be kept in mind when writing programs in IDL. Depending on the context, we will refer to an $n \times m$ IDL *array* and an $m \times n$ *matrix* as one and the same object.

When a matrix is multiplied with a vector, the result is another vector, e.g.,

$$\boldsymbol{A}\boldsymbol{x} = \begin{pmatrix} a_{11} & a_{12} \\ a_{21} & a_{22} \end{pmatrix} \begin{pmatrix} x_1 \\ x_2 \end{pmatrix} = \begin{pmatrix} a_{11}x_1 + a_{12}x_2 \\ a_{21}x_1 + a_{22}x_2 \end{pmatrix} = x_1 \begin{pmatrix} a_{11} \\ a_{21} \end{pmatrix} + x_2 \begin{pmatrix} a_{11} \\ a_{22} \end{pmatrix}.$$

In general, for $\boldsymbol{A} = (\boldsymbol{a}_1, \boldsymbol{a}_2 \ldots \boldsymbol{a}_N)$, where the vectors $\boldsymbol{a}_i$ are the columns of $\boldsymbol{A}$,

$$\boldsymbol{A}\boldsymbol{x} = x_1\boldsymbol{a}_1 + x_2\boldsymbol{a}_2 + \cdots + x_N\boldsymbol{a}_N. \tag{1.13}$$

The IDL operator for matrix and vector multiplication is `##`, for example:

```
1 IDL> A=[[1,2],[3,4]]
2 IDL> PRINT,A
3          1          2
4          3          4
5 IDL> PRINT,A##X
6                     5
7                    11
```

Numerical Python is similarly efficient in manipulating arrays, vectors and matrices. The scalar multiplication operator $*$ is interpreted as matrix multiplication for matrix operands:

```
1 >>> IMPORT numpy
2 >>> X = numpy.mat([[1],[2]])
3 >>> A = numpy.mat([[1,2],[3,4]])
4 >>> PRINT A
5 [[1 2]
6  [3 4]]
7 >>> PRINT A*X
8 [[ 5]
9  [11]]
```

The product of two 2×2 matrices is given by

$$\boldsymbol{AB} = \begin{pmatrix} a_{11} & a_{12} \\ a_{21} & a_{22} \end{pmatrix} \begin{pmatrix} b_{11} & b_{12} \\ b_{21} & b_{22} \end{pmatrix} = \begin{pmatrix} a_{11}b_{11} + a_{12}b_{21} & a_{11}b_{12} + a_{12}b_{22} \\ a_{21}b_{11} + a_{22}b_{21} & a_{21}b_{12} + a_{22}b_{22} \end{pmatrix}$$

and is another matrix. More generally, the matrix product $\boldsymbol{AB}$ is allowed whenever $\boldsymbol{A}$ has the same number of columns as $\boldsymbol{B}$ has rows. So if $\boldsymbol{A}$ has dimension $\ell \times m$ and $\boldsymbol{B}$ has dimension $m \times n$, then $\boldsymbol{AB}$ is $\ell \times n$ with elements

$$(\boldsymbol{AB})_{ij} = \sum_{k=1}^{m} a_{ik}b_{kj} \quad i = 1 \ldots \ell, \; j = 1 \ldots n. \tag{1.14}$$

Matrix multiplication is not *commutative*, i.e., $\boldsymbol{AB} \neq \boldsymbol{BA}$ in general. However it is *associative*:

$$(\boldsymbol{AB})\boldsymbol{C} = \boldsymbol{A}(\boldsymbol{BC}) \tag{1.15}$$

so we can write, for example,

$$(\boldsymbol{AB})\boldsymbol{C} = \boldsymbol{A}(\boldsymbol{BC}) = \boldsymbol{ABC} \tag{1.16}$$

without ambiguity. The *outer product* of two vectors of equal length, written $\boldsymbol{x}\boldsymbol{y}^\top$, is a matrix, e.g.,

$$\boldsymbol{x}\boldsymbol{y}^\top = \begin{pmatrix} x_1 \\ x_2 \end{pmatrix} (y_1, y_2) = \begin{pmatrix} x_1 & 0 \\ x_2 & 0 \end{pmatrix} \begin{pmatrix} y_1 & y_2 \\ 0 & 0 \end{pmatrix} = \begin{pmatrix} x_1y_1 & x_1y_2 \\ x_2y_1 & x_2y_2 \end{pmatrix}. \tag{1.17}$$

Matrices, like vectors, have a transposed form, obtained by interchanging their rows and columns:

$$A^\top = \begin{pmatrix} a_{11} & a_{21} \\ a_{12} & a_{22} \end{pmatrix}. \tag{1.18}$$

Transposition has the properties

$$\begin{aligned} (A+B)^\top &= A^\top + B^\top \\ (AB)^\top &= B^\top A^\top. \end{aligned} \tag{1.19}$$

The analogous operation to transposition for complex vectors and matrices is *conjugate transposition* (see Appendix A), for which a_{ij} is replaced by a^*_{ji}, where the asterisk denotes complex conjugation. (The complex conjugate of $a+\mathbf{i}b$ is $a-\mathbf{i}b$.) We will write the conjugate transpose of A as $A^\dagger$. Conjugate transposition has the same properties as given above for ordinary transposition.

1.3.2 Square matrices

A *square* matrix A has equal numbers of rows and columns. The *determinant* of a $p \times p$ square matrix is written $|A|$ and defined as

$$|A| = \sum_{(j_1 \dots j_p)} (-1)^{f(j_1 \dots j_p)} a_{1j_1} a_{2j_2} \cdots a_{pj_p}. \tag{1.20}$$

The sum is taken over all permutations $(j_1 \dots j_p)$ of the integers $(1 \dots p)$ and $f(j_1 \dots j_p)$ is the number of transpositions (interchanges of two integers) required to change $(1 \dots p)$ into $(j_1 \dots j_p)$. The determinant of a 2×2 matrix, for example, is given by

$$|A| = a_{11}a_{22} - a_{12}a_{21}.$$

The determinant has the properties

$$\begin{aligned} |AB| &= |A||B| \\ |A^\top| &= |A|. \end{aligned} \tag{1.21}$$

The *identity matrix* is a square matrix with ones along its diagonal and zeroes everywhere else. For example

$$I = \begin{pmatrix} 1 & 0 \\ 0 & 1 \end{pmatrix},$$

and for any A,

$$IA = AI = A. \tag{1.22}$$

The *matrix inverse* $\boldsymbol{A}^{-1}$ of a square matrix $\boldsymbol{A}$ is defined in terms of the identity matrix by the requirements

$$\boldsymbol{A}^{-1}\boldsymbol{A} = \boldsymbol{A}\boldsymbol{A}^{-1} = \boldsymbol{I}. \tag{1.23}$$

For example, it is easy to verify that a 2×2 matrix has inverse

$$\boldsymbol{A}^{-1} = \frac{1}{|\boldsymbol{A}|}\begin{pmatrix} a_{22} & -a_{12} \\ -a_{21} & a_{11} \end{pmatrix}.$$

In IDL, continuing the previous dialog,

```
IDL> PRINT, determ(A)
      -2.00000
IDL> PRINT, invert(A)
      -2.00000         1.00000
       1.50000       -0.500000
IDL> PRINT, A##invert(A)
       1.00000        0.000000
      0.000000         1.00000
```

The matrix inverse has the properties

$$\begin{aligned}(\boldsymbol{A}\boldsymbol{B})^{-1} &= \boldsymbol{B}^{-1}\boldsymbol{A}^{-1} \\ (\boldsymbol{A}^{-1})^\top &= (\boldsymbol{A}^\top)^{-1}.\end{aligned} \tag{1.24}$$

If the transpose of a square matrix is its inverse, i.e., if

$$\boldsymbol{A}^\top\boldsymbol{A} = \boldsymbol{I}, \tag{1.25}$$

then it is referred to as an *orthonormal matrix.*

A system of n linear equations of the form

$$y_i = \sum_{j=1}^{n} a_j x_j(i), \quad i = 1 \dots n, \tag{1.26}$$

can be written in matrix notation as

$$\boldsymbol{y} = \boldsymbol{A}\boldsymbol{a}, \tag{1.27}$$

where $\boldsymbol{y} = (y_1 \dots y_n)^\top$, $\boldsymbol{a} = (a_1 \dots a_n)^\top$ and $A_{ij} - x_j(i)$. Provided $\boldsymbol{A}$ is nonsingular (see below), the solution for the parameter vector $\boldsymbol{a}$ is given by

$$\boldsymbol{a} = \boldsymbol{A}^{-1}\boldsymbol{y}. \tag{1.28}$$

The *trace* of a square matrix is the sum of its diagonal elements, e.g., for a 2×2 matrix,

$$\mathrm{tr}(\boldsymbol{A}) = a_{11} + a_{22}. \tag{1.29}$$

The trace has the properties

$$\begin{aligned}\mathrm{tr}(\boldsymbol{A}+\boldsymbol{B}) &= \mathrm{tr}\boldsymbol{A} + \mathrm{tr}\boldsymbol{B} \\ \mathrm{tr}(\boldsymbol{A}\boldsymbol{B}) &= \mathrm{tr}(\boldsymbol{B}\boldsymbol{A}).\end{aligned} \tag{1.30}$$

1.3.3 Singular matrices

If $|\boldsymbol{A}| = 0$, then $\boldsymbol{A}$ has no inverse and is said to be a *singular matrix*. If $\boldsymbol{A}$ is nonsingular, then the equation

$$\boldsymbol{A}\boldsymbol{x} = \boldsymbol{0} \tag{1.31}$$

only has the so-called *trivial solution* $\boldsymbol{x} = \boldsymbol{0}$. To see this, multiply from the left with $\boldsymbol{A}^{-1}$. Then $\boldsymbol{A}^{-1}\boldsymbol{A}\boldsymbol{x} = \boldsymbol{I}\boldsymbol{x} = \boldsymbol{x} = \boldsymbol{0}$.

If $\boldsymbol{A}$ is singular, Equation (1.31) has at least one *nontrivial solution* $\boldsymbol{x} \neq \boldsymbol{0}$. This, again, is easy to see for a 2×2 matrix. Suppose $|\boldsymbol{A}| = 0$. Writing Equation (1.31) out fully:

$$a_{11}x_1 + a_{12}x_2 = 0$$
$$a_{21}x_1 + a_{22}x_2 = 0.$$

To get a nontrivial solution, assume without loss of generality that $a_{12} \neq 0$. Just choose $x_1 = 1$. Then the above two equations imply that

$$x_2 = -\frac{a_{11}}{a_{12}} \quad \text{and} \quad a_{21} - a_{22}\frac{a_{11}}{a_{12}} = 0.$$

The latter equality is satisfied because $|\boldsymbol{A}| = a_{11}a_{22} - a_{12}a_{21} = 0$.

1.3.4 Symmetric, positive definite matrices

The *variance–covariance matrix*, which we shall meet in the next chapter and which plays a central role in digital image analysis, is both *symmetric* and *positive definite.*

DEFINITION 1.1 *A square matrix is said to be* symmetric *if* $\boldsymbol{A}^\top = \boldsymbol{A}$. *The* $p \times p$ *matrix* $\boldsymbol{A}$ *is* positive definite *if*

$$\boldsymbol{x}^\top \boldsymbol{A}\boldsymbol{x} > 0 \tag{1.32}$$

for all p-dimensional vectors $\boldsymbol{x} \neq \boldsymbol{0}$.

The expression $\boldsymbol{x}^\top \boldsymbol{A}\boldsymbol{x}$ in the above definition is called a *quadratic form.* If $\boldsymbol{x}^\top \boldsymbol{A}\boldsymbol{x} \geq 0$ for all $\boldsymbol{x}$ then $\boldsymbol{A}$ is *positive semi-definite.* Definition 1.1 can be generalized to complex matrices; see Exercise 9.

Listing 1.5 shows how to extract the covariance matrix from an ENVI image. In line 21 the image is read into an array `G`, in which each pixel vector constitutes a row of the array. We will refer to such an array as a *data design matrix* or simply *data matrix*; see Section 2.3.1. Many of our IDL and Python programs will involve manipulations of the data matrix. Here, in line 22, it is passed to the IDL function `CORRELATE()` with keyword `COVARIANCE` set, which returns the covariance matrix. The result for the three VNIR bands 1, 2, and 3N of the Jülich image in Figure 1.1 is a symmetric 3×3 matrix:

Listing 1.5: Determining the covariance matrix of an image in ENVI/IDL.

```
PRO EX1_3

envi_select, title='Choose␣multispectral␣image', $
             fid=fid, dims=dims,pos=pos
IF (fid EQ -1) THEN BEGIN
   PRINT, 'cancelled'
   RETURN
ENDIF

envi_file_query, fid, fname=fname

cols = dims[2]-dims[1]+1
rows = dims[4]-dims[3]+1
bands = n_elements(pos)
pixels = cols*rows

; array with pixel vectors as rows
G = fltarr(bands,pixels)

FOR i=0,bands-1 DO $
   G[i,*]= envi_get_data(fid=fid,dims=dims,pos=pos[i])
C = correlate(G,/covariance,/double)

PRINT, 'Covariance␣matrix␣for␣image␣'+fname
PRINT, C

END
```

```
ENVI> EX1_3
'Covariance␣matrix␣for␣image␣E:\satellite␣images\juelich'
       87.117627       92.394480      -10.518867
       92.394480       115.83875      -77.420886
      -10.518867      -77.420886       374.61514
```

We will see how to show that this matrix is positive definite in Section 1.4.

1.3.5 Linear dependence and vector spaces

Vectors are said to be *linearly dependent* when any one can be expressed as a linear combination of the others. Here is a formal definition:

DEFINITION 1.2 *A set S of vectors $\boldsymbol{x}_1 \ldots \boldsymbol{x}_r$ is said to be* linearly dependent *if there exist scalars $c_1 \ldots c_r$, not all of which are zero, such that $\sum_{i=1}^r c_i \boldsymbol{x}_i = \boldsymbol{0}$. Otherwise they are* linearly independent.

A matrix is said to have *rank* r if the maximum number of linearly independent columns is r. If the $p \times p$ matrix $\boldsymbol{A} = (\boldsymbol{a}_1 \dots \boldsymbol{a}_p)$, where $\boldsymbol{a}_i$ is its ith column, is nonsingular, then it has full rank p. If this were not the case, then there must exist a set of scalars $c_1 \dots c_p$, not all of which are zero, for which

$$c_1\boldsymbol{a}_1 + \dots + c_p\boldsymbol{a}_p = \boldsymbol{A}\boldsymbol{c} = \boldsymbol{0}.$$

In other words, there would be a nontrivial solution to $\boldsymbol{A}\boldsymbol{c} = \boldsymbol{0}$, contradicting the fact that $\boldsymbol{A}$ is nonsingular.

The set S in Definition 1.2 is said to constitute a *basis* for a *vector space* V, comprising all vectors that can be expressed as a linear combination of the vectors in S. The number r of vectors in the basis is called the *dimension* of V. The vector space V is also an *inner product space* by virtue of the inner product definition Equation (1.8) in Subsection 1.3.1. Inner product spaces are elaborated upon in Appendix A.

1.4 Eigenvalues and eigenvectors

In image analysis, it is frequently necessary to solve an *eigenvalue problem*. In the simplest case, and the one which will concern us primarily, the eigenvalue problem consists of finding *eigenvectors* $\boldsymbol{u}$ and *eigenvalues* λ that satisfy the matrix equation

$$\boldsymbol{A}\boldsymbol{u} = \lambda\boldsymbol{u}, \tag{1.33}$$

where $\boldsymbol{A}$ is both symmetric and positive definite. Equation (1.33) can be written equivalently as

$$(\boldsymbol{A} - \lambda\boldsymbol{I})\boldsymbol{u} = \boldsymbol{0}, \tag{1.34}$$

so for a nontrivial solution we must have

$$|\boldsymbol{A} - \lambda\boldsymbol{I}| = 0. \tag{1.35}$$

This is known as the *characteristic equation* for the matrix $\boldsymbol{A}$. For instance, in the case of a 2×2 matrix eigenvalue problem,

$$\begin{pmatrix} a_{11} & a_{12} \\ a_{21} & a_{22} \end{pmatrix} \begin{pmatrix} u_1 \\ u_2 \end{pmatrix} = \lambda \begin{pmatrix} u_1 \\ u_2 \end{pmatrix}, \tag{1.36}$$

the characteristic equation is

$$(a_{11} - \lambda)(a_{22} - \lambda) - a_{12}^2 = 0,$$

which is a quadratic equation in λ with solutions

$$\begin{aligned} \lambda^{(1)} &= \frac{1}{2}\left(a_{11} + a_{22} + \sqrt{(a_{11} + a_{22})^2 - 4(a_{11}a_{22} - a_{12}^2)}\right) \\ \lambda^{(2)} &= \frac{1}{2}\left(a_{11} + a_{22} - \sqrt{(a_{11} + a_{22})^2 - 4(a_{11}a_{22} - a_{12}^2)}\right). \end{aligned} \tag{1.37}$$

Thus there are two eigenvalues and, correspondingly, two eigenvectors $\boldsymbol{u}^{(1)}$ and $\boldsymbol{u}^{(2)}$.* The eigenvectors can be obtained by first substituting $\lambda^{(1)}$ and then $\lambda^{(2)}$ into Equation (1.36) and solving for u_1 and u_2 each time. It is easy to show (Exercise 10) that the eigenvectors are orthogonal

$$(\boldsymbol{u}^{(1)})^\top \boldsymbol{u}^{(2)} = 0. \tag{1.38}$$

Moreover, since the left- and right-hand sides of Equation (1.33) can be multiplied by any constant, the eigenvectors can be always chosen to have unit length, $\|\boldsymbol{u}^{(1)}\| = \|\boldsymbol{u}^{(2)}\| = 1$. The matrix formed by two such eigenvectors, e.g.,

$$\boldsymbol{U} = (\boldsymbol{u}^{(1)}, \boldsymbol{u}^{(2)}) = \begin{pmatrix} u_1^{(1)} & u_1^{(2)} \\ u_2^{(1)} & u_2^{(2)} \end{pmatrix}, \tag{1.39}$$

is said to *diagonalize* the matrix $\boldsymbol{A}$. That is, if $\boldsymbol{A}$ is multiplied from the left by $\boldsymbol{U}^\top$ and from the right by $\boldsymbol{U}$, the result is a diagonal matrix with the eigenvalues along the diagonal:

$$\boldsymbol{U}^\top \boldsymbol{A} \boldsymbol{U} = \boldsymbol{\Lambda} = \begin{pmatrix} \lambda^{(1)} & 0 \\ 0 & \lambda^{(2)} \end{pmatrix}, \tag{1.40}$$

as can easily be verified. Note that $\boldsymbol{U}$ is an orthonormal matrix: $\boldsymbol{U}^\top \boldsymbol{U} = \boldsymbol{I}$ and therefore Equation (1.40) can be written as

$$\boldsymbol{A}\boldsymbol{U} = \boldsymbol{U}\boldsymbol{\Lambda}. \tag{1.41}$$

All of the above statements generalize in a straightforward fashion to square matrices of any size.

Suppose that $\boldsymbol{A}$ is a $p \times p$ symmetric matrix. Then its eigenvectors $\boldsymbol{u}^{(j)}$, $j = 1 \ldots p$, are orthogonal and any p-component vector $\boldsymbol{x}$ can be expressed as a linear combination of them,

$$\boldsymbol{x} = \sum_{j=1}^{p} \beta_j \boldsymbol{u}^{(j)}. \tag{1.42}$$

Multiplying from the right with $\boldsymbol{u}^{(i)\top}$,

$$\boldsymbol{u}^{(i)\top} \boldsymbol{x} \; (= \boldsymbol{x}^\top \boldsymbol{u}^{(i)}) = \sum_{j=1}^{p} \beta_j \boldsymbol{u}^{(i)\top} \boldsymbol{u}^{(j)} = \beta_i, \tag{1.43}$$

so that we have

$$\boldsymbol{x}^\top \boldsymbol{A} \boldsymbol{x} = \boldsymbol{x}^\top \sum_i \beta_i \lambda_i \boldsymbol{u}^{(i)} = \sum_i \beta_i^2 \lambda_i, \tag{1.44}$$

*The special case of degeneracy, which arises when the two eigenvalues are equal, will be ignored here.

Listing 1.6: Diagonalization of the covariance matrix in ENVI/IDL.

```
PRO EX1_4

envi_select, title='Choose␣multispectral␣image', $
             fid=fid, dims=dims,pos=pos
IF (fid EQ -1) THEN BEGIN
   PRINT, 'cancelled'
   RETURN
ENDIF

envi_file_query, fid, fname=fname

cols = dims[2]-dims[1]+1
rows = dims[4]-dims[3]+1
bands = n_elements(pos)
pixels = cols*rows

; data matrix
G = fltarr(bands,pixels)

FOR i=0,bands-1 DO $
 G[i,*]=envi_get_data(fid=fid,dims=dims,pos=pos[i])

C = correlate(G,/covariance)

PRINT, 'Eigenvalues'
PRINT, eigenql(C, eigenvectors=U, /double)
PRINT, 'Eigenvectors␣(as␣rows)'
PRINT, U

PRINT, 'Orthogonality'
PRINT, transpose(U)##U

END
```

where λ_i, $i = 1 \ldots p$, are the eigenvalues of $\boldsymbol{A}$. We can conclude, from the definition of positive definite matrices given in Section 1.2.4, that $\boldsymbol{A}$ is positive definite if and only if all of its eigenvalues λ_i, $i = 1 \ldots N$, are positive.

The eigenvalue problem can be solved in IDL with the built-in function `EIGENQL(A)`. This function returns an array of the eigenvalues of the symmetric matrix $\boldsymbol{A}$ and, optionally, the orthonormal matrix $\boldsymbol{U}^\top$ in the keyword `EIGENVECTORS`; see the program in Listing 1.6. Here is the output of the program with the VNIR bands 1, 2, and 3N of the Jülich image:

```
ENVI> Ex1_3
Eigenvalues
          934.08442        398.98373        4.0842497
```

```
Eigenvectors (as rows)
      -0.64092380        -0.71641836        0.27561100
      -0.22689155        -0.16619583       -0.95963492
       0.73330547        -0.67758666       -0.056030408
Orthogonality
        1.0000000         0.00000000    5.0306981e-017
       0.00000000          1.0000000    4.1633363e-017
   5.0306981e-017     4.1633363e-017         1.0000000
```

Notice that, due to rounding errors, $\boldsymbol{U}^\top \boldsymbol{U}$ has finite — but very small — off-diagonal elements. The eigenvalues are all positive, so the covariance matrix calculated by the program is positive definite.

In Python, the eigenvalues and eigenvectors of real symmetric (or complex Hermitian) matrices can be calculated with `numpy.linalg.eigh()`.

1.5 Singular value decomposition

Rewriting Equation (1.41), we get a special form of *singular value decomposition* (SVD) for symmetric matrices:

$$\boldsymbol{A} = \boldsymbol{U}\boldsymbol{\Lambda}\boldsymbol{U}^\top. \tag{1.45}$$

This says that any symmetric matrix $\boldsymbol{A}$ can be factored into the product of an orthonormal matrix $\boldsymbol{U}$ times a diagonal matrix $\boldsymbol{\Lambda}$ whose diagonal elements are the eigenvalues of $\boldsymbol{A}$ times the transpose of $\boldsymbol{U}$. This is also called the *eigen-decomposition* or *spectral decomposition* of $\boldsymbol{A}$ and can be written alternatively as a sum over outer products of the eigenvectors (Exercise 11),

$$\boldsymbol{A} = \sum_{i=1}^{p} \lambda_i \boldsymbol{u}^{(i)} \boldsymbol{u}^{(i)\top}. \tag{1.46}$$

For the general form for singular value decomposition of non-symmetric or non-square matrices, see Press et al. (2002).

SVD is a powerful tool for the solution of systems of linear equations, and is often used when a solution cannot be determined by other numerical algorithms. To invert a non-singular symmetric matrix $\boldsymbol{A}$, we simply write

$$\boldsymbol{A}^{-1} = \boldsymbol{U}\boldsymbol{\Lambda}^{-1}\boldsymbol{U}^\top, \tag{1.47}$$

since $\boldsymbol{U}^{-1} = \boldsymbol{U}^\top$. The matrix $\boldsymbol{\Lambda}^{-1}$ is the diagonal matrix whose diagonal elements are the inverses of the diagonal elements of $\boldsymbol{\Lambda}$. If $\boldsymbol{A}$ is singular (has no inverse), clearly at least one of its eigenvalues is zero.

The matrix $\boldsymbol{A}$ is said to be *ill-conditioned* or *nearly singular* if one or more of the diagonal elements of $\boldsymbol{\Lambda}$ is close to zero. SVD detects this situation effectively, since the factorization in Equation (1.45) is always possible, even if $\boldsymbol{A}$ is truly singular. The IDL procedure for singular value decomposition is `SVDC`. For example, the program

```
PRO svd_test
   b = [1,2,3]
; an almost singular matrix
   A = transpose(b)##b+randomu(seed,3,3)*0.001
; a symmetric almost singular matrix
   A = A + transpose(A)
   PRINT,'det(A) = ',determ(A)
; singular value decomposition A = U L V^T
   svdc,A,L,U,V,/double
   PRINT,'L ='
   PRINT, diag_matrix(L)
   PRINT,'U = V ='
   PRINT, U
END
```

generates the output

```
% Compiled module: SVD_TEST.
det(A) = -4.21913e-006
L =
        28.002635        0.00000000        0.00000000
       0.00000000      0.0010748116        0.00000000
       0.00000000        0.00000000     0.00014015930
U = V =
      -0.26730887        0.88174905       -0.38867026
      -0.53453525        0.19991515        0.82116137
      -0.80175934       -0.42726168       -0.41788685
```

indicating that $\boldsymbol{A}$ is ill-conditioned (two of the three diagonal elements of $\boldsymbol{\Lambda}$ are close to zero).

If $\boldsymbol{A}$ is singular and we order the eigenvalues and eigenvectors in order of decreasing eigenvalue, then the eigendecomposition reads

$$\boldsymbol{A} = \sum_{i=1}^{r} \lambda_i \boldsymbol{u}^{(i)} \boldsymbol{u}^{(i)\top}, \tag{1.48}$$

where r is the number of nonzero eigenvalues. Accordingly, Equation (1.45) becomes

$$\boldsymbol{A} = \boldsymbol{U}_r \boldsymbol{\Lambda}_r \boldsymbol{U}_r^\top, \tag{1.49}$$

where

$$\boldsymbol{U}_r = (\boldsymbol{u}^{(1)}, \dots, \boldsymbol{u}^{(r)}), \quad \boldsymbol{\Lambda} = \begin{pmatrix} \lambda_1 & 0 & \cdots & 0 \\ 0 & \lambda_2 & \cdots & 0 \\ \vdots & \vdots & \ddots & \vdots \\ 0 & 0 & \cdots & \lambda_r \end{pmatrix}.$$

The *pseudoinverse* of the symmetric, singular matrix $\boldsymbol{A}$ is defined as

$$\boldsymbol{A}^{+} = \boldsymbol{U}_r \boldsymbol{\Lambda}_r^{-1} \boldsymbol{U}_r^{\top}. \tag{1.50}$$

The pseudoinverse has the property that $\boldsymbol{A}^{+}\boldsymbol{A} - \boldsymbol{I}$, where $\boldsymbol{I}$ is the $p \times p$ identity matrix, is minimized (Press et al., 2002).

1.6 Finding minima and maxima

In order to enhance some desirable property of a remote sensing image, such as signal-to-noise ratio or spread in intensity, we often need to take derivatives with respect to vectors. A *vector partial derivative* operator is written in the form $\frac{\partial}{\partial \boldsymbol{x}}$. For instance, in two dimensions,

$$\frac{\partial}{\partial \boldsymbol{x}} = \begin{pmatrix} 1 \\ 0 \end{pmatrix} \frac{\partial}{\partial x_1} + \begin{pmatrix} 0 \\ 1 \end{pmatrix} \frac{\partial}{\partial x_2} = \boldsymbol{i}\frac{\partial}{\partial x_1} + \boldsymbol{j}\frac{\partial}{\partial x_2}.$$

Such an operator (also called a *gradient*) arranges the partial derivatives of any scalar function of the vector $\boldsymbol{x}$ with respect to each of its components into a column vector, e.g.,

$$\frac{\partial f(\boldsymbol{x})}{\partial \boldsymbol{x}} = \begin{pmatrix} \frac{\partial f(\boldsymbol{x})}{\partial x_1} \\ \frac{\partial f(\boldsymbol{x})}{\partial x_2} \end{pmatrix}.$$

Many of the operations with vector derivatives correspond exactly to operations with ordinary scalar derivatives (and can all be verified easily by writing out the expressions component by component):

$$\frac{\partial}{\partial \boldsymbol{x}}(\boldsymbol{x}^{\top}\boldsymbol{y}) = \boldsymbol{y} \quad \text{analogous to} \quad \frac{\partial}{\partial x}xy = y$$

$$\frac{\partial}{\partial \boldsymbol{x}}(\boldsymbol{x}^{\top}\boldsymbol{x}) = 2\boldsymbol{x} \quad \text{analogous to} \quad \frac{\partial}{\partial x}x^2 = 2x.$$

For quadratic forms, we have

$$\frac{\partial}{\partial \boldsymbol{x}}(\boldsymbol{x}^{\top}\boldsymbol{A}\boldsymbol{y}) = \boldsymbol{A}\boldsymbol{y}$$

$$\frac{\partial}{\partial \boldsymbol{y}}(\boldsymbol{x}^{\top}\boldsymbol{A}\boldsymbol{y}) = \boldsymbol{A}^{\top}\boldsymbol{x}$$

and

$$\frac{\partial}{\partial \boldsymbol{x}}(\boldsymbol{x}^{\top}\boldsymbol{A}\boldsymbol{x}) = \boldsymbol{A}\boldsymbol{x} + \boldsymbol{A}^{\top}\boldsymbol{x}.$$

Note that, if $\boldsymbol{A}$ is a symmetric matrix, this last equation can be written

$$\frac{\partial}{\partial \boldsymbol{x}}(\boldsymbol{x}^{\top}\boldsymbol{A}\boldsymbol{x}) = 2\boldsymbol{A}\boldsymbol{x}. \tag{1.51}$$

Suppose x^* is a stationary point of the function $f(x)$, by which is meant that its first derivative vanishes at that point:

$$\frac{d}{dx}f(x^*) = \frac{d}{dx}f(x)\Big|_{x=x^*} = 0; \tag{1.52}$$

see Figure 1.5. Then $f(x^*)$ is a local minimum if the second derivative at x^* is positive,

$$\frac{d^2}{dx^2}f(x^*) > 0.$$

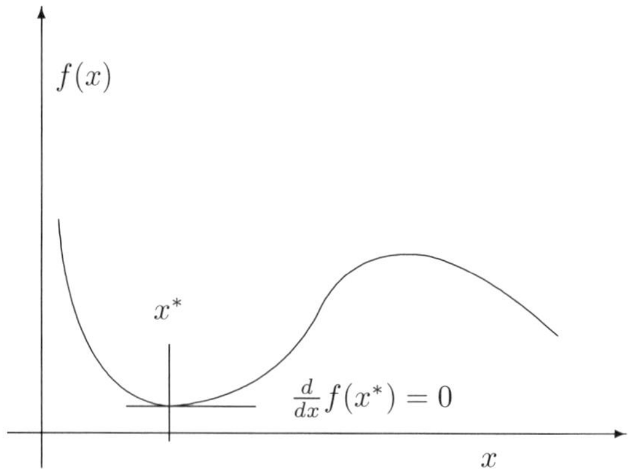

FIGURE 1.5
A function of one variable with a minimum at x^*.

This becomes obvious if $f(x)$ is expanded in a *Taylor series* about x^*,

$$f(x) = f(x^*) + (x - x^*)\frac{d}{dx}f(x^*) + \frac{1}{2}(x - x^*)^2\frac{d^2}{dx^2}f(x^*) + \dots . \tag{1.53}$$

The second term is zero, so for $|x - x^*|$ sufficiently small, $f(x)$ is approximately a quadratic function

$$f(x) \approx f(x^*) + \frac{1}{2}(x - x^*)^2\frac{d^2}{dx^2}f(x^*), \tag{1.54}$$

with a minimum at x^* when the second derivative is positive, a maximum when it is negative, and a point of inflection when it is zero.

The situation is similar for scalar functions of a vector. In this case the Taylor expansion is

$$f(\boldsymbol{x}) = f(\boldsymbol{x}^*) + (\boldsymbol{x} - \boldsymbol{x}^*)^\top\frac{\partial f(\boldsymbol{x}^*)}{\partial \boldsymbol{x}} + \frac{1}{2}(\boldsymbol{x} - \boldsymbol{x}^*)^\top\boldsymbol{H}(\boldsymbol{x} - \boldsymbol{x}^*) + \dots, \tag{1.55}$$

where $\boldsymbol{H}$ is called the *Hessian matrix*. Its elements are given by

$$(\boldsymbol{H})_{ij} = \frac{\partial^2 f(\boldsymbol{x}^*)}{\partial x_i \partial x_j}, \tag{1.56}$$

and it is a symmetric matrix.* At the stationary point, the vector derivative with respect to $\boldsymbol{x}$ vanishes,

$$\frac{\partial f(\boldsymbol{x}^*)}{\partial \boldsymbol{x}} = \boldsymbol{0},$$

so we get the second-order approximation

$$f(\boldsymbol{x}) \approx f(\boldsymbol{x}^*) + \frac{1}{2}(\boldsymbol{x} - \boldsymbol{x}^*)^\top \boldsymbol{H}(\boldsymbol{x} - \boldsymbol{x}^*). \tag{1.57}$$

Now the condition for a local minimum is clearly that the Hessian matrix be positive definite at the point $\boldsymbol{x}^*$. Note that the Hessian can be expressed in terms of an outer product of vector derivatives as

$$\frac{\partial}{\partial \boldsymbol{x}} \frac{\partial f(\boldsymbol{x})}{\partial \boldsymbol{x}^\top} = \frac{\partial^2 f(\boldsymbol{x})}{\partial \boldsymbol{x} \partial \boldsymbol{x}^\top}. \tag{1.58}$$

Suppose that we wish to find the position $\boldsymbol{x}^*$ of a minimum (or maximum) of a scalar function $f(\boldsymbol{x})$. If there are no constraints, then we solve the set of equations

$$\frac{\partial f(\boldsymbol{x})}{\partial x_i} = 0, \quad i = 1, 2 \ldots$$

or, in terms of our notation for vector derivatives,

$$\frac{\partial f(\boldsymbol{x})}{\partial \boldsymbol{x}} = \boldsymbol{0}, \tag{1.59}$$

and examine the Hessian matrix at the solution point. However, suppose that $\boldsymbol{x}$ is *constrained* by the condition

$$h(\boldsymbol{x}) = 0. \tag{1.60}$$

For example, in two dimensions we might have the constraint

$$h(\boldsymbol{x}) = x_1^2 + x_2^2 - 1 = 0,$$

which requires $\boldsymbol{x}$ to lie on a circle of radius 1. An obvious procedure would be to solve the above equation for x_1, say, and then substitute the result into Equation (1.59) to determine x_2. This method is not always practical, however. It may be the case, for example, that the constraint Equation (1.60) cannot be solved analytically.

*Because the order of partial differentiation does not matter.

A more convenient and generally applicable way to find an extremum of $f(\boldsymbol{x})$ subject to $h(\boldsymbol{x}) = 0$ is to determine an *unconstrained* minimum or maximum of the expression

$$L(\boldsymbol{x}) = f(\boldsymbol{x}) + \lambda h(\boldsymbol{x}). \tag{1.61}$$

This is called a *Lagrange function* and λ is a *Lagrange multiplier.* The Lagrange multiplier is treated as though it were an additional variable. To find the extremum, we solve the set of equations

$$\begin{aligned} \frac{\partial}{\partial \boldsymbol{x}}(f(\boldsymbol{x}) + \lambda h(\boldsymbol{x})) &= \boldsymbol{0}, \\ \frac{\partial}{\partial \lambda}(f(\boldsymbol{x}) + \lambda h(\boldsymbol{x})) &= 0 \end{aligned} \tag{1.62}$$

for $\boldsymbol{x}$ and λ. To see this, note that a minimum or maximum need not generally occur at a point $\boldsymbol{x}^*$ for which

$$\frac{\partial}{\partial \boldsymbol{x}} f(\boldsymbol{x}^*) = \boldsymbol{0}, \tag{1.63}$$

because of the presence of the constraint. This possibility is taken into account in the first of Equations (1.62), which at an extremum reads

$$\frac{\partial}{\partial \boldsymbol{x}} f(\boldsymbol{x}^*) = -\lambda \frac{\partial}{\partial \boldsymbol{x}} h(\boldsymbol{x}^*). \tag{1.64}$$

It implies that, if $\lambda \neq 0$, any small change in $\boldsymbol{x}$ away from $\boldsymbol{x}^*$ causing a change in $f(\boldsymbol{x}^*)$, i.e., any change in $\boldsymbol{x}$ which is not orthogonal to the gradient of $f(\boldsymbol{x}^*)$, would be accompanied by a proportional change in $h(\boldsymbol{x}^*)$. This would necessarily violate the constraint, so $\boldsymbol{x}^*$ is an extremum. The second equation is just the constraint $h(\boldsymbol{x}^*) = 0$ itself. For more detailed justifications of this procedure, see Bishop (1995), Appendix C; Milman (1999), Chapter 14; or Cristianini and Shawe–Taylor (2000), Chapter 5.

As a first example illustrating the Lagrange method, let $f(\boldsymbol{x}) = ax_1^2 + bx_2^2$ and $h(\boldsymbol{x}) = x_1 + x_2 - 1$. Then we get the three equations

$$\begin{aligned} \frac{\partial}{\partial x_1}(f(\boldsymbol{x}) + \lambda h(\boldsymbol{x})) &= 2ax_1 + \lambda = 0 \\ \frac{\partial}{\partial x_2}(f(\boldsymbol{x}) + \lambda h(\boldsymbol{x})) &= 2bx_2 + \lambda = 0 \\ \frac{\partial}{\partial \lambda}(f(\boldsymbol{x}) + \lambda h(\boldsymbol{x})) &= x_1 + x_2 - 1 = 0. \end{aligned}$$

The solution for $\boldsymbol{x}$ is

$$x_1 = \frac{b}{a+b}, \quad x_2 = \frac{a}{a+b}.$$

Listing 1.7: Principal components analysis in Python.

```
#!/usr/bin/env python
#Name:   ex1_3.py
IMPORT auxil.auxil as auxil
FROM numpy IMPORT *
FROM osgeo IMPORT gdal
FROM osgeo.gdalconst IMPORT GA_ReadOnly,GDT_Float32

DEF main():
    gdal.AllRegister()
    infile = auxil.select_infile()
    IF infile:
        inDataset = gdal.Open(infile,GA_ReadOnly)
        cols = inDataset.RasterXSize
        rows = inDataset.RasterYSize
        bands = inDataset.RasterCount
    ELSE:
        RETURN

#   spectral and spatial subsets
    pos =  auxil.select_pos(bands)
    bands = LEN(pos)
    x0,y0,rows,cols=auxil.select_dims([0,0,rows,cols])

#   data matrix
    G = zeros((rows*cols,LEN(pos)))
    k = 0
    FOR b IN pos:
        band = inDataset.GetRasterBand(b)
        tmp = band.ReadAsArray(x0,y0,cols,rows)\
                              .astype(FLOAT).ravel()
        G[:,k] = tmp - mean(tmp)
        k += 1

#   covariance matrix
    C = mat(G).T*mat(G)/(cols*rows-1)

#   diagonalize
    lams,U = linalg.eigh(C)

#   sort
    idx = argsort(lams)[::-1]
    lams = lams[idx]
    U = U[:,idx]

#   project
    PCs = reshape(array(G*U),(rows,cols,bands))
```

Listing 1.8: Principal components analysis in Python (continued).

```
#   write to disk
    outfile,fmt = auxil.select_outfilefmt()
    IF outfile:
        driver = gdal.GetDriverByName(fmt)
        outDataset = driver.Create(outfile,
                        cols,rows,bands,GDT_Float32)
        projection = inDataset.GetProjection()
        geotransform = inDataset.GetGeoTransform()
        IF geotransform IS NOT None:
            gt = LIST(geotransform)
            gt[0] = gt[0] + x0*gt[1]
            gt[3] = gt[3] + y0*gt[5]
            outDataset.SetGeoTransform(TUPLE(gt))
        IF projection IS NOT None:
            outDataset.SetProjection(projection)
        FOR k IN RANGE(bands):
            outBand = outDataset.GetRasterBand(k+1)
            outBand.WriteArray(PCs[:,:,k],0,0)
            outBand.FlushCache()
        outDataset = None
    inDataset = None

IF __name__ == '__main__':
    main()
```

A second — and very important — example: Let us find the maximum of

$$f(\boldsymbol{x}) = \boldsymbol{x}^\top \boldsymbol{C}\boldsymbol{x},$$

where $\boldsymbol{C}$ is symmetric positive definite, subject to the constraint

$$h(\boldsymbol{x}) = \boldsymbol{x}^\top \boldsymbol{x} - 1 = 0.$$

The Lagrange function is

$$L(\boldsymbol{x}) = \boldsymbol{x}^\top \boldsymbol{C}\boldsymbol{x} - \lambda(\boldsymbol{x}^\top \boldsymbol{x} - 1).$$

Any extremum of L must occur at a value of $\boldsymbol{x}$ for which

$$\frac{\partial L}{\partial \boldsymbol{x}} = 2\boldsymbol{C}\boldsymbol{x} - 2\lambda\boldsymbol{x} = 0,$$

which is the eigenvalue problem

$$\boldsymbol{C}\boldsymbol{x} = \lambda\boldsymbol{x}. \tag{1.65}$$

Let $\boldsymbol{u}$ be an eigenvector with eigenvalue λ. Then

$$f(\boldsymbol{u}) = \boldsymbol{u}^\top \boldsymbol{C}\boldsymbol{u} = \lambda\boldsymbol{u}^\top \boldsymbol{u} = \lambda.$$

Listing 1.9: Principal components analysis in ENVI/IDL.

```
PRO EX1_5
   COMPILE_OPT IDL2

   envi_select, title='Choose␣multispectral␣image', $
                fid=fid, dims=dims,pos=pos
   IF (fid EQ -1) THEN BEGIN
      PRINT, 'cancelled'
      RETURN
   ENDIF
   envi_file_query, fid, fname=fname
   cols = dims[2]-dims[1]+1
   rows = dims[4]-dims[3]+1
   bands = n_elements(pos)
   pixels = cols*rows
; data matrix
   G = fltarr(bands,pixels)
; read in image and subtract means
   FOR i=0,bands-1 DO BEGIN
     temp = envi_get_data(fid=fid,dims=dims,pos=pos[i])
     G[i,*] = temp-mean(temp)
   END
; principal components transformation
   C = correlate(G,/covariance,/double)
   void = eigenql(C, eigenvectors=U, /double)
   PCs = G##transpose(U)
; BSQ array
   image = fltarr(cols,rows,bands)
   FOR i = 0, bands-1 DO image[*,*,i] = $
      reform(PCs[i,*],cols,rows)
; map tie point
   map_info = envi_get_map_info(fid=fid)
   envi_convert_file_coordinates, fid, $
      dims[1], dims[3], e, n, /to_map
   map_info.mc = [0D,0D,e,n]
   envi_enter_data, image, map_info = map_info
END
```

So to maximize $f(\boldsymbol{x})$, we must choose the eigenvector of $\boldsymbol{C}$ with maximum eigenvalue.

If $\boldsymbol{C}$ is the covariance matrix of a multispectral image, then, as we shall see in Chapter 3, the above maximization corresponds to a *principal components analysis* (PCA). Listings 1.7 and 1.8 give a rudimentary script for doing PCA on a multispectral image in numerical Python. After estimating the covariance matrix in line 35 (see Chapter 2), the eigenvalue problem, Equation (1.65), is solved in line 38 using the `numpy.linalg.eigh()` function. The order of

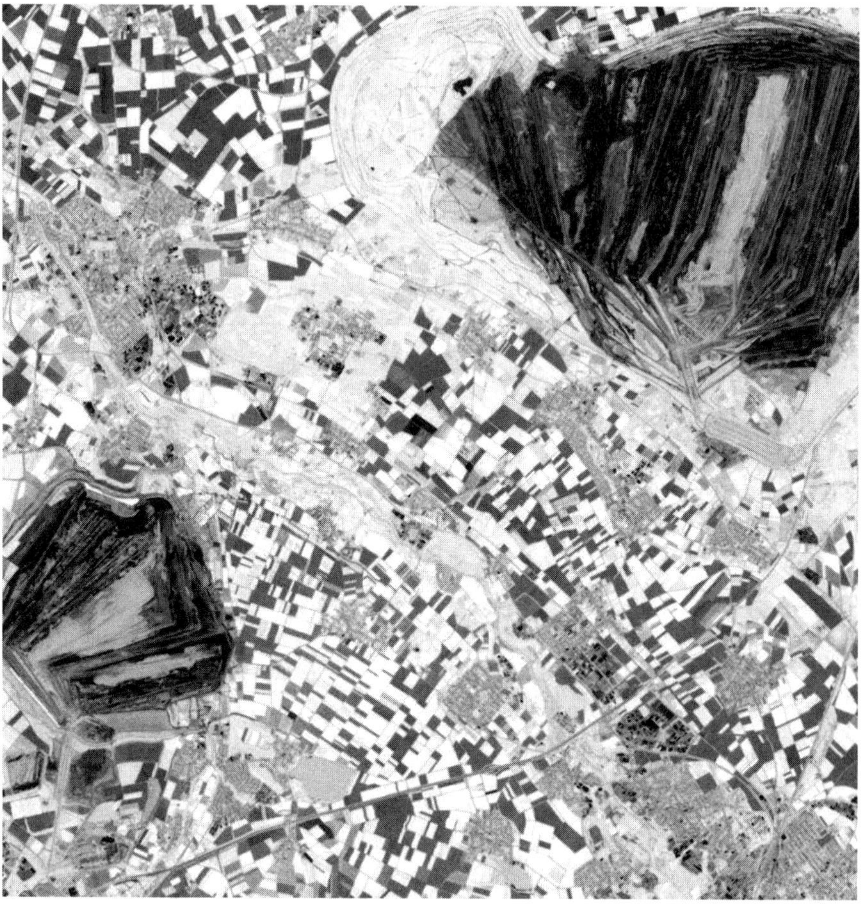

FIGURE 1.6
First principal component of the VNIR spectral bands for the Jülich scene calculated with the script in Listings 1.7 and 1.8 and displayed with the Python script `dispms.py`.

eigenvalues `lams` and eigenvectors (the columns of `U`) is undefined, so the eigenvalues and eigenvectors are sorted into decreasing order in lines 41–43. Then the principal components are calculated by projecting the original image bands along the eigenvectors and the resulting data matrix is rearranged in BIP format (line 46). Finally, the principal components are stored to disk, correcting the georeferencing information in case a spatial subset of the input image was chosen with a new origin `x0,y0`, (lines 9–20 in Listing 1.8). The first principal component of VNIR bands 1, 2, and 3N of the Jülich image determined with the Python script is shown in Figure 1.6. The Python script `dispms.py`, described in Appendix D, can be used to view histogram-enhanced

Listing 1.10: Principal components analysis with the new ENVI API.

```
 1 PRO EX1_6
 2    COMPILE_OPT IDL2
 3 ; start ENVI 5 GUI
 4    envi5 = ENVI()
 5 ; read image
 6    UI = envi5.UI
 7    inRaster = UI.SelectInputData(/raster, $
 8                /disable_sub_rect,bands=bands)
 9 ; get image bands in BIP format
10    data = inRaster.GetData(interleave='bip', $
11                                      bands=bands)
12    sz = size(data)
13    num_bands = sz[1]
14    cols = sz[2]
15    rows = sz[3]
16 ; data matrix
17    G = reform(data,num_bands,cols*rows)
18 ; subtract means
19    FOR i=0,num_bands-1 DO G[i,*] = G[i,*] $
20                           - mean(G[i,*])
21 ; principal components transformation
22    C = correlate(G,/covariance,/double)
23    void = eigenql(C, eigenvectors=U, /double)
24    PCs = G##transpose(U)
25    PCs = reform(PCs,num_bands,cols,rows)
26 ; create and save the PCs as a raster
27    tempFile = envi5.GetTemporaryFilename()
28    outRaster = envi5.CreateRaster(tempFile,PCs, $
29             inherits_from=inRaster,interleave='bip')
30    outRaster.Save
31 ; display the original and transformed images
32    view = envi5.GetView()
33    layer1 = view.CreateLayer(inRaster,bands=[0,1,2])
34    layer2 = view.CreateLayer(outRaster,bands=[0,1,2])
35 END
```

RGB composite band combinations of any multi-band image stored on disk.

Listing 1.9 codes a similar procedure for doing PCA with ENVI/IDL. After solving the eigenvalue problem in line 24, the principal components (PCs) are calculated by projecting the original image bands along the eigenvectors (line 25). Note that the eigenvector matrix has to be transposed in this case, because, unlike its Python equivalent, EIGENQL() returns the eigenvectors in the *rows* of the IDL array U. Then the transformed data matrix is rearranged in BSQ format and returned to ENVI with the ENVI_ENTER_DATA procedure, line 35. A map tie-point is recalculated with the ENVI_CONVERT_FILE_COORDINATES procedure (lines 31 to 34), again in case the user happens to choose a spatial subset of the original image with a new origin.

Finally, Listing 1.10 illustrates the use of the ENVI API introduced in ENVI Version 5.0. The principal components image along with the original image are displayed in so-called *views* in the new ENVI 5 GUI, lines 32–34. In lines 28–29 the *metadata* (the image dimensions and map projection information) of the PC image are simply inherited from the input image. For this reason, spatial subsetting is disabled (line 8). We will prefer to continue to code the ENVI/IDL examples in the traditional ENVI Classic style for compatibility with earlier ENVI versions.

1.7 Exercises

1. Demonstrate that the definition

$$\boldsymbol{x}^\top \boldsymbol{y} = \|\boldsymbol{x}\|\|\boldsymbol{y}\|\cos\theta$$

   is equivalent to

$$\boldsymbol{x}^\top \boldsymbol{y} = x_1 y_1 + x_2 y_2.$$

   Hint: Use the trigonometric identity $\cos(\alpha-\beta) = \cos\alpha\cos\beta + \sin\alpha\sin\beta$.

2. Show that the outer product of two two-dimensional vectors is a singular matrix.

3. Verify the matrix identity:

$$(\boldsymbol{AB})^\top = \boldsymbol{B}^\top \boldsymbol{A}^\top$$

   in IDL or Python. In Python, you must import the `numpy` package.

4. Show that three two-dimensional vectors representing three points, all lying on the same line, are linearly dependent.

5. Show that the determinant of a symmetric 2×2 matrix is given by the product of its eigenvalues.

6. Prove that the inverse of a symmetric nonsingular matrix is symmetric.

7. Prove that the eigenvectors of $\boldsymbol{A}^{-1}$ are the same as those of $\boldsymbol{A}$, but with reciprocal eigenvalues.

8. Prove the identity

$$\boldsymbol{x}^\top \boldsymbol{A}\boldsymbol{x} = \mathrm{tr}(\boldsymbol{A}\boldsymbol{x}\boldsymbol{x}^\top)$$

   with the aid of the second of Equations (1.30).

9. A square complex matrix $\boldsymbol{A}$ is said to be *Hermitian* if $\boldsymbol{A}^\dagger = \boldsymbol{A}$, which is a generalization of Definition 1.1 for a symmetric real matrix. It is positive semi-definite if $\boldsymbol{x}^\dagger \boldsymbol{A}\boldsymbol{x} \ge 0$ for all nonzero complex vectors $\boldsymbol{x}$. The matrix $\boldsymbol{B} = \boldsymbol{A}^\dagger \boldsymbol{A}$ is obviously Hermitian (why?). Demonstrate in IDL or Python, by generating some random complex matrices $\boldsymbol{A}$, that the eigenvalues of $\boldsymbol{B}$ are real and positive; in other words, that $\boldsymbol{B}$ is positive definite.

10. Prove that the eigenvectors or a 2×2 symmetric matrix are orthogonal.

11. Demonstrate the equivalence of Equations (1.45) and (1.46) for a symmetric 2×2 matrix.

12. Prove, from Equation (1.46), that the trace of a symmetric matrix is the sum of its eigenvalues.

13. Differentiate the function
$$\frac{1}{\boldsymbol{x}^\top \boldsymbol{A}\boldsymbol{y}}$$
with respect to $\boldsymbol{y}$.

14. Calculate the eigenvectors of the (nonsymmetric!) matrix
$$\begin{pmatrix} 1 & 2 & 3 \\ 4 & 5 & 6 \\ 7 & 8 & 9 \end{pmatrix}$$
with IDL or Python. In Python you require the `numpy.linalg` package.

15. Plot the function $f(\boldsymbol{x}) = x_1^2 - x_2^2$ with IDL. Find its minima and maxima subject to the constraint $h(\boldsymbol{x}) = x_1^2 + x_2^2 - 1 = 0$.

16. Modify the program in Listing 1.9 or 1.10 to make use of the IDL function `PCOMP()`.

2

Image Statistics

In a multispectral or a synthetic aperture radar image, a given pixel value $g(i, j)$, derived from the measured radiation field at a satellite sensor, is never exactly reproducible. It is the outcome of a complex measurement influenced by instrument noise, atmospheric conditions, changing illumination and so forth. It may be assumed, however, that there is an underlying random mechanism with an associated probability distribution which restricts the possible outcomes in some way. Each time we make an observation, we are sampling from that probability distribution or, put another way, we are observing a different possible *realization* of the random mechanism. In this chapter, some basic statistical concepts for multispectral and SAR images viewed as random mechanisms will be introduced.

2.1 Random variables

A *random variable* can be used to represent a quantity, in the present context an image gray-scale value, which changes in an unpredictable way each time it is observed. In order to make a precise definition, let us consider some chance experiment which has a set Ω of possible *outcomes*. This set is referred to as the *sample space* for the experiment. Subsets of Ω are called *events*. An event will be said to have *occurred* whenever the outcome of the experiment is contained within it.

To make this clearer, consider the random experiment consisting of the throw of two dice. The sample space is the set of 36 possible outcomes

$$\Omega = \{(1, 1), (1, 2), (2, 1) \dots (6, 6)\}.$$

An event is then, for example, that the sum of the points is 7. It is the subset

$$\{(1, 6), (2, 5), (3, 4), (4, 3), (5, 2), (6, 1)\}$$

of the sample space. If, for instance, $(3, 4)$ is thrown, then the event has occurred.

DEFINITION 2.1 *A random variable* $Z : \Omega \mapsto \mathbb{R}$ *is a function which maps all outcomes onto the set* $\mathbb{R}$ *of real numbers such that the set*

$$\{\omega \in \Omega \mid Z(\omega) \leq z\}$$

is an event, i.e., a subset of Ω. *This subset is usually abbreviated as* $\{Z \leq z\}$.

Thus, for the throw of two dice, the sum of points S is a random variable, since it maps all outcomes onto real numbers:

$$S(1,1) = 2,\ S(1,2) = S(2,1) = 3,\ \dots\ S(6,6) = 12,$$

and sets such as

$$\{S \leq 4\} = \{(1,1),(1,2),(2,1),(1,3),(3,1),(2,2)\}$$

are subsets of the sample space. The set $\{S \leq 1\}$ is the empty set, whereas $\{S \leq 12\} = \Omega$, the entire sample space.

On the basis of the *probabilities* for the individual outcomes, we can associate a function $P(z)$ with the random variable Z as follows:

$$P(z) = \Pr(Z \leq z).$$

This is the probability of observing the event that the random variable Z takes on a value less than or equal to z. The probability of an event may be thought of as the relative frequency with which it occurs in n repetitions of a random experiment in the limit as $n \to \infty$. (For the complete, axiomatic definition, see, e.g., Freund (1992).) In the dice example, the probability of throwing a four or less is

$$P(4) = \Pr(S \leq 4) = 6/36 = 1/6,$$

for instance.

DEFINITION 2.2 *Given the random variable* Z, *then*

$$P(z) = \Pr(Z \leq z), \quad -\infty < z < \infty, \tag{2.1}$$

is called its distribution function.

2.1.1 Discrete random variables

When, as in the case of the dice throw, a random variable Z is discrete and takes on values $z_1 < z_2 < z_3 < \dots$, then the probabilities of the separate outcomes

$$p(z_i) = \Pr(Z = z_i) = P(z_i) - P(z_{i-1}), \quad i = 1, 2 \dots$$

are said to constitute the *mass function* for the random variable. This is best illustrated with a practical example. As we shall see in Chapter 7, the evaluation of a land cover classification model involves repeated trials with a finite number of independent test observations, keeping track of the number of times the model fails to predict the correct land cover category. This will lead us to consideration of a discrete random variable having the so-called *binomial distribution*. We can derive its mass function as follows.

Let θ be the probability of failure in a single trial. The probability of getting y misclassifications (and hence $n-y$ correct classifications) in n trials *in a specific sequence* is

$$\theta^y(1-\theta)^{n-y}.$$

In this expression there is a factor θ for each of the y misclassifications and a factor $(1-\theta)$ for each of the $n-y$ correct classifications. Taking the product is justified by the assumption that the trials are independent of each other. The number of such sequences is just the number of ways of selecting y trials from n possible ones. This is given by the *binomial coefficient*

$$\binom{n}{y} = \frac{n!}{(n-y)!\,y!}, \tag{2.2}$$

so that the probability for y misclassifications in n trials is

$$\binom{n}{y}\theta^y(1-\theta)^{n-y}.$$

A discrete random variable Y is said to be *binomially distributed* with parameters n and θ if its mass function is given by

$$p_{n,\theta}(y) = \begin{cases} \binom{n}{y}\theta^y(1-\theta)^{n-y} & \text{for } y = 0,1,2\ldots n \\ 0 & \text{otherwise.} \end{cases} \tag{2.3}$$

Note that the values of $p_{n,\theta}(y)$ are the terms in the binomial expansion of

$$[\theta + (1-\theta)]^n = 1^n = 1,$$

so the sum over the probabilities equals 1, as it should.

2.1.2 Continuous random variables

In the case of continuous random variables, which we are in effect dealing with when we speak of pixel intensities, the distribution function is not expressed in terms of the discrete probabilities of a mass function, but rather in terms of a *probability density function* $p(z)$.

DEFINITION 2.3 *A function with values $p(z)$, defined over the set of all real numbers, is called a probability density function of the continuous random* *variable Z if and only if*

$$\Pr(a \leq Z \leq b) = \int_a^b p(z)dz \tag{2.4}$$

for any real numbers $a \leq b$.

The quantity $p(z)dz$ is the probability that the associated random variable Z lies within the infinitesimal interval $[z, z + dz]$. The integral (sum) over all such intervals is one:

$$\int_{-\infty}^{\infty} p(z)dz = 1. \tag{2.5}$$

The distribution function $P(z)$ can be written in terms of the density function and vice versa as

$$P(z) = \int_{-\infty}^{z} p(t)dt, \quad p(z) = \frac{d}{dz}P(z). \tag{2.6}$$

The distribution function has the limiting values

$$P(-\infty) = 0, \quad P(\infty) = \int_{-\infty}^{\infty} p(t)dt = 1.$$

The following theorem can often be used to determine the probability density of a function of some random variable whose density is known. For a proof, see Freund (1992).

THEOREM 2.1

Let $p_z(z)$ be the density function for random variable Z and

$$y = u(z) \tag{2.7}$$

a monotonic function of z for all values of z for which $p_z(z) \neq 0$. For these z values, Equation (2.7) can be solved for z to give $z = w(y)$. Then the density function of the random variable $Y = u(Z)$ is given by

$$p_y(y) = p_z(z)\left|\frac{dz}{dy}\right| = p_z(w(y))\left|\frac{dz}{dy}\right|. \tag{2.8}$$

As an example of the application of Theorem 2.1, suppose that a random variable Z has the exponential distribution (Section 2.1.5) with density function

$$p_z(z) = \begin{cases} e^{-z} & \text{for } z > 0 \\ 0 & \text{otherwise,} \end{cases}$$

and we wish to determine the probability density of the random variable $Y = \sqrt{Z}$. The monotonic function $y = u(z) = \sqrt{z}$ can be inverted to give

$$z = w(y) = y^2.$$

Thus

$$\left|\frac{dz}{dy}\right| = |2y|, \quad p_z(w(y)) = e^{-y^2},$$

and we obtain

$$p_y(y) = 2ye^{-y^2}, \quad y > 0.$$

For many practical applications it is sufficient to characterize a distribution function by a small number of its *moments*. The *mean* or *expected value* of a continuous random variable Z is commonly written $\langle Z \rangle$ or $E(Z)$.* It is defined in terms of its density function $p_z(z)$ according to

$$\langle Z \rangle = \int_{-\infty}^{\infty} z \cdot p_z(z) dg. \tag{2.9}$$

The mean has the important property that, for two random variables Z_1 and Z_2 and real numbers a_0, a_1 and a_2,

$$\langle a_0 + a_1 Z_1 + a_2 Z_2 \rangle = a_0 + a_1 \langle Z_1 \rangle + a_2 \langle Z_2 \rangle, \tag{2.10}$$

a fact which follows directly from Equation (2.9).

The *variance* of Z, written $\text{var}(Z)$, describes how widely the realizations scatter around the mean. It is defined as

$$\text{var}(Z) = \left\langle (Z - \langle Z \rangle)^2 \right\rangle, \tag{2.11}$$

that is, as the mean of the random variable $Y = (Z - \langle Z \rangle)^2$. In terms of the density function $p_y(y)$ of Y, the variance is given by

$$\text{var}(Z) = \int_{-\infty}^{\infty} y \cdot p_y(y) dy,$$

but in fact can be written (Freund (1992); Theorem 4.1) more conveniently as

$$\text{var}(Z) = \int_{-\infty}^{\infty} (z - \langle Z \rangle)^2 p_z(z) dz, \tag{2.12}$$

which is also referred to as the *second moment about the mean*. For discrete random variables, the integrals in Equations (2.9) and (2.12) are replaced by summations over the allowed values of Z and the probability density is replaced by the mass function.

---

*We will prefer to use the former.

As a simple example, consider a *uniformly distributed* random variable Z with density function

$$p(z) = \begin{cases} 1 & \text{if } 0 \le z \le 1 \\ 0 & \text{otherwise.} \end{cases}$$

We calculate the moments to be

$$\langle Z \rangle = \int_0^1 z \cdot 1 \; dz = 1/2$$
$$\text{var}(Z) = \int_0^1 (x - 1/2)^2 \cdot 1 \; dz = 1/12.$$

Since the populations we are dealing with in the case of actual measurements are infinite, it is clear that mean and variance can, in reality, never be known exactly. As we shall discuss Section 2.3, they must be estimated from the available data.

Two very useful identities follow from the definition of variance (Exercise 1):

$$\begin{aligned} \text{var}(Z) &= \langle Z^2 \rangle - \langle Z \rangle^2 \\ \text{var}(a_0 + a_1 Z) &= a_1^2 \; \text{var}(Z). \end{aligned} \tag{2.13}$$

2.1.3 Random vectors

The idea of a distribution function may be extended to more than one random variable. For convenience, we consider only two continuous random variables in the following discussion, but the generalization to any number of continuous or discrete random variables is straightforward.

Let $\boldsymbol{Z} = (Z_1, Z_2)^\top$ be a *random vector*, i.e., a vector the components of which are random variables. The *joint distribution function* of $\boldsymbol{Z}$ is defined by

$$P(\boldsymbol{z}) = P(z_1, z_2) = \Pr(Z_1 \le z_1 \text{ and } Z_2 \le z_2) \tag{2.14}$$

or, in terms of the *joint density function* $p(z_1, z_2)$,

$$P(z_1, z_2) = \int_{-\infty}^{z_1} \int_{-\infty}^{z_2} p(t_1, t_2) dt_1 dt_2 \tag{2.15}$$

and, conversely,

$$p(z_1, z_1) = \frac{\partial^2}{\partial z_1 \partial z_2} P(z_1, z_2). \tag{2.16}$$

The *marginal distribution function* for Z_1 is given by

$$P_1(z_1) = P(z_1, \infty) = \int_{-\infty}^{z_1} \left[\int_{-\infty}^{\infty} p(t_1, t_2) dt_2 \right] dt_1 \tag{2.17}$$

and similarly for Z_2. The *marginal density* is defined as

$$p_1(z_1) = \int_{-\infty}^{\infty} p(z_1, z_2) dz_2, \tag{2.18}$$

with a similar expression for $p_2(z_2)$. So to get a marginal density value at z_1 we integrate (sum) over all of the probabilities for z_2 at fixed z_1.

The mean of the random vector $\boldsymbol{Z}$ is the vector of mean values of Z_1 and Z_2,

$$\langle \boldsymbol{Z} \rangle = \begin{pmatrix} \langle Z_1 \rangle \\ \langle Z_2 \rangle \end{pmatrix},$$

where the vector components are calculated with Equation (2.9) using the corresponding marginal densities.

Next we formalize the concept of statistical independence.

DEFINITION 2.4 *Two random variables Z_1 and Z_2 are said to be* independent *when their joint distribution is the product of their marginal distributions:*

$$P(z_1, z_2) = P_1(z_1)P_2(z_2)$$

or, equivalently, when their joint density is the product of their marginal densities:

$$p(z_1, z_2) = p_1(z_1)p_2(z_2).$$

Thus we have, for the mean of the product of two independent random variables,

$$\begin{aligned} \langle Z_1 Z_2 \rangle &= \int_{-\infty}^{\infty} \int_{-\infty}^{\infty} z_1 z_2 p(z_1, z_2) dz_1 dz_2 \\ &= \int_{-\infty}^{\infty} z_1 p_1(z_1) dz_1 \int_{-\infty}^{\infty} z_2 p_2(z_2) dz_2 = \langle Z_1 \rangle \langle Z_2 \rangle. \end{aligned}$$

In particular, if Z_1 and Z_2 have the same distribution function with mean $\langle Z \rangle$, then

$$\langle Z_1 Z_2 \rangle = \langle Z \rangle^2. \tag{2.19}$$

The *covariance* of random variables Z_1 and Z_2 is a measure of how their realizations are dependent upon each other and is defined to be the mean of the random variable $(Z_1 - \langle Z_1 \rangle)(Z_2 - \langle Z_2 \rangle)$, i.e.,

$$\text{cov}(Z_1, Z_2) = \langle\, (Z_1 - \langle Z_1 \rangle)(Z_2 - \langle Z_2 \rangle)\, \rangle. \tag{2.20}$$

Their *correlation* is defined by

$$\rho_{12} = \frac{\text{cov}(Z_1, Z_2)}{\sqrt{\text{var}(Z_1)\text{var}(Z_2)}}. \tag{2.21}$$

The correlation is unitless and restricted to values $-1 \le \rho_{12} \le 1$. If $|\rho_{12}| = 1$, then Z_1 and Z_2 are *linearly dependent.* Two simple consequences of the definition of covariance are:

$$\begin{aligned} \text{cov}(Z_1, Z_2) &= \langle Z_1 Z_2 \rangle - \langle Z_1 \rangle \langle Z_2 \rangle \\ \text{cov}(Z_1, Z_1) &= \text{var}(Z_1). \end{aligned} \tag{2.22}$$

A convenient way of representing the variances and covariances of the components of a random vector is in terms of the *variance–covariance matrix.* Let $\boldsymbol{a} = (a_1, a_2)^\top$ be any constant vector. Then the variance of the random variable $\boldsymbol{a}^\top \boldsymbol{Z} = a_1 Z_1 + a_2 Z_2$ is, according to the preceding definitions,

$$\begin{aligned}
\mathrm{var}(\boldsymbol{a}^\top \boldsymbol{Z}) &= \mathrm{cov}(a_1 Z_1 + a_2 Z_2, a_1 Z_1 + a_2 Z_2) \\
&= a_1^2 \mathrm{var}(Z_1) + a_1 a_2 \mathrm{cov}(Z_1, Z_2) + a_1 a_2 \mathrm{cov}(Z_2, Z_1) + a_2^2 \mathrm{var}(Z_2) \\
&= (a_1, a_2) \begin{pmatrix} \mathrm{var}(Z_1) & \mathrm{cov}(Z_1, Z_2) \\ \mathrm{cov}(Z_2, Z_1) & \mathrm{var}(Z_2) \end{pmatrix} \begin{pmatrix} a_1 \\ a_2 \end{pmatrix}.
\end{aligned}$$

The matrix in the above equation is the variance–covariance matrix,* usually denoted by the symbol $\boldsymbol{\Sigma}$,

$$\boldsymbol{\Sigma} = \begin{pmatrix} \mathrm{var}(Z_1) & \mathrm{cov}(Z_1, Z_2) \\ \mathrm{cov}(Z_2, Z_1) & \mathrm{var}(Z_2) \end{pmatrix}. \tag{2.23}$$

Therefore we have

$$\mathrm{var}(\boldsymbol{a}^\top \boldsymbol{Z}) = \boldsymbol{a}^\top \boldsymbol{\Sigma} \boldsymbol{a}. \tag{2.24}$$

Note that, since $\mathrm{cov}(Z_1, Z_2) = \mathrm{cov}(Z_2, Z_1)$, $\boldsymbol{\Sigma}$ is a symmetric matrix. Moreover, since $\boldsymbol{a}$ is arbitrary and the variance of any random variable is positive, $\boldsymbol{\Sigma}$ is also positive definite, see Definition 1.1.

The covariance matrix can be written as an outer product:

$$\boldsymbol{\Sigma} = \langle\, (\boldsymbol{Z} - \langle \boldsymbol{Z} \rangle)(\boldsymbol{Z} - \langle \boldsymbol{Z} \rangle)^\top \,\rangle = \langle \boldsymbol{Z}\boldsymbol{Z}^\top \rangle - \langle \boldsymbol{Z} \rangle \langle \boldsymbol{Z} \rangle^\top, \tag{2.25}$$

as is easily verified. Indeed, if $\langle \boldsymbol{Z} \rangle = \boldsymbol{0}$, we can write simply

$$\boldsymbol{\Sigma} = \langle \boldsymbol{Z}\boldsymbol{Z}^\top \rangle. \tag{2.26}$$

The *correlation matrix* $\boldsymbol{R}$ is similar to the covariance matrix, except that each matrix element $(\boldsymbol{\Sigma})_{ij}$ is divided by $\sqrt{\mathrm{var}(Z_i)\mathrm{var}(Z_j)}$ as in Equation (2.21):

$$\boldsymbol{R} = \begin{pmatrix} 1 & \rho_{12} \\ \rho_{21} & 1 \end{pmatrix} = \begin{pmatrix} 1 & \frac{\mathrm{cov}(Z_1, Z_2)}{\sqrt{\mathrm{var}(Z_1)\mathrm{var}(Z_2)}} \\ \frac{\mathrm{cov}(Z_2, Z_1)}{\sqrt{\mathrm{var}(Z_1)\mathrm{var}(Z_2)}} & 1 \end{pmatrix}, \tag{2.27}$$

where $\rho_{12} = \rho_{21}$ is the correlation of Z_1 and Z_2.

2.1.4 The normal distribution

It is very often the case that random variables are well described by the *normal* or *Gaussian density function*

$$p(z) = \frac{1}{\sqrt{2\pi}\sigma} \exp\left(-\frac{1}{2\sigma^2}(z - \mu)^2 \right), \tag{2.28}$$

*For the sake of brevity, we will simply call it the *covariance matrix* from now on.

where $-\infty < \mu < \infty$ and $\sigma^2 > 0$. In that case, it follows from Equation (2.9) and Equation (2.12) that

$$\langle Z \rangle = \mu, \quad \mathrm{var}(Z) = \sigma^2.$$

This is commonly abbreviated by writing

$$Z \sim \mathcal{N}(\mu, \sigma^2).$$

If Z is normally distributed, then the *standardized* random variable $(Z - \mu)/\sigma$ has the *standard normal distribution* $\Phi(z)$ with zero mean and unit variance

$$\Phi(z) = \frac{1}{\sqrt{2\pi}} \int_{-\infty}^{z} \exp(-t^2/2) dt = \int_{-\infty}^{z} \phi(t) dt, \tag{2.29}$$

where the *standard normal density* $\phi(t)$ is given by

$$\phi(t) = \frac{1}{\sqrt{2\pi}} \exp(-t^2/2). \tag{2.30}$$

Since it is not possible to express the normal distribution function $\Phi(z)$ in terms of simple analytical functions, it is tabulated. From the symmetry of the density function it follows that

$$\Phi(-z) = 1 - \Phi(z), \tag{2.31}$$

so it is sufficient to give tables only for $z \geq 0$. Note that

$$P(z) = \Pr(Z \leq z) = \Pr\left(\frac{Z - \mu}{\sigma} \leq \frac{z - \mu}{\sigma}\right) = \Phi\left(\frac{z - \mu}{\sigma}\right), \tag{2.32}$$

so that values for any normally distributed random variable can be read from the table.

Proofs of the following two important theorems are given in Freund (1992).

THEOREM 2.2

(Additivity) *If the random variables $Z_1, Z_2 \ldots Z_m$ are independent (see Definition 2.4) and normally distributed, then the linear combination*

$$a_1 Z_1 + a_2 Z_2 + \ldots + a_m Z_m$$

is normally distributed with moments

$$\mu = a_1 \mu_1 + a_2 \mu_2 + \ldots + a_m \mu_m, \quad \sigma^2 = a_1^2 \sigma_1^2 + a_2^2 \sigma_2^2 + \ldots a_m^2 \sigma_m^2.$$

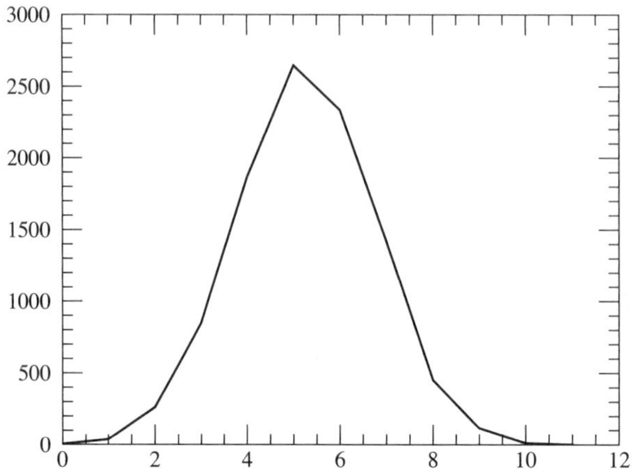

FIGURE 2.1
Histogram of sums of 12 uniformly distributed random numbers.

Listing 2.1: Illustrating the Central Limit Theorem.

```
PRO ex2_1
   thisDevice =!D.Name
   set_plot , 'PS'
   Device,Filename='fig2_1.eps',xsize=6,ysize=4,$
      /inches,/encapsulated
   PLOT, histogram(total(randomu(seed,10000,12),2), $
         nbins=12)
   device,/close_file
   set_plot , thisDevice
END
```

THEOREM 2.3
(Central Limit Theorem) *If random variables $Z_1, Z_2 \ldots Z_m$ are independent and have equal distributions with mean μ and variance σ^2, then the random variable*

$$\frac{1}{\sigma\sqrt{m}}\sum_{i=1}^{m}(Z_i-\mu)=\frac{\bar{Z}-\mu}{\sigma/\sqrt{m}}$$

with $\bar{Z}=(1/m)\sum_i Z_i$ is standard normally distributed in the limit $m\to\infty$.

Theorem 2.2 implies that, if Z_i, $i = 1\ldots m$, is a random sample drawn from a population which is distributed with mean μ and variance σ^2, then the *sample mean* (see Section 2.2),

$$\bar{Z} = \frac{1}{m}\sum_{i=1}^{m} Z_i \ ,$$

is normally distributed with mean μ and variance σ^2/m (Exercise 5). Theorem 2.3, on the other hand, justifies approximating the distribution of the mean $\bar{Z}$ with a normal distribution having mean μ and variance σ^2/m for large m, even when the Z_i are not normally distributed.

As an illustration of the Central Limit Theorem, the IDL code in Listing 2.1 calculates 10,000 sums of $m = 12$ random numbers uniformly distributed on the interval $[0, 1]$ and plots their histogram to a PostScript file; see Figure 2.1. Note the use of the IDL function `TOTAL(A,2)`, which sums a two-dimensional IDL array `A` along its second dimension, i.e., along its columns. The histogram closely approximates a normal distribution.

2.1.5 The gamma distribution and its derivatives

A random variable Z is said to have a *gamma distribution* if its probability density function is given by

$$p_\gamma(z) = \begin{cases} \frac{1}{\beta^\alpha\Gamma(\alpha)} z^{\alpha-1}e^{-z/\beta} & \text{for } z > 0 \\ 0 & \text{elsewhere,} \end{cases} \tag{2.33}$$

where $\alpha > 0$ and $\beta > 0$ and where the *gamma function* $\Gamma(\alpha)$ is given by

$$\Gamma(\alpha) = \int_0^\infty z^{\alpha-1}e^{-z}dz, \quad \alpha > 0. \tag{2.34}$$

The gamma function has the recursive property

$$\Gamma(\alpha) = (\alpha-1)\Gamma(\alpha-1), \quad \alpha > 1,$$

and generalizes the notion of a factorial; see Exercise 6. It is easy to show (Exercise 7(a)) that the gamma distribution has mean and variance

$$\mu = \alpha\beta, \quad \sigma^2 = \alpha\beta^2. \tag{2.35}$$

A special case of the gamma distribution arises for $\alpha = 1$. Since $\Gamma(1) = \int_0^\infty e^{-z}dz = 1$, we obtain the *exponential distribution* with density function

$$p_e(z) = \begin{cases} \frac{1}{\beta}e^{-z/\beta} & \text{for } z > 0 \\ 0 & \text{elsewhere,} \end{cases} \tag{2.36}$$

where $\beta > 0$. According to Equation (2.35), the exponential distribution has mean β and variance β^2. In addition we have (Exercise 7(b)) the following theorem:

THEOREM 2.4

If random variables $Z_1, Z_2 \dots Z_m$ are independent and exponentially distributed according Equation (2.36), then the random variable $Z = \sum_{i=1}^m Z_i$ is gamma distributed with $\alpha = m$.

The *chi-square distribution with m degrees of freedom* is another special case of the gamma distribution. We get its density function with $\beta = 2$ and $\alpha = m/2$, i.e.,

$$p_{\chi^2;m}(z) = \begin{cases} \frac{1}{2^{m/2}\Gamma(m/2)} z^{(m-2)/2} e^{-z/2} & \text{for } z > 0 \\ 0 & \text{otherwise.} \end{cases} \tag{2.37}$$

It follows that the chi-square distribution has mean $\mu = m$ and variance $\sigma^2 = 2m$. The reader is asked to prove a special case of the following theorem in Exercise 3.

THEOREM 2.5

If the random variables Z_i, $i = 1 \dots m$, are independent and standard normally distributed (i.e., with mean 0 and variance 1), then the random variable $Z = \sum_{i=1}^m Z_i^2$ is chi-square distributed with m degrees of freedom.

One can use the IDL function `CHISQR_PDF()` to calculate the chi-square probability distribution function or `CHISQR_CVF()` to calculate its percentiles (values of z for given $P_{\chi^2;m}(z)$). The Python equivalents are `chi2.cdf()` and `chi2.ppf()`. They may be imported with the `scipy.stats` package. The Scipy statistics package also provides the function `chi2.pdf()` to calculate the chi-square probability *density** function. We use it in the following Python script to generate the plots shown in Figure 2.2.

```
IMPORT numpy as np
IMPORT scipy.stats as st
IMPORT matplotlib.pyplot as plt

z = np.linspace(1,20,200)
FOR i IN RANGE(1,6):
    plt.plot(z,st.chi2.pdf(z,i))
plt.show()
```

The gamma and exponential distributions will be essential for the characterization of SAR speckle noise in Chapter 5. The chi-square distribution plays a central role in the iterative change detection algorithm IR-MAD of Chapter 9.

---

*Note the unfortunate ambiguity of the abbreviation "pdf"!

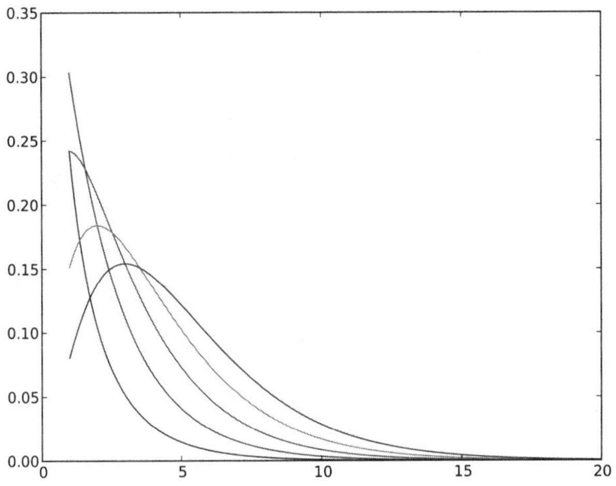

FIGURE 2.2
Plots of the chi-square probability density for $m = 1 \ldots 5$ degrees of freedom. **(See color insert.)**

2.2 Parameter estimation

Having introduced distribution functions for random variables, the question arises as to how to estimate the parameters which characterize those distributions, most importantly their means, variances and covariances, from observations.

2.2.1 Random samples

Consider a multispectral image and a specific land cover category within it. We might choose n pixels belonging to that category and use them to estimate the moments of the underlying distribution. That distribution will be determined not only by measurement noise, atmospheric disturbances, etc., but also by the spread in reflectances characterizing the land cover category itself. For example, Figure 2.3 shows the 3N band of the ASTER image of Figure 1.1, with region of interest (ROI) masks marking areas of mixed forest. The histogram of the observations under the masks is given in Figure 2.4 and is, roughly speaking, a normal distribution. The masked observations might thus be used to calculate an approximate mean and variance for a random variable describing mixed forest land cover.

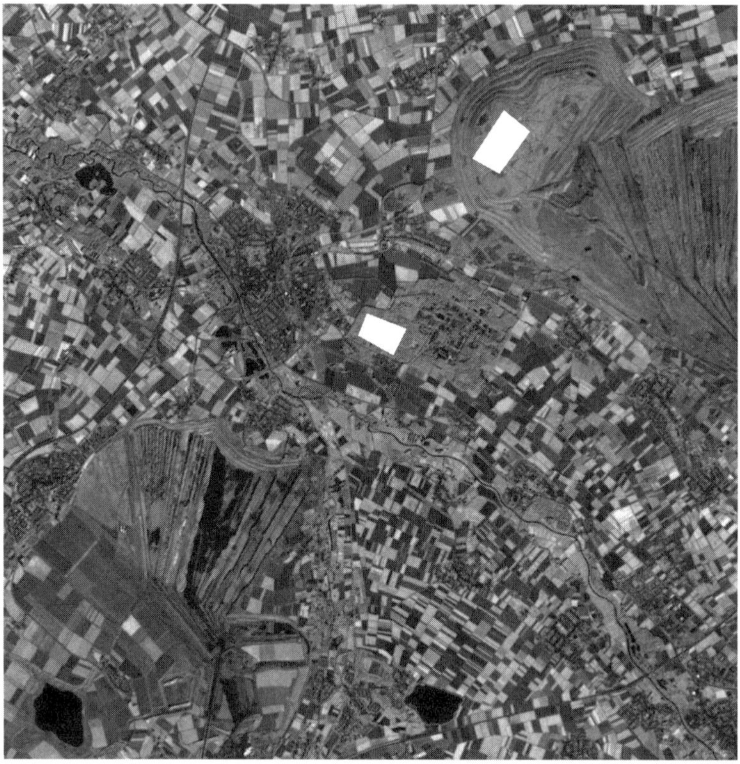

FIGURE 2.3
Spectral band 3N of the ASTER image over Jülich with ROIs covering areas of mixed forest (set with ENVI's ROI tool).

More formally, let $Z_1, Z_2 \ldots Z_m$ be independent random variables which all have the same distribution function $P(z)$ with mean $\langle Z \rangle$ and variance $\mathrm{var}(Z)$. These random variables are referred to as a *sample of the distribution* and are said to be *independent and identically distributed* (i.i.d.). Any function of them is called a *sample function* and is itself a random variable. The pixel intensities contributing to Figure 2.4 are a particular realization of some sample of the distribution corresponding to the land cover category mixed forest. For our present purposes, the sample functions of interest are those which can be used to estimate the mean and variance of the distribution $P(z)$. These are the *sample mean*

$$\bar{Z} = \frac{1}{m} \sum_{i=1}^{m} Z_i \tag{2.38}$$

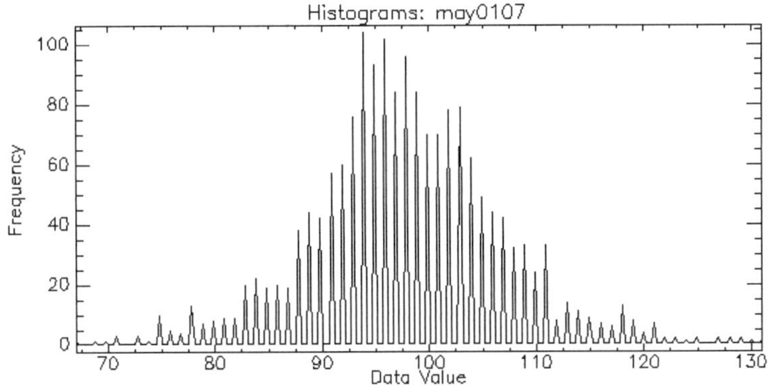

FIGURE 2.4
Histogram of the pixels in Figure 2.3 under the ROIs.

and the *sample variance*

$$S = \frac{1}{m-1}\sum_{i=1}^{m}(Z_i - \bar{Z})^2. \tag{2.39}$$

These two sample functions are also called *unbiased estimators* because their expected values are equal to the corresponding moments of $P(z)$, that is,

$$\langle \bar{Z} \rangle = \langle Z \rangle \tag{2.40}$$

and

$$\langle S \rangle = \mathrm{var}(Z). \tag{2.41}$$

The first result, Equation (2.40), follows immediately:

$$\langle \bar{Z} \rangle = \frac{1}{m}\sum_i \langle Z_i \rangle = \frac{1}{m}\, m\langle Z \rangle = \langle Z \rangle.$$

To see that Equation (2.41) holds, consider

$$\begin{aligned}(m-1)S &= \sum_i (Z_i - \bar{Z})^2 = \sum_i \left(Z_i^2 - 2Z_i\bar{Z} - \bar{Z}^2\right)\\ &= \sum_i Z_i^2 - 2m\bar{Z}^2 + m\bar{Z}^2.\end{aligned}$$

Therefore

$$(m-1)S = \sum_i Z_i^2 - m\bar{Z}^2 = \sum_i Z_i^2 - m\frac{1}{m^2}\left(\sum_i Z_i\right)\left(\sum_i Z_i\right).$$

Expanding the product of sums yields

$$(m-1)S = \sum_i Z_i^2 - \frac{1}{m}\sum_i Z_i^2 - \frac{1}{m}\sum_{i\neq i'} Z_i Z_{i'}.$$

But Z_i and $Z_{i'}$ are independent random variables; see Definition 2.4 and Equation (2.19), so that

$$\langle Z_i Z_{i'}\rangle = \langle Z\rangle^2.$$

Therefore, since the double sum above has $m(m-1)$ terms,

$$(m-1)\langle S\rangle = m\langle Z^2\rangle - \frac{1}{m}\, m\langle Z^2\rangle - \frac{1}{m}\, m(m-1)\langle Z\rangle^2$$

or

$$\langle S\rangle = \langle Z^2\rangle - \langle Z\rangle^2 = \text{var}(Z).$$

The denominator $(m-1)$ in the definition of the sample variance is thus seen to be required for unbiased estimation of the covariance matrix and to be due to the appearance of the sample mean $\bar{Z}$ rather than the distribution mean $\langle Z\rangle$ in the definition. The *maximum likelihood method*, which we will meet in Section 2.4, will lead to the same sample mean, but to a slightly different sample variance estimator.

2.2.2 Sample distributions and interval estimators

It follows from Theorem 2.2 that the sample mean $\bar{Z} = \sum_i Z_i/m$ for $Z_i \sim \mathcal{N}(\mu, \sigma^2)$ is normally distributed with mean μ and variance σ^2/m. For the sample variance we have (Freund, 1992) the following theorem:

THEOREM 2.6
If S is the sample variance

$$S = \frac{1}{m-1}\sum_{i=1}^{m}(Z_i - \bar{Z})^2$$

of a random sample Z_i, $i = 1\ldots m$, drawn from a normally distributed population with mean μ and variance σ^2, then the random variable

$$(m-1)S/\sigma^2$$

is independent of $\bar{Z}$ and has the chi-square distribution with $m-1$ degrees of freedom.

The estimators in Equations (2.38) and (2.39) are referred to as *point estimators*, since their realizations involve real numbers. The estimated value of any distribution parameter will of course differ from the true value. Generally, one prefers to quote an interval within which, to some specified degree of confidence, the true value will lie.

To give an important example, consider again a $\mathcal{N}(\mu, \sigma^2)$-distributed random sample $Z_1 \dots Z_m$. Then

$$\bar{Z} \sim \mathcal{N}(\mu, \sigma^2/m),$$

and the variance in the sample mean decreases inversely with sample size m. Moreover, $(\bar{Z} - \mu)/(\sigma/\sqrt{m})$ is standard normally distributed. Therefore, for any $t > 0$, and with Equation (2.31),

$$\Pr\left(-t < \frac{\bar{Z} - \mu}{\sigma/\sqrt{m}} \le t\right) = \Phi(t) - \Phi(-t) = 2\Phi(t) - 1.$$

This may be written in the form

$$\Pr\left(\bar{Z} - t\,\frac{\sigma}{\sqrt{m}} \le \mu < \bar{Z} + t\,\frac{\sigma}{\sqrt{m}}\right) = 2\Phi(t) - 1.$$

Thus, we can say that the probability that the *random interval*

$$\left(\bar{Z} - t\,\frac{\sigma}{\sqrt{m}}, \bar{Z} + t\,\frac{\sigma}{\sqrt{m}}\right) \tag{2.42}$$

covers the unknown mean value μ is $2\Phi(t) - 1$. Once a realization of the random interval has been obtained and reported, i.e., by plugging a realization of $\bar{Z}$ into Equation (2.42), then μ either lies within it or it doesn't. Therefore, one can no longer properly speak of probabilities. Instead, a *degree of confidence* for the reported interval is conventionally given and expressed in terms of a (usually small) quantity α defined by

$$1 - \alpha = 2\Phi(t) - 1.$$

This determines t in Equation (2.42) according to

$$\Phi(t) = 1 - \alpha/2. \tag{2.43}$$

For a given α, t can be read from the table of the normal distribution. One then says that the interval *covers the unknown parameter* μ *with confidence* $1 - \alpha$.

We can similarly derive a confidence interval for the estimated variance S. Define $\chi^2_{\alpha;m}$ by

$$\Pr(Z \ge \chi^2_{\alpha;m}) = \alpha$$

for a random variable Z having the chi-square distribution with m degrees of freedom. Then from Theorem 2.6,

$$\Pr\left(\chi^2_{1-\alpha/2;m-1} < \frac{(m-1)S}{\sigma^2} < \chi^2_{\alpha/2;m-1}\right) = 1 - \alpha$$

or

$$\Pr\left(\frac{(m-1)S}{\chi^2_{\alpha/2;m-1}} < \sigma^2 < \frac{(m-1)S}{\chi^2_{1-\alpha/2;m-1}}\right) = 1 - \alpha.$$

So we can say that, if s is the estimated variance (realization of S) of a random sample of size m from a normal population, then

$$\frac{(m-1)s}{\chi^2_{\alpha/2;m-1}} < \sigma^2 < \frac{(m-1)s}{\chi^2_{1-\alpha/2;m-1}}$$

is a $1-\alpha$ confidence interval for σ^2.

The following Python script calculates 95% mean and variance confidence intervals for a uniformly distributed random sample with $m = 1000$:

```
FROM math IMPORT sqrt
FROM scipy.stats IMPORT norm,chi2
IMPORT numpy as np

DEF x2(a,m):
    RETURN chi2.ppf(1-a,m)

m = 1000.0
a = 0.05
g = np.random.random(m)
gbar = np.SUM(g)/m
s = np.SUM((g-gbar)**2)/(m-1)
PRINT 'sample␣variance:␣%f'%s
lower = (m-1)*s/x2(a/2,m-1)
upper = (m-1)*s/x2(1-a/2,m-1)
PRINT '%i␣percent␣confidence␣interval:␣(%f,␣%f)'\
                %(INT((1-a)*100),lower,upper)
PRINT 'sample␣mean:␣%f'%gbar
t = norm.ppf(1-a/2)
sigma = sqrt(s)
lower = gbar-t*sigma/sqrt(m)
upper = gbar+t*sigma/sqrt(m)
PRINT '%i␣percent␣confidence␣interval:␣(%f,␣%f)'\
                %(INT((1-a)*100),lower,upper)
```

The result is

```
sample variance: 0.083069
95 percent confidence interval: (0.076240, 0.090864)
sample mean: 0.504085
95 percent confidence interval: (0.486221, 0.521948)
```

According to the Central Limit Theorem, the true mean is 0.5 and the true variance is $1/12 = 0.08333$; see Section 2.1.2.

2.3 Multivariate distributions

As with continuous random variables, the continuous random vector $\boldsymbol{Z} = (Z_1, Z_2)^\top$ is often assumed to be described by a *bivariate normal density function* $p(\boldsymbol{z})$ given by

$$p(\boldsymbol{z}) = \frac{1}{(2\pi)^{N/2}\sqrt{|\boldsymbol{\Sigma}|}} \exp\left(-\frac{1}{2}(\boldsymbol{z}-\boldsymbol{\mu})^\top \boldsymbol{\Sigma}^{-1}(\boldsymbol{z}-\boldsymbol{\mu})\right), \tag{2.44}$$

with $N = 2$. For $N > 2$, the same definition applies and we speak of a *multivariate normal density function.* The mean vector $\langle \boldsymbol{Z} \rangle$ is $\boldsymbol{\mu}$ and the covariance matrix is $\boldsymbol{\Sigma}$. This is indicated by writing

$$\boldsymbol{Z} \sim \mathcal{N}(\boldsymbol{\mu}, \boldsymbol{\Sigma}).$$

We have the following important result (see again Freund (1992)):

THEOREM 2.7
If two random variables Z_1, Z_2 have a bivariate normal distribution, they are independent if and only if $\rho_{12} = 0$, that is, if and only if they are uncorrelated.

In general, a zero correlation does not imply that two random variables are independent. However, this theorem says that it does if the variables are normally distributed.

A *complex Gaussian random variable* $Z = X + \mathbf{i}Y$ is a complex random variable whose real and imaginary parts are bivariate normally distributed. Under certain assumptions (Goodman, 1963), the real-valued form for the multivariate distribution, Equation (2.44), can be carried over straightforwardly to the domain of complex random vectors. In particular, it is assumed that the real and imaginary parts of each component of the complex random vector are uncorrelated and have equal variances, although the real and imaginary parts of different components can be correlated. As we shall see in Chapter 5, this corresponds closely to the properties of SAR amplitude data. The complex covariance matrix for a zero-mean complex random vector $\boldsymbol{Z}$ is, similar to Equation (2.26), given by

$$\boldsymbol{\Sigma} = \langle \boldsymbol{Z}\boldsymbol{Z}^\dagger \rangle, \tag{2.45}$$

where the $\dagger$ denotes conjugate transposition as defined in Chapter 1 and Appendix A. Note that $\boldsymbol{\Sigma}$ is Hermitian and positive semi-definite (see Exercise 9 in Chapter 1). The complex random vector $\boldsymbol{Z}$ is said to be *complex multivariate normally distributed* with zero mean and covariance matrix $\boldsymbol{\Sigma}$ if its density function is

$$p(\boldsymbol{z}) = \frac{1}{\pi^N|\boldsymbol{\Sigma}|} \exp(-\boldsymbol{z}^\dagger \boldsymbol{\Sigma}^{-1}\boldsymbol{z}). \tag{2.46}$$

This is indicated by writing $\boldsymbol{Z} \sim \mathcal{N}_C(\mathbf{0}, \boldsymbol{\Sigma})$.

2.3.1 Vector sample functions and the data matrix

Random vectors will, in our context, always represent pixel observation vectors in remote sensing images. The sample functions of interest are, as in the scalar case, those which can be used to estimate the vector mean and covariance matrix of a joint distribution $P(\boldsymbol{z})$, namely the *vector sample mean**

$$\bar{\boldsymbol{Z}} = \frac{1}{m}\sum_{\nu=1}^{m} \boldsymbol{Z}(\nu) \tag{2.47}$$

and the *sample covariance matrix*

$$\boldsymbol{S} = \frac{1}{m-1}\sum_{\nu=1}^{m}(\boldsymbol{Z}(\nu)-\bar{\boldsymbol{Z}})(\boldsymbol{Z}(\nu)-\bar{\boldsymbol{Z}})^\top. \tag{2.48}$$

These two sample functions are unbiased estimators because again, as in the scalar case, their expected values are equal to the corresponding parameters of $P(\boldsymbol{z})$, that is,

$$\langle\bar{\boldsymbol{Z}}\rangle = \langle\boldsymbol{Z}\rangle \tag{2.49}$$

and

$$\langle\boldsymbol{S}\rangle = \boldsymbol{\Sigma}. \tag{2.50}$$

Suppose that we have made i.i.d. observations $\boldsymbol{z}(\nu)$, $\nu = 1 \dots m$. They may be arranged into an $m \times N$ matrix $\boldsymbol{\mathcal{Z}}$ in which each N-component observation vector forms a row, i.e.,

$$\boldsymbol{\mathcal{Z}} = \begin{pmatrix} \boldsymbol{z}(1)^\top \\ \boldsymbol{z}(2)^\top \\ \vdots \\ \boldsymbol{z}(m)^\top \end{pmatrix}, \tag{2.51}$$

which, as mentioned in Section 1.3.4, is a very useful construct, and is called the *data matrix*. (Mardia et al. (1979) explain the theory of multivariate statistics entirely in terms of the data matrix!) The calligraphic font is chosen to avoid confusion with the random vector $\boldsymbol{Z}$. The unbiased estimate $\bar{\boldsymbol{z}}$ of the sample mean vector $\bar{\boldsymbol{Z}}$ is just the vector of the column means of the data matrix $\boldsymbol{\mathcal{Z}}$. It can be conveniently written as

$$\bar{\boldsymbol{z}} = \frac{1}{m}\boldsymbol{\mathcal{Z}}^\top \boldsymbol{1}_m, \tag{2.52}$$

where $\boldsymbol{1}_m$ denotes a column vector of m ones. If the column means have been subtracted out, then the data matrix is said to be *column centered*, in which case an unbiased estimate $\boldsymbol{s}$ for the covariance matrix $\boldsymbol{\Sigma}$ is given by

$$\boldsymbol{s} = \frac{1}{m-1}\boldsymbol{\mathcal{Z}}^\top\boldsymbol{\mathcal{Z}}. \tag{2.53}$$

*We shall prefer to use the Greek letter ν to index random vectors and their realizations from now on.

Note that the order of transposition is reversed relative to Equation (2.26) because the observations are stored as rows. The rules of matrix multiplication, Equation (1.14), take care of the sum over all observations, so it only remains to divide by $m-1$ to obtain the desired estimate. If $\boldsymbol{d}$ is the diagonal matrix having the diagonal elements of $\boldsymbol{s}$,

$$(\boldsymbol{d})_{ij} = \begin{cases} (\boldsymbol{s})_{ij} & \text{for } i = j \\ 0 & \text{otherwise,} \end{cases}$$

then an unbiased estimate $\boldsymbol{r}$ of the correlation matrix $\boldsymbol{R}$ is (Exercise 11)

$$\boldsymbol{r} = \boldsymbol{d}^{-1/2}\boldsymbol{s}\boldsymbol{d}^{-1/2}. \tag{2.54}$$

In terms of a column-centered data matrix stored in the numerical Python matrix `Z`, for instance, the covariance and correlation matrices can be coded as

```
IMPORT numpy as *
s = Z.T*Z/(Z.shape[0]-1)
d = mat(diag(sqrt(diag(s))))
r = d.I*s*d.I
```

For the corresponding data matrix in IDL we would write simply

```
s = correlate(Z,/covariance)
r = correlate(Z)
```

See, e.g., Listing 1.5.

2.3.2 Provisional means

In Listing 1.5 in Chapter 1 an IDL program was given which estimated the covariance matrix of an image by sampling *all* of its pixels. As we will see in subsequent chapters, this operation is required for many useful transformations of multispectral images. For large datasets, however, the use of the IDL function `CORRELATE()` (or of its Python equivalent) may become impractical since it requires that the image array be stored completely in memory. An alternative is to use an iterative algorithm and take advantage of the ENVI *tiling facility*, reading and processing small portions of the image at a time. The procedure we describe here is referred to as the *method of provisional means* and we give it in a form which includes the possibility of weighting each sample.

Let $\boldsymbol{g}(\nu) = (g_1(\nu), g_2(\nu) \dots g_N(\nu))^\top$ denote the νth sample from an N-band multispectral image with some distribution function $P(\boldsymbol{g})$. Set $\nu = 0$ and define the following quantities and their initial values:

$$\bar g_k(\nu = 0) = 0, \quad k = 1 \dots N$$
$$c_{k\ell}(\nu = 0) = 0, \quad k, \ell = 1 \dots N.$$

The $\nu+1$st observation is given weight $w_{\nu+1}$ and we define the update constant

$$r_{\nu+1} = \frac{w_{\nu+1}}{\sum_{\nu'=1}^{\nu+1} w_{\nu'}}.$$

Listing 2.2: An IDL object class for the method of provisional means.

```
FUNCTION CPM::Init, NN
   self.mn = ptr_new(dblarr(NN))
   self.cov= ptr_new(dblarr(NN,NN))
   self.sw = 0.0000001D
   RETURN, 1
END

PRO CPM::Cleanup
   Ptr_Free, self.mn
   Ptr_Free, self.cov
END

PRO CPM::Update, Xs, weights = Ws
   COMPILE_OPT STRICTARR
   sz = size(Xs)
   NN = sz[1]
   n = sz[2]
   IF n_elements(Ws) EQ 0 THEN Ws = fltarr(sz[2])+1.0
   sw  =  self.sw
   mn  = *self.mn
   cov = *self.cov
   void = provmeans(float(Xs),float(Ws),NN,n,sw,mn,cov)
   self.sw   = sw
   *self.mn  = mn
   *self.cov = cov
END

FUNCTION CPM::Covariance
   c = *self.cov/(self.sw-1)
   d = diag_matrix(diag_matrix(c))
   RETURN, c+transpose(c)-d
END

FUNCTION CPM::Means
   RETURN, *self.mn
END

PRO CPM__Define
class = {CPM, $
         sw: 0.0D, $      ;current sum of weights
         mn: ptr_new(), $;current mean
         cov: ptr_new()} ;current sum of cross products
END
```

Listing 2.3: Calculation of the covariance matrix of an image with the provisional means method and ENVI tiling.

```
PRO EX2_2

envi_select, title='Choose␣multispectral␣image', $
             fid=fid, dims=dims,pos=pos
IF (fid EQ -1) THEN BEGIN
   PRINT, 'cancelled'
   RETURN
ENDIF

envi_file_query,fid,fname=fname,interleave=interleave
IF interleave NE 2 THEN BEGIN
   PRINT, 'not␣BIP␣format'
   RETURN
ENDIF

num_cols = dims[2]-dims[1]+1
num_bands = n_elements(pos)

cpm = Obj_New("CPM",num_bands)

tile_id = envi_init_tile(fid,pos,num_tiles=num_tiles, $
       interleave=2,xs=dims[1],xe=dims[2], $
       ys=dims[3],ye=dims[4])

; spectral tiling
FOR tile_index = 0L, num_tiles-1 DO $
   cpm->update, envi_get_tile(tile_id,tile_index)

PRINT, 'Covariance␣matrix␣for␣image␣'+fname
PRINT, cpm->covariance()

Obj_Destroy, cpm
envi_tile_done, tile_id

END
```

Each new observation $\boldsymbol{g}(\nu+1)$ leads to the following updates:

$$\bar{g}_k(\nu+1) = \bar{g}_k(\nu) + (g_k(\nu+1) - \bar{g}_k(\nu))r_{\nu+1}$$
$$c_{k\ell}(\nu+1) = c_{k\ell}(\nu) + w_{\nu+1}(g_k(\nu+1) - \bar{g}_k(\nu))(g_\ell(\nu+1) - \bar{g}_\ell(\nu)(1 - r_{\nu+1}).$$

Then, after m observations, $\bar{\boldsymbol{g}}(m) = (\bar{g}_1(m)\ldots\bar{g}_N(m))^\top$ is a realization of the (weighted) sample mean, Equation (2.47), and $c_{k\ell}(m)$ is a realization of the (k,ℓ)th element of the (weighted) sample covariance matrix, Equation (2.48).

Listing 2.4: A Python object class for the method of provisional means.

```
CLASS Cpm(OBJECT):
    '''Provisional␣means␣algorithm'''
    DEF __init__(self,N):
        self.mn = np.zeros(N)
        self.cov = np.zeros((N,N))
        self.sw = 0.0000001

    DEF update(self,Xs,Ws=None):
        n,N = np.shape(Xs)
        IF Ws IS None:
            Ws = np.ones(n)
        sw = ctypes.c_double(self.sw)
        mn = self.mn
        cov = self.cov
        provmeans(Xs,Ws,N,n,ctypes.byref(sw),mn,cov)
        self.sw = sw.value
        self.mn = mn
        self.cov = cov

    DEF covariance(self):
        c = np.mat(self.cov/(self.sw-1.0))
        d = np.diag(np.diag(c))
        RETURN c + c.T - d

    DEF means(self):
        RETURN self.mn
```

To see that this prescription gives the desired result, consider the case in which $w_\nu = 1$ for all ν. The first two mean values are

$$\begin{aligned}\bar{g}_k(1) &= 0 + (g_k(1) - 0)\cdot 1 = g_k(1)\\ \bar{g}_k(2) &= g_k(1) + (g_k(2) - g_k(1))\cdot\frac{1}{2} = \frac{g_k(1)+g_k(2)}{2}\end{aligned}$$

as expected. The first two cross products are

$$\begin{aligned}c_{k\ell}(1) &= 0 + (g_k(1) - 0)(g_\ell(1) - 0)(1-1) = 0\\ c_{k\ell}(2) &= 0 + (g_k(2) - g_k(1))(g_\ell(2) - g_\ell(1))(1 - 1/2)\\ &= \frac{1}{2}(g_k(2) - g_k(1))(g_\ell(2) - g_\ell(1))\\ &= \frac{1}{2-1}\sum_{\nu=1}^{2}\left(g_k(\nu) - \frac{g_k(1)+g_k(2)}{2}\right)\left(g_\ell(\nu) - \frac{g_\ell(1)+g_\ell(2)}{2}\right)\end{aligned}$$

again as expected.

An IDL program for the provisional means method is given in Listing 2.2. It is implemented for convenience as an IDL *object class* `CPM` (see Fanning (2000) or Galloy (2011) for good introductions to object-oriented programming in IDL). In order to avoid an expensive IDL `FOR`-loop over the pixels, an external dynamically loadable module (DLM) `provmeans` written in C is called in line 22 to process an input array. Listing 2.3 is a modification of Listing 1.5 to calculate the covariance matrix of a multispectral image using the object class `CPM` and spectral tiling. The result for the VNIR bands of the Jülich image is, apart from some cumulative rounding errors, as before:

```
'Covariance␣matrix␣for␣image␣E:\satellite␣images\juelich'
       87.117540       92.394387      -10.518856
       92.394387       115.83864      -77.420809
      -10.518856      -77.420809       374.61476
```

Listing 2.4 shows the corresponding Python class `Cpm`, which is part of the `auxil.py` module. The `provmeans` function called in line 15 is also coded in C and accessed with the `ctypes` package. ENVI-style spectral tiling is easily implemented in Python simply by reading each raster band row-by-row and combining the pixels in each row into a data array (tile), for example:

```
    pos=auxil.select_pos(bands)
    tile=numpy.zeros((cols,bands))
    inDataset=gdal.Open(filename,GA_ReadOnly)
    rasterBands=[inDataset.GetRasterBand(b) FOR b IN pos]

    FOR row IN RANGE(rows):
        FOR k IN RANGE(bands):
            tile[:,k]=rasterBands[k]\
                              .ReadAsArray(0,row,cols,1)

#       process the tile
```

2.3.3 Real and complex multivariate sample distributions

For the purposes of interval estimation and hypothesis testing, and also for image classification and change detection, we require not only sample vector estimators, but also their distributions. We concluded, from Theorem 2.2 in Section 2.1.4, that the mean of m normally distributed samples $Z_i \sim \mathcal{N}(\mu, \sigma^2)$,

$$\bar{Z} = \frac{1}{m}\sum_{i=1}^{m} Z_i,$$

is normally distributed with mean μ and variance σ^2/m. Similarly, according to Theorem 2.6, $(m-1)S/\sigma^2$, where S is the sample variance, is chi-square distributed with $m-1$ degrees of freedom.

In the multivariate case we have a similar situation. The sample mean vector

$$\bar{\boldsymbol{Z}} = \frac{1}{m}\sum_{\nu=1}^{m} \boldsymbol{Z}(\nu)$$

is multivariate normally distributed with mean $\boldsymbol{\mu}$ and covariance matrix $\boldsymbol{\Sigma}/m$ for samples $\boldsymbol{Z}(\nu) \sim \mathcal{N}(\boldsymbol{\mu}, \boldsymbol{\Sigma})$, $\nu = 1 \dots m$. The sample covariance matrix $\boldsymbol{S}$ is described by a *Wishart distribution* in the following sense. Suppose $\boldsymbol{Z}(\nu) \sim \mathcal{N}(\boldsymbol{0}, \boldsymbol{\Sigma})$, $\nu = 1 \dots m$. We then have, with Equation (2.48),

$$(m-1)\boldsymbol{S} = \sum_{\nu=1}^{m} \boldsymbol{Z}(\nu)\boldsymbol{Z}(\nu)^\top =: \boldsymbol{X}. \tag{2.55}$$

Realizations $\boldsymbol{x}$ of the random sample matrix $\boldsymbol{X}$, namely

$$\boldsymbol{x} = \sum_{\nu=1}^{m} \boldsymbol{z}(\nu)\boldsymbol{z}(\nu)^\top,$$

are symmetric and, for sufficiently large m, positive definite.

THEOREM 2.8

(Anderson, 2003) *The probability density function of* $\boldsymbol{X}$ *given by Equation (2.55) is the* Wishart density with m degrees of freedom

$$p_W(\boldsymbol{x}) = \frac{|\boldsymbol{x}|^{(m-N-1)/2}\exp(-\mathrm{tr}(\boldsymbol{\Sigma}^{-1}\boldsymbol{x})/2)}{2^{Nm/2}\pi^{N(N-1)/4}|\boldsymbol{\Sigma}|^{m/2}\prod_{i=1}^{N}\Gamma[(m+1-i)/2]} \tag{2.56}$$

for $\boldsymbol{x}$ *positive definite, and* 0 *otherwise.*

We write $\boldsymbol{X} \sim \mathcal{W}(\boldsymbol{\Sigma}, N, m)$. Since the gamma function $\Gamma(\alpha)$, Equation (2.34), is not defined for $\alpha \leq 0$, the Wishart distribution is undefined for $m < N$. Equation (2.56) generalizes the chi-square density, Equation (2.37), as may easily be seen by setting $N \to 1$, $\boldsymbol{x} \to x$ and $\boldsymbol{\Sigma} \to 1$.

Now define, in analogy to Equation (2.55), the random complex sample matrix

$$\boldsymbol{X} = \sum_{\nu=1}^{m} \boldsymbol{Z}(\nu)\boldsymbol{Z}(\nu)^\dagger, \tag{2.57}$$

where $\boldsymbol{Z}(\nu) \sim \mathcal{N}_C(\boldsymbol{0}, \boldsymbol{\Sigma})$. Its realizations are Hermitian and, again for sufficiently large m, positive definite.

THEOREM 2.9

(Goodman, 1963) *The probability density function of* $\boldsymbol{X}$ *given by Equation (2.57) is the* complex Wishart density with m degrees of freedom

$$p_{W_c}(\boldsymbol{x}) = \frac{|\boldsymbol{x}|^{(m-N)}\exp(-\mathrm{tr}(\boldsymbol{\Sigma}^{-1}\boldsymbol{x}))}{\pi^{N(N-1)/2}|\boldsymbol{\Sigma}|^{m}\prod_{i=1}^{N}\Gamma(m+1-i)}. \tag{2.58}$$

This is denoted $\boldsymbol{X} \sim \mathcal{W}_C(\boldsymbol{\Sigma}, N, m)$. The complex Wishart density function plays an important role in the discussions of polarimetric SAR imagery in Chapters 5, 6 and 9. In particular we will need the following theorem:

THEOREM 2.10
If $\boldsymbol{X}_1$ and $\boldsymbol{X}_2$ are both complex Wishart distributed with covariance matrix $\boldsymbol{\Sigma}$ and m degrees of freedom, then the random matrix $\boldsymbol{X}_1 + \boldsymbol{X}_2$ is complex Wishart distributed with $2m$ degrees of freedom.

Proof. The theorem is an immediate consequence of Theorem 2.9. Let $\boldsymbol{Z}_1(\nu), \boldsymbol{Z}_2(\nu) \sim \mathcal{N}_C(\boldsymbol{0}, \boldsymbol{\Sigma})$, $\nu = 1 \ldots m$, and

$$\boldsymbol{X}_1 = \sum_{\nu=1}^{m} \boldsymbol{Z}_1(\nu)\boldsymbol{Z}_1(\nu)^\dagger, \quad \boldsymbol{X}_2 = \sum_{\nu=1}^{m} \boldsymbol{Z}_2(\nu)\boldsymbol{Z}_2(\nu)^\dagger.$$

But since $\boldsymbol{Z}_1$ and $\boldsymbol{Z}_2$ have the same distribution, the sum $\boldsymbol{X}_1 + \boldsymbol{X}_2$ has the same form as Equation (2.57) with m replaced by $2m$. □

2.4 Bayes' Theorem, likelihood and classification

If A and B are two events, i.e., two subsets of a sample space Ω, such that the probability of A and B occurring simultaneously is $\Pr(A, B)$, and if $\Pr(B) \neq 0$, then the *conditional probability of A occurring given that B occurs* is defined to be

$$\Pr(A \mid B) = \frac{\Pr(A, B)}{\Pr(B)}. \tag{2.59}$$

We have Theorem 2.11 (Freund, 1992):

THEOREM 2.11
(Theorem of Total Probability) *If $A_1, A_2 \ldots A_m$ are disjoint events associated with some random experiment and if their union is the set of all possible events, then for any event B*

$$\Pr(B) = \sum_{i=1}^{m} \Pr(B \mid A_i)\Pr(A_i) = \sum_{i=1}^{m} \Pr(B, A_i). \tag{2.60}$$

It should be noted that both Equations (2.59) and (2.60) have their counterparts for probability density functions:

$$p(x, y) = p(x \mid y)p(y) \tag{2.61}$$

and

$$p(x) = \int_{-\infty}^{\infty} p(x \mid y)p(y)dy = \int_{-\infty}^{\infty} p(x,y)dy. \tag{2.62}$$

Bayes' Theorem* follows directly from the definition of conditional probability and is the basic starting point for inference problems using probability theory as logic.

THEOREM 2.12

(Bayes' Theorem) *If $A_1, A_2 \dots A_m$ are disjoint events associated with some random experiment, their union is the set of all possible events, and if $Pr(A_i) \neq 0$ for $i = 1 \dots m$, then for any event B for which $Pr(B) \neq 0$*

$$\Pr(A_k \mid B) = \frac{\Pr(B \mid A_k)\Pr(A_k)}{\sum_{i=1}^m \Pr(B \mid A_i)\Pr(A_i)}. \tag{2.63}$$

Proof: From the definition of conditional probability, Equation (2.59),

$$\Pr(A_k \mid B)\Pr(B) = \Pr(A_k, B) = \Pr(B \mid A_k)\Pr(A_k),$$

and therefore

$$\Pr(A_k \mid B) = \frac{\Pr(B \mid A_k)\Pr(A_k)}{\Pr(B)} = \frac{\Pr(B \mid A_k)\Pr(A_k)}{\sum_{i=1}^m \Pr(B \mid A_i)\Pr(A_i)}$$

from the Theorem of Total Probability. □

We will use Bayes' Theorem primarily in the following form: Let $\boldsymbol{g}$ be a realization of a random vector associated with a distribution of multispectral image pixel intensities (gray values), and let $\{k \mid k = 1 \dots K\}$ be a set of possible *class labels* (e.g., land cover categories) for all of the pixels. Then the *a posteriori* conditional probability for class k, *given the observation* $\boldsymbol{g}$, may be written using Bayes' Theorem as

$$\Pr(k \mid \boldsymbol{g}) = \frac{\Pr(\boldsymbol{g} \mid k)\Pr(k)}{\Pr(\boldsymbol{g})}, \tag{2.64}$$

where $\Pr(k)$ is the *a priori* probability for class k, $\Pr(\boldsymbol{g} \mid k)$ is the conditional probability of observing the value $\boldsymbol{g}$ if it belongs to class k, and

$$\Pr(\boldsymbol{g}) = \sum_{k=1}^{K} \Pr(\boldsymbol{g} \mid k)\Pr(k) \tag{2.65}$$

*Named after Rev. Thomas Bayes, an 18th-century hobby mathematician who derived a special case.

is the total probability for $\boldsymbol{g}$.

One can formulate the problem of *classification* of multispectral imagery as the process of determining posterior conditional probabilities for all of the classes. This is accomplished by specifying prior probabilities $\Pr(k)$ (if prior information exists), modeling the class-specific probabilities $\Pr(\boldsymbol{g} \mid k)$ and then applying Bayes' Theorem. Thus we might choose a class-specific density function as our model:

$$\Pr(\boldsymbol{g} \mid k) \sim p(\boldsymbol{g} \mid \theta_k),$$

where θ_k is a set of parameters for the density function describing the kth class (for a multivariate normal distribution, for instance, just the mean vector $\boldsymbol{\mu}_k$ and covariance matrix $\boldsymbol{\Sigma}_k$). In this case Equation (2.64) takes the form

$$\Pr(k \mid \boldsymbol{g}) = \frac{p(\boldsymbol{g} \mid \theta_k)\Pr(k)}{p(\boldsymbol{g})}, \tag{2.66}$$

where $p(\boldsymbol{g})$ is the unconditional density function for $\boldsymbol{g}$,

$$p(\boldsymbol{g}) = \sum_{k=1}^{K} p(\boldsymbol{g} \mid \theta_k)\Pr(k), \tag{2.67}$$

which is constant when $\boldsymbol{g}$ is given. As we will see in Chapter 6, under reasonable assumptions the observation $\boldsymbol{g}$ should be assigned to the class k which maximizes $\Pr(k \mid \boldsymbol{g})$.

In order to find that maximum, we require estimates of the parameters θ_k, $k = 1 \dots K$. If we have access to measured values (realizations) that are known to be in class k, $\boldsymbol{g}(\nu)$, $\nu = 1 \dots m_k$, say, then we can form the product of probability densities

$$L(\theta_k) = \prod_{\nu=1}^{m_k} p(\boldsymbol{g}(\nu) \mid \theta_k), \tag{2.68}$$

which is called a *likelihood function*, and take its logarithm

$$\mathcal{L}(\theta_k) = \sum_{\nu=1}^{m_k} \log p(\boldsymbol{g}(\nu) \mid \theta_k), \tag{2.69}$$

which is the *log-likelihood.* Taking products is justified when the $\boldsymbol{g}(\nu)$ are realizations of independent random vectors. The parameter set $\hat\theta_k$ which maximizes the likelihood function or its logarithm, i.e., which gives the largest value for all of the realizations, is called the *maximum likelihood estimate* of θ_k. For normally distributed random vectors, maximum likelihood parameter estimators for $\theta_k = \{\boldsymbol{\mu}_k, \boldsymbol{\Sigma}_k\}$ turn out to correspond (almost) to the unbiased estimators, Equations (2.47) and (2.48).

To illustrate this for the class mean, write out Equation (2.69) for the multivariate normal distribution, Equation (2.44):

$$\mathcal{L}(\boldsymbol{\mu}_k, \boldsymbol{\Sigma}_k) = m_k \frac{N}{2}\log 2\pi - m_k \frac{1}{2}\log|\boldsymbol{\Sigma}_k| - \frac{1}{2}\sum_{\nu=1}^{m_k}(\boldsymbol{g}(\nu) - \boldsymbol{\mu}_k)^\top \boldsymbol{\Sigma}_k^{-1}(\boldsymbol{g}(\nu) - \boldsymbol{\mu}_k).$$

To maximize with respect to $\boldsymbol{\mu}_k$, we set

$$\frac{\partial \mathcal{L}(\boldsymbol{\mu}_k, \boldsymbol{\Sigma}_k)}{\partial \boldsymbol{\mu}_k} = \sum_{\nu=1}^{m_k} \boldsymbol{\Sigma}_k^{-1}(\boldsymbol{g}(\nu) - \boldsymbol{\mu}_k) = \boldsymbol{0},$$

giving

$$\hat{\boldsymbol{\mu}}_k = \frac{1}{m_k}\sum_{i=1}^{m_k} \boldsymbol{g}(\nu), \tag{2.70}$$

which is the realization $\bar{\boldsymbol{g}}_k$ of the unbiased estimator for the sample mean, Equation (2.47). In a similar way (see, e.g., Duda et al. (2001)) one can show that

$$\hat{\boldsymbol{\Sigma}}_k = \frac{1}{m_k}\sum_{\nu=1}^{m_k}(\boldsymbol{g}(\nu) - \hat{\boldsymbol{\mu}}_k)(\boldsymbol{g}(\nu) - \hat{\boldsymbol{\mu}}_k)^\top, \tag{2.71}$$

which is (almost) the realization $\boldsymbol{s}$ of the unbiased sample covariance matrix estimator, Equation (2.48), except that the denominator $m_k - 1$ is replaced by m_k, a fact which can be ignored for large m_k.

The observations, of course, must be chosen from the appropriate class k in each case. For *supervised classification* (Chapters 6 and 7), there exists a set of training data with known class labels. Therefore, maximum likelihood estimates can be obtained and posterior probability distributions for the classes can be calculated from Equation (2.66) and then used to generalize to all of the image data. In the case of *unsupervised classification*, the class memberships are not initially known. How this conundrum is solved will be discussed in Chapter 8.

2.5 Hypothesis testing

A *statistical hypothesis* is a conjecture about the distributions of one or more random variables. It might, for instance, be an assertion about the mean of a distribution, or about the equivalence of the variances of two different distributions. One distinguishes between *simple* hypotheses, for which the distributions are completely specified, for example: *the mean of a normal distribution with variance* σ^2 *is* $\mu = 0$, and *composite* hypotheses, for which this is not the case, e.g., *the mean is* $\mu \geq 0$.

In order to test such assertions on the basis of samples of the distributions involved, it is also necessary to formulate *alternative* hypotheses. To distinguish these from the original assertions, the latter are traditionally called *null* hypotheses. Thus we might be interested in testing the simple null hypothesis $\mu = 0$ against the composite alternative hypothesis $\mu \neq 0$. An appropriate sample function for deciding whether or not to reject the null hypothesis in favor of its alternative is referred to as a *test statistic*, often denoted by the symbol Q. An appropriate *test procedure* will partition the possible realizations of the test statistic into two subsets: an acceptance region for the null hypothesis and a rejection region. The latter is customarily referred to as the *critical region.*

DEFINITION 2.5 *Referring to the null hypothesis as H_0, there are two kinds of errors which can arise from any test procedure:*

1. *H_0 may be rejected when in fact it is true. This is called an* error of the first kind *and the probability that it will occur is denoted α.*

2. *H_0 may be accepted when in fact it is false, which is called an* error of the second kind *with probability of occurrence β.*

The probability of obtaining a value of the test statistic within the critical region when H_0 is true is thus α. The probability α is also referred to as the *level of significance* of the test. It is generally the case that the lower the value of α, the higher is the probability β of making a second kind error. Traditionally, significance levels of 0.01 or 0.05 are used. Such values are obviously arbitrary, and for exploratory data analysis it is common to avoid specifying them altogether. Instead, the *P-value* for the test is stated:

DEFINITION 2.6 *Given the observed value of a test statistic, the P-*value *is the lowest level of significance at which the null hypothesis could have been rejected.*

High P-values provide evidence in favor of accepting the null hypothesis, without actually forcing one to commit to a decision.

The theory of statistical testing specifies methods for determining the most appropriate test statistic for a given null hypothesis and its alternative. Fundamental to the theory is the *Neyman–Pearson Lemma*, which gives for simple hypotheses a prescription for finding the test procedure which maximizes the probability $1-\beta$ of rejecting the null hypothesis when it is false for a fixed level of significance α, see, e.g., Freund (1992). The following definition deals with the more general case of tests involving composite hypotheses H_0 and H_1.

DEFINITION 2.7 *Consider a vector random sample $\boldsymbol{z}(\nu)$, $\nu = 1 \dots m$, from a multivariate population whose density function is $p(\boldsymbol{z} \mid \theta)$. The likelihood function for the sample is (see Equation (2.68))*

$$L(\theta) = \prod_{\nu=1}^{m} p(\boldsymbol{g}(\nu) \mid \theta). \tag{2.72}$$

Let ω be space of all possible values of the parameter set θ and ω_0 be a subset of that space. The likelihood ratio test (LRT) *for the null hypothesis $\theta \in \omega_0$ against the alternative $\theta \in \omega - \omega_0$ has the critical region*

$$Q = \frac{\max_{\theta \in \omega_0} L(\theta)}{\max_{\theta \in \omega} L(\theta)} \le k. \tag{2.73}$$

.

This definition simply reflects the fact that, if H_0 is true, the maximum likelihood for θ when restricted to ω_0 should be close to the maximum likelihood for θ without that restriction. Therefore, if the likelihood ratio is small, (less than or equal to some small value k), then H_0 should be rejected.

To illustrate, consider scalar random samples $z(1), z(2) \dots z(m)$ from a normal distribution with mean μ and known variance σ^2. The likelihood ratio test at significance level α for the simple hypothesis $H_0 : \mu = \mu_0$ against the alternative composite hypothesis $H_1 : \mu \neq \mu_0$ has, according to Definition 2.7, the critical region

$$Q = \frac{L(\mu_0)}{L(\hat{\mu})} \le k_\alpha,$$

where $\hat{\mu}$ maximizes the likelihood function (2.72) and k_α depends on α. Therefore

$$\hat{\mu} = \bar{z} = \frac{1}{m} \sum_{\nu=1}^{m} z(\nu).$$

The critical region is then (Exercise 14)

$$Q = \frac{L(\mu_0)}{L(\hat{\mu})} = \exp\left(-\frac{1}{2\sigma^2/m} (\bar{z} - \mu_0)^2 \right) \le k_\alpha. \tag{2.74}$$

Equivalently,

$$e^{-x^2} \le k_\alpha, \quad \text{where} \quad x = \frac{\bar{z} - \mu_0}{\sigma/\sqrt{m}}.$$

Since the above exponential function is maximum at $x = 0$ and vanishes asymptotically for large values of $|x|$, the critical region can also be written in the form $|x| \ge \tilde{k}_\alpha$, where $\tilde{k}_\alpha$ also depends on α. But we know that $\bar{z}$ is normally distributed with mean μ_0 and variance σ^2/m. Therefore x has the standard normal distribution $\Phi(x)$ with probability density $\phi(x)$; see Equations (2.29) and (2.30). Thus $\tilde{k}_\alpha$ is determined by (see Figure 2.5)

$$\Phi(-\tilde{k}_\alpha) + 1 - \Phi(\tilde{k}_\alpha) = \alpha,$$

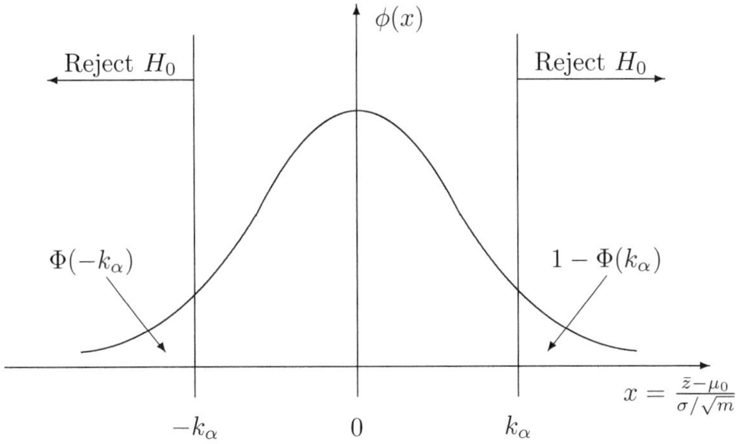

FIGURE 2.5
Critical region for rejecting the hypothesis $\mu = \mu_0$ at significance level α.

or

$$1 - \Phi(\tilde{k}_\alpha) = \alpha/2.$$

This example is straightforward because we assume that the variance σ^2 is known. Suppose, as is often the case, that the variance is unknown and that we wish nevertheless to make a statement about μ in terms of some realization of an appropriate test statistic. To treat this and similar problems, it is necessary to define some additional distribution functions.

If the random variables Z_i, $i = 0 \dots m$, are independent and standard normally distributed, then the random variable

$$T = \frac{Z_0}{\sqrt{\frac{1}{m}(Z_1^2 + \ldots + Z_m^2)}} \tag{2.75}$$

is said to be *Student-t distributed with m degrees of freedom.* The corresponding density is given by

$$p_{t;m}(z) = d_m \left(1 + \frac{z^2}{m}\right)^{-(m+1)/2}, \quad -\infty < z < \infty, \tag{2.76}$$

with d_m again being a normalization factor. The Student-t distribution converges to the standard normal distribution for $m \to \infty$. The IDL functions `T_PDF()` and `T_CVF()` or the Python functions `scipy.stats.t.cdf()` and `scipy.stats.t.ppf()` may be used to calculate the distribution and its percentiles.

The Student-t distribution is used to make statements regarding the mean when the variance is unknown. Thus if Z_i, $\nu = 1 \dots m$, is a sample from a normal distribution with mean μ and unknown variance, and

$$\bar{Z} = \frac{1}{m}\sum_{i=1}^{m} Z_i, \quad S = \frac{1}{m}\sum_{i=1}^{m}(Z_i - \bar{Z})^2,$$

then the random variable

$$T = \frac{\bar{Z} - \mu}{\sqrt{S/m}}$$

is Student-t distributed with $m-1$ degrees of freedom. $T \leq k$ is the critical region for a likelihood ratio test for $H_0 : \mu = \mu_0$ against $H_1 : \mu \neq \mu_0$.

If X_i and Y_i, $i = 1 \dots m$, are samples from normal distributions with equal variance, then the random variable

$$T_d = \frac{\bar{X} - \bar{Y} - (\mu_X - \mu_Y)}{\sqrt{(S_X + S_Y)/m}} \tag{2.77}$$

is Student-t distributed with $2m-2$ degrees of freedom. It may be used to test the hypothesis $\mu_X = \mu_Y$, for example.

Given independent and standard normally distributed random variables Y_i, $i = 1 \dots n$ and Y_i, $i = 1 \dots m$, the random variable

$$F = \frac{\frac{1}{m}(X_1^2 + \ldots + X_m^2)}{\frac{1}{n}(Y_1^2 + \ldots + Y_n^2)} \tag{2.78}$$

is *F-distributed with m and n degrees of freedom.* Its density function is

$$p_{f;m,n}(z) = \begin{cases} c_{mn} z^{(m-2)/2}\left(1 + \frac{m}{n}z\right)^{-(m+n)/2} & \text{for } z > 0 \\ 0 & \text{otherwise,} \end{cases} \tag{2.79}$$

with normalization factor c_{mn}. Note that F is the ratio of two chi-square distributed random variables. One can compute the F-distribution function and its percentiles with `F_PDF()` and `F_CVF()` in IDL and in Python with `scipy.stats.f.cdf()` and `scipy.stats.f.ppf()`.

The F-distribution can be used for hypothesis tests regarding two variances. If S_X and S_Y are sample variances for samples of size n and m, respectively, drawn from normally distributed populations with variances σ_X^2 and σ_Y^2, then

$$F = \frac{\sigma_Y^2 S_X}{\sigma_X^2 S_Y} \tag{2.80}$$

is a random variable having an F-distribution with $n-1$ and $m-1$ degrees of freedom. $F \leq k$ is the critical region for a likelihood ratio test for

$$H_0 : \sigma_X^2 = \sigma_Y^2, \quad H_1 : \sigma_X^2 < \sigma_Y^2.$$

In many cases the likelihood ratio test will lead to a test statistic whose distribution is unknown. The LRT has, however, an important asymptotic property (Mardia et al., 1979):

THEOREM 2.13

In the notation of Definition 2.7, if ω is a region of $\mathbb{R}^q$ and ω_0 is an r-dimensional subregion, then for each $\theta \in \omega_0$, $-2\log Q$ has an asymptotic chi-square distribution with $q - r$ degrees of freedom as $m \to \infty$.

Thus if we take minus twice the logarithm of the LRT statistic in Equation (2.74), we obtain

$$-2\log Q = -2 \cdot \left(-\frac{1}{2\sigma^2/m}(\bar{z} - \mu_0)^2 \right) = \frac{(\bar{z} - \mu_0)^2}{\sigma^2/m},$$

which is chi-square distributed with one degree of freedom (see Section 2.1.5). Since ω_0 consists of the single point μ_0, its dimension is $r = 0$, whereas ω is all real values of μ, so $q = 1$. Hence $q - r = 1$. In this simple case, Theorem 2.13 holds for all values of m.

2.6 Ordinary linear regression

Many image analysis tasks involve fitting a set of data points with a straight line*

$$y(x) = a + bx. \tag{2.81}$$

We review here the standard procedure for determining the parameters a and b and their uncertainties, namely *ordinary linear regression*, in which it is assumed that the dependent variable y has a random error but that the independent variable x is exact. The procedure is then generalized to more than one independent variable and the concepts of regularization and duality are introduced.

Linear regression can also be carried out sequentially by updating the best fit after each new observation. This topic, as well as *orthogonal linear regression*, where is assumed that both variables have random errors associated with them, are discussed in Appendix A.

2.6.1 One independent variable

Suppose that the dataset consists of m pairs $\{x(\nu), y(\nu) \mid \nu = 1 \ldots m\}$. An appropriate statistical model for linear regression is

$$Y(\nu) = a + bx(\nu) + R(\nu), \quad \nu = 1 \ldots m. \tag{2.82}$$

*Relative radiometric normalization, which we will meet in Chapter 9, as well as similarity warping, Chapter 5, are good examples.

$Y(\nu)$ is a random variable representing the νth measurement of the dependent variable and $R(\nu)$, referred to as the *residual error*, is a random variable representing the measurement uncertainty. The $x(\nu)$ are exact. We will assume that the individual measurements are uncorrelated and that they all have the same variance:

$$\mathrm{cov}(R(\nu), R(\nu')) = \begin{cases} \sigma^2 & \text{for } \nu = \nu' \\ 0 & \text{otherwise.} \end{cases} \tag{2.83}$$

The realizations of $R(\nu)$ are $y(\nu) - a - bx(\nu)$, $\nu = 1 \ldots m$, from which we define a least squares *goodness-of-fit* function

$$z(a, b) = \sum_{\nu=1}^{m} \left(\frac{y(\nu) - a - bx(\nu)}{\sigma} \right)^2. \tag{2.84}$$

If the residuals $R(\nu)$ are normally distributed, then we recognize Equation (2.84) as a realization of a chi-square distributed random variable. For the "best" values of a and b, Equation (2.84) is in fact chi-square distributed with $m - 2$ degrees of freedom (Press et al., 2002). The best values for the parameters a and b are obtained by minimizing $z(a, b)$, that is, by solving the equations

$$\frac{\partial z}{\partial a} = \frac{\partial z}{\partial b} = 0$$

for a and b. The solution is (Exercise 15)

$$\hat{b} = \frac{s_{xy}}{s_{xx}}, \quad \hat{a} = \bar{y} - \hat{b}\bar{x}, \tag{2.85}$$

where

$$\begin{aligned} s_{xy} &= \frac{1}{m} \sum_{\nu=1}^{m} (x(\nu) - \bar{x})(y(\nu) - \bar{y}) \\ s_{xx} &= \frac{1}{m} \sum_{\nu=1}^{m} (x(\nu) - \bar{x})^2 \\ \bar{x} &= \frac{1}{m} \sum_{\nu=1}^{m} x(\nu), \quad \bar{y} = \frac{1}{m} \sum_{\nu=1}^{m} y(\nu). \end{aligned} \tag{2.86}$$

The uncertainties in the estimates $\hat{a}$ and $\hat{b}$ are given by (Exercise 16)

$$\begin{aligned} \sigma_a^2 &= \frac{\sigma^2 \sum x(\nu)^2}{m \sum x(\nu)^2 - (\sum x(\nu))^2} \\ \sigma_b^2 &= \frac{\sigma^2}{\sum x(\nu)^2 - (\sum x(\nu))^2}. \end{aligned} \tag{2.87}$$

The goodness of fit can be determined by substituting the estimates $\hat{a}$ and $\hat{b}$ into Equation (2.84) which, as we have said, will then be chi-square distributed with $m-2$ degrees of freedom. The probability of finding a value $z = z(\hat{a}, \hat{b})$ *or higher* by chance is

$$Q = 1 - P_{\chi^2;m-2}(z),$$

where $P_{\chi^2;m-2}(z)$ is given by Equation (2.37). If $Q < 0.001$, one would typically reject the fit as being unsatisfactory.

If σ^2 is not known *a priori*, then it can be estimated by

$$\hat{\sigma}^2 = \frac{1}{m-2}\sum_{\nu=1}^{m}(y(\nu) - \hat{a} - \hat{b}x(\nu))^2, \tag{2.88}$$

in which case the goodness-of-fit procedure cannot be applied, since we *assume* the fit to be good in order to estimate σ^2 with Equation (2.88) (Press et al., 2002).

2.6.2 Coefficient of determination (R^2)

The fitted or predicted values are

$$\hat{y}(\nu) = \hat{a} + \hat{b}x(\nu) \quad \nu = 1 \ldots m.$$

Consider the total variation of the observed variables $y(\nu)$, $\nu = 1 \ldots m$, about their mean value $\bar{y}$,

$$\begin{aligned}\sum_\nu (y(\nu) - \bar{y})^2 &= \sum_\nu (y(\nu) - \hat{y}(\nu) + \hat{y}(\nu) - \bar{y})^2 \\ &= \sum_\nu (y(\nu) - \hat{y}(\nu))^2 + (\hat{y}(\nu) - \bar{y})^2 + 2(y(\nu) - \hat{y}(\nu))(\hat{y}(\nu) - \bar{y}).\end{aligned}$$

The last term in the summation is

$$\sum_\nu 2r(\nu)(\hat{y}(\nu) - \bar{y}) = 2\sum_\nu r(\nu)\hat{y}(\nu) - 2\bar{y}\sum_\nu r(\nu) = 0,$$

since the errors are uncorrelated with the predicted values $\hat{y}(\nu)$ and have mean zero. Therefore we have

$$\sum_\nu (y(\nu) - \bar{y})^2 = \sum_\nu (y(\nu) - \hat{y}(\nu))^2 + \sum_\nu (\hat{y}(\nu) - \bar{y})^2$$

or

$$1 = \frac{\sum_\nu (y(\nu) - \hat{y}(\nu))^2}{\sum_\nu (y(\nu) - \bar{y})^2} + \frac{\sum_\nu (\hat{y}(\nu) - \bar{y})^2}{\sum_\nu (y(\nu) - \bar{y})^2}$$

or

$$1 = \frac{\sum_\nu (y(\nu) - \hat{y}(\nu))^2}{\sum_\nu (y(\nu) - \bar{y})^2} + R^2,$$

where the *coefficient of determination* R^2,

$$R^2 = \frac{\sum_\nu (\hat{y}(\nu) - \bar{y})^2}{\sum_\nu (y(\nu) - \bar{y})^2}, \quad 0 \le R^2 \le 1, \tag{2.89}$$

is the fraction of the variance in Y that is explained by the regression model. It is easy to show (Exercise 19) that

$$R = \text{corr}(y, \hat{y}) = \frac{\sum_\nu (y(\nu) - \bar{y})(\hat{y}(\nu) - \bar{y})}{\sqrt{\sum_\nu (y(\nu) - \bar{y})^2 \sum_\nu (\hat{y}(\nu) - \bar{y})^2}}. \tag{2.90}$$

In words: The coefficient of determination is the square of the sample correlation between the observed and predicted values of y.

2.6.3 More than one independent variable

The statistical model of the preceding Section may be written more generally in the form

$$Y(\nu) = w_o + \sum_{i=1}^{N} w_i x_i(\nu) + R(\nu), \quad \nu = 1 \dots m, \tag{2.91}$$

relating m measurements of the N independent variables $x_1 \dots x_N$ to a measured dependent variable Y via the parameters $w_0, w_1 \dots w_N$. Equivalently, in vector notation we can write

$$Y(\nu) = \boldsymbol{w}^\top \boldsymbol{x}(\nu) + R(\nu), \quad \nu = 1 \dots m, \tag{2.92}$$

where $\boldsymbol{x} = (x_0 = 1, x_1 \dots x_N)^\top$ and $\boldsymbol{w} = (w_0, w_1 \dots w_N)^\top$. The random variables $R(\nu)$ again represent the measurement uncertainty in the realizations $y(\nu)$ of $Y(\nu)$. We assume that they are independent and identically distributed with zero mean and variance σ^2, whereas the values $\boldsymbol{x}(\nu)$ are, as before, assumed to be exact. Now we wish to determine the best value for parameter vector $\boldsymbol{w}$.

Introducing the $m \times (N+1)$ data matrix

$$\mathcal{X} = \begin{pmatrix} \boldsymbol{x}(1)^\top \\ \vdots \\ \boldsymbol{x}(m)^\top \end{pmatrix},$$

we can express Equation (2.91) or (2.92) in the form

$$\boldsymbol{Y} = \mathcal{X}\boldsymbol{w} + \boldsymbol{R}, \tag{2.93}$$

where $\boldsymbol{Y} = (Y(1) \dots Y(m))^\top$, $\boldsymbol{R} = (R(1) \dots R(m))^\top$ and, by assumption,

$$\boldsymbol{\Sigma}_R = \langle \boldsymbol{R}\boldsymbol{R}^\top \rangle = \sigma^2 \boldsymbol{I}.$$

The identity matrix $\boldsymbol{I}$ is $m \times m$. The goodness-of-fit function analog to Equation (2.84) is

$$z(\boldsymbol{w}) = \sum_{\nu=1}^{m} \left[\frac{y(\nu) - \boldsymbol{w}^\top \boldsymbol{x}(\nu)}{\sigma} \right]^2 = \frac{1}{\sigma^2} (\boldsymbol{y} - \mathcal{X}\boldsymbol{w})^\top (\boldsymbol{y} - \mathcal{X}\boldsymbol{w}). \tag{2.94}$$

This is minimized by solving the equations

$$\frac{\partial z(\boldsymbol{w})}{\partial w_k} = 0, \quad k = 0 \ldots N.$$

Using the rules for vector differentiation we obtain (Exercise 17)

$$\boldsymbol{\mathcal{X}}^\top \boldsymbol{y} = (\boldsymbol{\mathcal{X}}^\top \boldsymbol{\mathcal{X}})\boldsymbol{w}\ . \tag{2.95}$$

Equation (2.95) is referred to as the *normal equation.* The estimated parameters of the model are obtained by solving for $\boldsymbol{w}$,

$$\hat{\boldsymbol{w}} = (\boldsymbol{\mathcal{X}}^\top \boldsymbol{\mathcal{X}})^{-1}\boldsymbol{\mathcal{X}}^\top \boldsymbol{y} =: \boldsymbol{\mathcal{X}}^+ \boldsymbol{y}. \tag{2.96}$$

The matrix

$$\boldsymbol{\mathcal{X}}^+ = (\boldsymbol{\mathcal{X}}^\top \boldsymbol{\mathcal{X}})^{-1}\boldsymbol{\mathcal{X}}^\top \tag{2.97}$$

is the pseudoinverse of the data matrix $\boldsymbol{\mathcal{X}}$.*

In order to obtain the uncertainty in the estimate $\hat{\boldsymbol{w}}$, Equation (2.96), we can think of $\boldsymbol{w}$ as a random vector with mean value $\langle \boldsymbol{w} \rangle$. Its covariance matrix is then given by

$$\begin{aligned}
\boldsymbol{\Sigma}_w &= \big\langle (\boldsymbol{w} - \langle \boldsymbol{w} \rangle)(\boldsymbol{w} - \langle \boldsymbol{w} \rangle)^\top \big\rangle \\
&\approx \big\langle (\boldsymbol{w} - \hat{\boldsymbol{w}})(\boldsymbol{w} - \hat{\boldsymbol{w}})^\top \big\rangle \\
&= \big\langle (\boldsymbol{w} - \boldsymbol{\mathcal{X}}^+ \boldsymbol{y})(\boldsymbol{w} - \boldsymbol{\mathcal{X}}^+ \boldsymbol{y})^\top \big\rangle \\
&= \big\langle (\boldsymbol{w} - \boldsymbol{\mathcal{X}}^+ (\boldsymbol{\mathcal{X}}\boldsymbol{w} + \boldsymbol{r}))(\boldsymbol{w} - \boldsymbol{\mathcal{X}}^+ (\boldsymbol{\mathcal{X}}\boldsymbol{w} + \boldsymbol{r}))^\top \big\rangle.
\end{aligned}$$

But from Equation (2.97) we see that $\boldsymbol{\mathcal{X}}^+ \boldsymbol{\mathcal{X}} = \boldsymbol{I}$, so

$$\begin{aligned}
\boldsymbol{\Sigma}_w &\approx \langle (-\boldsymbol{\mathcal{X}}^+ \boldsymbol{r})(-\boldsymbol{\mathcal{X}}^+ \boldsymbol{r})^\top \rangle = \boldsymbol{\mathcal{X}}^+ \langle \boldsymbol{r}\boldsymbol{r}^\top \rangle \boldsymbol{\mathcal{X}}^{+\top} \\
&= \sigma^2 \boldsymbol{\mathcal{X}}^+ \boldsymbol{\mathcal{X}}^{+\top}.
\end{aligned}$$

Again with Equation (2.97) we have finally

$$\boldsymbol{\Sigma}_w \approx \sigma^2 (\boldsymbol{\mathcal{X}}^\top \boldsymbol{\mathcal{X}})^{-1} \tag{2.98}$$

To check that this is indeed a generalization of ordinary linear regression on a single independent variable, identify the parameter vector $\boldsymbol{w}$ with the straight line parameters a and b, i.e.,

$$\boldsymbol{w} = \begin{pmatrix} w_0 \\ w_1 \end{pmatrix} = \begin{pmatrix} a \\ b \end{pmatrix}.$$

*In terms of the singular value decomposition (SVD) of $\boldsymbol{\mathcal{X}}$, namely $\boldsymbol{\mathcal{X}} = \boldsymbol{U}\boldsymbol{\Lambda}\boldsymbol{V}^\top$, the pseudoinverse is $\boldsymbol{\mathcal{X}}^+ = \boldsymbol{V}\boldsymbol{\Lambda}^{-1}\boldsymbol{U}^\top$, generalizing the definition for a symmetric square matrix given by Equation (1.50).

The matrix $\boldsymbol{\mathcal{X}}$ and vector $\boldsymbol{y}$ are similarly

$$\boldsymbol{\mathcal{X}} = \begin{pmatrix} 1 & x(1) \\ 1 & x(2) \\ \vdots & \vdots \\ 1 & x(m) \end{pmatrix}, \quad \boldsymbol{y} = \begin{pmatrix} y(1) \\ y(2) \\ \vdots \\ y(m) \end{pmatrix}.$$

Thus the best estimates for the parameters are

$$\hat{\boldsymbol{w}} = \begin{pmatrix} \hat{a} \\ \hat{b} \end{pmatrix} = (\boldsymbol{\mathcal{X}}^\top \boldsymbol{\mathcal{X}})^{-1}(\boldsymbol{\mathcal{X}}^\top \boldsymbol{y}).$$

Evaluating:

$$(\boldsymbol{\mathcal{X}}^\top \boldsymbol{\mathcal{X}})^{-1} = \begin{pmatrix} m & \sum x(\nu) \\ \sum x(\nu) & \sum x(\nu)^2 \end{pmatrix}^{-1} = \begin{pmatrix} m & m\bar{x} \\ m\bar{x} & \sum x(\nu)^2 \end{pmatrix}^{-1}.$$

Recalling the expression for the inverse of a 2×2 matrix in Chapter 1, we then have

$$(\boldsymbol{\mathcal{X}}^\top \boldsymbol{\mathcal{X}})^{-1} = \frac{1}{m \sum x(\nu)^2 + m^2 \bar{x}^2} \begin{pmatrix} \sum x(\nu)^2 & -m\bar{x} \\ -m\bar{x} & m \end{pmatrix}.$$

Furthermore,

$$\boldsymbol{\mathcal{X}}^\top \boldsymbol{y} = \begin{pmatrix} m\bar{y} \\ \sum x(\nu) y(\nu) \end{pmatrix}.$$

Therefore the estimate for b is

$$\hat{b} = \frac{1}{m \sum x(\nu)^2 + m^2 \bar{x}^2} \left(-m^2 \bar{x}\bar{y} + m \sum x(\nu) y(\nu)\right) = \frac{-m\bar{x}\bar{y} + \sum x(\nu) y(\nu)}{m \sum x(\nu)^2 + m^2 \bar{x}^2}. \tag{2.99}$$

From Equation (2.98), the uncertainty in b is given by σ^2 times the (2,2) element of $(\boldsymbol{\mathcal{X}}^\top \boldsymbol{\mathcal{X}})^{-1}$,

$$\sigma_b^2 = \sigma^2 \frac{m}{m \sum_\nu x(\nu)^2 + m^2 \bar{x}^2}. \tag{2.100}$$

Equations (2.99) and (2.100) correspond to Equations (2.85) and (2.87).

2.6.4 Regularization, duality and the Gram matrix

For *ill-conditioned* regression problems (e.g., large amount of noise, insufficient data or $\boldsymbol{\mathcal{X}}^\top \boldsymbol{\mathcal{X}}$ nearly singular) the solution $\hat{\boldsymbol{w}}$ in Equation (2.96) may be unreliable. A remedy is to restrict $\boldsymbol{w}$ in some way, the simplest one being to favor a small length or, equivalently, a small squared norm $\|\boldsymbol{w}\|^2$. In the modified goodness-of-fit function

$$z(\boldsymbol{w}) = (\boldsymbol{y} - \boldsymbol{\mathcal{X}}\boldsymbol{w})^\top (\boldsymbol{y} - \boldsymbol{\mathcal{X}}\boldsymbol{w}) + \lambda \|\boldsymbol{w}\|^2, \tag{2.101}$$

where we have assumed $\sigma^2 = 1$ for simplicity, the parameter λ defines a trade-off between minimum residual error and minimum norm. Equating the vector derivative with respect to $\boldsymbol{w}$ with zero as before then leads to the normal equation

$$\mathcal{X}^\top \boldsymbol{y} = (\mathcal{X}^\top \mathcal{X} + \lambda \boldsymbol{I}_{N+1})\boldsymbol{w}, \tag{2.102}$$

where the identity matrix $\boldsymbol{I}_{N+1}$ has dimensions $(N+1) \times (N+1)$. The least squares estimate for the parameter vector $\boldsymbol{w}$ is now

$$\hat{\boldsymbol{w}} = (\mathcal{X}^\top \mathcal{X} + \lambda \boldsymbol{I}_{N+1})^{-1} \mathcal{X}^\top \boldsymbol{y}. \tag{2.103}$$

For $\lambda > 0$, the matrix $\mathcal{X}^\top \mathcal{X} + \lambda \boldsymbol{I}_{N+1}$ can always be inverted. This procedure is known as *ridge regression* and Equation (2.103) may be referred to as its *primal solution.* Regularization will be encountered in Chapter 9 in the context of change detection.

With a simple manipulation, Equation (2.103) can be put in the form

$$\hat{\boldsymbol{w}} = \mathcal{X}^\top \boldsymbol{\alpha} = \sum_{\nu=1}^{m} \alpha_\nu \boldsymbol{x}(\nu), \tag{2.104}$$

where $\boldsymbol{\alpha}$ is given by

$$\boldsymbol{\alpha} = \frac{1}{\lambda}(\boldsymbol{y} - \mathcal{X}\hat{\boldsymbol{w}}); \tag{2.105}$$

see Exercise 20. Equation (2.104) expresses the unknown parameter vector $\hat{\boldsymbol{w}}$ as a linear combination of the observation vectors $\boldsymbol{x}(\nu)$. It remains to find a suitable expression for the vector $\boldsymbol{\alpha}$. We can eliminate $\hat{\boldsymbol{w}}$ from Equation (2.105) by substituting Equation (2.104) and solving for $\boldsymbol{\alpha}$,

$$\boldsymbol{\alpha} = (\mathcal{X}\mathcal{X}^\top + \lambda \boldsymbol{I}_m)^{-1} \boldsymbol{y}, \tag{2.106}$$

where $\boldsymbol{I}_m$ is the $m \times m$ identity matrix. Equations (2.104) and (2.106) taken together constitute the *dual solution* of the ridge regression problem, and the components of $\boldsymbol{\alpha}$ are called the *dual parameters.* Once they have been determined from Equation (2.106), the solution for the original parameter vector $\hat{\boldsymbol{w}}$ is recovered from Equation (2.104). Note that in the primal solution, Equation (2.103), we are inverting a $(N+1) \times (N+1)$ matrix,

$$\mathcal{X}^\top \mathcal{X} + \lambda \boldsymbol{I}_{N+1}$$

whereas in the dual solution we must invert the (often much larger) $m \times m$ matrix

$$\mathcal{X}\mathcal{X}^\top + \lambda \boldsymbol{I}_m.$$

The matrix $\mathcal{X}\mathcal{X}^\top$ is called a *Gram matrix.* Its elements consist of all inner products of the observation vectors

$$\boldsymbol{x}(\nu)^\top \boldsymbol{x}(\nu'), \quad \nu, \nu' = 1 \ldots m,$$

and it is obviously a symmetric matrix. Moreover, it is positive semi-definite since, for any vector $\boldsymbol{z}$ with m components,

$$\boldsymbol{z}^\top \boldsymbol{\mathcal{X}} \boldsymbol{\mathcal{X}}^\top \boldsymbol{z} = \|\boldsymbol{\mathcal{X}}^\top \boldsymbol{z}\|^2 \geq 0.$$

The dual solution is interesting because the dual parameters are expressed entirely in terms of inner products of observation vectors $\boldsymbol{x}(\nu)$. Moreover, predicting values of y from new observations $\boldsymbol{x}$ can be expressed purely in terms of inner products as well. Thus

$$y = \hat{\boldsymbol{w}}^\top \boldsymbol{x} = \left(\sum_{\nu=1}^{m} \alpha_\nu \boldsymbol{x}(\nu)^\top \right) \boldsymbol{x} = \sum_{\nu=1}^{m} \alpha_\nu \, \boldsymbol{x}(\nu)^\top \boldsymbol{x}. \tag{2.107}$$

Later we will see that the inner products can be substituted by so-called *kernel functions*, which allow very elegant and powerful nonlinear generalizations of linear methods such as ridge regression.

2.7 Entropy and information

Suppose we make an observation on a discrete random variable X with mass function

$$p\big(X = x(i)\big) = p(i), \quad i = 1 \ldots n.$$

Qualitatively speaking, the *amount of information* we receive on observing a particular realization $x(i)$ may be thought of as the "amount of surprise" associated with the result. The information content of the observation should be a monotonically decreasing function of the probability $p(i)$ for that observation: if the probability is unity, there is no surprise; if $p(i) \ll 1$, the surprise is large. The function chosen to express the information content of $x(i)$ is

$$h(x(i)) = -\log p(i). \tag{2.108}$$

This function is monotonically decreasing and zero when $p(i) = 1$. It also has the desirable property that, for two independent observations $x(i), x(j)$,

$$\begin{aligned} h(x(i), x(j)) &= -\log p(X = x(i), X = x(j)) \\ &= -\log \left[p(X = x(i))p(X = x(j))\right] \\ &= -\log \left[p(i)p(j)\right] = h(x(i)) + h(x(j)), \end{aligned}$$

that is, information gained from two independent observations is additive.

The *average amount of information* that we expect to receive on observing the random variable X is called the *entropy* of X and is given by

$$H(X) = -\sum_{i=1}^{n} p(i) \log p(i). \tag{2.109}$$

The entropy can also be interpreted as the *average amount of information required to specify the random variable.*

The discrete distribution with maximum entropy can be determined by maximizing the Lagrange function

$$L\left(p(1)\dots p(n)\right) = -\sum_i p(i)\log p(i) + \lambda\left(\sum_i p(i) - 1\right).$$

Equating the derivatives to zero,

$$\frac{\partial L}{\partial p(i)} = -\log p(i) - 1 + \lambda = 0,$$

so $p(i)$ is independent of i. The condition $\sum_i p(i) = 1$ then requires that

$$p(i) = 1/n.$$

The Hessian matrix is easily seen to have diagonal elements given by

$$(\boldsymbol{H})_{ii} = \frac{\partial^2 L}{\partial p(i)^2} = -\frac{1}{p(i)},$$

and off-diagonal elements zero. It is therefore negative definite, i.e., for any $\boldsymbol{x} > \boldsymbol{0}$,

$$\boldsymbol{x}^\top \boldsymbol{H}\boldsymbol{x} = -\frac{1}{p_1}x_1^2 - \dots - \frac{1}{p_n}x_n^2 < 0.$$

Thus the uniform distribution indeed maximizes the entropy.

If X is a continuous random variable with probability density function $p(x)$, then its entropy* is defined analogously to Equation (2.109) as

$$H(X) = -\int p(x)\log[p(x)]dx. \tag{2.110}$$

The continuous distribution function which has maximum entropy is the normal distribution; see Bishop (2006), Chapter 1, for a derivation of this fact.

If $p(x,y)$ is a joint density function for random variables X and Y, then the *conditional entropy* of Y given X is

$$\begin{aligned} H(Y \mid X) &= -\int p(x)\left(\int p(y \mid x)\log[p(y \mid x)]dy\right)dx \\ &= -\int\int p(x,y)\log[p(y \mid x)]dydx. \end{aligned} \tag{2.111}$$

For the second equality we have used Equation (2.61). This is just the information $-\log[p(y \mid x)]$ gained on observing y given x, averaged over the joint

*More correctly, *differential entropy* (Bishop, 2006).

probability for x and y. If Y is independent of X, then $p(y \mid x) = p(y)$ and $H(Y \mid X) = H(Y)$.

We can express the entropy $H(X,Y)$ associated with the random vector $(X,Y)^\top$ in terms of conditional entropy as follows:

$$\begin{aligned} H(X,Y) &= -\int\int p(x,y)\log[p(x,y)]dxdy \\ &= -\int\int p(x,y)ln[p(y \mid x)p(x)]dxdy \\ &= -\int\int p(x,y)\log[p(y \mid x)]dxdy - \int\int p(x,y)\log[p(x)]dxdy \\ &= H(Y \mid X) - \int\left(\int p(x,y)dy\right)\log[p(x)]dx \\ &= H(Y \mid X) - \int p(x)\log[p(x)]dx \end{aligned}$$

or

$$H(X,Y) = H(Y \mid X) + H(X). \tag{2.112}$$

2.7.1 Kullback–Leibler divergence

Let $p(x)$ be some unknown density function for a random variable X, and let $q(x)$ represent an approximation of that density function. Then the information required to specify X when using $q(x)$ as an approximation for $p(x)$ is given by

$$-\int p(x)\log[q(x)]dx.$$

The *additional information* required relative to that for the correct density function is called the *Kullback–Leibler (KL) divergence* between density functions $p(x)$ and $q(x)$ and is given by

$$\begin{aligned} \mathrm{KL}(p,q) &= -\int p(x)\log[q(x)]dx - \left(-\int p(x)\log[p(x)]dx\right) \\ &= -\int p(x)\log\left[\frac{q(x)}{p(x)}\right]dx. \end{aligned} \tag{2.113}$$

The KL divergence can be shown to satisfy (Exercise 21)

$$\mathrm{KL}(p,q) > 0,\ p(x) \neq q(x), \quad \mathrm{KL}(p,p) = 0,$$

and is thus a measure of the dissimilarity between $p(x)$ and $q(x)$.

2.7.2 Mutual information

Consider two gray-scale images represented by random variables X and Y. Their joint probability distribution is $p(x,y)$. If the images are completely

Listing 2.5: Mutual information using IDL histogram functions.

```
FUNCTION MI, image1, image2
; returns the mutual information of
; two grayscale byte images
   p12 = hist_2d(image1,image2,min1=0,max1=255, $
                 min2=0,max2=255)
   p12 = float(p12)/total(p12)
   p1  = histogram(image1,min=0,max=255)
   p1  = float(p1)/total(p1)
   p2  = histogram(image2,min=0,max=255)
   p2  = float(p2)/total(p2)
   p1p2 = transpose(p2)##p1
   i = where(p1p2 GT 0 AND p12 GT 0)
   RETURN, total(p12[i]*alog(p12[i]/p1p2[i]))
END

PRO ex2_3
  envi_select, title='Choose␣first␣single␣band␣image',$
             fid=fid1, dims=dims,pos=pos1, /band_only
  IF (fid1 EQ -1) THEN RETURN
  envi_select,title='Choose␣second␣single␣band␣image',$
             fid=fid2, dims=dims,pos=pos2, /band_only
  IF (fid2 EQ -1) THEN RETURN
  image1 = $
   (envi_get_data(fid=fid1,dims=dims,pos=pos1))[*]
  image2 = $
   (envi_get_data(fid=fid2,dims=dims,pos=pos2))[*]
  PRINT, MI(bytscl(image1),bytscl(image2))
END
```

independent, then

$$p(x,y) = p(x)p(y).$$

Thus the extent $I(X,Y)$ to which they are *not* independent can be measured by the KL divergence between $p(x,y)$ and $p(x)p(y)$:

$$I(X,Y) = \mathrm{KL}(p(x,y),p(x)p(y)) = -\int\int p(x,y)\log\left[\frac{p(x)p(y)}{p(x,y)}\right]dxdy, \tag{2.114}$$

which is called the *mutual information* between X and Y. Expanding:

$$\begin{aligned} I(X,Y) &= -\int\int p(x,y)\big[\log[p(x)] + \log[p(y)] - \log[p(x,y)]\big]dxdy \\ &= H(X) + H(Y) + \int\int p(x,y)\log[p(x \mid y)p(y)]dxdy \\ &= H(X) + H(Y) - H(X \mid Y) - H(Y), \end{aligned}$$

and thus

$$I(X,Y) = H(X) - H(X \mid Y). \tag{2.115}$$

Mutual information measures the degree of dependence between the two images, a value of zero indicating statistical independence. This is to be contrasted with correlation, where a value of zero implies statistical independence only for normally distributed quantities; see Theorem 2.7.

In practice the images are quantized, so that if p_1 and p_2 are their normalized histograms (i.e., $\sum_i p_1(i) = \sum_i p_2(i) = 1$) and p_{12} is the normalized two-dimensional histogram, $\sum_{ij} p_{12}(i,j) = 1$, then the mutual information is

$$\begin{aligned} I(1,2) &= -\sum_{ij} p_{12}(i,j)\big(\log[p_1(i)] + \log[p_2(j)] - \log[p_{12}(i,j)]\big) \\ &= \sum_{ij} p_{12}(i,j) \log \frac{p_{12}(i,j)}{p_1(i)p_2(j)}. \end{aligned} \tag{2.116}$$

The IDL program in Listing 2.5 calculates the mutual information between two image bands.

2.8 Exercises

1. Derive Equations (2.13).

2. Let the random variable X be standard normally distributed with density function

$$\phi(x) = \frac{1}{\sqrt{2\pi}} \exp(-x^2/2), \quad -\infty < x < \infty.$$

   Show that the random variable $|X|$ has the density function

$$p(x) = \begin{cases} 2\phi(x) & \text{for } x > 0 \\ 0 & \text{otherwise.} \end{cases}$$

3. Use Equation (2.8) and the result of Exercise 2 to show that the random variable $Y = X^2$, where X is standard normally distributed, has the chi-square density function, Equation (2.37), with $m = 1$ degree of freedom.

4. If X_1 and X_2 are independent random variables, both standard normally distributed, show that $X_1 + X_2$ is normally distributed with mean 0 and variance 2. (*Hint*: Write down the joint density function $f(x_1, x_2)$ for X_1 and X_2. Then treat x_1 as fixed and apply Theorem 2.1.)

5. Show from Theorem 2.2 that the sample mean

$$\bar{Z} = \frac{1}{m}\sum_{i=1}^{m} Z_i ,$$

is normally distributed with mean μ and variance σ^2/m.

6. Prove that, for $\alpha > 1$, $\Gamma(\alpha) = (\alpha-1)\Gamma(\alpha-1)$ and hence that, for positive integers n, $\Gamma(n) = (n-1)!$ (Hint: Use integration by parts.)

7. (a) Show that the mean and variance of a random variable Z with the gamma probability density, Equation (2.33), are $\mu = \alpha\beta$ and $\sigma^2 = \alpha\beta^2$.

   (b) (Proof of Theorem 2.4 for $m = 2$) Suppose that Z_1 and Z_2 are independent and exponentially distributed random variables with density functions as in Equation (2.36). Then we can write the probability distribution of $Z = Z_1 + Z_2$ in the form

$$P(z) = \Pr(Z_1 + Z_2 < z) = \int_0^z \int_0^{z-z_2} \frac{1}{\beta}e^{-z_1/\beta}\frac{1}{\beta}e^{-z_2/\beta}dz_1 dz_2.$$

   Evaluate this double integral and then take its derivative with respect to z to show that the probability density function for Z is the gamma density with $\alpha = 2$, i.e.,

$$p(z) = \begin{cases} \frac{1}{\beta^2\Gamma(2)} z e^{-z/\beta} & \text{for } z > 0 \\ 0 & \text{elsewhere.} \end{cases}$$

8. For constant vectors $\boldsymbol{a}$ and $\boldsymbol{b}$ and random vector $\boldsymbol{G}$ with covariance matrix $\boldsymbol{\Sigma}$, demonstrate that $\text{cov}(\boldsymbol{a}^\top \boldsymbol{G}, \boldsymbol{b}^\top \boldsymbol{G}) = \boldsymbol{a}^\top \boldsymbol{\Sigma} \boldsymbol{b}$.

9. Write down the multivariate normal probability density function $p(\boldsymbol{z})$ for the case $\boldsymbol{\Sigma} = \sigma^2 \boldsymbol{I}$. Show that probability density function $p(z)$ for a one-dimensional random variable Z is a special case. Using the fact that $\int_{-\infty}^{\infty} p(z)dz = 1$, demonstrate that $\langle Z \rangle = \mu$.

10. Given the $m \times N$ (uncentered) data matrix $\mathcal{Z}$, show that the covariance matrix estimate can be written in the form

$$(m-1)\boldsymbol{s} = \mathcal{Z}^\top \boldsymbol{H} \mathcal{Z},$$

   where the *centering matrix* $\boldsymbol{H}$ is given by

$$\boldsymbol{H} = \boldsymbol{I}_{mm} - \frac{1}{m}\boldsymbol{1}_m \boldsymbol{1}_m^\top.$$

   Show that $\boldsymbol{H}$ is not only symmetric, but also *idempotent* ($\boldsymbol{H}\boldsymbol{H} = \boldsymbol{H}$). Use this fact to prove that $\boldsymbol{s}$ is positive semi-definite.

11. Demonstrate Equation (2.54).

12. In the game *Lets Make a Deal!* a contestant is asked to choose between one of three doors. Behind one of the doors the prize is an automobile. After the contestant has chosen, the quizmaster opens one of the other two doors to show that the automobile is not there. He then asks the contestant if she wishes to change her mind and switch from her original choice to the other unopened door. Use Bayes' Theorem to prove that her correct answer is "yes."

13. Write an IDL script to generate two normal distributions with the random number generator `RANDOMU()` and test them for equal means with the Student-t test (IDL function `TM_TEST()`) and for equal variance with the F-test (IDL function `FV_TEST()`). Alternatively, do the same in Python with `scipy.stats.ttest_ind()` for equal means and `scipy.stats.bartlett()` for equal variances.

14. Show that the critical region for the likelihood ratio test for $\mu = \mu_0$ against $\mu \neq \mu_0$ for known variance σ^2 and m samples can be written as

$$\exp\left(-\frac{1}{2\sigma^2/m}(\bar{z} - \mu_0)^2\right) \leq k.$$

15. Prove Equations (2.85) for the regression parameter estimates $\hat{a}$ and $\hat{b}$. Demonstrate that these values correspond to a minimum of the goodness-of-fit function, Equation (2.84), and not to a maximum.

16. Derive the uncertainty for a in Equation (2.87) from the formula for error propagation for uncorrelated errors

$$\sigma_a^2 = \sum_{i=1}^{n} \sigma^2 \left(\frac{\partial a}{\partial y(i)}\right)^2.$$

17. Derive Equation (2.95) by applying the rules for vector differentiation to minimize the goodness-of-fit function, Equation (2.94).

18. Write an IDL or Python program to calculate the regression coefficients of spectral band 2 on spectral band 1 of a multispectral image. (The built-in IDL function for ordinary linear regression is `REGRESS()`. In Python, use `numpy.linalg.lstsq()`.)

19. Prove Equation (2.90) by replacing $(y(\nu) - \bar{y})$ in the numerator by $(y(\nu) + \hat{y}(\nu) - \hat{y}(\nu) + \bar{y})$ and expanding.

20. Show that Equations (2.103) and (2.104) are equivalent to Equation (2.105).

21. A *convex function* $f(x)$ satisfies $f(\lambda a + (1-\lambda)b) \leq \lambda f(a) + (1-\lambda)f(b)$. *Jensen's inequality* states that, for any convex function $f(x)$, any function $g(x)$ and any probability density $p(x)$,

$$\int f(g(x))p(x)dx \geq f\left(\int g(x)p(x)dx\right). \tag{2.117}$$

Use this to show that the KL divergence satisfies

$$\mathrm{KL}(p,q) > 0,\ p(x) \neq q(x), \quad \mathrm{KL}(p,p) = 0.$$

3

Transformations

Thus far we have thought of multispectral and SAR images as three-dimensional arrays of pixel intensities (columns × rows × bands) representing, more or less directly, measured radiances. In the present chapter we consider other, more abstract representations which are useful in image interpretation and analysis and which will play an important role in later chapters.

The discrete Fourier and wavelet transforms that we treat in Sections 3.1 and 3.2 convert the pixel values in a given spectral band to linear combinations of orthogonal functions of spatial frequency and distance. They may therefore be classified as *spatial transformations*. The principal components, minimum noise fraction and maximum autocorrelation factor transformations (Sections 3.3 to 3.5), on the other hand, create at each pixel location new linear combinations of the pixel intensities from all of the spectral bands and can properly be called *spectral transformations* (Schowengerdt, 1997).

3.1 The discrete Fourier transform

Let the function $g(x)$ represent the radiance at a point x focused along a row of pushbroom-geometry sensors, and $g(j)$ be the corresponding pixel intensities stored in a row of a digital image. We can think of $g(j)$ approximately as a discrete sample* of the function $g(x)$, sampled c times at some sampling interval Δ, c being the number of columns in the image, i.e.,

$$g(j) = g(x = j\Delta), \quad j = 0 \dots c - 1.$$

For convenience, the pixels are numbered from zero, a convention that will be adhered to in the remainder of the book. The interval Δ is the sensor width or, projected back to the Earth's surface, the across-track ground sample distance (GSD).

The theory of Fourier analysis states that the function $g(x)$ can be expressed in the form

$$g(x) = \int_{-\infty}^{\infty} \hat{g}(f) e^{\mathrm{i}2\pi f x} df, \tag{3.1}$$

*More correctly, $g(j)$ is a result of convolutions of the spatial and spectral response functions of the detector with the focused signal; see Chapter 4.

where $\hat{g}(f)$ is called the *Fourier transform* of $g(x)$. Equation (3.1) describes a continuous superposition of periodic complex functions of x,

$$e^{\mathbf{i}2\pi fx} = \cos(2\pi fx) + \mathbf{i}\sin(2\pi fx),$$

having frequency* f. In general, we require a continuum of periodic functions $e^{\mathbf{i}2\pi fx}$ to represent $g(x)$ in this way. However, if $g(x)$ is in fact itself periodic with period T, that is, if $g(x+T) = g(x)$, then the integral in Equation (3.1) can be replaced by an infinite sum of *discrete* periodic functions of frequency kf for $-\infty < k < \infty$, where f is the *fundamental frequency* $f = 1/T$:

$$g(x) = \sum_{k=-\infty}^{\infty} \hat{g}(k)e^{\mathbf{i}2\pi(kf)x}. \tag{3.2}$$

If we think of the sampled series of pixels $g(j)$, $j = 0 \dots c-1$, as also being periodic with period $T = c\Delta$, that is, repeating itself to infinity in both positive and negative directions, then we can replace x in Equation (3.2) by $j\Delta$ and express $g(j)$ in a similar way:

$$g(j) = g(j\Delta) = \sum_{k=-\infty}^{\infty} \hat{g}(k)e^{\mathbf{i}2\pi(kf)j\Delta} = \sum_{k=-\infty}^{\infty} \hat{g}(k)e^{\mathbf{i}2\pi kj/c}, \tag{3.3}$$

where in the last equality we have used $f\Delta = \Delta/T = 1/c$.

The limits in the summation in Equation (3.3) must, however, be truncated. This is due to the fact that there is a limit to the highest frequency $k_{max}f$ that can be measured by sampling at the interval Δ. The limit is called the *Nyquist critical frequency* f_N. It may be determined simply by observing that the minimum number of samples needed to describe a sine wave completely is two per period (e.g., at the maximum and minimum values). Therefore, the shortest period measurable is 2Δ and the Nyquist frequency is

$$f_N = \frac{1}{2\Delta} = \frac{cf}{2}.$$

Hence $k_{max} = c/2$. Taking this into account in Equation (3.3) we obtain

$$g(j) = \sum_{k=-c/2}^{c/2} \hat{g}(k)e^{\mathbf{i}2\pi kj/c}, \quad j = 0 \dots c-1. \tag{3.4}$$

The effect of truncation depends upon the nature of the function $g(x)$ being sampled. According to the *Sampling Theorem*, see, e.g., Press et al. (2002), $g(x)$ is completely determined by the samples $g(j)$ in Equation (3.4) if it

*Most often, frequency is associated with inverse time (cycles per second). Here, of course, we are speaking of *spatial frequency*, or cycles per meter.

is *bandwidth limited* to frequencies smaller than f_N, i.e., provided that, in Equation (3.1), $\hat{g}(f) = 0$ for all $|f| \geq f_N$. If this is not the case, then any frequency component outside the interval $(-f_N, f_N)$ is spuriously moved into that range, a phenomenon referred to as *aliasing*.

To bring Equation (3.4) into a more convenient form, we have to make a few simple manipulations. To begin with, note that the exponents in the first and last terms in the summation are equal, i.e.,

$$e^{\mathbf{i}2\pi(-c/2)j/c} = e^{-\mathbf{i}\pi j} = (-1)^j = e^{\mathbf{i}\pi j} = e^{\mathbf{i}2\pi(c/2)j/c},$$

so we can lump those two terms together and write Equation (3.4) equivalently as

$$g(j) = \sum_{k=-c/2}^{c/2-1} \hat{g}(k) e^{\mathbf{i}2\pi kj/c}, \quad j = 0 \ldots c-1.$$

Rearranging further,

$$\begin{aligned} g(j) &= \sum_{k=0}^{c/2-1} \hat{g}(k) e^{\pi 2\pi kj/c} + \sum_{k=-c/2}^{-1} \hat{g}(k) e^{\mathbf{i}2\pi kj/c} \\ &= \sum_{k=0}^{c/2-1} \hat{g}(k) e^{\mathbf{i}2\pi kj/c} + \sum_{k'=c/2}^{c-1} \hat{g}(k'-c) e^{\mathbf{i}2\pi(k'-c)j/c} \\ &= \sum_{k=0}^{c/2-1} \hat{g}(k) e^{\mathbf{i}2\pi kj/c} + \sum_{k'=c/2}^{c-1} \hat{g}(k'-c) e^{\mathbf{i}2\pi k'j/c}. \end{aligned}$$

Thus we have finally

$$g(j) = \sum_{k=0}^{c-1} \hat{g}(k) e^{\mathbf{i}2\pi kj/c}, \quad j = 0 \ldots c-1, \tag{3.5}$$

provided that we interpret $\hat{g}(k)$ as meaning $\hat{g}(k-c)$ when $k \geq c/2$.*

Equation (3.5) is a set of c equations in the c unknown frequency components $\hat{g}(k)$. Its solution is called the *discrete Fourier transform* and is given by

$$\hat{g}(k) = \frac{1}{c} \sum_{j=0}^{c-1} g(j) e^{-\mathbf{i}2\pi kj/c}, \quad k = 0 \ldots c-1. \tag{3.6}$$

This follows (Exercise 2) from the orthogonality property of the exponentials:

$$\sum_{j=0}^{c-1} e^{\mathbf{i}2\pi(k-k')j/c} = c\delta_{k,k'}, \tag{3.7}$$

*This is the convention adopted in IDL in the fast Fourier transform function `FFT()`.

Listing 3.1: Displaying the power spectrum of an image band in ENVI.

```
PRO EX3_1
envi_select, title='Choose␣multispectral␣band', $
             fid=fid, dims=dims,pos=pos, /band_only
IF (fid EQ -1) THEN BEGIN
   PRINT, 'cancelled'
   RETURN
ENDIF
cols = dims[2]-dims[1]+1
rows = dims[4]-dims[3]+1
image = envi_get_data(fid=fid,dims=dims,pos=pos[0])
; arrays of i and j values
a = lindgen(cols,rows)
i = a MOD cols
j = a/cols
; shift Fourier transform to center
image = (-1)^(i+j)*image
; compute power spectrum an return to ENVI
envi_enter_data, alog((abs(FFT(image)))^2)
END
```

where $\delta_{k,k'}$ is the *delta function*,

$$\delta_{k,k'} = \begin{cases} 1 & \text{if } k = k' \\ 0 & \text{otherwise.} \end{cases}$$

Equation (3.5) itself is the *discrete inverse Fourier transform*. We write

$$g(j) \Leftrightarrow \hat{g}(k),$$

to signify that $g(j)$ and $\hat{g}(k)$ constitute a *discrete Fourier transform pair*.

Determining the frequency components in Equation (3.6) from the original pixel intensities would appear to involve, in all, c^2 floating point multiplication operations. The *fast Fourier transform* (FFT) exploits the structure of the complex e-functions to reduce this to order $c \log c$, a very considerable saving in computation time for large arrays. For good explanations of the FFT algorithm see, e.g., Press et al. (2002) or Gonzalez and Woods (2002).

The discrete Fourier transform is easily generalized to two dimensions. Let $g(i,j)$, $i = 0 \ldots c-1$, $j = 0, r-1$, represent a gray-scale image. Its discrete inverse Fourier transform is

$$g(i,j) = \sum_{k=0}^{c-1} \sum_{\ell=0}^{r-1} \hat{g}(k,\ell) e^{\mathbf{i}2\pi(ik/c + j\ell/r)} \tag{3.8}$$

and the discrete Fourier transform is

$$\hat{g}(k,\ell) = \frac{1}{cr} \sum_{i=0}^{c-1} \sum_{j=0}^{r-1} g(i,j) e^{-\mathbf{i}2\pi(ik/c + j\ell/r)}. \tag{3.9}$$

Listing 3.2: Displaying the power spectrum of an image band in Python.

```
#!/usr/bin/env python
#  Name:     ex3_1.py
IMPORT auxil.auxil as auxil
FROM numpy IMPORT *
FROM numpy IMPORT fft
FROM osgeo IMPORT gdal
FROM osgeo.gdalconst IMPORT GA_ReadOnly
IMPORT matplotlib.pyplot as plt

DEF main():
    gdal.AllRegister()
    infile = auxil.select_infile()
    IF infile:
        inDataset = gdal.Open(infile,GA_ReadOnly)
        cols = inDataset.RasterXSize
        rows = inDataset.RasterYSize
    ELSE:
        RETURN
    band = inDataset.GetRasterBand(1)
    image = band.ReadAsArray(0,0,cols,rows) \
                                     .astype(FLOAT)
#  arrays of i and j values
    a = reshape(RANGE(rows*cols),(rows,cols))
    i = a % cols
    j = a / cols
#  shift Fourier transform to center
    image = (-1)**(i+j)*image
#  compute power spectrum and display
    image = log(ABS(fft.fft2(image))**2)
    mn = amin(image)
    mx = amax(image)
    plt.imshow((image-mn)/(mx-mn), cmap='gray' )
    plt.show()

IF __name__ == '__main__':
    main()
```

The frequency coefficients $\hat{g}(k,\ell)$ in Equations (3.8) and (3.9) are complex numbers. In order to represent an image in the frequency domain as a raster, one can calculate its *power spectrum*, which is defined as*

$$P(k,\ell) = |\hat{g}(k,\ell)|^2 = \hat{g}(k,\ell)\hat{g}^*(k,\ell). \tag{3.10}$$

*The magnitude $|z|$ of a complex number $z = x + \mathbf{i}y$ is $\sqrt{x^2+y^2} = \sqrt{zz^*}$, where $z^* = x - \mathbf{i}y$ is the complex conjugate of z; see Appendix A.

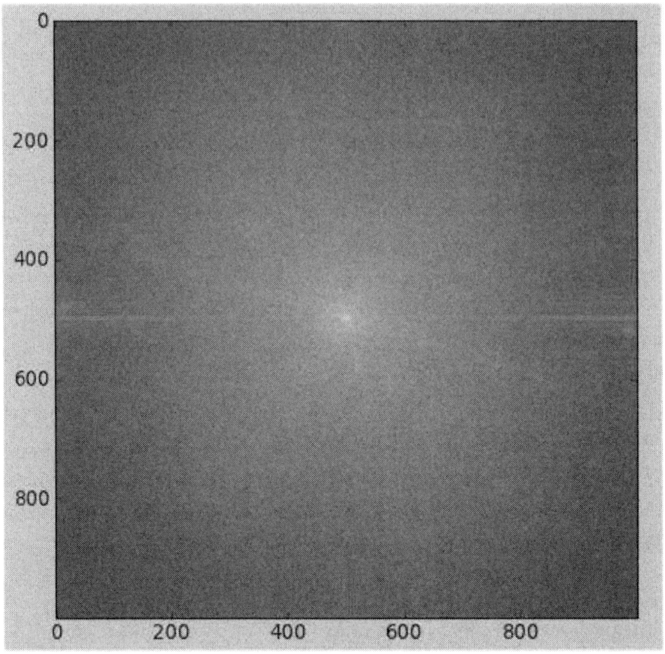

FIGURE 3.1
Logarithm of the power spectrum for the 3N band of the Jülich ASTER image with the Python script in Listing 3.2.

Rather than displaying $P(k,\ell)$ directly, which, according to ENVI's display convention, would place zero frequency components $k=0, \ell=0$ in the upper left-hand corner, use can be made of the *translation property* (Exercise 4) of the Fourier transform:

$$g(i,j)e^{\mathbf{i}2\pi(k_0 i/c+\ell_0 j/r)} \Leftrightarrow \hat{g}(k-k_0, \ell-\ell_0). \tag{3.11}$$

In particular, for $k_0 = c/2$ and $\ell_0 = r/2$, we can write

$$e^{\mathbf{i}2\pi(k_0 i/c+\ell_0 j/r)} = e^{\mathbf{i}\pi(i+j)} = (-1)^{i+j}.$$

Therefore

$$g(i,j)(-1)^{i+j} \Leftrightarrow \hat{g}(k-c/2, \ell-r/2),$$

so if we multiply an image by $(-1)^{i+j}$ before transforming, zero frequency will be at the center. This is illustrated in Listing 3.1, which performs a fast Fourier transform of an image band using the IDL function `FFT()` and displays the logarithm of the power spectrum with zero frequency at the center. Listing 3.2 shows the same procedure in Python; see Figure 3.1. We shall return to discrete Fourier transforms in the next chapter when we discuss convolutions and filters.

3.2 The discrete wavelet transform

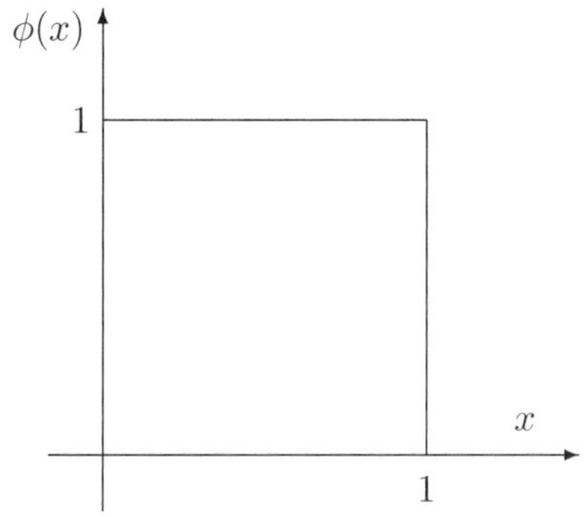

FIGURE 3.2
The Haar scaling function.

Unlike the Fourier transform, which represents an array of pixel intensities in terms of pure frequency functions, the wavelet transform expresses an image array in terms of functions which are restricted both in terms of frequency and spatial extent. In many image processing applications, this turns out to be particularly efficient and useful. The traditional (and most intuitive) way of introducing wavelets is in terms of the Haar scaling function (Strang, 1989), Aboufadel and Schlicker (1999), Gonzalez and Woods (2002), and we will adopt this approach here as well, in particular following the development in Aboufadel and Schlicker (1999).

Fundamental to the definition of wavelet transforms is the concept of an *inner product* of real-valued functions and the associated *inner product space* (Appendix A).

DEFINITION 3.1 *If f and g are two real functions on the set of real numbers $\mathbb{R}$, then their* inner product *is given by*

$$\langle f, g\rangle = \int_{-\infty}^{\infty} f(x)g(x)dx. \tag{3.12}$$

The inner product space $L_2(\mathbb{R})$ *is the collection of all functions $f : \mathbb{R} \mapsto \mathbb{R}$ with the property that*

$$\langle f, f\rangle = \int_{-\infty}^{\infty} f(x)^2 dx \quad \textit{is finite.} \tag{3.13}$$

3.2.1 Haar wavelets

The *Haar scaling function* is the function

$$\phi(x) = \begin{cases} 1 & \text{if } 0 \le x \le 1 \\ 0 & \text{otherwise} \end{cases} \tag{3.14}$$

shown in Figure 3.2. We shall use it to represent pixel intensities.

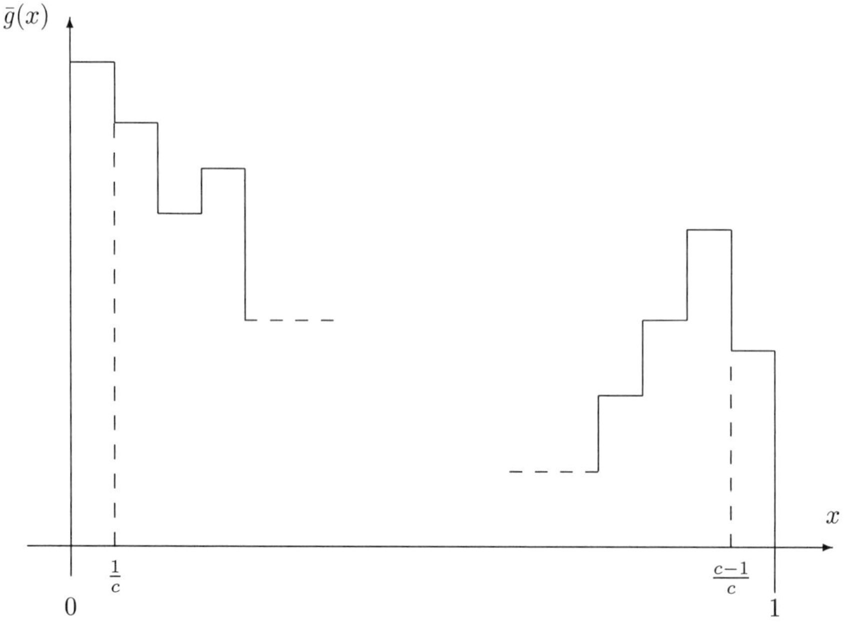

FIGURE 3.3
A row of c pixel intensities on the interval $[0, 1]$ as a piecewise constant function $\bar{g}(x) \in L_2(\mathbb{R})$. In the text it is assumed that $c = 2^n$ for some integer n.

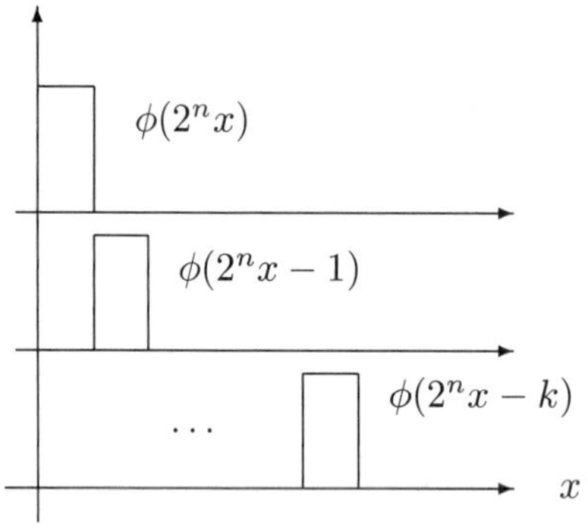

FIGURE 3.4
Basis functions C_n for space V_n.

The quantities $g(j)$, $j = 0, c-1$, representing a row of pixel intensities, can be thought of as a piecewise constant function of x. This is indicated in Figure 3.3. The abscissa has been normalized to the interval $[0, 1]$, so that j measures the distance in increments of $1/c$ along the pixel row, with the last pixel occupying the interval $[\frac{c-1}{c}, 1]$. We have called this piecewise constant function $\bar{g}(x)$ to distinguish it from $g(j)$. According to Definition 3.1 it is in $L_2(\mathbb{R})$.

Now let V_n be the collection of *all* piecewise constant functions on the interval $[0, 1]$ that have possible discontinuities at the rational points $j \cdot 2^{-n}$, where j and n are nonnegative integers. If the number of pixels c in Figure 3.3 is a power of two, $c = 2^n$ say, then $\bar{g}(x)$ clearly is a function which belongs to V_n, i.e., $\bar{g}(x) \in V_n$. The (possible) discontinuities occur at

$$x = 1 \cdot 2^{-n}, 2 \cdot 2^{-n} \ldots (c-1) \cdot 2^{-n}.$$

Certainly all members of V_n also belong to the function space $L_2(\mathbb{R})$, so that $V_n \subset L_2(\mathbb{R})$. Any function in V_n confined to the interval $[0, 1]$ in this way can be expressed as a linear combination of the *standard Haar basis functions.* These are scaled and shifted versions of the Haar scaling function of Figure 3.2 and comprise the set

$$C_n = \{\phi_{n,k}(x) = \phi(2^n x - k) \mid k = 0, 1 \ldots 2^n - 1\}, \tag{3.15}$$

see Figure 3.4.

Note that $\phi_{0,0}(x) = \phi(x)$. The index n corresponds to a compression or change of scale by a factor of 2^{-n}, whereas the index k shifts the basis function across the interval $[0, 1]$. The row of pixels in Figure 3.3 can be expanded in terms of the standard Haar basis trivially as

$$\begin{aligned} \bar{g}(x) &= g(0)\phi_{n,0}(x) + g(1)\phi_{n,1}(x) + \ldots + g(c-1)\phi_{n,c-1}(x) \\ &= \sum_{j=0}^{c-1} g(j)\phi_{n,j}(x). \end{aligned} \tag{3.16}$$

The Haar basis functions are clearly orthogonal:

$$\langle \phi_{n,k}, \phi_{n,k'} \rangle = \int_0^1 \phi_{n,k}(x)\phi_{n,k'}(x)dx = \frac{1}{2^n}\delta_{k,k'}. \tag{3.17}$$

The expansion coefficients $g(j)$ are therefore given formally by

$$g(j) = \frac{\langle \bar{g}, \phi_{n,j} \rangle}{\langle \phi_{n,j}, \phi_{n,j} \rangle} = 2^n \langle \bar{g}, \phi_{n,j} \rangle. \tag{3.18}$$

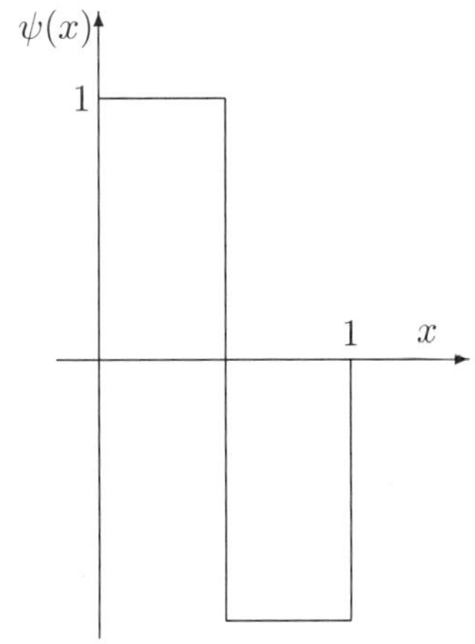

FIGURE 3.5
The Haar mother wavelet.

We will now derive a new and more interesting orthogonal basis for V_n. Consider, first of all, the function spaces V_0 and V_1 with standard Haar bases $\{\phi_{0,0}(x)\}$ and $\{\phi_{1,0}(x), \phi_{1,1}(x)\}$, respectively. According to the Orthogonal Decomposition Theorem (Appendix A, Theorem A.4), any function in V_1 can be expressed as a linear combination of the basis for V_0 plus some function in a *residual space* $V_0^\perp$ which is orthogonal to V_0 (i.e., any function in $V_0^\perp$ is orthogonal to any function in V_0). This is denoted formally by writing

$$V_1 = V_0 \oplus V_0^\perp. \tag{3.19}$$

For example, the basis function $\phi_{1,0}(x)$ of V_0 is also in V_1, and so can be written in the form

$$\phi_{1,0}(x) = \frac{\langle \phi_{1,0}, \phi_{0,0} \rangle}{\langle \phi_{0,0}, \phi_{0,0} \rangle}\, \phi_{0,0}(x) + r(x) = \frac{1}{2}\, \phi_{0,0}(x) + r(x).$$

The function $r(x)$ is in the residual space $V_0^{\perp}$. We see that, in this case,

$$r(x) = \phi_{1,0}(x) - \frac{1}{2}\, \phi_{0,0}(x) = \phi(2x) - \frac{1}{2}\, \phi(x).$$

But we can express $\phi(x)$ as $\phi(x) = \phi(2x) + \phi(2x-1)$ so that

$$r(x) = \phi(2x) - \frac{1}{2}(\phi(2x) + \phi(2x-1)) = \frac{1}{2}(\phi(2x) - \phi(2x-1)) =: \frac{1}{2}\, \psi(x).$$

The function

$$\psi(x) = \phi(2x) - \phi(2x-1) \tag{3.20}$$

is shown in Figure 3.5. It is orthogonal to $\phi(x)$ and is called the *Haar mother wavelet.* Thus an *alternative basis* for V_1 is

$$B_1 = \{\phi_{0,0}, \psi_{0,0}\},$$

where for consistency we have defined $\psi_{0,0}(x) = \psi(x)$.

This argument can be repeated (Exercise 6) for $V_2 = V_1 \oplus V_1^{\perp}$ to obtain the basis

$$B_2 = \{\phi_{0,0}, \psi_{0,0}, \psi_{1,0}, \psi_{1,1}\}$$

for V_2, where now $\{\psi_{1,0}, \psi_{1,1}\}$ is an orthogonal basis for $V_1^{\perp}$ given by

$$\psi_{1,0} = \psi(2x),\ \psi_{1,1} = \psi(2x-1).$$

Indeed, the argument can be continued indefinitely, so in general the *Haar wavelet basis* for V_n is

$$B_n = \{\phi_{0,0}, \psi_{0,0}, \psi_{1,0}, \psi_{1,1} \dots \psi_{n-1,0}, \psi_{n-1,1} \dots \psi_{n-1,2^n-1}\},$$

where $\{\psi_{m,k} = \psi(2^m x - k) \mid k = 0 \dots 2^m - 1\}$ is an orthogonal basis for $V_m^{\perp}$, and

$$V_n = V_{n-1} \oplus V_{n-1}^{\perp} = V_0 \oplus V_0^{\perp} \oplus \dots \oplus V_{n-2}^{\perp} \oplus V_{n-1}^{\perp}.$$

In terms of this new basis, the function $\bar{g}(x)$ in Figure 3.3 can now be expressed as

$$\bar{g}(x) = \hat{g}(0)\phi_{0,0}(x) + \hat{g}(1)\psi_{0,0}(x) + \dots + \hat{g}(c-1)\psi_{n-1,c-1}(x), \tag{3.21}$$

where $c = 2^n$. The expansion coefficients $\hat{g}(j)$ are called the *wavelet coefficients.* They are still to be determined.

In the case of the Haar wavelets, their determination turns out to be quite easy because there is a simple correspondence between the basis functions (ϕ, ψ) and the space of 2^n-component vectors (Strang, 1989). Consider for instance $n = 2$. Then the correspondence is

$$\phi_{0,0} = \begin{pmatrix} 1 \\ 1 \\ 1 \\ 1 \end{pmatrix}, \ \phi_{1,0} = \begin{pmatrix} 1 \\ 1 \\ 0 \\ 0 \end{pmatrix}, \ \phi_{1,1} = \begin{pmatrix} 0 \\ 0 \\ 1 \\ 1 \end{pmatrix}, \ \phi_{2,0} = \begin{pmatrix} 1 \\ 0 \\ 0 \\ 0 \end{pmatrix}, \ \ldots$$

and

$$\psi_{0,0} = \begin{pmatrix} 1 \\ 1 \\ -1 \\ -1 \end{pmatrix}, \ \psi_{1,0} = \begin{pmatrix} 1 \\ -1 \\ 0 \\ 0 \end{pmatrix}, \ \psi_{1,1} = \begin{pmatrix} 0 \\ 0 \\ 1 \\ -1 \end{pmatrix}.$$

Thus the orthogonal basis B_2 may be represented equivalently by the mutually orthogonal vectors

$$B_2 = \left\{ \begin{pmatrix} 1 \\ 1 \\ 1 \\ 1 \end{pmatrix}, \begin{pmatrix} 1 \\ 1 \\ -1 \\ -1 \end{pmatrix}, \begin{pmatrix} 1 \\ -1 \\ 0 \\ 0 \end{pmatrix}, \begin{pmatrix} 0 \\ 0 \\ 1 \\ -1 \end{pmatrix} \right\}.$$

This gives us a more convenient representation of the expansion in Equation (3.21), namely

$$\bar{\boldsymbol{g}} = \boldsymbol{B}_n \hat{\boldsymbol{g}}, \tag{3.22}$$

where $\bar{\boldsymbol{g}} = \big(g(0) \ldots g(c-1)\big)^\top$ is a column vector of the original pixel intensities, $\hat{\boldsymbol{g}} = \big(\hat{g}(0) \ldots \hat{g}(c-1)\big)^\top$ is a column vector of the wavelet coefficients, and $\boldsymbol{B}_n$ is a transformation matrix whose columns are the basis vectors of B_n. The wavelet coefficients for the pixel vector are then given by inverting the representation:

$$\hat{\boldsymbol{g}} = \boldsymbol{B}_n^{-1} \bar{\boldsymbol{g}}. \tag{3.23}$$

A full gray-scale image is transformed by first applying Equation (3.23) to its columns and then to its rows. Wavelet coefficients for gray-scale images thus obtained tend to have "simple statistics" (Gonzalez and Woods, 2002), e.g., they might be approximately Gaussian with zero mean.

3.2.2 Image compression

The fact that many of the wavelet coefficients are close to zero makes the wavelet transformation useful for image compression. This is illustrated in Listing 3.2, which is adapted from an example in the IDL Reference Guide. First the auxiliary functions `PSI_M()` and `PSI()` are used to generate the Haar wavelet transformation matrix $\boldsymbol{B}_n$ for $n = 8$ (lines 17 to 23). Then a 256×256 image subset is read from ENVI into the IDL variable `G`, converted to

Listing 3.3: Image compression with the Haar wavelet transform.

```
FUNCTION psi_m, x
   IF x LT 0.0 THEN RETURN, 0.0
   IF x LT 0.5 THEN RETURN, 1.0
   IF x LT 1.0 THEN RETURN, -1.0
   RETURN, 0.0
END
FUNCTION psi, m, k, n
   c = 2^n
   result = fltarr(c)
   x = findgen(c)/c
   FOR i=0,c-1 DO result[i]=psi_m(2^m*x[i]-k)
   RETURN, result
END

PRO ex3_2
; generate wavelet basis B_8
   n = 8L
   B = fltarr(2^n,2^n)+1.0
   i = 1
   FOR m=0,n-1 DO FOR k=0,2^m-1 DO BEGIN
      B[i,*] = psi(m,k,n)
      i++
   ENDFOR
; get a 256x256 grayscale image
   envi_select, title='Choose␣multispectral␣band', $
               fid=fid, dims=dims,pos=pos, /band_only
   G = envi_get_data(fid=fid,dims=dims,pos=pos)
   G = float(G[0:255,0:255])
   PRINT, 'Size␣original␣image:',256L*256L*4L,'␣bytes'
   envi_enter_data, G + 0.0
; transform the columns and rows
   FOR i=0,255 DO G[i,*] = invert(B)##G[i,*]
   FOR j=0,255 DO G[*,j] = invert(B)##transpose(G[*,j])
   envi_enter_data, G + 0.0
; convert to sparse format
   G = sprsin(G,thresh=1.0)
   write_spr, G, 'sparse.dat'
   OPENR, 1,'sparse.dat' & status = fstat(1) & close,1
   PRINT, 'Size␣compressed␣image:',status.size,'␣bytes'
; invert the transformation
   G = fulstr(G)
   FOR j=0,255 DO G[*,j] = B##transpose(G[*,j])
   FOR i=0,255 DO G[i,*]  = B##G[i,*]
   envi_enter_data, G
END
```

floating point (4-byte) format and its size is printed out (lines 25 to 29). After the wavelet transformation in Equation (3.23) has been applied to the rows and columns (lines 32 to 34, the result of which is returned as an image to ENVI), the IDL function `SPRSIN()` is invoked to convert the wavelet coefficient image into row-indexed sparse storage format, overwriting `G` and retaining only elements with an absolute value greater than a specified threshold, in this instance 1.0 (line 36). The sparse image is then written to disk with the `WRITE_SPR` procedure so that its file size can be queried and printed out. Finally, the compressed image is restored to normal format with the function `FULSTR()` and the wavelet transform is inverted, again overwriting `G` (lines 41 to 43). The result is returned to ENVI for comparison with the original; see Figure 3.6. Here is the printout for spectral band 1 of the Jülich image:

```
ENVI> ex3_2
Size original image:       262144 bytes
Size compressed image:      60424 bytes
```

A compression of about a factor 4.3 is achieved at the cost of the (considerable) loss of image quality visible in Figure 3.6. (The program is very slow due to the many floating point operations involved in the transformations. This will be remedied in Chapter 4 when we treat filters and the fast wavelet transform.) The wavelet coefficients themselves and their histogram are illustrated in Figure 3.7, where it is apparent that they are tightly distributed about zero and that, in this case, most have absolute magnitudes less than 1.

FIGURE 3.6
Left: a 256×256 spatial subset of band 1 of the Jülich ASTER image of Figure 1.1. Right: the same subset after transformation to the Haar wavelet basis B_8, compression with `SPRSIN()` and restoration with the inverse transformation; see Listing 3.3.

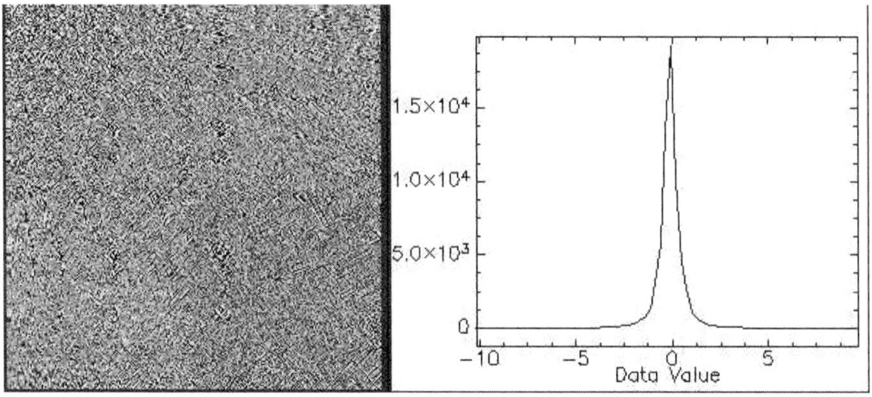

FIGURE 3.7
Left: the Haar wavelet coefficients for the image of Figure 3.6. Right: histogram of the wavelet coefficient image.

3.2.3 Multiresolution analysis

So far we have represented only functions on the interval $[0, 1]$ with the standard basis $\phi_{n,k}(x) = \phi(2^n x - k)$, $k = 1 \ldots 2^n - 1$. We can extend this to functions defined on all real numbers in a straightforward way, still restricting ourselves, however, to functions with *compact support.* These are zero everywhere outside a closed, bounded interval. Thus

$$\{\phi(x - k) \mid k \in \mathbb{Z}\},$$

where $\mathbb{Z}$ is the set of *all* integers, is a basis for the space V_0 of all piecewise constant functions with compact support having possible breaks at integer values. Note that $\phi(x - k) = \phi_{0,k}$ is an *orthonormal* basis for V_0, that is,

$$\langle \phi(x - k), \phi(x - k') \rangle = \delta_{k,k'},$$

whereas $\phi(2^n x - k) = \phi_{n,k}$ with $n > 0$ is only an orthogonal basis for V_n, since the inner products for equal k are not unity. Quite generally then, an orthogonal basis for the set V_n of piecewise constant functions with possible breaks at $j \cdot 2^{-n}$ and compact support is

$$\{\phi(2^n x - k) \mid k \in \mathbb{Z}\}. \tag{3.24}$$

One can even allow $n < 0$. For example, $n = -1$ means that the possible breaks are at even integer values.

We can think of the collection of nested subspaces of piecewise constant functions as being *generated* by the Haar scaling function ϕ. Such a collection is an example of a *multiresolution analysis* (MRA). There are many other possible scaling functions that define or generate an MRA. Although the subspaces will no longer consist of simple piecewise constant functions, nevertheless, based on our experience with the Haar wavelets, we can appreciate the following definition (Aboufadel and Schlicker, 1999):

DEFINITION 3.2 *An MRA is a collection of nested subspaces*

$$\ldots \subseteq V_{-1} \subseteq V_0 \subseteq V_1 \subseteq V_2 \subseteq \ldots \subseteq L_2(\mathbb{R}),$$

with the following properties:

1. *For any function $f \in L_2(\mathbb{R})$, there exists a series of functions, one in each V_n, which converges to f.*
2. *The only function common to all V_n is $f(x) = 0$.*
3. *The function $f(x) \in V_n$ if and only if $f(2^{-n}x) \in V_0$.*
4. *The scaling function ϕ is an orthonormal basis for the function space V_0, i.e., $\langle \phi(x-k), \phi(x-k') \rangle = \delta_{kk'}$.*

Clearly, property 1 is met for the Haar MRA, since any function in $L_2(\mathbb{R})$ can be approximated to arbitrary accuracy with successively finer piecewise constant functions. Being common to all V_n means being piecewise constant on all intervals. The only function in $L_2(\mathbb{R})$ with this property and compact support is $f(x) = 0$, so property 2 is also satisfied for the Haar MRA. If $f(x) \in V_1$ then it is piecewise constant on intervals of length 1/2. Therefore, the function $f(2^{-1}x)$ is piecewise constant on intervals of length 1, that is, $f(2^{-1}x) \in V_0$, etc., and so property 3 is satisfied as well. Finally, property 4 also holds for the Haar scaling function.

3.2.3.1 The dilation equation and refinement coefficients

In the following, we will think of $\phi(x)$ as any scaling function which generates an MRA in the sense of Definition 3.2. Since $\{\phi(x-k) \mid k \in \mathbb{Z}\}$ is an orthonormal basis for V_0, it follows that $\{\phi(2x-k) \mid k \in \mathbb{Z}\}$ is an orthogonal basis for V_1. That is, let $f(x) \in V_1$. Then by property 3, $f(x/2) \in V_0$, hence

$$f(x/2) = \sum_k a_k \phi(x-k),$$

which implies that

$$f(x) = \sum_k a_k \phi(2x-k).$$

In particular, since $\phi(x) \in V_0 \subset V_1$, we have the *dilation equation*

$$\phi(x) = \sum_k c_k \phi(2x-k). \tag{3.25}$$

The constants c_k are called the *refinement coefficients.* For example, the dilation equation for the Haar scaling function is

$$\phi(x) = \phi(2x) + \phi(2x-1),$$

so that the refinement coefficients are $c_0 = c_1 = 1$, $c_k = 0$ otherwise. Note that $c_0^2 + c_1^2 = 2$. This is a general property of the refinement coefficients:

$$1 = \langle \phi(x), \phi(x) \rangle = \langle \sum_k c_k \phi(2x-k), \sum_{k'} c_{k'} \phi(2x-k') \rangle = \frac{1}{2} \sum_k c_k^2$$

and therefore,

$$\sum_{k=-\infty}^{\infty} c_k^2 = 2, \tag{3.26}$$

which is also called *Parseval's formula.* In a similar way, one can show (Exercise 7)

$$\sum_{k=-\infty}^{\infty} c_k c_{k-2j} = 0 \quad \text{for all} \quad j \neq 0. \tag{3.27}$$

3.2.3.2 The cascade algorithm

Some of the scaling functions which generate an MRA cannot be expressed as simple, analytical functions. Nevertheless, we can work with an MRA even when there is no simple representation for the scaling function which

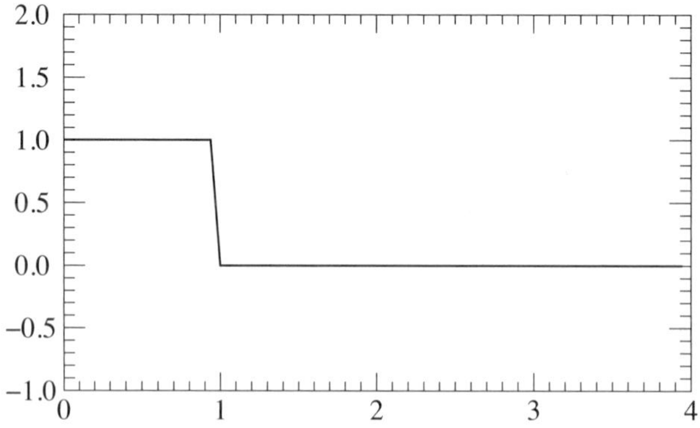

FIGURE 3.8
Approximation to the Haar scaling function with the program of Listing 3.4 after $n = 4$ iterations.

Listing 3.4: Cascade algorithm approximation to the scaling function.

```
FUNCTION F, x, i
   COMMON refinement, c0,c1,c2,c3,c4
   IF (i EQ 0) THEN IF (x EQ 0) THEN $
      RETURN, 1.0 ELSE RETURN, 0.0 ELSE $
      RETURN, c0*f(2*x,i-1)+c1*f(2*x-1,i-1)+ $
        c2*f(2*x-2,i-1)+c3*f(2*x-3,i-1)+c4*f(2*x-4,i-1)
END

PRO ex3_3

   COMMON refinement, c0,c1,c2,c3,c4
; refinement coefficients for Haar scaling function
   c0=1 & c1=1 & c2=0 & c3=0 & c4=0
; refinement coefficients for D4 scaling function
;  c0=(1+sqrt(3))/4 & c1=(3+sqrt(3))/4
;  c2=(3-sqrt(3))/4 & c3=(1-sqrt(3))/4 & c4=0

; fourth order approximation
   n=4
   x = findgen(4*2^n)
   ff=fltarr(4*2^n)
   FOR i=0,4*2^n-1 DO ff[i]=F(x[i]/2^n,n)

; output as EPS file
   thisDevice =!D.Name
   set_plot, 'PS'
   Device, Filename='fig3_8.eps',xsize=3,ysize=2, $
            /inches,/encapsulated
   PLOT,x/2^n,ff,yrange=[-1,2]
   device,/close_file
   set_plot,thisDevice

END
```

generates it. For instance, once we have the refinement coefficients for a scaling function, it can be approximated to any desired degree of accuracy using the dilation equation. The idea is to iterate the refinement equation with a so-called *cascade algorithm* until it converges to a sequence of points which approximates $\phi(x)$.

The following recursive scheme can be used to estimate a scaling function with up to five nonzero refinement coefficients $c_0, c_1 \dots c_4$:

$$
\begin{aligned}
f_0(x) &= \delta_{x,0} \\
f_i(x) &= c_0 f_{i-1}(2x) + c_1 f_{i-1}(2x-1) + c_2 f_{i-1}(2x-2) + c_3 f_{i-1}(2x-3) \\
&\quad + c_4 f_{i-1}(2x-4).
\end{aligned}
$$

In this scheme, x takes on values $j/2^n$, where j, n are any integers. The first definition is the termination condition for the recursion and approximates the scaling function to zeroth order as the delta function

$$\delta_{x,0} = \begin{cases} 1 & \text{if } x = 0 \\ 0 & \text{otherwise.} \end{cases}$$

The second relation defines the ith approximation to the scaling function in terms of the $(i-1)$th approximation using the dilation equation. We can calculate the set of values

$$\phi \approx f_n\left(\frac{j}{2^n}\right) \quad \text{for } j = 0 \ldots 4 \cdot 2^n$$

for some $n > 1$ as a point-wise approximation of ϕ on the interval $[0, 4]$. Listing 3.4 uses a recursive IDL function `F(x,i)` to approximate a scaling function in this way. The result using the Haar refinement coefficients is shown in Figure 3.8 and is seen to be an approximation of Figure 3.2.

3.2.3.3 The mother wavelet

For a general MRA, we also require a generalization of Equation (3.20), which relates the Haar mother wavelet to the scaling function. Let some MRA have a scaling function ϕ defined by the dilation Equation (3.25). Since

$$\langle \phi(2x-k), \phi(2x-k) \rangle = \frac{1}{2} \cdot \langle \phi(x), \phi(x) \rangle = \frac{1}{2},$$

the functions $\sqrt{2}\phi(2x-k)$ are both normalized and orthogonal. We can write Equation (3.25) in the form

$$\phi(x) = \sum_k h_k \sqrt{2}\phi(2x-k), \tag{3.28}$$

where

$$h_k = \frac{c_k}{\sqrt{2}}.$$

It follows from Equation (3.26) that

$$\sum_k h_k^2 = 1. \tag{3.29}$$

Now we assume, in analogy to Equation (3.28), that the mother wavelet ψ can also be expressed in terms of the scaling function as*

$$\psi(x) = \sum_k g_k \sqrt{2}\phi(2x-k). \tag{3.30}$$

*The coefficients g_k should not be confused with pixel intensities.

Since $\phi \in V_0$ and $\psi \in V_0^\perp$, we have

$$\langle \phi, \psi \rangle = \sum_k h_k g_k = 0. \tag{3.31}$$

Similarly, with some simple index manipulations,

$$\langle \psi(x-k), \psi(x-m) \rangle = \sum_i g_i g_{i-2(k-m)} = \delta_{k,m}. \tag{3.32}$$

A set of coefficients that satisfies Equations (3.31) and (3.32) is given by

$$g_k = (-1)^k h_{1-k}. \tag{3.33}$$

So we obtain, finally, the general relationship between the mother wavelet and the scaling function:

$$\psi(x) = \sum_k (-1)^k h_{1-k} \sqrt{2} \phi(2x-k) = \sum_k (-1)^k c_{1-k} \phi(2x-k). \tag{3.34}$$

3.2.3.4 The Daubechies D4 scaling function

A family of MRAs which is very useful in digital image analysis is generated by the Daubechies scaling functions and their associated wavelets (Daubechies, 1988). The Daubechies D4 scaling function, for example, can be derived by placing the following two additional requirements on an MRA (Aboufadel and Schlicker, 1999):

1. *Compact support:* The scaling function $\phi(x)$ is required to be zero outside the interval $0 < x < 3$. This means that the refinement coefficients c_k vanish for $k < 0$ and for $k > 3$. To see this, note that

$$c_{-3} = 2\langle \phi(x), \phi(2x+3) \rangle = \int_0^3 \phi(x)\phi(2x+3)dx = 0$$

and similarly for $k \le -4$ and for $k \ge 6$. Therefore, from the dilation equation,

$$\phi(-1/2) = 0 = c_{-2}\phi(-1+2) + c_{-1}\phi(-1+1) + \ldots \quad \text{implying } c_{-2} = 0$$

and similarly for $k = -1, 4, 5$. Thus from Equation (3.26), we can conclude that

$$c_0^2 + c_1^2 + c_2^2 + c_3^2 = 2 \tag{3.35}$$

and from Equation(3.27) with $j = 1$, that

$$c_0 c_2 + c_1 c_3 = 0. \tag{3.36}$$

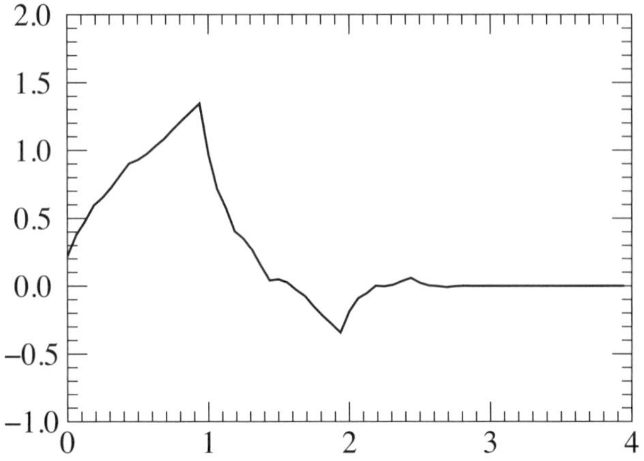

FIGURE 3.9
Approximation to the Daubechies D4 scaling function with the program of Listing 3.4 after $n = 4$ iterations.

2. *Regularity:* All constant and linear polynomials can be written as a linear combination of the basis $\{\phi(x-k) \mid k \in \mathbb{Z}\}$ for V_0. This implies that there is no residual in the orthogonal decomposition of $f(x) = 1$ and $f(x) = x$ onto the basis, that is,

$$\int_{-\infty}^{\infty} 1 \cdot \psi(x)dx = \int_{-\infty}^{\infty} x \cdot \psi(x)dt = 0. \tag{3.37}$$

With Equation (3.34) the mother wavelet is

$$\begin{aligned}\psi(x) &= -c_0\phi(2x-1) + c_1\phi(2x) - c_2\phi(2x+1) + c_3\phi(2x+2) \\ &= \sum_{k=0}^{3}(-1)^{k+1}c_k\phi(2x-1+k).\end{aligned} \tag{3.38}$$

The first requirement in Equation (3.37) gives immediately

$$-c_0 + c_1 - c_2 + c_3 = 0. \tag{3.39}$$

From the second requirement we have

$$0 = \int_{-\infty}^{\infty} x\psi(x)dx = \sum_{k=0}^{3}(-1)^{k+1}c_k \int_{-\infty}^{\infty} x\phi(2x-1+k)dx$$
$$= \sum_{k=0}^{3}(-1)^{k+1}c_k \int_{-\infty}^{\infty} \frac{u+1-k}{4}\phi(u)du$$
$$= \frac{0}{4}\cdot \int_{-\infty}^{\infty} u\phi(u)du + \frac{-c_0+c_2-2c_3}{4}\int_{-\infty}^{\infty}\phi(u)du,$$

using Equation (3.39). Thus

$$-c_0 + c_2 - 2c_3 = 0. \tag{3.40}$$

Equations (3.35), (3.36), (3.39), and (3.40) comprise a system of four (nonlinear) equations in four unknowns. A solution is given by

$$c_0 = \frac{1+\sqrt{3}}{4}, \quad c_1 = \frac{3+\sqrt{3}}{4}, \quad c_2 = \frac{3-\sqrt{3}}{4}, \quad c_3 = \frac{1-\sqrt{3}}{4},$$

which are known as the D4 refinement coefficients. Figure 3.9 shows the corresponding scaling function, determined with the cascade algorithm described earlier (Listing 3.4).

The D4 scaling function and the subspaces that it generates are thus anything but simple. The scaling function is continuous but not everywhere differentiable and also self-similar (the tail is an exact but re-scaled copy of the entire function). Nevertheless, the D4 wavelets provide a much more useful representation of digital images than the Haar wavelets.* We will return to them in Chapter 4 when we treat the fast wavelet transform and examine pyramid algorithms for image processing.

3.3 Principal components

The principal components transformation, also called *principal components analysis* (PCA), generates linear combinations of multispectral pixel intensities which are mutually uncorrelated and which have maximum variance. Specifically, consider a multispectral image represented by the random vector $\boldsymbol{G}$ (for vector of gray-scale values) and assume that $\langle \boldsymbol{G} \rangle = \boldsymbol{0}$, so that the covariance matrix is given by $\boldsymbol{\Sigma} = \langle \boldsymbol{G}\boldsymbol{G}^\top \rangle$. Let us seek a linear combination $Y = \boldsymbol{w}^\top \boldsymbol{G}$ whose variance $\boldsymbol{w}^\top \boldsymbol{\Sigma} \boldsymbol{w}$ is maximum. This quantity can

*The IDL function `WTN()` implements the discrete wavelet transform with various wavelets from the Daubechies family; see Exercise 5 in Chapter 4.

trivially be made as large as we like by choosing $\boldsymbol{w}$ sufficiently large, so that the maximization only makes sense if we restrict $\boldsymbol{w}$ in some way. A convenient constraint is $\boldsymbol{w}^\top \boldsymbol{w} = 1$. According to the discussion in Section 1.6, we can solve this problem by maximizing the unconstrained Lagrange function

$$L = \boldsymbol{w}^\top \boldsymbol{\Sigma} \boldsymbol{w} - \lambda(\boldsymbol{w}^\top \boldsymbol{w} - 1).$$

This leads directly, see Equation (1.65), to the eigenvalue problem

$$\boldsymbol{\Sigma} \boldsymbol{w} = \lambda \boldsymbol{w}. \tag{3.41}$$

Denote the orthogonal and normalized eigenvectors of $\boldsymbol{\Sigma}$ obtained by solving the above problem by $\boldsymbol{w}_1 \dots \boldsymbol{w}_N$, sorted according to decreasing eigenvalue $\lambda_1 \geq \dots \geq \lambda_N$. These eigenvectors are the *principal axes* and the corresponding linear combinations $\boldsymbol{w}_i^\top \boldsymbol{G}$ are projections along the principal axes, called the *principal components* of $\boldsymbol{G}$. The individual principal components

$$Y_1 = \boldsymbol{w}_1^\top \boldsymbol{G},\ Y_2 = \boldsymbol{w}_2^\top \boldsymbol{G},\ \dots,\ Y_N = \boldsymbol{w}_N^\top \boldsymbol{G}$$

can be expressed more compactly as a random vector $\boldsymbol{Y}$ by writing

$$\boldsymbol{Y} = \boldsymbol{W}^\top \boldsymbol{G}, \tag{3.42}$$

where $\boldsymbol{W}$ is the matrix whose columns comprise the eigenvectors, that is,

$$\boldsymbol{W} = (\boldsymbol{w}_1 \dots \boldsymbol{w}_N).$$

Since the eigenvectors are orthogonal and normalized, $\boldsymbol{W}$ is an orthonormal matrix:

$$\boldsymbol{W}^\top \boldsymbol{W} = \boldsymbol{I}.$$

If the covariance matrix of the principal components vector $\boldsymbol{Y}$ is called $\boldsymbol{\Sigma}'$, then we have

$$\begin{aligned}\boldsymbol{\Sigma}' &= \langle \boldsymbol{Y}\boldsymbol{Y}^\top \rangle = \langle \boldsymbol{W}^\top \boldsymbol{G}\boldsymbol{G}^\top \boldsymbol{W} \rangle \\ &= \boldsymbol{W}^\top \boldsymbol{\Sigma} \boldsymbol{W} = \begin{pmatrix} \lambda_1 & 0 & \cdots & 0 \\ 0 & \lambda_2 & \cdots & 0 \\ \vdots & \vdots & \ddots & \vdots \\ 0 & 0 & \cdots & \lambda_N \end{pmatrix} =: \boldsymbol{\Lambda}.\end{aligned} \tag{3.43}$$

The eigenvalues are thus seen to be the variances of the principal components, and all of the covariances are zero. The first principal component Y_1 has maximum variance $\text{var}(Y_1) = \lambda_1$, the second principal component Y_2 has maximum variance $\text{var}(Y_2) = \lambda_2$ subject to the condition that it is uncorrelated with Y_1, and so on. The fraction of the total variance in the original multispectral image which is accounted for by the first i principal components is

$$\frac{\lambda_1 + \dots + \lambda_i}{\lambda_1 + \dots + \lambda_i + \dots + \lambda_N}.$$

FIGURE 3.10
RGB color composites (2% linear histogram stretch) of the first three (left) and last three (right) principal components of the six nonthermal bands of a LANDSAT 7 ETM+ image over Jülich, acquired August 29, 2001. **(See color insert.)**

3.3.1 Image compression and reconstruction

If the original multispectral channels are highly correlated, as is often the case, then the first few principal components will usually account for a very high percentage of the total variance in the image. For example, a color composite of the first three principal components of a LANDSAT 7 ETM+ scene displays essentially all of the information contained in the six non-thermal spectral components in one single image; see Figure 3.10. The principal components transformation is therefore often used for dimensionality reduction of imagery prior, for instance, to land cover classification (Chapters 6, 7, and 8).

Alternatively, one can think of the first few principal components as being the main contributing factors to the observed image and then reconstruct the image from those factors. With Equation (3.42), we can recover the original image losslessly by inverting,

$$\boldsymbol{G} = (\boldsymbol{W}^\top)^{-1}\boldsymbol{Y} = \boldsymbol{W}\boldsymbol{Y}.$$

Suppose that the first r principal components account for (explain) most of the variance in the data and let $\boldsymbol{W} = (\boldsymbol{W}_r, \boldsymbol{W}_{r-})$, where

$$\boldsymbol{W}_r = (\boldsymbol{w}_1, \ldots \boldsymbol{w}_r), \quad \boldsymbol{W}_{r-} = (\boldsymbol{w}_{r+1}, \ldots \boldsymbol{w}_N),$$

and similarly

$$\boldsymbol{Y} = \begin{pmatrix} \boldsymbol{Y}_r \\ \boldsymbol{Y}_{r-} \end{pmatrix}.$$

Now reconstruct $\boldsymbol{G}$ from $\boldsymbol{Y}_r = (Y_1 \dots Y_r)^\top$ and $\boldsymbol{W}_r$,

$$\boldsymbol{G} \approx \boldsymbol{G}_r = \boldsymbol{W}_r \boldsymbol{Y}_r,$$

so we can write

$$\boldsymbol{G} = \boldsymbol{W}_r \boldsymbol{Y}_r + \boldsymbol{\epsilon}, \tag{3.44}$$

where the *reconstruction error* $\boldsymbol{\epsilon}$ is

$$\boldsymbol{\epsilon} = \boldsymbol{G} - \boldsymbol{G}_r = \boldsymbol{G} - \boldsymbol{W}_r \boldsymbol{W}_r^\top \boldsymbol{G} = (\boldsymbol{I} - \boldsymbol{W}_r \boldsymbol{W}_r^\top)\boldsymbol{G} = \boldsymbol{W}_{r-} \boldsymbol{W}_{r-}^\top \boldsymbol{G}.$$

The covariance matrix for the error term is given by

$$\begin{aligned} \langle \boldsymbol{\epsilon}\boldsymbol{\epsilon}^\top \rangle &= \langle \boldsymbol{W}_{r-} \boldsymbol{W}_{r-}^\top \boldsymbol{G}\boldsymbol{G}^\top \boldsymbol{W}_{r-} \boldsymbol{W}_{r-}^\top \rangle \\ &= \langle \boldsymbol{W}_{r-} \boldsymbol{Y}_{r-} \boldsymbol{Y}_{r-}^\top \boldsymbol{W}_{r-} \rangle \\ &= \boldsymbol{W}_{r-} \boldsymbol{\Lambda}_{r-} \boldsymbol{W}_{r-}^\top, \end{aligned} \tag{3.45}$$

where

$$\boldsymbol{\Lambda}_{r-} = \begin{pmatrix} \lambda_{r+1} & 0 & \cdots & 0 \\ 0 & \lambda_{r+2} & \cdots & 0 \\ \vdots & \vdots & \ddots & \vdots \\ 0 & 0 & \cdots & \lambda_N \end{pmatrix}.$$

Thus, the smaller the eigenvalues (variances) of the disregarded principal components are, the smaller is the reconstruction error. Obviously the dataset $\{\boldsymbol{W}_r, \boldsymbol{Y}_r\}$ can be considerably smaller than the original image $\boldsymbol{G}$. Reconstruction is easily carried out in the ENVI environment with

`Transform/Principal Components/Inverse PC Rotation.`

The Python script in Listing 3.5 illustrates the reconstruction process for $r = 3$, without actually saving the reconstructed image; see Exercise 11. For clarity, a transposed data matrix is used so that the code corresponds closely with the above equations. In line 39 the reconstruction error covariance matrix is printed. It is identical to Equation (3.45), printed for comparison in line 41.

Equation (3.44) is in the form of a *factor analysis model*; see Mardia et al. (1979). In this case the "factors" are the first r principal components. The elements of $\boldsymbol{W}_r$ are called the "factor loadings."

3.3.2 Primal solution

In order to perform PCA in practice, one calculates the estimate $\boldsymbol{s}$ of the covariance matrix $\boldsymbol{\Sigma}$ in terms of a data matrix $\boldsymbol{\mathcal{G}}$, see Equation (2.53), and then solves the eigenvalue problem, Equation (3.41), in the form

$$\boldsymbol{s}\boldsymbol{w} = \frac{1}{m-1}\boldsymbol{\mathcal{G}}^\top \boldsymbol{\mathcal{G}}\boldsymbol{w} = \lambda \boldsymbol{w}. \tag{3.46}$$

Listing 3.5: Image reconstruction from principal components.

```
#!/usr/bin/env python
#Name:   ex3_2.py
IMPORT auxil.auxil as auxil
FROM numpy IMPORT *
FROM osgeo IMPORT gdal
FROM osgeo.gdalconst IMPORT GA_ReadOnly

DEF main():
    gdal.AllRegister()
    infile = auxil.select_infile()
    IF infile:
        inDataset = gdal.Open(infile,GA_ReadOnly)
        cols = inDataset.RasterXSize
        rows = inDataset.RasterYSize
        bands = inDataset.RasterCount
    ELSE:
        RETURN
#   transposed data matrix
    m = rows*cols
    G = zeros((bands,m))
    FOR b IN RANGE(bands):
        band = inDataset.GetRasterBand(b+1)
        tmp = band.ReadAsArray(0,0,cols,rows)\
                                  .astype(FLOAT).ravel()
        G[b,:] = tmp - mean(tmp)
    G = mat(G)
#   covariance matrix
    S = G*G.T/(m-1)
#   diagonalize and sort eigenvectors
    lamda,W = linalg.eigh(S)
    idx = argsort(lamda)[::-1]
    lamda = lamda[idx]
    W = W[:,idx]
#   get principal components and reconstruct
    r = 3
    Y = W.T*G
    G_r = W[:,:r]*Y[:r,:]
#   reconstruction error covariance matrix
    PRINT  (G-G_r)*(G-G_r).T/(m-1)
#   Equation (3.45)
    PRINT  W[:,r:]*diag(lamda[r:])*W[:,r:].T
    inDataset = None

IF __name__ == '__main__':
    main()
```

This is the *primal* problem for PCA (see the discussion of primal and dual formulations for ridge regression in Section 2.6.4). Solution of the primal problem can be performed directly from the ENVI main menu:

```
Transform/Principal Components/Forward PC Rotation
```

The Python and IDL programs in Listings 1.7 through 1.10 in Chapter 1 illustrated the procedure explicitly.

3.3.3 Dual solution

The normalized eigenvectors of the estimated covariance matrix $\boldsymbol{s}$ are $\boldsymbol{w}_i$, $i = 1 \dots N$. These, as explained, are the principal vectors, and we now show how to express them in terms of the eigenvectors of the Gram matrix $\mathcal{G}\mathcal{G}^\top$, which was initially introduced in Section 2.6.4 in connection with ridge regression. Recall that the Gram matrix is symmetric, positive semi-definite.

Assume that $m > N$ and consider an eigenvector-eigenvalue pair $(\boldsymbol{v}_i, \lambda_i)$ for the Gram matrix $\mathcal{G}\mathcal{G}^\top$. Then we can write

$$\boldsymbol{s}(\mathcal{G}^\top \boldsymbol{v}_i) = \frac{1}{m-1}(\mathcal{G}^\top \mathcal{G})(\mathcal{G}^\top \boldsymbol{v}_i) = \frac{1}{m-1}\mathcal{G}^\top(\mathcal{G}\mathcal{G}^\top)\boldsymbol{v}_i = \frac{1}{m-1}\lambda_i(\mathcal{G}^\top \boldsymbol{v}_i), \tag{3.47}$$

so that $(\mathcal{G}^\top \boldsymbol{v}_i, \lambda_i/(m-1))$ is an eigenvector-eigenvalue pair for $\boldsymbol{s}$. The norm of the eigenvector $\mathcal{G}^\top \boldsymbol{v}_i$ is

$$\|\mathcal{G}^\top \boldsymbol{v}_i\| = \sqrt{\boldsymbol{v}_i^\top \mathcal{G}\mathcal{G}^\top \boldsymbol{v}_i} = \sqrt{\lambda_i}. \tag{3.48}$$

For $\lambda_1 > \lambda_2 > \dots \lambda_N > 0$, the normalized principal vectors $\boldsymbol{w}_i$ can thus be expressed equivalently in the form

$$\boldsymbol{w}_i = \lambda_i^{-1/2} \mathcal{G}^\top \boldsymbol{v}_i, \quad i = 1 \dots N.$$

In fact, the Gram matrix $\mathcal{G}\mathcal{G}^\top$ has exactly N positive eigenvalues, the rest being zero ($\mathcal{G}\mathcal{G}^\top$ has rank N). Informally, every positive eigenvalue for $\mathcal{G}\mathcal{G}^\top$ generates, via Equation (3.47), an eigenvector-eigenvalue pair for $\boldsymbol{s}$, and $\boldsymbol{s}$ has only N eigenvectors. For example, the IDL program

```
PRO dual_pca
; column centered design matrix for random 2D data
   m = 100
   G = 2*randomu(seed,2,m,/double)-1
; covariance matrix
   S = transpose(G)##G/(m-1)
; Gram matrix
   K = G##transpose(G)
   PRINT, eigenql(S)
   PRINT, (eigenql(K))[0:3]/(m-1)
END
```

generates the output

```
1 ENVI> dual_pca
2 0.38706425     0.30646066
3 0.38706425     0.30646066   8.1496153e-017   6.5566325e-017
```

The first $N = 2$ eigenvalues of $\mathcal{G}\mathcal{G}^\top$ are equal to $(m-1)\times$ the eigenvalues of $\boldsymbol{s}$. The remaining $m - N$ eigenvalues of $\mathcal{G}\mathcal{G}^\top$ are zero.

In terms of m-dimensional *dual vectors* $\boldsymbol{\alpha}_i = \lambda_i^{-1/2}\boldsymbol{v}_i$, we have

$$\boldsymbol{w}_i = \mathcal{G}^\top \boldsymbol{\alpha}_i = \sum_{\nu=1}^{m} (\boldsymbol{\alpha}_i)_\nu \boldsymbol{g}(\nu), \quad i = 1 \ldots N. \tag{3.49}$$

So, just as for ridge regression, we get the dual form by expressing the parameter vector as a linear combination of the observations. The projection of any observation $\boldsymbol{g}$ along a principal axis is then

$$\boldsymbol{w}_i^\top \boldsymbol{g} = \sum_{\nu=1}^{m} (\boldsymbol{\alpha}_i)_\nu (\boldsymbol{g}(\nu)^\top \boldsymbol{g}), \quad i = 1 \ldots N. \tag{3.50}$$

Thus we can alternatively perform PCA by finding eigenvalues and eigenvectors of the Gram matrix. The observations $\boldsymbol{g}(\nu)$ appear only in the form of inner products, both in the determination of the Gram matrix as well as in the projection of any new observations, Equation (3.50). This forms the starting point for nonlinear, or *kernel* PCA. Nonlinear kernels will be introduced in Chapter 4 with application to nonlinear PCA and appear again in connection with support vector machine classification in Chapter 6, hyperspectral anomaly detection in Chapter 7, and unsupervised classification in Chapter 8. Chapter 9 gives an example of kernel PCA for change detection.

3.4 Minimum noise fraction

Principal components analysis maximizes variance. This doesn't always lead to images of the desired quality (e.g., having minimal noise). The *minimum noise fraction* (MNF) transformation (Green et al., 1988) can be used to maximize the *signal-to-noise ratio* (SNR) rather than maximizing variance, so, if this is the desired criterion, it is to be preferred over PCA. In the following, we derive the MNF transformation directly by maximizing the ratio of signal variance to noise variance. Then we relate our procedure to the algorithm used in ENVI to perform the MNF transformation.

3.4.1 Additive noise

A noisy multispectral image $\boldsymbol{G}$ may often be represented in terms of an *additive noise model*, i.e., as a sum of signal and noise contributions*

$$\boldsymbol{G} = \boldsymbol{S} + \boldsymbol{N}. \tag{3.51}$$

The signal $\boldsymbol{S}$ (not to be confused here with the covariance matrix estimator) is understood as the component carrying the information of interest. Noise, introduced most often by the sensors, corrupts the signal and masks that information. If both components are assumed to be normally distributed with respective covariance matrices $\boldsymbol{\Sigma}_S$ and $\boldsymbol{\Sigma}_N$, to have zero mean and, furthermore, to be uncorrelated, then the covariance matrix $\boldsymbol{\Sigma}$ for the image $\boldsymbol{G}$ is given by

$$\boldsymbol{\Sigma} = \langle \boldsymbol{G}\boldsymbol{G}^\top \rangle = \langle (\boldsymbol{S}+\boldsymbol{N})(\boldsymbol{S}+\boldsymbol{N})^\top \rangle = \langle \boldsymbol{S}\boldsymbol{S}^\top \rangle + \langle \boldsymbol{N}\boldsymbol{N}^\top \rangle,$$

the covariance $\langle \boldsymbol{N}\boldsymbol{S}^\top \rangle$ being zero by assumption. Thus the image covariance matrix is simply the sum of signal and noise contributions,

$$\boldsymbol{\Sigma} = \boldsymbol{\Sigma}_S + \boldsymbol{\Sigma}_N. \tag{3.52}$$

The signal-to-noise ratio in the ith band of a multispectral image is usually expressed as the ratio of the variance of the signal and noise components,

$$\text{SNR}_i = \frac{\text{var}(S_i)}{\text{var}(N_i)}, \quad i = 1 \ldots N.$$

Let us now seek a linear combination $Y = \boldsymbol{a}^\top \boldsymbol{G}$ of image bands for which this ratio is maximum. That is, we wish to maximize

$$\text{SNR} = \frac{\text{var}(\boldsymbol{a}^\top \boldsymbol{S})}{\text{var}(\boldsymbol{a}^\top \boldsymbol{N})} = \frac{\boldsymbol{a}^\top \boldsymbol{\Sigma}_S \boldsymbol{a}}{\boldsymbol{a}^\top \boldsymbol{\Sigma}_N \boldsymbol{a}}. \tag{3.53}$$

The ratio of quadratic forms on the right is referred to as a *Rayleigh quotient.*

With Equation (3.52) we can write the Equation (3.53) equivalently in the form

$$\text{SNR} = \frac{\boldsymbol{a}^\top \boldsymbol{\Sigma} \boldsymbol{a}}{\boldsymbol{a}^\top \boldsymbol{\Sigma}_N \boldsymbol{a}} - 1. \tag{3.54}$$

Setting its vector derivative with respect to $\boldsymbol{a}$ equal to zero, we get

$$\frac{\partial}{\partial \boldsymbol{a}}\text{SNR} = \frac{1}{\boldsymbol{a}^\top \boldsymbol{\Sigma}_N \boldsymbol{a}} 2\boldsymbol{\Sigma}\boldsymbol{a} - \frac{\boldsymbol{a}^\top \boldsymbol{\Sigma} \boldsymbol{a}}{(\boldsymbol{a}^\top \boldsymbol{\Sigma}_N \boldsymbol{a})^2} 2\boldsymbol{\Sigma}_N \boldsymbol{a} = \boldsymbol{0}$$

*The phenomenon of *speckle* in SAR imagery, on the other hand, can be treated as a form of multiplicative noise; see Chapter 5.

or, equivalently,

$$(\boldsymbol{a}^\top \boldsymbol{\Sigma}_N \boldsymbol{a})\boldsymbol{\Sigma}\boldsymbol{a} = (\boldsymbol{a}^\top \boldsymbol{\Sigma}\boldsymbol{a})\boldsymbol{\Sigma}_N \boldsymbol{a}.$$

This condition is met when $\boldsymbol{a}$ solves the *symmetric generalized eigenvalue problem*

$$\boldsymbol{\Sigma}_N \boldsymbol{a} = \lambda \boldsymbol{\Sigma}\boldsymbol{a}, \tag{3.55}$$

as can easily be seen by substitution. In Equation (3.55), both $\boldsymbol{\Sigma}_N$ and $\boldsymbol{\Sigma}$ are symmetric and positive definite. The equation can be reduced to the standard eigenvalue problem that we are familiar with by performing a *Cholesky decomposition* of $\boldsymbol{\Sigma}$. As explained in Appendix A, Cholesky decomposition will factor $\boldsymbol{\Sigma}$ as $\boldsymbol{\Sigma} = \boldsymbol{L}\boldsymbol{L}^\top$, where $\boldsymbol{L}$ is a lower triangular matrix. Substituting this into Equation (3.55) gives

$$\boldsymbol{\Sigma}_N \boldsymbol{a} = \lambda \boldsymbol{L}\boldsymbol{L}^\top \boldsymbol{a}$$

or, multiplying both sides of the equation from the left by $\boldsymbol{L}^{-1}$ and inserting the identity $(\boldsymbol{L}^\top)^{-1}\boldsymbol{L}^\top$,

$$\boldsymbol{L}^{-1}\boldsymbol{\Sigma}_N (\boldsymbol{L}^\top)^{-1}\boldsymbol{L}^\top \boldsymbol{a} = \lambda \boldsymbol{L}^\top \boldsymbol{a}.$$

Now let $\boldsymbol{b} = \boldsymbol{L}^\top \boldsymbol{a}$. From the commutativity of the operations of inverse and transpose, it follows that

$$[\boldsymbol{L}^{-1}\boldsymbol{\Sigma}_N (\boldsymbol{L}^{-1})^\top]\boldsymbol{b} = \lambda \boldsymbol{b}, \tag{3.56}$$

a standard eigenvalue problem for the symmetric matrix $\boldsymbol{L}^{-1}\boldsymbol{\Sigma}_N(\boldsymbol{L}^{-1})^\top$. Let its (orthogonal and normalized) eigenvectors be $\boldsymbol{b}_i$, $i = 1 \dots N$. Then

$$\boldsymbol{b}_i^\top \boldsymbol{b}_j = \boldsymbol{a}_i^\top \boldsymbol{L}\boldsymbol{L}^\top \boldsymbol{a}_j = \boldsymbol{a}_i^\top \boldsymbol{\Sigma}\boldsymbol{a}_j = \delta_{ij}.$$

Therefore we see that the variances of the transformed components $Y_i = \boldsymbol{a}_i^\top \boldsymbol{G}$ are all unity:

$$\mathrm{var}(Y_i) = \boldsymbol{a}_i^\top \boldsymbol{\Sigma}\boldsymbol{a}_i = 1, \quad i = 1 \dots N.$$

and that they are mutually uncorrelated:

$$\mathrm{cov}(Y_i, Y_j) = \boldsymbol{a}_i^\top \boldsymbol{\Sigma}\boldsymbol{a}_j = 0, \quad i, j = 1 \dots N,\ i \neq j.$$

Listings 3.6 and 3.7 show, respectively, IDL and Python routines for solving the generalized eigenvalue problem with Cholesky decomposition. In the latter script, the Cholesky algorithm is programmed explicitly,* but we could just as well have used the `numpy.linalg.cholesky()` function.

With the definition $\boldsymbol{A} = (\boldsymbol{a}_1, \boldsymbol{a}_2 \dots \boldsymbol{a}_N)$, the complete minimum noise fraction (MNF) transformation can be represented as

$$\boldsymbol{Y} = \boldsymbol{A}^\top \boldsymbol{G}, \tag{3.57}$$

*As described in `http://en.wikipedia.org/wiki/Cholesky_decomposition`.

Listing 3.6: Solving the generalized eigenvalue problem in IDL by Cholesky decomposition.

```
PRO gen_eigenproblem , C,B,A,lambda
; solve the generalized eigenproblem C##a = lambda*B##a
   choldc , B, P, /double
   FOR i=1L,(size(B))[1]-1 DO B[i,0:i]=[fltarr(i),P[i]]
   B[0,0]=P[0]
   Li  = invert(B,/double)
   D = Li ## C ## transpose(Li)
; ensure symmetry after roundoff errors
   D = (D+transpose(D))/2
   lambda = eigenql(D,/double,eigenvectors=A)
; eigenvectors are in columns of A
   A = transpose(A##Li)
END
```

Listing 3.7: Solving the generalized eigenvalue problem in Python by Cholesky decomposition.

```
DEF choldc(A):
# Cholesky-Banachiewicz algorithm,
# A is a numpy matrix
    L = A - A
    FOR i IN RANGE(LEN(L)):
        FOR j IN RANGE(i):
            sm = 0.0
            FOR k IN RANGE(j):
                sm += L[i,k]*L[j,k]
            L[i,j] = (A[i,j]-sm)/L[j,j]
        sm = 0.0
        FOR k IN RANGE(i):
            sm += L[i,k]*L[i,k]
        L[i,i] = math.sqrt(A[i,i]-sm)
    RETURN L

DEF geneiv(A,B):
# solves A*x = lambda*B*x for numpy matrices A and B,
# returns eigenvectors in columns
    Li = np.linalg.inv(choldc(B))
    C = Li*A*(Li.transpose())
    C = np.asmatrix((C + C.transpose())*0.5,np.float32)
    eivs,V = np.linalg.eig(C)
    RETURN eivs, Li.transpose()*V
```

in a manner similar to the principal components transformation, Equation (3.42). The covariance matrix of $\boldsymbol{Y}$ is (compare with Equation (3.43))

$$\boldsymbol{\Sigma}' = \boldsymbol{A}^\top \boldsymbol{\Sigma} \boldsymbol{A} = \boldsymbol{I}, \tag{3.58}$$

where $\boldsymbol{I}$ is the $N \times N$ identity matrix.

It follows from Equation (3.54) that the SNR for eigenvalue λ_i is just

$$\mathrm{SNR}_i = \frac{\boldsymbol{a}_i^\top \boldsymbol{\Sigma} \boldsymbol{a}_i}{\boldsymbol{a}_i^\top (\lambda_i \boldsymbol{\Sigma} \boldsymbol{a}_i)} - 1 = \frac{1}{\lambda_i} - 1. \tag{3.59}$$

Thus the projection $Y_i = \boldsymbol{a}_i^\top \boldsymbol{G}$ corresponding to the *smallest* eigenvalue λ_i will have largest signal-to-noise ratio. Note that with Equation (3.55) we can write

$$\boldsymbol{\Sigma}_N \boldsymbol{A} = \boldsymbol{\Sigma} \boldsymbol{A} \boldsymbol{\Lambda}, \tag{3.60}$$

where $\boldsymbol{\Lambda} = \mathrm{Diag}(\lambda_1 \ldots \lambda_N)$, a diagonal matrix with diagonal elements $\lambda_1 \ldots \lambda_N$.

3.4.2 Minimum noise fraction in ENVI

The MNF transformation is available in the ENVI environment:

```
Transform/MNF Rotation/Forward MNF
```

It is carried out somewhat differently in two steps which are, as we shall now show, equivalent to the above derivation.

In the first step the noise contribution to the observation $\boldsymbol{G}$ is "whitened," that is, a transformation is performed after which the noise component $\boldsymbol{N}$ has covariance matrix $\boldsymbol{\Sigma}_N = \boldsymbol{I}$, the identity matrix. This can be accomplished by first doing a transformation which diagonalizes $\boldsymbol{\Sigma}_N$. Suppose that the transformation matrix for this operation is $\boldsymbol{C}$ and that $\boldsymbol{Z}$ is the resulting random vector. Then

$$\boldsymbol{Z} = \boldsymbol{C}^\top \boldsymbol{G}, \quad \boldsymbol{C}^\top \boldsymbol{\Sigma}_N \boldsymbol{C} = \boldsymbol{\Lambda}_N, \ \boldsymbol{C}^\top \boldsymbol{C} = \boldsymbol{I}, \tag{3.61}$$

where $\boldsymbol{\Lambda}_N$ is a diagonal matrix, the diagonal elements of which are the variances of the transformed noise component $\boldsymbol{C}^\top \boldsymbol{N}$. Next apply the transformation $\boldsymbol{\Lambda}_N^{-1/2}$ (the diagonal matrix whose diagonal elements are the square roots of the diagonal elements of $\boldsymbol{\Lambda}_N$) to give a new random vector $\boldsymbol{X}$,

$$\boldsymbol{X} = \boldsymbol{\Lambda}_N^{-1/2} \boldsymbol{Z} = \boldsymbol{\Lambda}_N^{-1/2} \boldsymbol{C}^\top \boldsymbol{G}.$$

Then the covariance matrix of the noise component $\boldsymbol{\Lambda}_N^{-1/2} \boldsymbol{C}^\top \boldsymbol{N}$ is given by

$$\boldsymbol{\Lambda}_N^{-1/2} \boldsymbol{\Lambda}_N \boldsymbol{\Lambda}_N^{-1/2} = \boldsymbol{I},$$

as desired and the noise contribution has been "whitened." At this stage of affairs, the covariance matrix of the transformed random vector $\boldsymbol{X}$ is

$$\boldsymbol{\Sigma}_X = \boldsymbol{\Lambda}_N^{-1/2} \boldsymbol{C}^\top \boldsymbol{\Sigma} \boldsymbol{C} \boldsymbol{\Lambda}_N^{-1/2}. \tag{3.62}$$

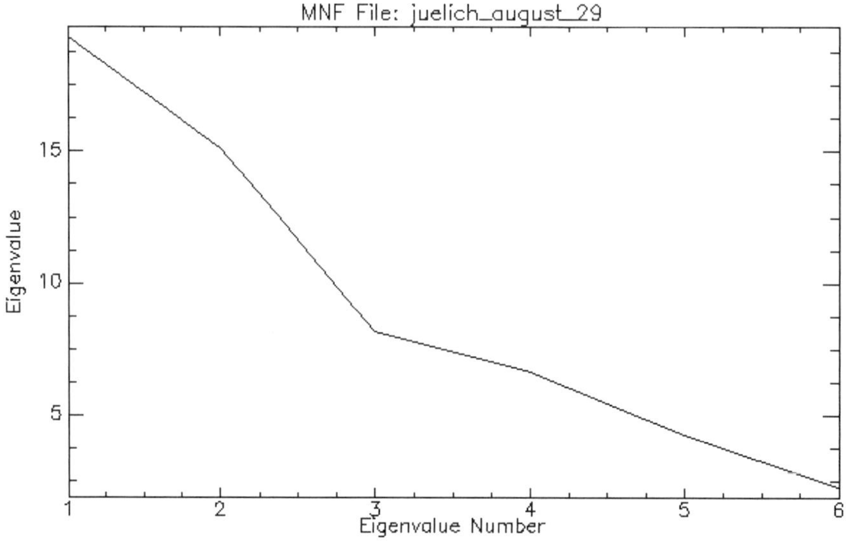

FIGURE 3.11
Eigenvalues of the minimum noise fraction (MNF) transformation calculated with ENVI for the LANDSAT 7 ETM+ image of Figure 3.10.

In the second step, an ordinary principal components transformation is performed on $\boldsymbol{X}$, leading finally to the random vector $\boldsymbol{Y}$ representing the MNF components:

$$\boldsymbol{Y} = \boldsymbol{B}^\top \boldsymbol{X}, \quad \boldsymbol{B}^\top \boldsymbol{\Sigma}_X \boldsymbol{B} = \boldsymbol{\Lambda}_X, \ \boldsymbol{B}^\top \boldsymbol{B} = \boldsymbol{I}. \tag{3.63}$$

The overall transformation is thus

$$\boldsymbol{Y} = \boldsymbol{B}^\top \boldsymbol{\Lambda}_N^{-1/2} \boldsymbol{C}^\top \boldsymbol{G} = \boldsymbol{A}^\top \boldsymbol{G}, \tag{3.64}$$

where $\boldsymbol{A} = \boldsymbol{C}\boldsymbol{\Lambda}_N^{-1/2}\boldsymbol{B}$. To see that this transformation is indeed equivalent to solving the generalized eigenvalue problem, Equation (3.60), consider

$$\begin{aligned}
\boldsymbol{\Sigma}_N \boldsymbol{A} &= \boldsymbol{\Sigma}_N \boldsymbol{C}\boldsymbol{\Lambda}_N^{-1/2}\boldsymbol{B} \\
&= \boldsymbol{C}\boldsymbol{\Lambda}_N\boldsymbol{\Lambda}_N^{-1/2}\boldsymbol{B} \quad \text{from Equation (3.61)} \\
&= \boldsymbol{C}\boldsymbol{\Lambda}_N^{1/2}\boldsymbol{B} \\
&= \boldsymbol{C}\boldsymbol{\Lambda}_N^{1/2}(\boldsymbol{\Sigma}_X\boldsymbol{B}\boldsymbol{\Lambda}_X^{-1}) \quad \text{from Equation (3.63)} \\
&= \boldsymbol{C}\boldsymbol{\Lambda}_N^{1/2}\boldsymbol{\Lambda}_N^{-1/2}\boldsymbol{C}^\top\boldsymbol{\Sigma}\boldsymbol{C}\boldsymbol{\Lambda}_N^{-1/2}\boldsymbol{B}\boldsymbol{\Lambda}_X^{-1} \quad \text{from Equation (3.62)} \\
&= \boldsymbol{\Sigma}\boldsymbol{A}\boldsymbol{\Lambda}_X^{-1}.
\end{aligned} \tag{3.65}$$

This is the same as Equation (3.60) with $\boldsymbol{\Lambda}$ replaced by $\boldsymbol{\Lambda}_X^{-1}$, that is,

$$\lambda_{Xi} = \frac{1}{\lambda_i} = \text{SNR}_i + 1, \qquad i = 1 \dots N,$$

using Equation (3.59). The eigenvalues returned by ENVI's MNF transformation are those of the second (i.e., the principal components) transformation, namely the λ_{Xi} above.* They are equal to the SNR plus one, so that values equal to one correspond to "pure noise." Figure 3.11 shows an example.

3.5 Spatial correlation

Before the MNF transformation can be performed, it is of course necessary to estimate both the image and noise covariance matrices $\boldsymbol{\Sigma}$ and $\boldsymbol{\Sigma}_N$. The former poses no problem, but how does one estimate the noise covariance matrix? The spatial characteristics of the image can be used to estimate $\boldsymbol{\Sigma}_N$, taking advantage of the fact that the intensity of neighboring pixels is usually approximately constant. This property is quantified as the *autocorrelation* of an image. We shall first find a spectral transformation that maximizes the autocorrelation and then see how to relate it to the image noise statistics.

3.5.1 Maximum autocorrelation factor

Let $\boldsymbol{x} = (x_1, x_2)^\top$ represent the coordinates of a pixel within image $\boldsymbol{G}$ and assume that $\langle \boldsymbol{G} \rangle = 0$. The *spatial covariance* $\boldsymbol{C}(\boldsymbol{x}, \boldsymbol{h})$ is defined as the covariance of the original image, represented by $\boldsymbol{G}(\boldsymbol{x})$, with itself, but shifted by the amount $\boldsymbol{h} = (h_1, h_2)^\top$,

$$\boldsymbol{C}(\boldsymbol{x}, \boldsymbol{h}) = \langle \boldsymbol{G}(\boldsymbol{x}) \boldsymbol{G}(\boldsymbol{x} + \boldsymbol{h})^\top \rangle. \tag{3.66}$$

We make the so-called *second-order stationarity assumption*, namely that $\boldsymbol{C}(\boldsymbol{x}, \boldsymbol{h}) = \boldsymbol{C}(\boldsymbol{h})$ is independent of $\boldsymbol{x}$. Then $\boldsymbol{C}(\boldsymbol{0}) = \langle \boldsymbol{G}\boldsymbol{G}^\top \rangle = \boldsymbol{\Sigma}$, and furthermore

$$\begin{aligned} \boldsymbol{C}(-\boldsymbol{h}) &= \langle \boldsymbol{G}(\boldsymbol{x}) \boldsymbol{G}(\boldsymbol{x} - \boldsymbol{h})^\top \rangle \\ &= \langle \boldsymbol{G}(\boldsymbol{x} + \boldsymbol{h}) \boldsymbol{G}(\boldsymbol{x})^\top \rangle \\ &= \langle (\boldsymbol{G}(\boldsymbol{x}) \boldsymbol{G}(\boldsymbol{x} + \boldsymbol{h})^\top)^\top \rangle \\ &= \boldsymbol{C}(\boldsymbol{h})^\top. \end{aligned} \tag{3.67}$$

*Note that the MNF components returned by ENVI, unlike those of Section 3.4.1, do not have unit variance. Their variances are the eigenvalues λ_{Xi}.

The *multivariate variogram*, $\boldsymbol{\Gamma}(\boldsymbol{h})$, is defined as the covariance matrix of the difference image $\boldsymbol{G}(\boldsymbol{x}) - \boldsymbol{G}(\boldsymbol{x}+\boldsymbol{h})$,

$$
\begin{aligned}
\boldsymbol{\Gamma}(\boldsymbol{h}) &= \langle(\boldsymbol{G}(\boldsymbol{x}) - \boldsymbol{G}(\boldsymbol{x}+\boldsymbol{h}))(\boldsymbol{G}(\boldsymbol{x}) - \boldsymbol{G}(\boldsymbol{x}+\boldsymbol{h})^\top\rangle \\
&= \langle\boldsymbol{G}(\boldsymbol{x})\boldsymbol{G}(\boldsymbol{x})^\top\rangle + \langle\boldsymbol{G}(\boldsymbol{x}+\boldsymbol{h})\boldsymbol{G}(\boldsymbol{x}+\boldsymbol{h})^\top\rangle \\
&\quad - \langle\boldsymbol{G}(\boldsymbol{x})\boldsymbol{G}(\boldsymbol{x}+\boldsymbol{h})^\top\rangle - \langle\boldsymbol{G}(\boldsymbol{x}+\boldsymbol{h})\boldsymbol{G}(\boldsymbol{x})^\top\rangle \\
&= 2\boldsymbol{\Sigma} - \boldsymbol{C}(\boldsymbol{h}) - \boldsymbol{C}(-\boldsymbol{h}).
\end{aligned}
\tag{3.68}
$$

Now let us look at the covariance of projections $Y = \boldsymbol{a}^\top\boldsymbol{G}$ of the original and shifted images. This is given by

$$
\begin{aligned}
\mathrm{cov}(\boldsymbol{a}^\top\boldsymbol{G}(\boldsymbol{x}), \boldsymbol{a}^\top\boldsymbol{G}(\boldsymbol{x}+\boldsymbol{h})) &= \boldsymbol{a}^\top\langle\boldsymbol{G}(\boldsymbol{x})\boldsymbol{G}(\boldsymbol{x}+\boldsymbol{h})^\top\rangle\boldsymbol{a} \\
&= \boldsymbol{a}^\top\boldsymbol{C}(\boldsymbol{h})\boldsymbol{a} \\
&= \boldsymbol{a}^\top\boldsymbol{C}(-\boldsymbol{h})\boldsymbol{a} \\
&= \frac{1}{2}\boldsymbol{a}^\top(\boldsymbol{C}(\boldsymbol{h}) + \boldsymbol{C}(-\boldsymbol{h}))\boldsymbol{a},
\end{aligned}
\tag{3.69}
$$

where we have used Equation (3.67). From Equation (3.68), $\boldsymbol{C}(\boldsymbol{h})+\boldsymbol{C}(-\boldsymbol{h}) = 2\boldsymbol{\Sigma} - \boldsymbol{\Gamma}(\boldsymbol{h})$, and so we can write Equation (3.69) in the form

$$
\mathrm{cov}(\boldsymbol{a}^\top\boldsymbol{G}(\boldsymbol{x}), \boldsymbol{a}^\top\boldsymbol{G}(\boldsymbol{x}+\boldsymbol{h})) = \boldsymbol{a}^\top\boldsymbol{\Sigma}\boldsymbol{a} - \frac{1}{2}\boldsymbol{a}^\top\boldsymbol{\Gamma}(\boldsymbol{h})\boldsymbol{a}. \tag{3.70}
$$

The *spatial autocorrelation* of the projections is therefore given by

$$
\begin{aligned}
\mathrm{corr}(\boldsymbol{a}^\top\boldsymbol{G}(\boldsymbol{x}), \boldsymbol{a}^\top\boldsymbol{G}(\boldsymbol{x}+\boldsymbol{h})) &= \frac{\boldsymbol{a}^\top\boldsymbol{\Sigma}\boldsymbol{a} - \frac{1}{2}\boldsymbol{a}^\top\boldsymbol{\Gamma}(\boldsymbol{h})\boldsymbol{a}}{\sqrt{\mathrm{var}(\boldsymbol{a}^\top\boldsymbol{G}(\boldsymbol{x}))\mathrm{var}(\boldsymbol{a}^\top\boldsymbol{G}(\boldsymbol{x}+\boldsymbol{h}))}} \\
&= \frac{\boldsymbol{a}^\top\boldsymbol{\Sigma}\boldsymbol{a} - \frac{1}{2}\boldsymbol{a}^\top\boldsymbol{\Gamma}(\boldsymbol{h})\boldsymbol{a}}{\sqrt{(\boldsymbol{a}^\top\boldsymbol{\Sigma}\boldsymbol{a})(\boldsymbol{a}^\top\boldsymbol{\Sigma}\boldsymbol{a})}} \\
&= 1 - \frac{1}{2}\frac{\boldsymbol{a}^\top\boldsymbol{\Gamma}(\boldsymbol{h})\boldsymbol{a}}{\boldsymbol{a}^\top\boldsymbol{\Sigma}\boldsymbol{a}}.
\end{aligned}
\tag{3.71}
$$

The *maximum autocorrelation factor* (MAF) transformation determines the vector $\boldsymbol{a}$ which maximizes Equation (3.71). We obtain it by minimizing the Rayleigh quotient

$$
R(\boldsymbol{a}) = \frac{\boldsymbol{a}^\top\boldsymbol{\Gamma}(\boldsymbol{h})\boldsymbol{a}}{\boldsymbol{a}^\top\boldsymbol{\Sigma}\boldsymbol{a}}.
$$

Setting the vector derivative equal to zero gives

$$
\frac{\partial R}{\partial \boldsymbol{a}} = \frac{1}{\boldsymbol{a}^\top\boldsymbol{\Sigma}\boldsymbol{a}}\frac{1}{2}\boldsymbol{\Gamma}(\boldsymbol{h})\boldsymbol{a} - \frac{\boldsymbol{a}^\top\boldsymbol{\Gamma}(\boldsymbol{h})\boldsymbol{a}}{(\boldsymbol{a}^\top\boldsymbol{\Sigma}\boldsymbol{a})^2}\frac{1}{2}\boldsymbol{\Sigma}\boldsymbol{a} = \boldsymbol{0}
$$

or

$$
(\boldsymbol{a}^\top\boldsymbol{\Sigma}\boldsymbol{a})\boldsymbol{\Gamma}(\boldsymbol{h})\boldsymbol{a} = (\boldsymbol{a}^\top\boldsymbol{\Gamma}(\boldsymbol{h})\boldsymbol{a})\boldsymbol{\Sigma}\boldsymbol{a}.
$$

This condition is met when $\boldsymbol{a}$ solves the generalized eigenvalue problem

$$\boldsymbol{\Gamma}(\boldsymbol{h})\boldsymbol{a} = \lambda\boldsymbol{\Sigma}\boldsymbol{a}, \tag{3.72}$$

which is seen to have the same form as Equation (3.55), with $\boldsymbol{\Gamma}(\boldsymbol{h})$ replacing the noise covariance matrix $\boldsymbol{\Sigma}_N$. Again, both $\boldsymbol{\Gamma}(\boldsymbol{h})$ and $\boldsymbol{\Sigma}$ are symmetric, and the latter is also positive definite. We obtain as before, via Cholesky decomposition, the standard eigenvalue problem

$$[\boldsymbol{L}^{-1}\boldsymbol{\Gamma}(\boldsymbol{h})(\boldsymbol{L}^{-1})^\top]\boldsymbol{b} = \lambda\boldsymbol{b}, \tag{3.73}$$

for the symmetric matrix $\boldsymbol{L}^{-1}\boldsymbol{\Gamma}(\boldsymbol{h})(\boldsymbol{L}^{-1})^\top$ with $\boldsymbol{b} = \boldsymbol{L}^\top\boldsymbol{a}$.

Let the eigenvalues of Equation (3.73) be ordered from smallest to largest, $\lambda_1 \leq \ldots \leq \lambda_N$, and the corresponding (orthogonal) eigenvectors be $\boldsymbol{b}_i$. We have

$$\boldsymbol{b}_i^\top\boldsymbol{b}_j = \boldsymbol{a}_i^\top\boldsymbol{L}\boldsymbol{L}^\top\boldsymbol{a}_j = \boldsymbol{a}_i^\top\boldsymbol{\Sigma}\boldsymbol{a}_j = \delta_{ij} \tag{3.74}$$

so that, like the components of the MNF transformation, the MAF components $Y_i = \boldsymbol{a}_i^\top\boldsymbol{G},\ \ i = 1\ldots N$, are orthogonal (uncorrelated) with unit variance. Moreover, with Equation (3.71),

$$\text{corr}(\boldsymbol{a}_i^\top\boldsymbol{G}(\boldsymbol{x}), \boldsymbol{a}_i^\top\boldsymbol{G}(\boldsymbol{x}+\boldsymbol{h})) = 1 - \frac{1}{2}\lambda_i, \quad i = 1\ldots N, \tag{3.75}$$

and the first MAF component has maximum autocorrelation.

3.5.2 Noise estimation

The similarity of Equation (3.72) and Equation (3.55) is a result of the fact that $\boldsymbol{\Gamma}(\boldsymbol{h})$ is, under fairly general circumstances, proportional to $\boldsymbol{\Sigma}_N$. We can demonstrate this as follows (Green et al., 1988). Let

$$\boldsymbol{G}(\boldsymbol{x}) = \boldsymbol{S}(\boldsymbol{x}) + \boldsymbol{N}(\boldsymbol{x})$$

and assume

$$\begin{aligned}\langle\boldsymbol{S}(\boldsymbol{x})\boldsymbol{N}(\boldsymbol{x})^\top\rangle &= \boldsymbol{0}\\ \langle\boldsymbol{S}(\boldsymbol{x})\boldsymbol{S}(\boldsymbol{x}\pm\boldsymbol{h})^\top\rangle &= b_h\boldsymbol{\Sigma}_S\\ \langle\boldsymbol{N}(\boldsymbol{x})\boldsymbol{N}(\boldsymbol{x}\pm\boldsymbol{h})^\top\rangle &= c_h\boldsymbol{\Sigma}_N,\end{aligned} \tag{3.76}$$

where b_h and c_h are constants. Under these assumptions, $\boldsymbol{C}(\boldsymbol{h}) = \boldsymbol{C}(-\boldsymbol{h})$ from Equation (3.67) and, from Equation (3.68), we can conclude that

$$\boldsymbol{\Gamma}(\boldsymbol{h}) = 2(\boldsymbol{\Sigma} - \boldsymbol{C}(\boldsymbol{h})). \tag{3.77}$$

But with Equation (3.66)

$$\boldsymbol{C}(\boldsymbol{h}) = \langle(\boldsymbol{S}(\boldsymbol{x}) + \boldsymbol{N}(\boldsymbol{x}))(\boldsymbol{S}(\boldsymbol{x}+\boldsymbol{h}) + \boldsymbol{N}(\boldsymbol{x}+\boldsymbol{h}))^\top\rangle$$

Listing 3.8: Estimation of the noise covariance matrix in IDL from the difference of one-pixel shifts.

```
PRO ex3_4
envi_select, title='Choose␣multispectral␣image', $
                fid=fid, dims=dims,pos=pos
IF (fid EQ -1) THEN BEGIN
   PRINT, 'Canceled'
   RETURN
ENDIF
IF n_elements(pos) LT 2 THEN BEGIN
   PRINT, 'Aborted'
   Message, 'Spectral␣subset␣size␣must␣be␣at␣least␣2'
ENDIF
envi_file_query, fid, fname=fname
cols = dims[2]-dims[1]+1
rows = dims[4]-dims[3]+1
bands = n_elements(pos)
; data matrix for difference image
D = fltarr(bands,cols*rows)
FOR i=0,bands-1 DO BEGIN
   temp=float(envi_get_data(fid=fid,dims=dims, $
              pos=pos[i]))
   D[i,*]=temp-(shift(temp,1,0)+shift(temp,0,1))/2.0
ENDFOR
; noise covariance
S_N = correlate(D,/covariance,/double)
PRINT,'Noise␣covariance␣matrix,␣file␣', $
     file_basename(fname)
PRINT,S_N
END
```

or, with Equation (3.76),

$$\boldsymbol{C}(\boldsymbol{h}) = b_h \boldsymbol{\Sigma}_S + c_h \boldsymbol{\Sigma}_N. \tag{3.78}$$

Finally, combining Equations (3.77), (3.78), and (3.52) gives

$$\frac{1}{2}\boldsymbol{\Gamma}(\boldsymbol{h}) = (1 - b_h)\boldsymbol{\Sigma} + (b_h - c_h)\boldsymbol{\Sigma}_N. \tag{3.79}$$

For a signal with high spatial coherence and for random ("salt and pepper") noise, we expect that in Equation (3.76)

$$b_h \approx 1 \gg c_h$$

and therefore, from Equation (3.79), that

$$\boldsymbol{\Sigma}_N \approx \frac{1}{2}\boldsymbol{\Gamma}(\boldsymbol{h}). \tag{3.80}$$

Listing 3.9: Estimation of the noise covariance matrix in Python from the difference of one-pixel shifts.

```
#!/usr/bin/env python
#Name:   ex3_3.py
IMPORT auxil.auxil as auxil
FROM numpy IMPORT *
FROM osgeo IMPORT gdal
FROM osgeo.gdalconst IMPORT GA_ReadOnly

DEF main():
    gdal.AllRegister()
    infile = auxil.select_infile()
    IF infile:
        inDataset = gdal.Open(infile,GA_ReadOnly)
        cols = inDataset.RasterXSize
        rows = inDataset.RasterYSize
        bands = inDataset.RasterCount
    ELSE:
        RETURN
#   spectral and spatial subsets
    pos =  auxil.select_pos(bands)
    bands = LEN(pos)
    x0,y0,rows,cols=auxil.select_dims([0,0,rows,cols])
#   data matrix for difference images
    D = zeros((rows*cols,bands))
    i = 0
    FOR b IN pos:
        band = inDataset.GetRasterBand(b)
        tmp = band.ReadAsArray(x0,y0,cols,rows)\
                                    .astype(FLOAT)
        D[:,i] = (tmp-(roll(tmp,1,axis=0)+\
                    roll(tmp,1,axis=1))/2).ravel()
        i += 1
#   noise covariance matrix
    S_N = mat(D).T*mat(D)/(rows*cols-1)
    PRINT 'Noise␣covariance␣matrix,␣file␣%s'%infile
    PRINT S_N

IF __name__ == '__main__':
    main()
```

Thus we can obtain an estimate for the noise covariance matrix by estimating the multivariate variogram

$$\mathbf{\Gamma}(\boldsymbol{h}) = \langle(\boldsymbol{G}(\boldsymbol{x}) - \boldsymbol{G}(\boldsymbol{x}+\boldsymbol{h}))(\boldsymbol{G}(\boldsymbol{x}) - \boldsymbol{G}(\boldsymbol{x}+\boldsymbol{h})^\top\rangle$$

and dividing the result by 2. This is illustrated in IDL and Python in Listings 3.8 and 3.9, respectively. There, the matrix $\mathbf{\Gamma}(\boldsymbol{h})$ is determined by averaging the correlations from two difference images, one for a horizontal shift of one pixel, the other for a vertical shift of one pixel.* Here is the IDL calculation for the first 3 bands of the LANDSAT 7 ETM+ image of Figure 3.10:

```
Noise covariance matrix, file juelich_august_29
         7.3732025        7.5563052        11.989391
         7.5563052        10.240908        15.273025
         11.989391        15.273025        28.044577
```

3.6 Exercises

1. Show for $g(x) = \sin(2\pi x)$ in Equation (3.1), that the corresponding frequency coefficients in Equation (3.4) are given by

$$\hat{g}(-1) = -\frac{1}{2\mathbf{i}}, \quad \hat{g}(1) = \frac{1}{2\mathbf{i}},$$

   and $\hat{g}(k) = 0$ otherwise.

2. Demonstrate Equation (3.6) with the help of Equation (3.7).

3. Calculate the discrete Fourier transform of the sequence $2, 4, 6, 8$ from Equation (3.5). You have to solve four simultaneous equations, the first of which is

$$2 = \hat{g}(0) + \hat{g}(1) + \hat{g}(2) + \hat{g}(3).$$

   Verify your result in IDL with the command

```
IDL> PRINT, FFT([2,4,6,8])
```

   or in Python with

```
>>> fft.fft([2,4,6,8])/4
```

   (Note that the IDL and Python Fourier transform conventions are different!)

4. Prove the Fourier translation property, Equation (3.11).

5. Derive the discrete form of *Parseval's Theorem*,

$$\sum_{k=0}^{c-1} |\hat{g}(k)|^2 = \frac{1}{c}\sum_{j=0}^{c-1} |g(j)|^2,$$

*This will tend to overestimate the noise in an image in which the signal itself varies considerably. The noise determination should be restricted to regions having as little detailed structure as possible. The MNF procedure in ENVI (and in the IDL and Python examples listed) allow spatial subsetting for precisely this reason.

using the orthogonality property, Equation (3.7).

6. Show that
$$B_2 = \{\phi_{0,0}(x), \psi_{0,0}(x), \psi_{1,0}(x), \psi_{1,1}(x)\},$$
where
$$\psi_{1,0}(x) = \psi(2x),\ \ \psi_{1,1}(x) = \psi(2x - 1)$$
is an orthogonal basis for the subspace $V_1^{\perp}$.

7. Prove Equation (3.27).

8. It can be shown that, for any MRA, $\int_{-\infty}^{\infty} \phi(x)dx \neq 0$. Show that this implies that the refinement coefficients satisfy
$$\sum_k c_k = 2.$$

9. The cubic B-spline wavelet has the refinement coefficients $c_0 = c_4 = 1/8$, $c_1 = c_3 = 1/2$, $c_2 = 3/4$. Use the cascade algorithm of Listing 3.4 to display the scaling function.

10. (a) (Strang and Nguyen, 1997) Given the dilation Equation (3.25) with n nonzero refinement coefficients $c_0 \dots c_{n-1}$, argue on the basis of the cascade algorithm, that the scaling function $\phi(x)$ must be zero outside the interval $[0, n-1]$.

    (b) Prove that $\phi(x)$ is supported on (extends over) the entire interval $[0, n-1]$.

11. Complete the Python script in Listing 3.5 to have it store the reconstructed image to disk.

12. As discussed in Section 3.3.2, for PCA, one estimates the covariance matrix $\boldsymbol{\Sigma}$ of an image in terms of the $m \times N$ data matrix $\boldsymbol{\mathcal{G}}$, solving the eigenvalue problem
$$\frac{1}{m-1}\boldsymbol{\mathcal{G}}^\top \boldsymbol{\mathcal{G}} \boldsymbol{w} = \lambda \boldsymbol{w}.$$
Alternatively, consider the singular value decomposition of $\boldsymbol{\mathcal{G}}$ itself:
$$\boldsymbol{\mathcal{G}} = \boldsymbol{U}\boldsymbol{W}\boldsymbol{V}^\top,$$
where $\boldsymbol{U}$ is $m \times N$, $\boldsymbol{V}$ is $N \times m$, $\boldsymbol{W}$ is a diagonal $N \times N$ matrix, and
$$\boldsymbol{U}^\top \boldsymbol{U} = \boldsymbol{V}^\top \boldsymbol{V} = \boldsymbol{I}.$$
Explain why the columns of $\boldsymbol{V}$ are the principal axes (eigenvectors) of the transformation and the corresponding variances (eigenvalues) are proportional to the squares of the diagonal elements of $\boldsymbol{W}$.

Listing 3.10: Simulation of two classes of observations.

```
PRO ex3_5
; generate two classes
   n1 = randomu(seed,1000,/normal)
   n2 = n1 + randomu(seed,1000,/normal)
   B1 = [[n1],[n2]]
   B2 = [[n1+4],[n2]]
   image = [B1,B2]
   center_x = mean(image[*,0])
; principal components analysis
   C = correlate(transpose(image),/covariance,/double)
   void = eigenql(C, eigenvectors=U, /double)
; slopes of the principal axes
   a1 = U[1,0]/U[0,0]
   a2 = U[1,1]/U[0,1]
; scatterplot and principal axes
   thisDevice =!D.Name
   set_plot, 'PS'
   Device, Filename='D:\temp\fig3_12.eps',xsize=3, $
         ysize=3,/inches,/encapsulated
   PLOT, image[*,0],image[*,1],psym=4,/isotropic
   oplot,[center_x-10,center_x+10], [a1*(-10),a1*(10)]
   oplot,[center_x-10,center_x+10], [a2*(-10),a2*(10)]
   device,/close_file
   set_plot,thisDevice
END
```

13. The IDL routine in Listing 3.10 simulates two classes of observations B_1 and B_2 in a two-dimensional feature space and calculates the principal axes of the combined data, see Figure 3.12. While the classes are nicely separable in two dimensions, their one-dimensional projections along the x- or y-axes or along either of the principal axes are obviously not. A dimensionality reduction with PCA would thus result in a considerable loss of information about the class structure. *Fisher's linear discriminant* projects the observations $\boldsymbol{g}$ onto a direction $\boldsymbol{w}$, $v = \boldsymbol{w}^\top \boldsymbol{g}$, such that the ratio $J(\boldsymbol{w})$ of the squared difference of the class means of the projections v to their overall variance is maximized (Duda and Hart, 1973). Specifically, define

$$\boldsymbol{m}_i = \frac{1}{n_i}\sum_{\boldsymbol{g}\in B_i}\boldsymbol{g}, \quad \boldsymbol{C}_i = \frac{1}{n_i}\sum_{\boldsymbol{g}\in B_i}(\boldsymbol{g}-\boldsymbol{m}_i)(\boldsymbol{g}-\boldsymbol{m}_i)^\top, \quad i = 1,2.$$

(a) Show that the objective function can be written in the form

$$J(\boldsymbol{w}) = \frac{\boldsymbol{w}^\top \boldsymbol{C}_B \boldsymbol{w}}{\boldsymbol{w}^\top(\boldsymbol{C}_1+\boldsymbol{C}_2)\boldsymbol{w}}, \tag{3.81}$$

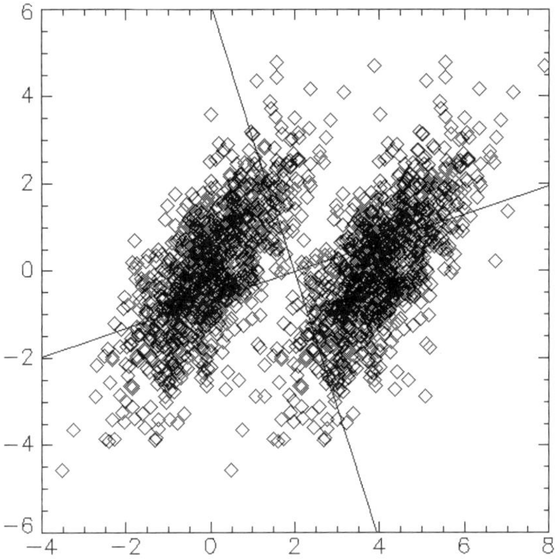

FIGURE 3.12
Two classes of observations in a two-dimensional feature space. The solid lines are the principal axes; see Listing 3.10.

where $\boldsymbol{C}_B = (\boldsymbol{m}_1 - \boldsymbol{m}_2)(\boldsymbol{m}_1 - \boldsymbol{m}_2)^\top$.

(b) Show that the desired projection direction is given by

$$\boldsymbol{w} = (\boldsymbol{C}_1 + \boldsymbol{C}_2)^{-1}(\boldsymbol{m}_1 - \boldsymbol{m}_2). \tag{3.82}$$

(c) Modify the program in Listing 3.10 to calculate and plot the projection direction $\boldsymbol{w}$.

14. Using the code in Listings 3.6 and 3.8 (IDL) or Listings 3.7 and 3.9 (Python) as a starting point, write a procedure to perform the MNF transformation on a multispectral image. If you are using ENVI/IDL, compare your output with ENVI's MNF procedure.

15. Show from Equation (3.50) that the variance of the principal components is given in terms of the eigenvalues λ_i of the Gram matrix by

$$\text{var}(\boldsymbol{w}_i^\top \boldsymbol{g}) = \frac{\lambda_i}{m-1}.$$

16. Formulate the primal and dual problems for the MNF transformation. (*Hint:* Similarly to the case for PCA, Equation (3.49), a dual problem can be obtained by writing $\boldsymbol{a} \propto \mathcal{G}^\top \boldsymbol{\alpha}$.) Write the dual formulation

in the form of a symmetric generalized eigenvalue problem. Can it be solved with Cholesky decomposition?

4

Filters, Kernels and Fields

This chapter is somewhat of a catch-all, intended mainly to consolidate and extend material presented in the preceding chapters and to help lay the foundation for the rest of the book. In Sections 4.1 and 4.2, building on the discrete Fourier transform introduced in Chapter 3, the concept of discrete convolution is introduced and filtering, both in the spatial and in the frequency domain, is discussed. Frequent reference to filtering will be made in Chapter 5 when we treat enhancement and geometric and radiometric correction of multispectral and SAR imagery. In Section 4.3 it is shown that the discrete wavelet transform of Chapter 3 is equivalent to a recursive application of low- and high-pass filters (a filter bank) and a pyramid algorithm for multiscale image representation is described and programmed in IDL and Python. Wavelet pyramid representations are applied in Chapter 5 for panchromatic sharpening and in Chapter 8 for contextual clustering. Section 4.4 introduces so-called *kernelization*, in which the dual representations of linear problems described in Chapters 2 and 3 can be modified to treat nonlinear data. Kernel methods are illustrated with a nonlinear version of the principal components transformation, for which both an ENVI/IDL extension and a Python script are provided. Kernel methods will be met again in Chapter 6 when we consider support vector machines for supervised classification, in Chapter 7 in connection with anomaly detection, in Chapter 8 in the form of a kernel K-means clustering algorithm and in Chapter 9 to illustrate nonlinear change detection. The present chapter closes in Section 4.5 with a brief introduction to Gibbs-Markov random fields, which are invoked in Chapter 8 in order to include spatial context in unsupervised image classification.

4.1 The Convolution Theorem

The *convolution* of two continuous functions $g(x)$ and $h(x)$, denoted by $h * g$, is defined by the integral

$$(h * g)(x) = \int_{-\infty}^{\infty} h(t)g(x-t)dt. \tag{4.1}$$

This definition is symmetric, i.e., $h * g = g * h$, but often one function, $g(x)$ for example, is considered to be a *signal* and the other, $h(x)$, an *instrument response* or *kernel* which is more local than $g(x)$ and which "smears" the signal according to the above prescription.

In the analysis of digital images, of course, we are dealing mainly with discrete signals. In order to define the discrete analog of Equation (4.1) we will again make reference to a signal consisting of a row of pixels $g(j)$, $j = 0 \dots c-1$. The discrete convolution kernel is any array of values $h(\ell)$, $\ell = 0 \dots m-1$, where $m < c$. The array h is referred to as a *finite impulse response* (FIR) filter kernel having duration m. The discrete convolution $f = h * g$ is then defined as

$$f(j) = \begin{cases} \sum_{\ell=0}^{m-1} h(\ell)g(j-\ell) & \text{for } m-1 \le j \le c-1 \\ 0 & \text{otherwise.} \end{cases} \tag{4.2}$$

Discrete convolution can be performed in IDL with the function `CONVOL()` with the keyword `CENTER` explicitly set to zero. The restriction on j in Equation (4.2) is necessary because of edge effects: $g(j)$ is not defined for $j < 0$. This can be circumvented in `CONVOL()` by setting the keywords `EDGE_WRAP` or `EDGE_TRUNCATE`. In the former case, the subscripts are wrapped as if the signal were periodic; in the latter case the endpoints are repeated for $j < 0$ as often as necessary, thereby allowing calculation $f(j)$ for all values of $j = 0 \dots c-1$. The function `convolve()` in the Python Numpy package has a similar behavior.

Let us extend the kernel $h(\ell)$ to have the same length $m = c$ as the signal $g(j)$ by padding it with zeroes, that is, $h(\ell) = 0, \quad \ell = m \dots c-1$. Then we can write Equation (4.2), assuming edge effects have been accommodated, simply as

$$f(j) = \sum_{\ell=0}^{c-1} h(\ell)g(j-\ell), \quad j = 0 \dots c-1. \tag{4.3}$$

The following theorem provides us with a useful alternative to performing this calculation explicitly.

THEOREM 4.1

(Convolution Theorem) *In the frequency domain, convolution is replaced by multiplication, that is,* $h * g \Leftrightarrow c \cdot \hat{h} \cdot \hat{g}$.

Proof: Taking the Fourier transform of Equation (4.3) we have

$$\hat{f}(k) = \frac{1}{c}\sum_{j=0}^{c-1} f(j)e^{-\mathrm{i}2\pi kj/c} = \frac{1}{c}\sum_{\ell=0}^{c-1} h(\ell)\sum_{j=0}^{c-1} g(j-\ell)e^{-\mathrm{i}2\pi kj/c}.$$

But from the translation property, Equation (3.11),

$$\frac{1}{c}\sum_{j=0}^{c-1} g(j-\ell)e^{-\mathrm{i}2\pi kj/c} = \hat{g}(k)e^{-\mathrm{i}2\pi k\ell/c},$$

Listing 4.1: Illustrating convolution in the spatial and frequency domains.

```
1  PRO EX4_1
2
3  ; get an image band from ENVI
4      envi_select, title='Choose␣multispectral␣band', $
5               fid=fid, dims=dims, pos=pos, /band_only
6      IF (fid EQ -1) THEN RETURN
7      num_cols = (c = dims[2]-dims[1]+1)
8      num_rows = dims[4]-dims[3]+1
9  ; pick out the center row of pixels
10     image = envi_get_data(fid=fid,dims=dims,pos=pos[0])
11     g = float(image[*,num_rows/2])
12 ; define a FIR kernel of length m = 5
13     h = [1,2,3,2,1]
14 ; setup for postscript
15     thisDevice = !D.Name
16     set_plot, 'PS'
17     Device, FileName = 'fig4_1.eps',xsize=5,ysize=3, $
18             /inches,/encapsulated
19 ; convolve in the spatial domain and plot
20     PLOT, convol(g,h,center=0)
21 ; pad the arrays to c + m - 1
22     g = [g,[0,0,0,0]]
23     hp = g*0
24     hp[0:4] = h
25 ; convolve in the frequency domain and plot
26     oplot, fft(c*fft(g)*fft(hp),1)-200
27     Device, /close_file
28     set_plot,thisDevice
29
30 END
```

therefore

$$\hat{f}(k) - \sum_{\ell=0}^{c-1} h(\ell)e^{-\mathrm{i}2\pi k\ell/c} \cdot \hat{g}(k) = c \cdot \hat{h}(k) \cdot \hat{g}(k).$$

$\square$

A full statement of the theorem includes the fact that $h \cdot g \Leftrightarrow c \cdot \hat{h} * \hat{g}$, but that needn't concern us here. Theorem 4.1 says that we can carry out the convolution operation, Equation (4.3), by performing the following steps:

1. doing a Fourier transform on the signal and on the (padded) filter,

2. multiplying the two transforms together (and multiplying with c), and

3. performing the inverse Fourier transform on the result.

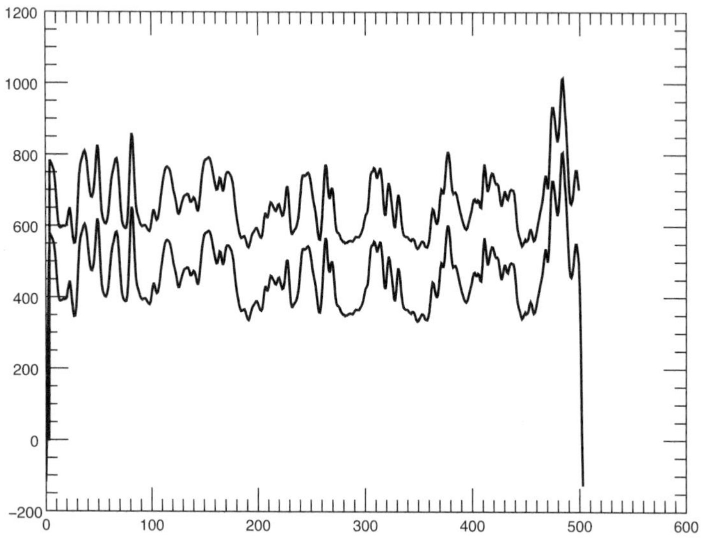

FIGURE 4.1
Illustrating the equivalence of convolution in the spatial (upper curve) and frequency (lower curve) domains; see Listing 4.1.

The FFT, as its name implies, is very fast and ordinary array multiplication is much faster than convolution. So, depending on the size of the arrays involved, convolving them in the frequency domain may be the better alternative. A pitfall when doing convolution in this fashion has to do with so-called *wraparound error*. The discrete Fourier transform assumes that both arrays are periodic. That means that the signal might overlap at the edges with a preceding or following period of the kernel, thus falsifying the result. The problem can be avoided by padding *both* arrays to $c+m-1$, see, e.g., Gonzalez and Woods (2002), Chapter 4. The two alternative convolution procedures are illustrated in Listing 4.1 and Figure 4.1, where a single row of image pixels is convolved with a smoothing filter kernel. They are seen to be completely equivalent.

A second example, programmed in Listing 4.2 and shown Figure 4.2, illustrates the use of convolution for *radar ranging*, which is also part of the SAR imaging process (Richards, 2009). In order to resolve ground features in the range direction (transverse to the direction of flight of the antenna) frequency modulated bursts (chirps), emitted and then received by the antenna after reflection from the Earth's surface, are convolved with the original signal. This

Listing 4.2: A Python program to illustrate radar ranging.

```
#!/usr/bin/env python
#  Name:       ex4_1.py
FROM numpy IMPORT *
IMPORT matplotlib.pyplot as plt

DEF chirp(t,t0):
    result = 0.0*t
    idx = array(RANGE(2000))+t0
    tt = t[idx] - t0
    result[idx] = sin(2*math.pi*2e-3*(tt+1e-3*tt**2))
    RETURN result

DEF main():
    t = array(RANGE(5000))
    plt.plot(t,chirp(t,400)+9)
    plt.plot(t,chirp(t,800)+6)
    plt.plot(t,chirp(t,1400)+3)
    signal = chirp(t,400)+chirp(t,800)+chirp(t,1400)
    kernel = chirp(t,0)[:2000]
    kernel = kernel[::-1]
    plt.plot(t,signal)
    plt.plot(0.003*convolve(signal,kernel,\
                                 mode='same')-5)
    plt.xlabel('Time')
    plt.ylim((-8,12))
    plt.show()

IF __name__ == '__main__':
    main()
```

allows discrimination of features on the ground even when the reflected bursts are not resolved from one another.

Convolution of a two-dimensional array is a straightforward extension of Equation (4.3). For a two-dimensional kernel $h(k,\ell)$ which has been appropriately padded, the convolution with a $c \times r$ pixel array $g(i,j)$ is given by

$$f(i,j) = \sum_{k=0}^{c-1} \sum_{\ell=0}^{r-1} h(k,\ell) g(i-k, j-\ell). \tag{4.4}$$

The Convolution Theorem now reads

$$h * g \Leftrightarrow c \cdot r \cdot \hat{h} \cdot \hat{g}, \tag{4.5}$$

so that convolution can be carried out in the frequency domain using the Fourier and inverse Fourier transforms in two dimensions, Equations (3.9) and (3.8).

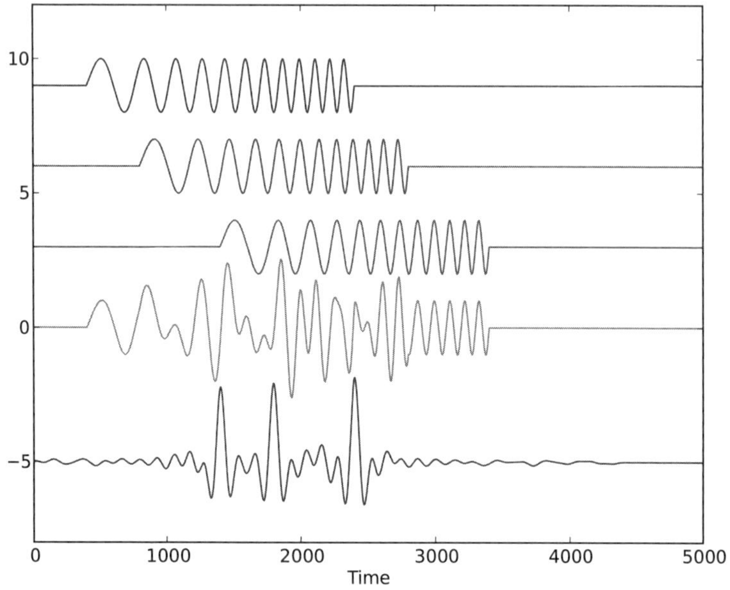

FIGURE 4.2
Radar ranging, see Listing 4.2. The upper three signals represent reflections of a frequency modulated radar pulse (chirp) from three ground points lying close to one another. Their separation in time is proportional to the distances separating the ground features along the direction of pulse emission, that is, transverse to the flight direction. The fourth signal is the superposition actually received. By convolving it with the emitted signal waveform, the arrival times are resolved (bottom signal). **(See color insert.)**

4.2 Linear filters

Linear filtering of images in the spatial domain generally involves moving a template across the image array, forming some specified linear combination of the pixel intensities within the template and associating the result with the coordinates of the pixel at the template's center. Specifically, for a rectangular template h of dimension $(2m+1) \times (2n+1)$,

$$f(i,j) = \sum_{k=-m}^{m} \sum_{\ell=-n}^{n} h(k,\ell) g(i+k, j+\ell), \tag{4.6}$$

Listing 4.3: Illustrating filtering in the frequency domain.

```
PRO EX4_3

   envi_select, title='Choose␣multispectral␣band', $
             fid=fid, dims=dims,pos=pos, /band_only
   IF (fid EQ -1) THEN BEGIN
      PRINT, 'cancelled' & RETURN
   ENDIF
   cols = dims[2]-dims[1]+1
   rows = dims[4]-dims[3]+1
; get the image band from ENVI
   g = envi_get_data(fid=fid,dims=dims,pos=pos[0])
; transform it
   g_hat = fft(g)
; create a Gauss filter in the frequency domain
   sigma = 50
   d = dist(cols,rows)
   h_hat =  exp(-d^2/sigma^2)
; output surface plot of filter as EPS file
   thisDevice =!D.Name
   set_plot, 'PS'
   Device, Filename='fig4_2.eps',xsize=3,ysize=3, $
        /inches,/encapsulated
   shade_surf, shift(h_hat,cols/2,rows/2)
   device,/close_file
   set_plot,thisDevice
; multiply, do inverse FFT and return result to ENVI
   envi_enter_data, fft(g_hat*h_hat,1)     ; low-pass
   envi_enter_data, fft(g_hat*(1-h_hat),1); high-pass

END
```

where g represents the original image array and f the filtered result. The similarity of Equation (4.6) to convolution, Equation (4.4), is readily apparent and spatial filtering can be carried out in the frequency domain if desired. Whether or not the Convolution Theorem should be used to evaluate Equation (4.6) depends again on the size of the arrays involved. Richards and Jia (2006) give a detailed discussion and calculate a cost factor

$$F = \frac{m^2}{2\log_2 c + 1} \cdot e$$

for an $m \times m$ template on a $c \times c$ image. If $F > 1$, it is economical to convolve in the frequency domain.

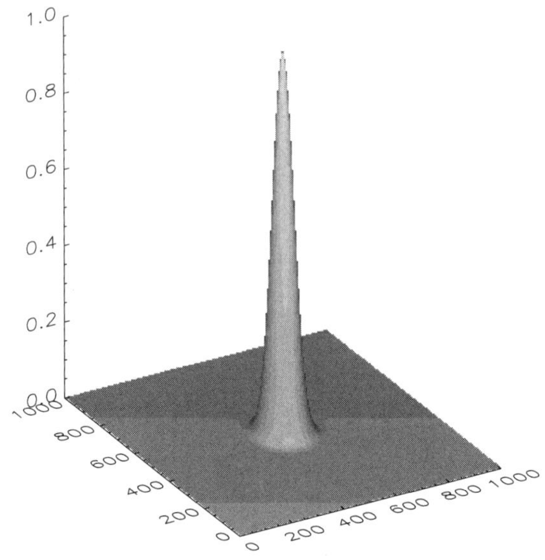

FIGURE 4.3
Gaussian filter in the frequency domain with $\sigma = 50$. Zero frequency is at the center; see Listing 4.3.

Linear smoothing templates are usually normalized so that $\sum_{k,\ell} h(k,\ell) = 1$. For example, the 3×3 "weighted average" filter

$$h = \frac{1}{16}\begin{pmatrix} 1 & 2 & 1 \\ 2 & 4 & 2 \\ 1 & 2 & 1 \end{pmatrix} \tag{4.7}$$

might be used to suppress uninteresting small details or random noise in an image prior to intensity thresholding in order to identify larger objects. However the Convolution Theorem suggests the alternative approach of designing filters in the frequency domain right from the beginning. This is often more intuitive, since suppressing fine details in an image $g(i,j)$ is equivalent to attenuating high spatial frequencies in its Fourier representation $\hat{g}(k,\ell)$ (low-pass filtering). Conversely, enhancing detail, for instance for performing edge detection, can be done by attenuating the low frequencies in $\hat{g}(k,\ell)$ (high-pass filtering). Both effects are achieved by transforming $g(i,j)$ to $\hat{g}(k,\ell)$, choosing an appropriate form for $\hat{h}(k,\ell)$ in Equation (4.5) without reference to a spatial filter h, multiplying the two together and then doing the inverse transformation. Listing 4.3 illustrates the procedure in the case of a Gaussian filter. Figure 4.3 shows the Gaussian low-pass filter

$$\hat{h}(k,\ell) = \exp(d^2/\sigma^2), \quad d^2 = \left((k-c/2)^2 + (\ell - r/2)^2\right)$$

generated by the program. The high-pass filter is its complement $1 - \hat{h}(k,\ell)$. Figures 4.4 and 4.5 display the result of applying them to an image band. We will return to the subject of low- and high-pass filters in Chapter 5, where we discuss image enhancement.

4.3 Wavelets and filter banks

Using the Haar scaling function of Section 3.2.1 we were able to carry out the wavelet transformation in an equivalent vector space. In general, as was pointed out there, one can't represent scaling functions and wavelets in this way. In fact, usually all that we have to work with are the refinement coefficients of Equation (3.25) or (3.28). So how does one perform the wavelet transformation in this case?

4.3.1 One-dimensional arrays

To answer this question, consider again a row of pixel intensities in a satellite image, which we now write in the form of a row vector $\boldsymbol{f} = (f_0, f_1, \ldots f_{c-1})$,* where we assume that $c = 2^n$ for some integer n.

In the multiresolution analysis (MRA) (see Definition 3.2) generated by a scaling function ϕ, such as the Daubechies D_4 scaling function of Figure 3.9, the signal $\boldsymbol{f}$ defines a function $f_n(x)$ in the subspace V_n on the interval $[0,1]$ according to

$$f_n(x) = \sum_{j=0}^{c-1} f_j \phi_{n,j}(x) = \sum_{j=0}^{c-1} f_j \phi(2^n x - j). \tag{4.8}$$

Assume that the V_n basis functions $\phi_{n,j}(x)$ are appropriately normalized:

$$\langle \phi_{n,j}(x), \phi_{n,j'}(x) \rangle = \delta_{j,j'}.$$

Now let us project $f_n(x)$ onto the subspace V_{n-1}, which has a factor of two coarser resolution. The projection is given by

$$f_{n-1}(x) = \sum_{k=0}^{c/2-1} \langle f_n(x), \phi(2^{n-1}x-k)\rangle \phi(2^{n-1}x-k) = \sum_{k=0}^{c/2-1} (H\boldsymbol{f})_k \phi(2^{n-1}x-k),$$

*Rather than our usual $\boldsymbol{g}$, in order to avoid confusion with the wavelet coefficients g_k.

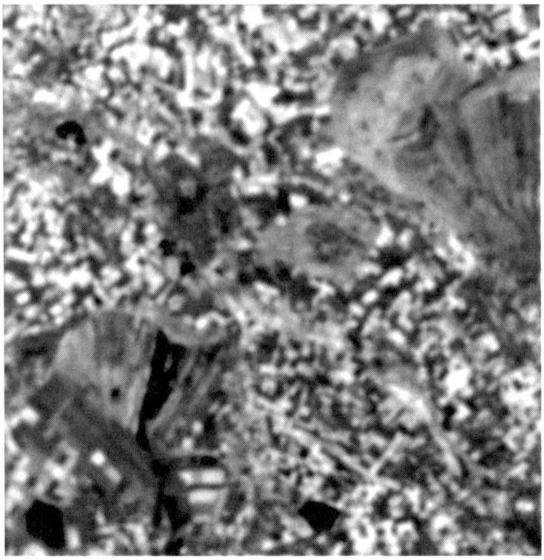

FIGURE 4.4
The 3N band of the Jülich ASTER image after filtering with the low-pass Gaussian filter of Figure 4.3.

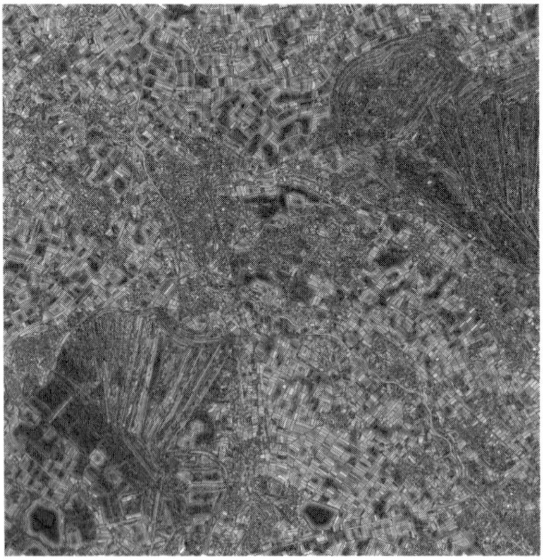

FIGURE 4.5
The 3N band of the Jülich ASTER image after filtering with a high-pass Gaussian filter (complement of Figure 4.3).

where we have introduced the quantity

$$(H\boldsymbol{f})_k = \langle f_n(x), \phi(2^{n-1}x - k)\rangle, \quad k = 0 \dots c/2 - 1. \tag{4.9}$$

This is the kth component of a vector $H\boldsymbol{f}$ representing the row of pixels in V_{n-1}. The notation implies that H is an operator. Its effect is to average the pixel vector $\boldsymbol{f}$ and to reduce its length by a factor of two. It is thus a kind of low-pass filter. More specifically, we have from Equation (4.8),

$$(H\boldsymbol{f})_k = \sum_{j=0}^{c-1} f_j \langle \phi(2^n x - j), \phi(2^{n-1}x - k)\rangle. \tag{4.10}$$

The dilation Equation (3.28) with normalized basis functions can be written in the form

$$\phi(2^{n-1}x - k) = \sum_{k'} h_{k'} \phi(2^n x - 2k - k'). \tag{4.11}$$

Substituting this into Equation (4.10), we have

$$\begin{aligned}(H\boldsymbol{f})_k &= \sum_{j=0}^{c-1} f_j \sum_{k'} h_{k'} \langle \phi(2^n x - j), \phi(2^n x - 2k - k')\rangle \\ &= \sum_{j=0}^{c-1} f_j \sum_{k'} h_{k'} \delta_{j,k'+2k}.\end{aligned}$$

Therefore the filtered pixel vector has components

$$(H\boldsymbol{f})_k = \sum_{j=0}^{c-1} h_{j-2k} f_j, \quad k = 0 \dots \frac{c}{2} - 1 \; (= 2^{n-1} - 1). \tag{4.12}$$

Let us examine this vector in the case of the four non-vanishing refinement coefficients

$$h_0, h_1, h_2, h_3$$

of the Daubechies D4 wavelet. The elements of the filtered signal are

$$\begin{aligned}(H\boldsymbol{f})_0 &= h_0 f_0 + h_1 f_1 + h_2 f_2 + h_3 f_3 \\ (H\boldsymbol{f})_1 &= h_0 f_2 + h_1 f_3 + h_2 f_4 + h_3 f_5 \\ (H\boldsymbol{f})_2 &= h_0 f_4 + h_1 f_5 + h_2 f_6 + h_3 f_7 \\ &\vdots\end{aligned}$$

We recognize the above as the convolution $H * \boldsymbol{f}$ of the low-pass filter kernel $H = (h_3, h_2, h_1, h_0)$ (note the order!) with the vector $\boldsymbol{f}$ (see Equation (4.3)), except that *only every second result is retained.* This is referred to as *downsampling* or *decimation* and is illustrated in Figure 4.6.

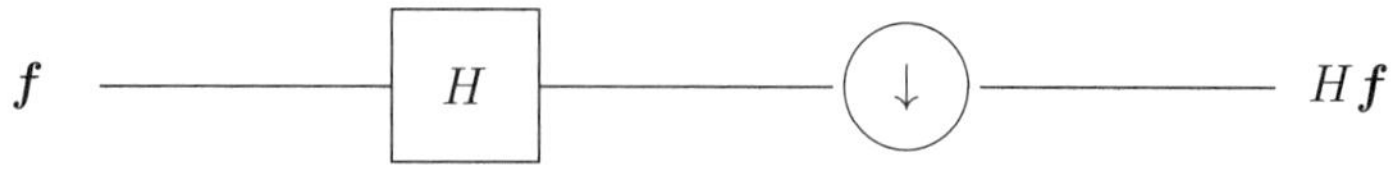

FIGURE 4.6
Schematic representation of Equation (4.12). The symbol ↓ indicates down-sampling by a factor of two.

The residual, that is, the difference between the original vector $\boldsymbol{f}$ and its projection $H\boldsymbol{f}$ onto V_{n-1}, resides in the orthogonal subspace $V_{n-1}^{\perp}$. It is the projection of $f_n(x)$ onto the wavelet basis

$$\psi_{n-1,j}(x), \quad j = 0 \ldots 2^{n-1} - 1.$$

A similar argument (Exercise 4) then leads us to the high-pass filter G, which projects $f_n(x)$ onto $V_{n-1}^{\perp}$ according to

$$(G\boldsymbol{f})_k = \sum_{j=0}^{c-1} g_{j-2k} f_j, \quad k = 0 \ldots \frac{c}{2} - 1 = 2^{n-1} - 1. \tag{4.13}$$

The g_k are related to the h_k by Equation (3.33), i.e.,

$$g_k = (-1)^k h_{1-k}$$

so that the nonzero high-pass filter coefficients are actually

$$g_{-2} = h_3, \; g_{-1} = -h_2, \; g_0 = h_1, \; g_1 = -h_0. \tag{4.14}$$

The *concatenated vector*

$$(H\boldsymbol{f}, G\boldsymbol{f}) = (\boldsymbol{f}^1, \boldsymbol{d}^1) \tag{4.15}$$

is thus the projection of $f_n(x)$ onto $V_{n-1} \oplus V_{n-1}^{\perp}$. It has the same length as the original vector $\boldsymbol{f}$ and is an alternative representation of that vector. Its generation is illustrated in Figure 4.7 as a *filter bank*.

The projections can be repeated on $\boldsymbol{f}^1 = H\boldsymbol{f}$ to obtain the projection

$$(H\boldsymbol{f}^1, G\boldsymbol{f}^1, G\boldsymbol{f}) = (\boldsymbol{f}^2, \boldsymbol{d}^2, \boldsymbol{d}^1) \tag{4.16}$$

onto $V_{n-2} \oplus V_{n-2}^{\perp} \oplus V_{n-1}^{\perp}$, and so on until the complete wavelet transformation has been obtained. Since the filtering process is applied recursively to arrays which are, at each application, reduced by a factor of two, the procedure is very fast. It constitutes *Mallat's algorithm* (Mallat, 1989) and is also referred to as the *fast wavelet transform, discrete wavelet transform* or *pyramid algorithm.*

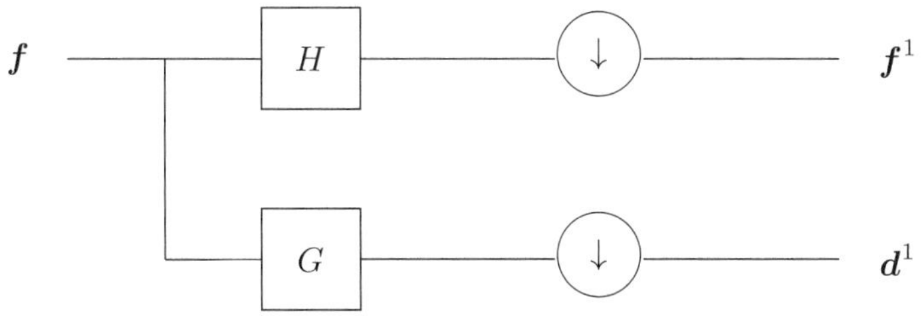

FIGURE 4.7
Schematic representation of the filter bank H, G.

The original vector can be reconstructed at any stage by applying the inverse operators H^* and G^*. For recovery from $(\boldsymbol{f}^1, \boldsymbol{d}^1)$, these are defined by

$$(H^*\boldsymbol{f}^1)_k = \sum_{j=0}^{c/2-1} h_{k-2j}\boldsymbol{f}_j^1, \quad k = 0 \ldots c-1 = 2^n - 1, \tag{4.17}$$

$$(G^*\boldsymbol{d}^1)_k = \sum_{j=0}^{c/2-1} g_{k-2j}\boldsymbol{d}_j^1, \quad k = 0 \ldots c-1 = 2^n - 1, \tag{4.18}$$

with analogous definitions for the other stages. To understand what's happening, consider the elements of the filtered vector in Equation (4.17). These are

$$\begin{aligned}
(H^*\boldsymbol{f}^1)_0 &= h_0 f_0^1 \\
(H^*\boldsymbol{f}^1)_1 &= h_1 f_0^1 \\
(H^*\boldsymbol{f}^1)_2 &= h_2 f_0^1 + h_0 f_1^1 \\
(H^*\boldsymbol{f}^1)_3 &= h_3 f_0^1 + h_1 f_1^1 \\
(H^*\boldsymbol{f}^1)_4 &= h_2 f_1^1 + h_0 f_2^1 \\
(H^*\boldsymbol{f}^1)_5 &= h_3 f_1^1 + h_1 f_2^1 \\
&\vdots
\end{aligned}$$

This is just the convolution of the filter $H^* = (h_0, h_1, h_2, h_3)$ with the vector

$$f_0^1, 0, f_1^1, 0, f_2^1, 0 \ldots f_{c/2-1}^1, 0,$$

which is called an *upsampled* array. The filter of Equation (4.17) is represented schematically in Figure 4.8.

Equation (4.18) is interpreted in a similar way. Finally we add the two results to get the original pixel vector:

$$H^*\boldsymbol{f}^1 + G^*\boldsymbol{d}^1 = \boldsymbol{f}. \tag{4.19}$$

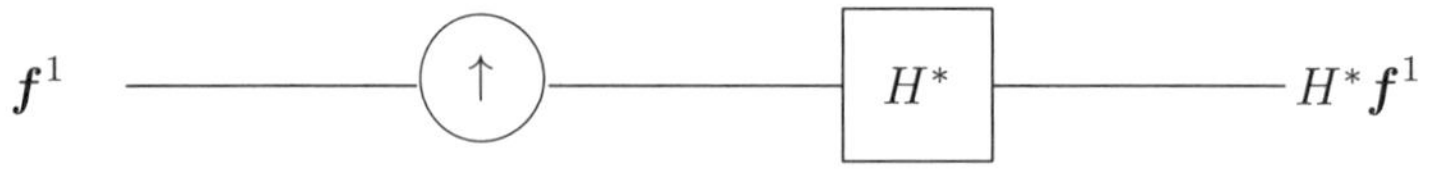

FIGURE 4.8
Schematic representation of the filter H^*. The symbol ↑ indicates upsampling by a factor of two.

To see this, write the equation out for a particular value of k:

$$(H^*\boldsymbol{f}^1)_k + (G^*\boldsymbol{d}^1)_k = \sum_{j=0}^{c/2-1} h_{k-2j}\left[\sum_{j'=0}^{c-1} h_{j'-2j}f_{j'} + g_{k-2j}\sum_{j'=0}^{c-1} g_{j'-2j}f_{j'}\right].$$

Combining terms and interchanging the summations, we get

$$(H^*\boldsymbol{f}^1)_k + (G^*\boldsymbol{d}^1)_k = \sum_{j'=0}^{c-1} f_{j'}\sum_{j=0}^{c/2-1}[h_{k-2j}h_{j'-2j} + g_{k-2j}g_{j'-2j}].$$

Now, using $g_k = (-1)^k h_{1-k}$,

$$(H^*\boldsymbol{f}^1)_k + (G^*\boldsymbol{d}^1)_k = \sum_{j'=0}^{c-1} f_{j'}\sum_{j=0}^{c/2-1}[h_{k-2j}h_{j'-2j} + (-1)^{k+j'}h_{1-k+2j}h_{1-j'+2j}].$$

With the help of Equations (3.26) and (3.27) it is easy to show that the second summation above is just $\delta_{j'k}$. For example, suppose k is even. Then

$$\sum_{j=0}^{c/2-1}[h_{k-2j}h_{j'-2j} + (-1)^{k+j'}h_{1-k+2j}h_{1-j'+2j}] =$$

$$h_0h_{j'-k} + h_2h_{j'-k+2} + (-1)^{j'}[h_1h_{1-j'+k} + h_3h_{3-j'+k}].$$

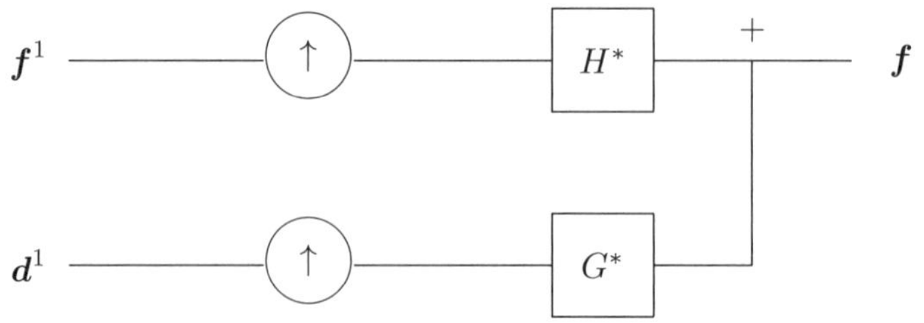

FIGURE 4.9
Schematic representation of the synthesis bank H^*, G^*.

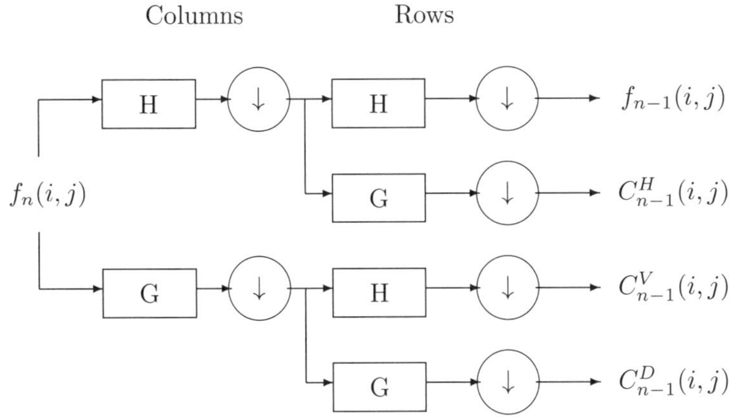

FIGURE 4.10
Wavelet filter bank. H is a low-pass filter and G a high-pass filter derived from the refinement coefficients of the wavelet transformation. The symbol $\downarrow$ indicates downsampling by a factor of two.

If $j' = k$, the right-hand side reduces to

$$h_0^2 + h_1^2 + h_2^2 + h_3^2 = 1,$$

from Equation (3.26) and the fact that $h_k = c_k/\sqrt{2}$. For any other value of j', the expression is zero. Therefore we can write

$$(H^* \boldsymbol{f}^1)_k + (G^* \boldsymbol{d}^1)_k = \sum_{j'=0}^{c-1} f_{j'} \delta_{j'k} = f_k, \quad k = 0 \dots c-1, \tag{4.20}$$

as claimed. The reconstruction of the original vector from $\boldsymbol{f}^1$ and $\boldsymbol{d}^1$ is shown in Figure 4.9 as a *synthesis bank*.

4.3.2 Two dimensional arrays

The extension of the procedure to two-dimensional arrays is straightforward. Figure 4.10 shows an application of the filters H and G to the rows and columns of an image array $f_n(i,j)$ at scale n. The image is filtered and downsampled into four quadrants:

- f_{n-1}, the result of applying the low-pass filter H to both rows and columns plus downsampling;
- C^H_{n-1}, the result of applying the low-pass filter H to the columns and then the high-pass filter G to the rows plus downsampling;

Listing 4.4: Application of two iterations of the discrete wavelet transform to an image band with the object class `DWT`.

```
PRO EX4_4

envi_select , title='Choose␣multispectral␣band', $
              fid=fid, dims=dims,pos=pos, /band_only
IF (fid EQ -1) THEN BEGIN
   PRINT, 'cancelled'
   RETURN
ENDIF

; get the image band from ENVI
g = envi_get_data(fid=fid,dims=dims,pos=pos[0])
; create an instance of the DWT class with image band
aDWT = Obj_New('DWT', g)
; apply the filter bank once
aDWT->filter
; return the result to ENVI
envi_enter_data, aDWT->get_image()
; apply again
aDWT->filter
; return the result to ENVI
envi_enter_data, aDWT->get_image()
; remove the class instance from the heap
Obj_Destroy, aDWT

END
```

- C^V_{n-1}, the result of applying the high-pass filter G to the columns and then the low-pass filter H to the rows plus downsampling; and
- C^D_{n-1}, the result of applying the high-pass filter G to both the columns and the rows plus downsampling.

The original image can be losslessly recovered by inverting the filter as in Figure 4.8.

The filter bank of Figure 4.10 (and the corresponding synthesis filter bank) are implemented in IDL as the object class `DWT`, which is described in Appendix C. Listing 4.4 illustrates its application to an image band using the Daubechies D4 MRA. The images generated are shown in Figures 4.11 and 4.12. Thus `DWT` produces *pyramid representations* of image bands, which occupy the same amount of storage as the original image array. An excerpt from the program is shown in Listing 4.5 in which the forward transformation (the method `FILTER`) is implemented. The object class provides similar functionality to the IDL built-in function `WV_DWT()` (discrete wavelet transform) with the `N_LEVELS`

FIGURE 4.11
Application of the filter bank of Figure 4.10 to the 3N band of the Jülich ASTER image; see Listing 4.4. The original 1000×1000 image, padded out to 1024×1024 ($1024 = 2^{10}$), is a two-dimensional function $f(x, y)$ in $V_{10} \otimes V_{10}$. The result of low-pass filtering rows and columns is in the upper left-hand quadrant, the projection of f onto $V_9 \otimes V_9$. The other three quadrants represent the projections onto the orthogonal subspaces $V_9^{\perp} \otimes V_9$ (upper right), $V_9 \otimes V_9^{\perp}$ (lower left), and $V_9^{\perp} \otimes V_9^{\perp}$ (lower right).

FIGURE 4.12
Recursive application of the filter bank of Figure 4.10 to the upper left quadrant in Figure 4.11; see Listing 4.4.

Listing 4.5: The FILTER method for the IDL object class DWT.

```
PRO DWT::Filter
IF self.compressions LT self.max_compressions $
THEN BEGIN
; single application of filter bank
   H = *self.H & G = *self.G
   m = self.num_rows/2^(self.compressions)
   n = self.num_cols/2^(self.compressions)
   f0 = (*self.image)[0:n-1,0:m-1]
; temporary arrays for wavelet coefficients
   f1 = fltarr(n,m/2)
   g1 = f1*0
   ff1 = fltarr(n/2,m/2)
   fg1 = ff1*0 & gf1 = ff1*0 & gg1 = ff1*0
; filter columns and downsample
   ds = indgen(m/2)*2+1
   FOR i=0,n-1 DO BEGIN
      temp = convol(transpose(f0[i,*]),H,/edge_wrap)
      f1[i,*] = temp[ds]
      temp = convol(transpose(f0[i,*]),G,/edge_wrap)
      g1[i,*] = temp[ds]
   ENDFOR
; filter rows and downsample
   ds = indgen(n/2)*2+1
   FOR i=0,m/2-1 DO BEGIN
      temp = convol(f1[*,i],H,/edge_wrap)
      ff1[*,i] = temp[ds]
      temp = convol(f1[*,i],G,/edge_wrap)
      fg1[*,i] = temp[ds]
      temp = convol(g1[*,i],H,/edge_wrap)
      gf1[*,i] = temp[ds]
      temp = convol(g1[*,i],G,/edge_wrap)
      gg1[*,i] = temp[ds]
   ENDFOR
   f0[0:n/2-1,0:m/2-1] = ff1[*,*]
   f0[n/2:n-1,0:m/2-1] = fg1[*,*]
   f0[0:n/2-1,m/2:m-1] = gf1[*,*]
   f0[n/2:n-1,m/2:m-1] = gg1[*,*]
   (*self.image)[0:n-1,0:m-1] = f0
   self.compressions = self.compressions+1
ENDIF
END
```

Listing 4.6: The `FILTER` method for the Python object `DWTArray` (excerpt from `auxil.py`).

```
    DEF FILTER(self):
#       single application of filter bank
        IF self.num_iter == self.max_iter:
            RETURN 0
#       get upper left quadrant
        m = self.lines/2**self.num_iter
        n = self.samples/2**self.num_iter
        f0 = self.data[:m,:n]
#       temporary arrays
        f1 = np.zeros((m/2,n))
        g1 = np.zeros((m/2,n))
        ff1 = np.zeros((m/2,n/2))
        fg1 = np.zeros((m/2,n/2))
        gf1 = np.zeros((m/2,n/2))
        gg1 = np.zeros((m/2,n/2))
#       filter columns and downsample
        ds = np.asarray(RANGE(m/2))*2+1
        FOR i IN RANGE(n):
            temp = np.convolve(f0[:,i].ravel(),\
                                           self.H,'same')
            f1[:,i] = temp[ds]
            temp = np.convolve(f0[:,i].ravel(),\
                                           self.G,'same')
            g1[:,i] = temp[ds]
#       filter rows and downsample
        ds = np.asarray(RANGE(n/2))*2+1
        FOR i IN RANGE(m/2):
            temp = np.convolve(f1[i,:],self.H,'same')
            ff1[i,:] = temp[ds]
            temp = np.convolve(f1[i,:],self.G,'same')
            fg1[i,:] = temp[ds]
            temp = np.convolve(g1[i,:],self.H,'same')
            gf1[i,:] = temp[ds]
            temp = np.convolve(g1[i,:],self.G,'same')
            gg1[i,:] = temp[ds]
        f0[:m/2,:n/2] = ff1
        f0[:m/2,n/2:] = fg1
        f0[m/2:,:n/2] = gf1
        f0[m/2:,n/2:] = gg1
        self.data[:m,:n] = f0
        self.num_iter = self.num_iter+1
```

keyword, but includes methods for normalization and substitution of wavelet coefficients.

The analogous class method in the Python implementation (see Appendix D) is shown in Listing 4.6. The discrete wavelet transformation will be made use of in Chapter 5 for image sharpening and in Chapter 8 for unsupervised classification.

4.4 Kernel methods

In Section 2.6.4 it was shown that regularized linear regression (ridge regression) possesses a dual formulation in which observation vectors $\boldsymbol{x}(\nu)$, $\nu = 1 \dots m$, only enter in the form of inner products $\boldsymbol{x}(\nu)^\top \boldsymbol{x}(\nu')$. A similar dual representation was found in Section 3.3.3 for the principal components transformation. In both cases the symmetric, positive semi-definite Gram matrix

$$\mathcal{X}\mathcal{X}^\top \quad (\mathcal{X} = \text{ data matrix})$$

played a central role. It turns out that many linear methods in pattern recognition have dual representations, ones which can be exploited to extend well-established linear theories to treat nonlinear data. One speaks in this context of *kernel methods* or *kernelization.* An excellent reference for kernel methods is Shawe–Taylor and Cristianini (2004). In the following we outline the basic ideas and then illustrate them with a nonlinear, or kernelized, version of principal components analysis.

4.4.1 Valid kernels

Suppose that $\phi(\boldsymbol{g})$ is some nonlinear function which maps the original N-dimensional Euclidean input space of the observation vectors $\boldsymbol{g}$ (image pixels) to some nonlinear (usually higher-dimensional) inner product space $\mathcal{H}$,

$$\phi : \mathbb{R}^N \mapsto \mathcal{H}. \tag{4.21}$$

The mapping takes a data matrix $\mathcal{G}$ into a new data matrix $\boldsymbol{\Phi}$ given by

$$\boldsymbol{\Phi} = \begin{pmatrix} \phi(\boldsymbol{g}(1))^\top \\ \phi(\boldsymbol{g}(2))^\top \\ \vdots \\ \phi(\boldsymbol{g}(m))^\top \end{pmatrix}. \tag{4.22}$$

This matrix has dimension $m \times p$, where $p \geq N$ is the (possibly infinite) dimension of the nonlinear feature space $\mathcal{H}$.

DEFINITION 4.1 *A* valid kernel *is a function* κ *that, for all* $\boldsymbol{g}, \boldsymbol{g}' \in \mathbb{R}^N$, *satisfies*

$$\kappa(\boldsymbol{g}, \boldsymbol{g}') = \boldsymbol{\phi}(\boldsymbol{g})^\top \boldsymbol{\phi}(\boldsymbol{g}'), \tag{4.23}$$

where $\boldsymbol{\phi}$ *is given by Equation (4.21). For a set of observations* $\boldsymbol{g}(\nu)$, $\nu = 1 \dots m$, *the* $m \times m$ *matrix* $\mathcal{K}$ *with elements* $\kappa(\boldsymbol{g}(\nu), \boldsymbol{g}(\nu'))$ *is called the* kernel matrix.

Note that with Equation (4.22) we can write the kernel matrix equivalently in the form

$$\mathcal{K} = \boldsymbol{\Phi}\boldsymbol{\Phi}^\top. \tag{4.24}$$

In Section 2.6.4 we saw that the Gram matrix is symmetric and positive semi-definite. This is also the case for kernel matrices since, for any m-component vector $\boldsymbol{x}$,

$$\begin{aligned}
\boldsymbol{x}^\top \mathcal{K} \boldsymbol{x} &= \sum_{\nu,\nu'} x_\nu x_{\nu'} \mathcal{K}_{\nu\nu'} \\
&= \sum_{\nu,\nu'} x_\nu x_{\nu'} \boldsymbol{\phi}_\nu^\top \boldsymbol{\phi}_{\nu'} \\
&= (\sum_\nu x_\nu \boldsymbol{\phi}_\nu)^\top (\sum_{\nu'} x_{\nu'} \boldsymbol{\phi}_{\nu'}) \\
&= \|\sum_\nu x_\nu \boldsymbol{\phi}_\nu\|^2 \geq 0.
\end{aligned}$$

Positive semi-definiteness of the kernel matrix is in fact a necessary and sufficient condition for any symmetric function $\kappa(\boldsymbol{g}, \boldsymbol{g}')$ to be a valid kernel in the sense of Definition 4.1. This is stated in the following Theorem (Shawe–Taylor and Cristianini, 2004):

THEOREM 4.2
Let $k(\boldsymbol{g}, \boldsymbol{g}')$ *be a symmetric function on the space of observations, that is,* $k(\boldsymbol{g}, \boldsymbol{g}') = k(\boldsymbol{g}', \boldsymbol{g})$. *Let* $\{\boldsymbol{g}(\nu) \mid \nu = 1 \dots m\}$ *be any finite subset of the input space* $\mathbb{R}^N$ *and define the matrix* $\mathcal{K}$ *with elements*

$$(\mathcal{K})_{\nu\nu'} = k(\boldsymbol{g}(\nu), \boldsymbol{g}(\nu')), \quad \nu, \nu' = 1 \dots m.$$

Then $k(\boldsymbol{g}, \boldsymbol{g}')$ *is a valid kernel if and only if* $\mathcal{K}$ *is positive semi-definite.*

The motivation for using valid kernels is that it allows us to apply known linear methods to nonlinear data simply by replacing the inner products in the dual formulation by an appropriate nonlinear, valid kernel. One implication of Theorem 4.2 is that one can obtain valid kernels without even specifying the nonlinear mapping $\boldsymbol{\phi}$ at all. In fact, it is possible to build up whole families of valid kernels in which the associated mappings are defined implicitly and are

Listing 4.7: Calculating a kernel matrix.

```
FUNCTION gausskernel_matrix ,G1 ,G2 ,GMA=gma ,NSCALE=nscale
   IF n_params () EQ 1 THEN G2 = G1
   IF n_elements (nscale) EQ 0 THEN nscale = 1.0
   m = n_elements (G1 [0 ,*])
   n = n_elements (G2 [0 ,*])
   K = transpose (total (G1^2 ,1))##(intarr (n)+1)
   K = K + transpose (intarr (m)+1)##total (G2^2 ,1)
   K = K - 2*G1##transpose (G2)
   IF n_elements (gma) EQ 0 THEN BEGIN
     scale = total(sqrt (abs (K)))/(m^2-m)
     gma = 1/(2*(nscale*scale)^2)
   ENDIF
   RETURN , exp (-gma*K)
END
```

otherwise unknown. An important example, one which we will use shortly to illustrate kernel methods, is the *Gaussian kernel* given by

$$\kappa_{\rm rbf}(\boldsymbol{g}, \boldsymbol{g}') = \exp(-\gamma \|\boldsymbol{g} - \boldsymbol{g}'\|^2). \tag{4.25}$$

This is an instance of a *homogeneous kernel*, also called *radial basis kernel*, one which depends only on the Euclidean distance between the observations. It is equivalent to the inner product of two infinite-dimensional feature vector mappings $\boldsymbol{\phi}(\boldsymbol{g})$ (Exercise 7). Listing 4.7 shows an IDL function to calculate a Gaussian kernel matrix (found in the `AUXIL` directory, see Appendix C).

If we wish to make use of kernels which, like the Gaussian kernel, imply nonlinear mappings $\boldsymbol{\phi}(\boldsymbol{g})$ to which we have no direct access, then clearly we must work exclusively with inner products

$$\kappa(\boldsymbol{g}, \boldsymbol{g}') = \boldsymbol{\phi}(\boldsymbol{g})^\top \boldsymbol{\phi}(\boldsymbol{g}').$$

Apart from dual formulations which only involve these inner products, it turns out that many other properties of the mapped observations $\boldsymbol{\phi}(\boldsymbol{g})$ can also be expressed purely in terms of the elements $\kappa(\boldsymbol{g}, \boldsymbol{g}')$ of the kernel matrix. Thus the norm or length of the mapping of $\boldsymbol{g}$ in the nonlinear feature space $\mathcal{H}$ is

$$\|\boldsymbol{\phi}(\boldsymbol{g})\| = \sqrt{\boldsymbol{\phi}(\boldsymbol{g})^\top \boldsymbol{\phi}(\boldsymbol{g})} = \sqrt{\kappa(\boldsymbol{g}, \boldsymbol{g})}. \tag{4.26}$$

The squared distance between any two points in $\mathcal{H}$ is given by

$$\begin{aligned} \|\boldsymbol{\phi}(\boldsymbol{g}) - \boldsymbol{\phi}(\boldsymbol{g}')\|^2 &= (\boldsymbol{\phi}(\boldsymbol{g}) - \boldsymbol{\phi}(\boldsymbol{g}'))^\top (\boldsymbol{\phi}(\boldsymbol{g}) - \boldsymbol{\phi}(\boldsymbol{g}')) \\ &= \boldsymbol{\phi}(\boldsymbol{g})^\top \boldsymbol{\phi}(\boldsymbol{g}) + \boldsymbol{\phi}(\boldsymbol{g}')^\top \boldsymbol{\phi}(\boldsymbol{g}') - 2\boldsymbol{\phi}(\boldsymbol{g})^\top \boldsymbol{\phi}(\boldsymbol{g}') \\ &= \kappa(\boldsymbol{g}, \boldsymbol{g}) + \kappa(\boldsymbol{g}', \boldsymbol{g}') - 2\kappa(\boldsymbol{g}, \boldsymbol{g}'). \end{aligned} \tag{4.27}$$

The $1 \times p$ row vector $\bar{\boldsymbol{\phi}}^\top$ of column means of the mapped data matrix $\boldsymbol{\Phi}$, Equation (4.22), can be written in matrix form as, see Equation (2.52),

$$\bar{\boldsymbol{\phi}}^\top = \mathbf{1}_m^\top \boldsymbol{\Phi}/m, \tag{4.28}$$

where $\mathbf{1}_m$ is a column vector of m ones. The norm of the mean vector $\bar{\boldsymbol{\phi}}$ is thus

$$\|\bar{\boldsymbol{\phi}}\| = \sqrt{\bar{\boldsymbol{\phi}}^\top \bar{\boldsymbol{\phi}}} = \frac{1}{m}\sqrt{\mathbf{1}_m^\top \boldsymbol{\Phi}\boldsymbol{\Phi}^\top \mathbf{1}_m} = \frac{1}{m}\sqrt{\mathbf{1}_m^\top \boldsymbol{\mathcal{K}} \mathbf{1}_m}.$$

We are even able to determine the elements of the kernel matrix $\tilde{\boldsymbol{\mathcal{K}}}$ which corresponds to a column centered data matrix $\tilde{\boldsymbol{\Phi}}$ (see Section 2.3.1). The rows of $\tilde{\boldsymbol{\Phi}}$ are

$$\tilde{\boldsymbol{\phi}}(\boldsymbol{g}(\nu)) = (\boldsymbol{\phi}(\boldsymbol{g}(\nu)) - \bar{\boldsymbol{\phi}})^\top = \boldsymbol{\phi}(\boldsymbol{g}(\nu))^\top - \bar{\boldsymbol{\phi}}^\top, \quad \nu = 1 \ldots m. \tag{4.29}$$

With Equation (4.28) the $m \times p$ matrix of m repeated column means of $\boldsymbol{\Phi}$ is given by

$$\begin{pmatrix} \bar{\boldsymbol{\phi}}^\top \\ \bar{\boldsymbol{\phi}}^\top \\ \vdots \\ \bar{\boldsymbol{\phi}}^\top \end{pmatrix} = \mathbf{1}_{mm}\boldsymbol{\Phi}/m,$$

where $\mathbf{1}_{mm}$ is an $m \times m$ matrix of ones, so that Equation (4.29) can be written in matrix form:

$$\tilde{\boldsymbol{\Phi}} = \boldsymbol{\Phi} - \mathbf{1}_{mm}\boldsymbol{\Phi}/m. \tag{4.30}$$

Therefore

$$\begin{aligned} \tilde{\boldsymbol{\mathcal{K}}} = \tilde{\boldsymbol{\Phi}}\tilde{\boldsymbol{\Phi}}^\top &= (\boldsymbol{\Phi} - \mathbf{1}_{mm}\boldsymbol{\Phi}/m)(\boldsymbol{\Phi} - \mathbf{1}_{mm}\boldsymbol{\Phi}/m)^\top \\ &= \boldsymbol{\Phi}\boldsymbol{\Phi}^\top - \boldsymbol{\Phi}\boldsymbol{\Phi}^\top \mathbf{1}_{mm}/m - \mathbf{1}_{mm}\boldsymbol{\Phi}\boldsymbol{\Phi}^\top/m - 1_{mm}\boldsymbol{\Phi}\boldsymbol{\Phi}^\top \mathbf{1}_{mm}/m^2 \\ &= \boldsymbol{\mathcal{K}} - \boldsymbol{\mathcal{K}}\mathbf{1}_{mm}/m - \mathbf{1}_{mm}\boldsymbol{\mathcal{K}}/m + \mathbf{1}_{mm}\boldsymbol{\mathcal{K}}\mathbf{1}_{mm}/m^2. \end{aligned} \tag{4.31}$$

Examining this equation component-wise we see that, to center the kernel matrix, we subtract from each element $(\boldsymbol{\mathcal{K}})_{ij}$ the mean of the ith row and the mean of the jth column and add to that the mean of the entire matrix. The following IDL and Python functions take a kernel matrix $\boldsymbol{\mathcal{K}}$ as input and returns the column-centered kernel matrix $\tilde{\boldsymbol{\mathcal{K}}}$:

```
FUNCTION center, K
   m = (size(K))[1]
   Imm =dblarr(m,m) + 1
   RETURN K - (Imm##K + K##Imm - total(K)/m)/m
```

```
DEF center(K):
    m = K.shape[0]
    Imm = mat(ones((m,m)))
    RETURN K - (Imm*K + K*Imm - SUM(K)/m)/m
```

4.4.2 Kernel PCA

In Section 3.3.3 it was shown that, in the dual formulation of the principal components transformation, the projection $P_i[\boldsymbol{g}]$ of an observation $\boldsymbol{g}$ along a principal axis $\boldsymbol{w}_i$ can be expressed as

$$P_i[\boldsymbol{g}] = \boldsymbol{w}_i^\top \boldsymbol{g} = \sum_{\nu=1}^{m} (\boldsymbol{\alpha}_i)_\nu \, \boldsymbol{g}(\nu)^\top \boldsymbol{g}, \quad i = 1 \dots N. \tag{4.32}$$

In this equation the dual vectors $\boldsymbol{\alpha}_i$ are determined by the eigenvectors $\boldsymbol{v}_i$ and eigenvalues λ_i of the Gram matrix $\boldsymbol{\mathcal{G}}\boldsymbol{\mathcal{G}}^\top$ according to

$$\boldsymbol{\alpha}_i = \lambda^{-1/2} \boldsymbol{v}_i, \quad i = 1 \dots m. \tag{4.33}$$

Kernelization of principal components analysis simply involves replacing the Gram matrix $\boldsymbol{\mathcal{G}}\boldsymbol{\mathcal{G}}^\top$ by the (centered) kernel matrix $\tilde{\boldsymbol{\mathcal{K}}}$, Equation (4.31), and the inner products $\boldsymbol{g}(\nu)^\top \boldsymbol{g}$ in Equation (4.32) by the corresponding kernel function. The projection along the ith principal axes in the nonlinear feature space $\mathcal{H}$ (the ith nonlinear principal component) is then

$$P_i[\boldsymbol{\phi}(\boldsymbol{g})] = \sum_{\nu=1}^{m} (\boldsymbol{\alpha}_i)_\nu \kappa(\boldsymbol{g}(\nu), \boldsymbol{g}), \quad i = 1 \dots m, \tag{4.34}$$

where the dual vectors $\boldsymbol{\alpha}_i$ are still given by Equation (4.33), but $\boldsymbol{v}_i$ and λ_i, $i = 1 \dots m$, are the eigenvectors and eigenvalues of the kernel matrix $\tilde{\boldsymbol{\mathcal{K}}}$. The variance of the projections is given by (see Chapter 3, Exercise 15)

$$\text{var}(P_i[\boldsymbol{\phi}(\boldsymbol{g})]) = \frac{\lambda_i}{m-1}, \quad i = 1 \dots m.$$

Schölkopf et al. (1998) were the first to introduce kernel PCA and Shawe–Taylor and Cristianini (2004) analyze the method in detail. Canty and Nieslsen (2012) discuss it in the context of linear and kernel change detection methods.

In the analysis of remote sensing imagery, the number of observations is in general very large (order $10^6 - 10^8$), so that diagonalization of the kernel matrix is only feasible if the image is sampled. The sampled pixel vectors $\boldsymbol{g}(\nu)$, $\nu = 1 \dots m$, are then referred to as *training data* and the calculation of the sampled kernel matrix and its centering/diagonalization constitute the *training phase.* A realistic upper limit on m would appear to be about 2000. Diagonalization of a 2000×2000 symmetric matrix on, e.g., a PC workstation with IDL will generally not require page swapping and takes the order of minutes. However, this is a very small sample ($\lessapprox 10^{-3}$). Kwon and Nasrabadi (2005) suggest extracting the most representative samples with a clustering algorithm such as k-means (see Chapter 8). The cluster mean vectors serve as the training data and may just number a few hundred.

After diagonalization, the *generalization phase* involves the projection of each image pixel vector according to Equation (4.34). This means, for every

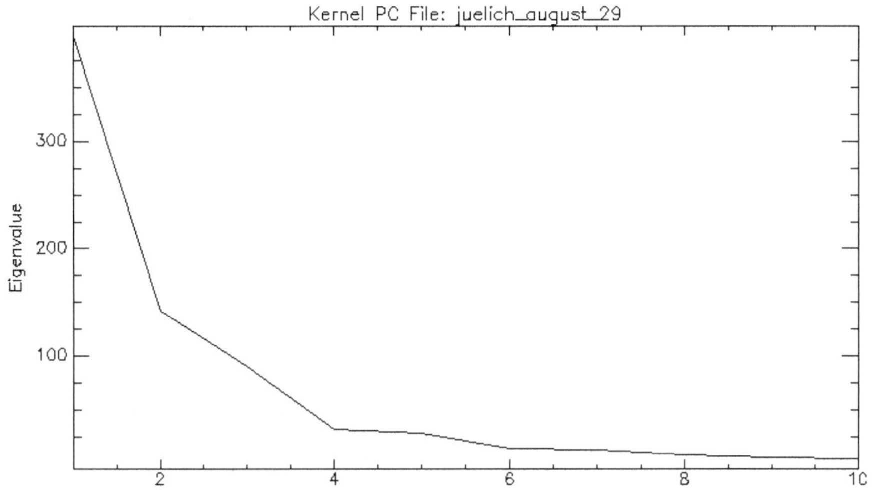

FIGURE 4.13
Eigenvalues for kernel PCA on the image of Figure 3.10. The size of the training dataset was $m = 2000$.

pixel $\boldsymbol{g}$, recalculation of the kernels $\kappa(\boldsymbol{g}(\nu), \boldsymbol{g})$ for $\nu = 1 \ldots m$.* For an image with n pixels there are, therefore, $m \times n$ kernels involved, generally much too large an array to be held in memory, so that it is advisable to read in and project the image pixels row-by-row.

Appendix C describes an ENVI extension `KPCA_RUN` for performing kernel PCA on multispectral imagery using the Gaussian kernel, Equation (4.25). The parameter γ defaults to $1/2\sigma^2$, where σ is equal to the mean distance between the observations in input space, i.e., the average over all test observations of $\|\boldsymbol{g}(\nu) - \boldsymbol{g}(\nu')\|$, $\nu \neq \nu'$. There are two equivalent code blocks in this extension, one of which is programmed in standard ENVI/IDL. The other makes use of Tech-X Corporation's GPULib,† taking advantage of the parallel computing power available in graphics hardware to speed up the projection step by an order of magnitude. A Python implementation of kernel PCA, `kpca.py`, is also provided and is described in Appendix D.

Figure 4.13 shows the logarithms of the eigenvalues for kernel PCA performed on the LANDSAT 7 ETM+ image of Figure 3.10 with a sample of 2000 pixels. In this example, only 85.8% of the variance is explained by the first three eigenvalues. In contrast, linear PCA concentrates 98.9% of the

*This is not the case for linear PCA, where we don't require the training data to project new observations. For this reason, kernel PCA is said to be *memory-based.*

†GPULib is a library of high-level IDL bindings to the so-called Compute Unified Device Architecture (CUDA), which allows access to graphics processors (GPUs); see Appendix C and `http://www.txcorp.com/home/gpulib` for details.

variance into the first three principal components. Chapter 9 will illustrate the use of kernel PCA for change detection with (simulated) nonlinear data.

4.5 Gibbs–Markov random fields

Random fields are frequently invoked to describe prior expectations in a Bayesian approach to image analysis (Winkler, 1995; Li, 2001). The following brief introduction will serve to make their use more plausible in the unsupervised land cover classification context that we will meet in Chapter 8. The development adheres closely to Li (2001), but in a notation specific to that used later in the treatment of image classification.

Image classification is a problem of *labeling*: Given an observation, that is, a pixel intensity vector, we ask: "Which class label should be assigned to it?" If the observations are assumed to have no spatial context, then the labeling will consist of partitioning the pixels into K disjoint subsets according to some decision criterion, K being the number of land cover categories present. If spatial context within the image is also to be taken into account, the labeling will take place on what is referred to as a *regular lattice*.

A regular lattice representing an image with c columns and r rows is the discrete set of sites

$$\mathcal{I} = \{(i,j) \mid 0 \le i \le c-1, 0 \le j \le r-1\}.$$

By re-indexing, we can write this in a more convenient, linear form

$$\mathcal{I} = \{i \mid 1 \le i \le n\},$$

where $n = rc$ is the number of pixels. The interrelationship between the sites is governed by a *neighborhood system*

$$\mathcal{N} = \{\mathcal{N}_i \mid i \in \mathcal{I}\},$$

where $\mathcal{N}_i$ is the set of pixels neighboring site i. Two frequently used neighborhood systems are shown in Figure 4.14.

The pair $(\mathcal{I}, \mathcal{N})$ may be thought of as constituting an *undirected graph* in which the pixels are nodes and the neighborhood system determines the edges between the nodes. Thus any two neighboring pixels are represented by two nodes connected by an edge. A *clique* is a single node or a subset of nodes which are all directly connected to one another in the graph by edges, i.e., they are mutual neighbors. Figure 4.15 shows the possible cliques for the neighborhoods of Figure 4.14. (Note that one also distinguishes their orientation.) We will denote by the symbol $\mathcal{C}$ the set of all cliques in $\mathcal{I}$.

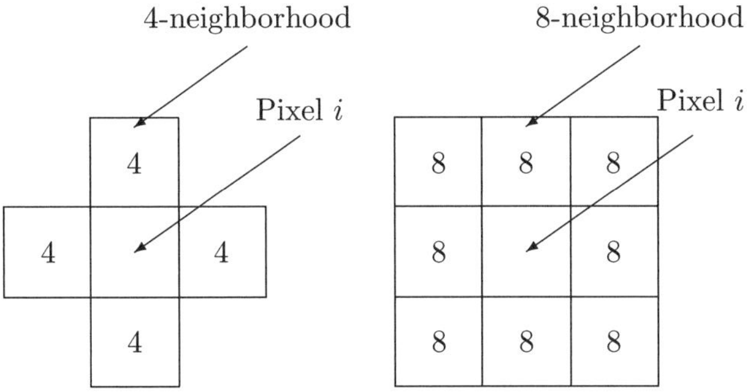

FIGURE 4.14
Pixel neighborhoods $\mathcal{N}_i$.

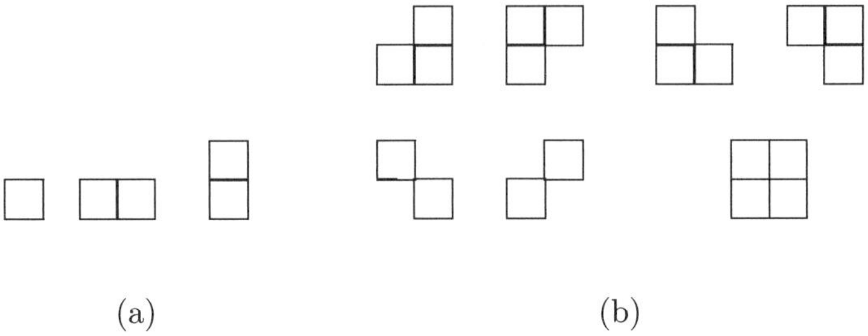

FIGURE 4.15
(a) Cliques for a 4-neighborhood. (a) and (b) Cliques for an 8-neighborhood.

Next, let us introduce a set of *class labels*

$$\mathcal{K} = \{k \mid 1 \le k \le K\},$$

K being the number of possible classes present in the image. Assigning a class label to a site on the basis of measurement is a random experiment, so we associate with the ith site a discrete random variable L_i representing its label. The set L of all such random variables

$$L = \{L_1 \dots L_n\}$$

is called a *random field* on $\mathcal{I}$. The possible realizations of L_i are labels $\ell_i \in \mathcal{K}$. A specific realization for the entire lattice, for example a classified image or thematic map, is a *configuration* ℓ, given by

$$\ell = \{\ell_1 \dots \ell_n\}, \quad \ell_i \in \mathcal{K},$$

and the space of all configurations is the Cartesian product set

$$\mathcal{L} = \overbrace{\mathcal{K} \otimes \mathcal{K} \ldots \otimes \mathcal{K}}^{n \text{ times}}.$$

Thus there are K^n possible configurations. For each site i, the probability that the site has label ℓ_i is

$$\Pr(L_i = \ell_i) = \Pr(\ell_i),$$

and the joint probability for configuration ℓ is

$$\Pr(L_1 = \ell_1, L_2 = \ell_2 \ldots L_n = \ell_n) = \Pr(L = \ell) = \Pr(\ell).$$

DEFINITION 4.2 *The random field L is said to be a* Markov random field (MRF) *on $\mathcal{I}$ with respect to neighborhood system $\mathcal{N}$ if and only if*

$$Pr(\ell) > 0 \quad \text{for all} \quad \ell \in \mathcal{L}, \tag{4.35}$$

which is referred to as the positivity *condition, and*

$$Pr(\ell_i \mid \ell_1 \ldots \ell_{i-1}, \ell_{i+1} \ldots \ell_n) = Pr(\ell_i \mid \ell_j, j \in \mathcal{N}_i), \tag{4.36}$$

called the Markovianity *condition.*

The Markovianity condition in the above definition simply says that a label assignment can be influenced only by neighboring pixels.

DEFINITION 4.3 *The random field L constitutes a* Gibbs random field (GRF) *on $\mathcal{I}$ with respect to neighborhood system $\mathcal{N}$ if and only if it obeys a* Gibbs distribution. *A Gibbs distribution has density function*

$$p(\ell) = \frac{1}{Z} \exp(-\beta U(\ell)), \tag{4.37}$$

where β is a constant and Z is the normalization factor

$$Z = \sum_{\ell \in \mathcal{L}} \exp(-\beta U(\ell)), \tag{4.38}$$

and where the energy function *$U(\ell)$ is represented as a sum of contributing terms for each clique*

$$U(\ell) = \sum_{c \in \mathcal{C}} V_c(\ell). \tag{4.39}$$

$V_c(\ell)$ is called a clique potential *for clique c in configuration ℓ. If clique potentials are independent of the location of the clique within the lattice, then*

the GRF is said to be homogeneous. *If V_c is independent of the orientation of c, then the GRF is* isotropic.

According to this definition, configurations in a GRF with low clique potentials are more probable than those with high clique potentials. The parameter β is an "inverse temperature." For small β (high temperature) all configurations become equally probable, irrespective of their associated energy.

A MRF is characterized by its local property (Markovianity) and a GRF by its global property (Gibbs distribution). It turns out that the two are equivalent:

THEOREM 4.3

(Hammersley–Clifford Theorem) *A random field L is an MRF on $\mathcal{I}$ with respect to $\mathcal{N}$ if and only if L is a GRF on $\mathcal{I}$ with respect to $\mathcal{N}$.*

Proof: The proof of the "if" part of the theorem, that a random field is an MRF if it is a GRF, is straightforward.* Write the conditional probability on the left-hand side of Equation (4.36) as

$$\Pr(\ell_i \mid \ell_1 \ldots \ell_{i-1}, \ell_{i+1} \ldots \ell_n) = \Pr(\ell_i \mid \ell_{\mathcal{I}-\{i\}}).$$

Here $\mathcal{I}-\{i\}$ is the set if all sites except the ith one. According to the definition of conditional probability, Equation (2.59),

$$\Pr(\ell_i \mid \ell_{\mathcal{I}-\{i\}}) = \frac{\Pr(\ell_i, \ell_{\mathcal{I}-\{i\}})}{\Pr(\ell_{\mathcal{I}-\{i\}})}.$$

Equivalently,

$$\Pr(\ell_i \mid \ell_{\mathcal{I}-\{i\}}) = \frac{\Pr(\ell)}{\sum_{\ell_i \in \mathcal{K}} \Pr(\ell')},$$

where $\ell' = \{\ell_1 \ldots \ell_{i-1}, \ell'_i, \ell_{i+1} \ldots \ell_n\}$ is any configuration which agrees with ℓ at all sites except possibly i. With Equations (4.37) and (4.39),

$$\Pr(\ell_i \mid \ell_{\mathcal{I}-\{i\}}) = \frac{\exp(-\sum_{c\in\mathcal{C}} V_c(\ell))}{\sum_{\ell_i \in \mathcal{K}} \exp(-\sum_{c\in\mathcal{C}} V_c(\ell'))}.$$

Now divide the set of cliques $\mathcal{C}$ into two sets, namely $\mathcal{A}$, consisting of those cliques which contain site i, and $\mathcal{B}$, consisting of the rest. Then

$$\Pr(\ell_i \mid \ell_{\mathcal{I}-\{i\}}) = \frac{[\exp(-\sum_{c\in\mathcal{A}} V_c(\ell))][\exp(-\sum_{c\in\mathcal{B}} V_c(\ell))]}{\sum_{\ell_i \in \mathcal{K}}[\exp(-\sum_{c\in\mathcal{A}} V_c(\ell'))][\exp(-\sum_{c\in\mathcal{B}} V_c(\ell'))]}.$$

---

*The "only if" part is more difficult, see Li (2001).

But for all cliques $c \in \mathcal{B}$, $V_c(\ell) = V_c(\ell')$ and the second factors in numerator and denominator cancel. Thus

$$\Pr(\ell_i \mid \ell_{\mathcal{I}-\{i\}}) = \frac{\exp(-\sum_{c\in\mathcal{A}} V_c(\ell))}{\sum_{\ell_i\in\mathcal{K}} \exp(-\sum_{c\in\mathcal{A}} V_c(\ell'))}.$$

This shows that the probability for ℓ_i is conditional only on the potentials of the cliques containing site i; in other words, that the random field is an MRF.

□

The above theorem allows one to express the joint probability for a configuration ℓ of image labels in terms of the local clique potentials $V_c(\ell)$. We shall encounter an example in Chapter 8 in connection with unsupervised image classification.

4.6 Exercises

1. Give an explicit expression for the convolution of the array $(g(0)\dots g(5))$ with the array $(h(0), h(1), h(2))$ as it would be calculated with the IDL function `CONVOL()` with the `EDGE_WRAP` keyword set.

2. Modify the program in Listing 4.1 to convolve a two-dimensional array with the filter of Equation (4.7), both in the spatial and frequency domains.

3. Modify the program in Listing 4.3 to implement the *Butterworth filter*

$$h(k,\ell) = \frac{1}{1 + (d/d_0)^{2n}}, \tag{4.40}$$

   where d_0 is a width parameter and $n = 1, 2\dots$.

4. Demonstrate Equation (4.13).

5. (Press et al., 2002) Consider a row of $c = 8$ pixels represented by the row vector

$$\boldsymbol{f} = (f_0, f_1 \dots f_7).$$

   Assuming that $\boldsymbol{f}$ is periodic (repeats itself), the application of low- and high-pass filter Equations (4.9) and (4.13) can be accomplished by multiplication of the column vector $\boldsymbol{f}^\top$ with the matrix

$$\boldsymbol{W} = \begin{pmatrix} h_0 & h_1 & h_2 & h_3 & 0 & 0 & 0 & 0 \\ h_3 & -h_2 & h_1 & -h_0 & 0 & 0 & 0 & 0 \\ 0 & 0 & h_0 & h_1 & h_2 & h_3 & 0 & 0 \\ 0 & 0 & h_3 & -h_2 & h_1 & -h_0 & 0 & 0 \\ 0 & 0 & 0 & 0 & h_0 & h_1 & h_2 & h_3 \\ 0 & 0 & 0 & 0 & h_3 & -h_2 & h_1 & -h_0 \\ h_2 & h_3 & 0 & 0 & 0 & 0 & h_0 & h_1 \\ h_1 & -h_0 & 0 & 0 & 0 & 0 & h_3 & -h_2 \end{pmatrix}.$$

(a) Prove that $\boldsymbol{W}$ is an orthonormal matrix (its inverse is equal to its transpose).

(b) In the transformed vector $\boldsymbol{W}\boldsymbol{f}^\top$, the components of $\boldsymbol{f}^1 = H\boldsymbol{f}$ and $\boldsymbol{d}^1 = G\boldsymbol{f}$ are interleaved. They can be sorted to give the vector $(\boldsymbol{f}^1, \boldsymbol{d}^1)$. When the matrix

$$\boldsymbol{W}^1 = \begin{pmatrix} h_0 & h_1 & h_2 & h_3 \\ h_3 & -h_2 & h_1 & -h_0 \\ h_2 & h_3 & h_0 & h_1 \\ h_1 & -h_0 & h_3 & -h_2 \end{pmatrix}$$

is then applied to the smoothed vector $(\boldsymbol{f}^1)^\top$ and the result again sorted, we obtain the complete discrete wavelet transformation of $\boldsymbol{f}$, namely $(\boldsymbol{f}^2, \boldsymbol{d}^2, \boldsymbol{d}^1)$. Precisely this transformation can be invoked in IDL with the function `WTN(F,4,inverse=inverse)` for the Daubechies D4 wavelet and for any vector `F` of length $c = 2^n$. Moreover, by applying the inverse transformation to a unit vector, the D4 wavelet itself can be generated. Write an IDL routine to plot the D4 wavelet by performing the inverse transformation on the vector $\underbrace{(0, 0, 0, 1, 0 \ldots 0)}_{1024}$.

6. Most filtering operations with the Fourier transform have their wavelet counterparts. Modify the IDL program in Listing 4.4, or write a Python equivalent, to perform high-pass filtering with the discrete wavelet transformation. See Appendix C or D and the IDL listing for `DWT__DEFINE` or the Python class `DWTArray` in the `auxil.auxil` module to understand how these classes work, then proceed as follows:

   - Perform a single wavelet transform with the IDL object method `FILTER` (Python method `FILTER`).
   - Zero the upper left quadrant in the transformed image with a null array. Use IDL object method `INJECT` or Python method `put_quadrant`
   - Then invert the transformation with the IDL object method `INVERT` (Python: `invert`).

- Display the result in ENVI or with `dispms.py`.

7. Show that the Gaussian kernel, Equation (4.25), is equivalent to the inner product of infinite-dimensional mappings $\phi(\boldsymbol{g})$.

8. Modify the program in Listing 4.7 to calculate the kernel matrix for the *polynomial kernel function*

$$\kappa_{\text{poly}}(\boldsymbol{g}_i, \boldsymbol{g}_j) = (\gamma \boldsymbol{g}_i^\top \boldsymbol{g}_j + r)^d. \tag{4.41}$$

   The parameter r is called the *bias*, d the *degree* of the kernel function.

9. Shawe–Taylor and Cristianini (2004) prove (in their Proposition 5.2) that the centering operation minimizes the average eigenvalue of the kernel matrix. Write an IDL or Python script to do the following:

   (a) Generate m random N-dimensional observation vectors and, with the IDL routine in Listing 4.7 or its Python equivalent, calculate the $m \times m$ Gaussian kernel matrix.

   (b) Determine its eigenvalues in order to confirm that the kernel matrix is positive semi-definite.

   (c) Center it (see the code following Equation (4.31)).

   (d) Re-determine the eigenvalues and verify that their average value has decreased.

   (e) Do the same using the polynomial kernel matrix from the preceding exercise.

10. A kernelized version of the dual formulation for the MNF transformation (see Chapter 3, Exercise 16) leads to a generalized eigenvalue problem having the form

$$\boldsymbol{A}\boldsymbol{x} = \lambda \boldsymbol{B}\boldsymbol{x}, \tag{4.42}$$

   where $\boldsymbol{A}$ and $\boldsymbol{B}$ are symmetric but not full rank. Let the symmetric $m \times m$ matrix $\boldsymbol{B}$ have rank $r < m$, eigenvalues $\lambda_1 \dots \lambda_m$, and eigenvectors $\boldsymbol{u}_1 \dots \boldsymbol{u}_m$. Show that Equation (4.42) can be re-formulated as an ordinary symmetric eigenvalue problem by writing $\boldsymbol{B}$ as a product of *matrix square roots* $\boldsymbol{B} = \boldsymbol{B}^{1/2}\boldsymbol{B}^{1/2}$. The square root is defined as

$$\boldsymbol{B}^{1/2} = \boldsymbol{P}\boldsymbol{\Lambda}^{1/2}\boldsymbol{P}^\top,$$

   where $\boldsymbol{P} = (\boldsymbol{u}_1 \dots \boldsymbol{u}_r)$ and $\boldsymbol{\Lambda} = \text{Diag}(\lambda_1 \dots \lambda_r)$.

11. Does a 24-neighborhood (the 5×5 array centered at a location i) have more clique types than are shown in Figure 4.14?

5

Image Enhancement and Correction

In preparation for the treatment of supervised/unsupervised classification and change detection, the subjects of the last four chapters of this book, the present chapter focuses on preprocessing methods. These fall into the two general categories of *image enhancement* (Sections 5.1 through 5.4) and *geometric correction* (Sections 5.5 and 5.6). Discussion mainly focuses on the processing of optical/infrared image data. However, Section 5.4 introduces polarimetric SAR imagery and treats the problem of speckle removal.

5.1 Lookup tables and histogram functions

Gray-level enhancements of an image are easily accomplished by means of *lookup tables*. For byte-encoded data, for example, the pixel intensities $g(i,j)$ are used to index into the array

$$LUT[k], \quad k = 0 \ldots 255,$$

the entries of which also lie between 0 and 255. These entries can be chosen to implement simple histogram processing, such as linear stretching, saturation, equalization, etc. The pixel values $g(i,j)$ are replaced by

$$f(i,j) = LUT[g(i,j)], \quad 0 \le i \le r-1,\ 0 \le j \le c-1. \tag{5.1}$$

In deriving the appropriate transformations it is convenient to think of the normalized histogram of pixel intensities g as a probability density $p_g(g)$ of a continuous random variable G, and the lookup table itself as a continuous transformation function, i.e.,

$$f(g) = LUT(g),$$

where both g and f are restricted to the interval $[0,1]$. We will illustrate the technique for the case of *histogram equalization*. First of all, we claim that the function

$$f(g) = LUT(g) = \int_0^g p_g(t)dt \tag{5.2}$$

Listing 5.1: Histogram equalization in Python (excerpt from auxil.py)

```
DEF histeqstr(x):
#   histogram equalization stretch of input ndarray
    hist,bin_edges = np.histogram(x,256,(0,256))
    cdf = hist.cumsum()
    lut = 255*cdf/FLOAT(cdf[-1])
    RETURN np.interp(x,bin_edges[:-1],lut)
```

corresponds to histogram equalization in the sense that a random variable $F = LUT(G)$ has uniform probability density. To see this, note that the function LUT in Equation (5.2) satisfies the monotonicity condition of Theorem 2.1. Therefore, with Equation (2.13), the PDF for F is

$$p_f(f) = p_g(g)\left|\frac{dg}{df}\right|.$$

Differentiating Equation (5.2),

$$\frac{df}{dg} = p_g(g), \quad \frac{dg}{df} = \frac{1}{p_g(g)}$$

and, since all probabilities are positive on the interval [0,1],

$$p_f(f) = p_g(g)\left|\frac{1}{p_g(g)}\right| = 1,$$

so F indeed has a uniform density. Histogram equalization can be approximated for byte-encoded data by first replacing $p_g(g)$ by the normalized histogram

$$p_g(g_k) = \frac{n_k}{n}, \quad k = 0\dots 255, \quad n = \sum_{j=0}^{255} n_j,$$

where n_k is the number of pixels with gray value g_k. This leads to the lookup table

$$LUT(g_k) = 255\cdot\sum_{j=0}^{k} p_g(g_k) = 255\cdot\sum_{j=0}^{k}\frac{n_j}{n}, \quad k = 0\dots 255. \tag{5.3}$$

The result is rounded down to the nearest integer. Because of the quantization, the resulting histogram will in general not be perfectly uniform, however the desired effect of spreading the intensities to span the full range of gray-scale values will be achieved.

The closely related procedure of *histogram matching*, which is the transformation of an image histogram to match the histogram of another image or some specified function, can be similarly derived using the probability density approximation; see Gonzalez and Woods (2002) for a detailed discussion.

Both histogram equalization and histogram matching are standard functions within the ENVI environment.

Listing 5.1 shows a straightforward Python implementation of histogram equalization stretching. It makes efficient use of the `numpy.interp()` function to interpolate each intensity in the input array between the histogram bin edges `bin_edges[:-1]` (just the sequence $[0, 1 \dots 255]$) and the normalized lookup table values `lut`, Equation (5.3).

5.2 High-pass spatial filtering and feature extraction

In Chapter 4, Section 4.2, we introduced filtering in the spatial domain, giving a simple example of a low-pass filter; see Equation (4.7). We shall now examine high-pass filtering for edge and contour detection, techniques that will be used later in the present chapter to implement low-level feature matching for image co-registration. We will also see how to access the highly optimized image processing algorithms of the Open Source Computer Vision Library (OpenCV) from the Python interpreter. Edge detection is often used in conjunction with feature extraction for scene analysis and for image segmentation, one of the subjects treated in Chapter 8.

FIGURE 5.1
Power spectrum image of the Sobel filter h_1 of Equation (5.5). Low spatial frequencies are at the center of the image.

Localized image features and/or segments can often be conveniently characterized by their *geometric moments*. A short description of geometric moments together with IDL and Python scripts to calculate them will also be presented in this Section.

5.2.1 Sobel filter

We begin by introducing the *gradient operator*

$$\nabla = \frac{\partial}{\partial \boldsymbol{x}} = \boldsymbol{i}\frac{\partial}{\partial x_1} + \boldsymbol{j}\frac{\partial}{\partial x_2}, \tag{5.4}$$

where $\boldsymbol{i}$ and $\boldsymbol{j}$ are unit vectors in the image plane in the horizontal and vertical directions, respectively. In a two-dimensional image represented by

Listing 5.2: Calculating the power spectrum of the filter h_1 in Equation (5.5).

```
PRO EX5_1
; create filter
   g = fltarr(512,512)
   g[0:2,0:2] = [[1,0,-1],[2,0,-2],[1,0,-1]]
; shift Fourier transform to center
   a = lindgen(512,512)
   i = a MOD 512
   j = a/512
   g = (-1)^(i+j)*g
; output Fourier power spectrum as EPS file
   thisDevice =!D.Name
   set_plot, 'PS'
   Device, Filename='fig5_1.eps',xsize=3,ysize=3, $
           /inches,/encapsulated
   tvscl, (abs(FFT(g)))^2
   device,/close_file
   set_plot,thisDevice
END
```

the continuous scalar function $g(x_1, x_2) = g(\boldsymbol{x})$, $\nabla g(\boldsymbol{x})$ is a vector in the direction of the maximum rate of change of gray-scale intensity.

Of course, the intensity values are actually discrete, so the partial derivatives must be approximated. For example, we can use the *Sobel operators*:

$$\begin{aligned}\frac{\partial g(\boldsymbol{x})}{\partial x_1}\bigg|_{\boldsymbol{x}=(i,j)} &\approx [g(i-1,j-1)+2g(i-1,j)+g(i-1,j+1)]\\ &\quad -[g(i+1,j-1)+2g(i+1,j)+g(i+1,j+1)] =: \nabla_1(\boldsymbol{x})\end{aligned}$$

$$\begin{aligned}\frac{\partial g(\boldsymbol{x})}{\partial x_2}\bigg|_{\boldsymbol{x}=(i,j)} &\approx [g(i-1,j-1)+2g(i,j-1)+g(i+1,j-1)]\\ &\quad -[g(i-1,j+1)+2g(i,j+1)+g(i+1,j+1)] =: \nabla_2(\boldsymbol{x}),\end{aligned}$$

which are equivalent to the two-dimensional spatial filters

$$h_1 = \begin{pmatrix} 1 & 0 & -1 \\ 2 & 0 & -2 \\ 1 & 0 & -1 \end{pmatrix} \quad \text{and} \quad h_2 = \begin{pmatrix} 1 & 2 & 1 \\ 0 & 0 & 0 \\ -1 & -2 & -1 \end{pmatrix}, \tag{5.5}$$

respectively; see Equation (4.6). The magnitude of the gradient at pixel $x = (i, j)$ is

$$\|\nabla g(i,j))\| = \sqrt{\nabla_1(\boldsymbol{x})^2 + \nabla_2(\boldsymbol{x})^2}.$$

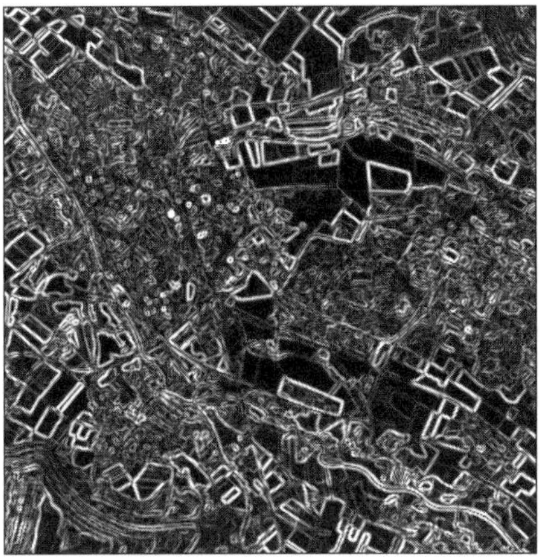

FIGURE 5.2
Sobel edge detection on the 3N band of the Jülich ASTER image of Figure 1.1.

Edge detection can be achieved by calculating the filtered image

$$f(i,j) = \|\nabla(g(\boldsymbol{x}))\|$$

and setting an appropriate threshold. The Sobel filters, Equations (5.5), involve *differences*, a characteristic of high-pass filters. They have the property of returning near-zero values when traversing regions of constant intensity, and positive or negative values in regions of changing intensity. Listing 5.2 calculates the Fourier power spectrum of h_1 and the result is shown in Figure 5.1. The Fourier spectrum of h_2 is the same, but rotated by 90 degrees. The Sobel filter, invoked from the ENVI main menu and applied to a spatial subset of the 3N spectral band from the ASTER image, Figure 1.1, generates the image shown in Figure 5.2. Note that the edge widths vary considerably, depending upon the contrast and abruptness (strength) of the edge.

5.2.2 Laplacian-of-Gaussian filter

FIGURE 5.3
Power spectrum image of the Laplacian filter, Equation (5.6). Low spatial frequencies are at the center of the image.

The magnitude of the gradient reaches a maximum at an edge in a gray-scale image. The second derivative, on the other hand, is zero at that maximum and has opposite signs immediately on either side. This offers the possibility to determine edge positions to the accuracy of one pixel by using second derivative filters. Thin edges, as we will see later, are useful in automatic determination of invariant features for image registration.

The second derivatives of the image intensities can be calculated with the *Laplacian* operator

$$\nabla^2 = \nabla^\top \nabla = \frac{\partial^2}{\partial x_1^2} + \frac{\partial^2}{\partial x_2^2}.$$

The Laplacian $\nabla^2 g(\boldsymbol{x})$ is an isotropic scalar function which is zero whenever the gradient magnitude is maximum. Like the gradient operator, the Laplacian operator can also be approximated by a spatial filter, for example,

$$h = \begin{pmatrix} 0 & 1 & 0 \\ 1 & -4 & 1 \\ 0 & 1 & 0 \end{pmatrix}. \tag{5.6}$$

Its power spectrum is depicted in Figure 5.3. This filter has the desired property of returning zero in regions of constant intensity and in regions of constantly varying intensity (e.g., ramps), but nonzero at their onset or termination. Laplacian filters tend to be very sensitive to image noise. Often a low-pass Gaussian filter is first used to smooth the image before the Laplacian filter is applied. This is equivalent to calculating the Laplacian of the Gaussian itself and then using the result to derive a high-pass filter. Recall that the normalized Gauss function in two dimensions is given by

$$\frac{1}{2\pi\sigma^2}\exp\left(-\frac{1}{2\sigma^2}(x_1^2+x_2^2)\right),$$

where the parameter σ determines its extent. Taking the second partial derivatives gives the *Laplacian-of-Gaussian* (LoG) filter

$$\frac{1}{2\pi\sigma^6}(x_1^2+x_2^2-2\sigma^2)\exp\left(-\frac{1}{2\sigma^2}(x_1^2+x_2^2)\right). \tag{5.7}$$

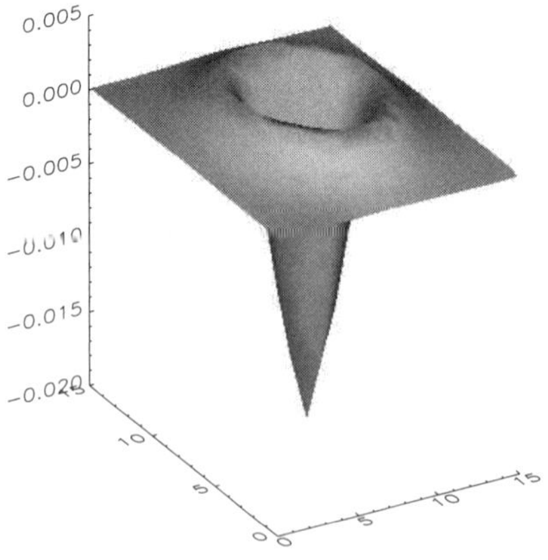

FIGURE 5.4
Laplacian-of-Gaussian filter on a 16 × 16 grid, with $\sigma = 2$.

Listing 5.3 illustrates the creation and use of a LoG filter together with determination of sign change to generate thin edges or contours from a gray-scale image. The filtering is carried out, after appropriate padding, in the frequency domain (lines 24 to 27), and the zero crossings in the horizontal and vertical directions are determined from the products of the image with a copy of itself shifted by one pixel to the right and upward, respectively (lines 29 and 30). Sign changes correspond to negative values in the product arrays and these define the contours. The contour pixel intensities are set equal to the magnitude of the

Listing 5.3: LoG filtering with sign change detection in IDL.

```
1 PRO EX5_2
2 ; create a LoG filter (and plot it)
3    sigma = 2.0
4    filter = fltarr(16,16)
5    FOR i=0L,15 DO FOR j=0L,15 DO filter[i,j] = $
6    (1/(2*!pi*sigma^6))*((i-8)^2+(j-8)^2-2*sigma^2) $
7     *exp(-((i-8)^2+(j-8)^2)/(2*sigma^2))
8 ;    thisDevice =!D.Name
9 ;    set_plot, 'PS'
10 ;   Device, Filename='fig5_4.eps',xsize=3,ysize=3, $
11 ;               /encapsulated
12   shade_surf,filter
13 ;   device,/close_file
14 ;   set_plot,thisDevice
15 ; get an image band and pad it
16   envi_select, title='Choose␣multispectral␣band', $
17               fid=fid, dims=dims,pos=pos, /band_only
18   num_cols = dims[2]-dims[1]+1
19   num_rows = dims[4]-dims[3]+1
20   image = fltarr(num_cols+16,num_rows+16)
21   image[0:num_cols-1,0:num_rows-1] = $
22       envi_get_data(fid=fid,dims=dims,pos=0)
23 ; pad the filter as well
24   filt = image*0
25   filt[0:15,0:15] = filter
26 ; perform filtering in frequency domain
27   image= float(fft(fft(image)*fft(filt),1))
28 ; get zero-crossing positions
29   indices = where( (image*shift(image,1,0) LT 0) $
30                OR (image*shift(image,0,1) LT 0) )
31 ; perform Sobel filter for edge strengths
32   edge_strengths = sobel(image)
33 ; create the contour image and return it to ENVI
34   image = image*0
35   image[indices]=edge_strengths[indices]
36   envi_enter_data, image
37 END
```

local gradient as determined by a Sobel filter (lines 32 to 35). This allows subsequent thresholding to identify the more significant contours. The program generates the surface plot of the two-dimensional LoG filter shown in Figure 5.4; the filtered image is displayed in Figure 5.5. The "spaghetti" effect is characteristic.

FIGURE 5.5
Image contours from a spatial subset of the 3N band of the Jülich ASTER image (Figure 1.1) calculated with the program in Listing 5.3.

5.2.3 OpenCV functions

The Open Source Computer Vision Library is a treasure chest of useful imaging processing routines, all of which are easily accessible from Python. The following is a quote from the website `opencv.org`:

> OpenCV is released under a BSD license and hence it's free for both academic and commercial use. It has C++, C, Python and Java interfaces and supports Windows, Linux, Mac OS, iOS and Android. OpenCV was designed for computational efficiency and with a strong focus on real-time applications.

In the following we will demonstrate the use of OpenCV for remote sensing imagery analysis with two well-known feature detection algorithms, and in the remainder of the text we will continue to make free use of the library for many of our Python scripts.

5.2.3.1 Corner detection

In a gray-scale image, the local (scalar) variogram in a neighborhood of the pixel position $\boldsymbol{x}$ can be estimated as

$$\gamma(\boldsymbol{h}) = \left\langle (g(\boldsymbol{x}) - g(\boldsymbol{x}+\boldsymbol{h}))^2 \right\rangle, \tag{5.8}$$

where $\langle\cdot\rangle$ signifies the average over the pixel's neighborhood; see Equation (3.68). A corner, or in general some interesting feature such as a localized bright or dark spot, will be characterized by a large value for $\gamma(\boldsymbol{h})$ in all directions of the displacement vector $\boldsymbol{h}$. An edge, on the other hand, would exhibit a large variation only in the directions nearly orthogonal to itself, and featureless regions would have small variations in all directions. Following Harris and Stephens (1988), we first of all obtain the first-order Taylor series approximation to $g(\boldsymbol{x}+\boldsymbol{h})$ in Equation (5.8):

$$g(\boldsymbol{x}+\boldsymbol{h}) \approx g(\boldsymbol{x}) + \boldsymbol{h}^\top \frac{\partial g(\boldsymbol{x})}{\partial \boldsymbol{x}} = g(\boldsymbol{x}) + \boldsymbol{h}^\top \nabla g(\boldsymbol{x});$$

see Equation (1.55). With this, the variogram $\gamma(\boldsymbol{h})$ can be approximately written as

$$\gamma(\boldsymbol{h}) \approx \left\langle \left(\boldsymbol{h}^\top \nabla g(\boldsymbol{x})\right)^2 \right\rangle,$$

or, expanding the quadratic term,

$$\begin{aligned}
\gamma(\boldsymbol{h}) &\approx \boldsymbol{h}^\top \left\langle \nabla g(\boldsymbol{x}) \nabla g(\boldsymbol{x})^\top \right\rangle \boldsymbol{h} \\
&= \boldsymbol{h}^\top \left\langle \begin{pmatrix} \left(\frac{\partial g(\boldsymbol{x})}{\partial x_1}\right)^2 & \frac{\partial g(\boldsymbol{x})}{\partial x_1}\frac{\partial g(\boldsymbol{x})}{\partial x_2} \\ \frac{\partial g(\boldsymbol{x})}{\partial x_1}\frac{\partial g(\boldsymbol{x})}{\partial x_2} & \left(\frac{\partial g(\boldsymbol{x})}{\partial x_2}\right)^2 \end{pmatrix} \right\rangle \boldsymbol{h} \\
&\approx \boldsymbol{h}^\top \begin{pmatrix} \langle \nabla_1(\boldsymbol{x})^2 \rangle & \langle \nabla_1(\boldsymbol{x})\nabla_2(\boldsymbol{x}) \rangle \\ \langle \nabla_1(\boldsymbol{x})\nabla_2(\boldsymbol{x}) \rangle & \langle \nabla_2(\boldsymbol{x})^2 \rangle \end{pmatrix} \boldsymbol{h} = \boldsymbol{h}^\top \boldsymbol{A} \boldsymbol{h},
\end{aligned} \tag{5.9}$$

where $\nabla_i(\boldsymbol{x})$, $i = 1, 2$, are the Sobel gradient operators introduced in Section 5.2.1. The matrix $\boldsymbol{A}$ in Equation (5.9) is symmetric and also positive definite for sufficiently large neighborhoods. In its principal axis coordinate system, therefore, we can write Equation (5.9) as

$$\gamma(\boldsymbol{h}) \approx \boldsymbol{h}^\top \begin{pmatrix} \lambda_1 & 0 \\ 0 & \lambda_2 \end{pmatrix} \boldsymbol{h} = \lambda_1 h_1^2 + \lambda_2 h_2^2,$$

where λ_1 and λ_2 are the real positive eigenvalues of $\boldsymbol{A}$. If the variogram $\gamma(\boldsymbol{h})$ is to be large for all directions $\boldsymbol{h}$, then clearly both eigenvalues must be

Listing 5.4: Corner detection with OpenCV.

```
1 #!/usr/bin/env python
2 #Name:  ex5_1.py
3 IMPORT auxil.auxil as auxil
4 FROM numpy IMPORT *
5 FROM osgeo IMPORT gdal
6 FROM osgeo.gdalconst IMPORT GA_ReadOnly,GDT_Float32
7 IMPORT cv2 as cv
8
9 DEF main():
10     gdal.AllRegister()
11     infile = auxil.select_infile()
12     IF infile:
13         inDataset = gdal.Open(infile,GA_ReadOnly)
14         cols = inDataset.RasterXSize
15         rows = inDataset.RasterYSize
16     ELSE:
17         RETURN
18 #  read first image band
19     rasterBand = inDataset.GetRasterBand(1)
20     band = rasterBand.ReadAsArray(0,0,cols,rows)\
21                                 .astype(uint8)
22 #  corner detection, window size 7x7
23     result = cv.cornerMinEigenVal(band, 7)
24 #  write to disk
25     outfile,fmt = auxil.select_outfilefmt()
26     IF outfile:
27         driver = gdal.GetDriverByName(fmt)
28         outDataset = driver.Create(outfile,
29                         cols,rows,1,GDT_Float32)
30         outBand = outDataset.GetRasterBand(1)
31         outBand.WriteArray(result,0,0)
32         outBand.FlushCache()
33         outDataset = None
34     inDataset = None
35
36 IF __name__ == '__main__':
37     main()
```

large. The OpenCV filter function `cornerMinEigenVal()` calculates the minimum eigenvalue of $\boldsymbol{A}$ for a square neighborhood of each pixel in a scene, returning the result as a floating point image. This image can then be thresholded to distinguish significant corners or other features.

In Listing 5.4, the first spectral band of an image is read in (lines 11–15) and processed with `cornerMinEigenVal()` using a 7×7 window, line 23. An example is shown on the left in Figure 5.6.

FIGURE 5.6
Corners and edges detected from a small spatial subset of the ASTER image of Figure 1.1. Center: a subset of the first spectral band. Left: the minimum eigenvalue intensities in a linear 2% histogram stretch. Right: the Canny edges.

5.2.3.2 Canny edge detector

The Canny edge detector (Canny, 1986),* which is similar to the corner detection algorithm, is based upon a gradient filter such as the Sobel filter. First a Gaussian smoothing filter is applied to suppress noise, followed by the gradient filter, whereby both the magnitude of the gradient

$$\|\nabla g(i,j))\| = \sqrt{\nabla_1(\boldsymbol{x})^2 + \nabla_2(\boldsymbol{x})^2}$$

as well as its direction

$$\theta = \arctan\left(\frac{\nabla_2(\boldsymbol{x})}{\nabla_1(\boldsymbol{x})}\right)$$

are calculated. From here on we quote the excellent Wikipedia article:

> Given estimates of the image gradients, a search is then carried out to determine if the gradient magnitude assumes a local maximum in the gradient direction. ... From this stage, referred to as non-maximum suppression, a set of edge points, in the form of a binary image, is obtained. These are sometimes referred to as "thin edges."
>
> Large intensity gradients are more likely to correspond to edges than small intensity gradients. It is in most cases impossible to specify a threshold at which a given intensity gradient switches from corresponding to an edge into not doing so. Therefore Canny uses thresholding with hysteresis.†
>
> Thresholding with hysteresis requires two thresholds high and low. Making the assumption that important edges should be along

*See also `http://en.wikipedia.org/wiki/Canny_edge_detector`.

†In Section 5.6 we will be applying a similar hysteresis approach for contour detection.

> continuous curves in the image allows us to follow a faint section of a given line and to discard a few noisy pixels that do not constitute a line but have produced large gradients. Therefore we begin by applying a high threshold. This marks out the edges we can be fairly sure are genuine. Starting from these, using the directional information derived earlier, edges can be traced through the image. While tracing an edge, we apply the lower threshold, allowing us to trace faint sections of edges as long as we find a starting point.
>
> Once this process is complete we have a binary image where each pixel is marked as either an edge pixel or a non-edge pixel. From complementary output from the edge tracing step, the binary edge map obtained in this way can also be treated as a set of edge curves.

The code demonstrating the Canny edge detector (in the file `ex5_2.py` included with the software for this book) is almost identical to that shown in Listing 5.4 except that line 23 is replaced by

```
result = cv.canny(band,50,150)
```

and the GDAL data type `GDT_Float32` in lines 6 and 29 is replaced by `GDT_Byte`. The lower and upper hysteresis thresholds are 50 and 150, respectively. An example is shown on the right in Figure 5.6.

5.2.4 Invariant moments

Moments and functions of moments are employed extensively in image classification, target identification and image scene analysis (Prokop and Reeves, 1992). A feature such as, for instance, a closed contour extracted from the image in Figure 5.5 can be described in terms of its *geometric moments*. Let S denote the set of pixels belonging to the feature. The (discrete) geometric moment of order p, q of the feature is defined by (see, e.g., Haberächer (1995); Gonzalez and Woods (2002))

$$m_{pq} = \sum_{i,j\in S} g(i,j)i^p j^q, \quad p,q = 0,1,2\dots. \tag{5.10}$$

Thus m_{00} is the total intensity of the feature or, in a binary representation in which $g(i,j) = 1$ if $(i,j) \in S$ and $g(i,j) = 0$ otherwise, m_{00} is the number of pixels. The center of gravity $(\bar{x}_1, \bar{x}_2)$ of the feature is

$$\bar{x}_1 = \frac{m_{10}}{m_{00}}, \quad \bar{x}_2 = \frac{m_{01}}{m_{00}}. \tag{5.11}$$

The translation-invariant *centralized moments* μ_{pq} are obtained by shifting the origin to the center of gravity,

$$\mu_{pq} = \sum_{i,j\in S} g(i-\bar{x}_1, j-\bar{x}_2)(i-\bar{x}_1)^p(j-\bar{x}_2)^q \tag{5.12}$$

Listing 5.5: Illustrating the Hu invariant moments.

```
PRO Ex5_3

; Airplane
   A = float([[0,0,0,0,0,1,0,0,0,0,0], $
              [0,0,0,0,1,1,1,0,0,0,0], $
              [0,0,0,0,1,1,1,0,0,0,0], $
              [0,0,0,1,1,1,1,1,0,0,0], $
              [0,0,1,1,0,1,0,1,1,0,0], $
              [0,1,1,0,0,1,0,0,1,1,0], $
              [1,0,0,0,0,1,0,0,0,0,1], $
              [0,0,0,0,0,1,0,0,0,0,0], $
              [0,0,0,0,1,1,1,0,0,0,0], $
              [0,0,0,0,0,1,0,0,0,0,0]])
   Im = fltarr(200,200)
   A =  rebin(A,55,50)
   Im[25:79,75:124]=A
; Rotate through 60 deg, magnify by 1.5 and translate
   Im1 = shift(rot(Im,60,1.5,cubic=-0.5),90,-60)
; Rotate through 180 deg, shrink by 0.5 and translate
   Im2 = shift(rot(Im,180,0.5,cubic=-0.5),-90,-60)
; Plot them
   tvscl,Im+Im1+Im2
; Invariant moments
   PRINT, hu_moments(Im,/log), format='(7F8.3)'
   PRINT, hu_moments(Im1,/log), format='(7F8.3)'
   PRINT, hu_moments(Im2,/log), format='(7F8.3)'

END
```

and the *normalized centralized moments* η_{pq} by the following normalization; see Exercise 5,

$$\eta_{pq} = \frac{1}{\mu_{00}^{(p+q)/2+1}} \mu_{pq}. \tag{5.13}$$

The normalized centralized moments are, apart from effects of digital quantization, invariant under both translations and scale changes. For example, in a binary representation the moment η_{20} is

$$\eta_{20} = \frac{1}{\mu_{00}^2}\mu_{20} = \frac{1}{n^2}\sum_{i,j\in S}(i-\bar{x}_1)^2 = \frac{1}{n}\sum_{i\in S}(i-\bar{x}_1)^2,$$

where $n = |S|$, the cardinality of S. This is just the variance of the feature in the x_1 direction.

Finally, we can define *Hu moments*, which are functions of the normalized centralized moments of orders $p+q \leq 3$ and which are invariant under rotations; see Hu (1962) or Gonzalez and Woods (2002). There are seven such moments in all, the first four of which are given by

$$\begin{aligned}
h_1 &= \eta_{20} + \eta_{02} \\
h_2 &= (\eta_{20} - \eta_{02})^2 + 4\eta_{11}^2 \\
h_3 &= (\eta_{30} - 3\eta_{12})^2 + (\eta_{03} - 3\eta_{21})^2 \\
h_4 &= (\eta_{30} + \eta_{12})^2 + (\eta_{03} + \eta_{21})^2.
\end{aligned} \tag{5.14}$$

To illustrate their rotational invariance, consider a rotation of the coordinate axes through the angle θ with origin at the center of gravity of a feature. A point (i, j) transforms according to

$$\begin{pmatrix} i' \\ j' \end{pmatrix} = \begin{pmatrix} \cos\theta & \sin\theta \\ -\sin\theta & \cos\theta \end{pmatrix} \begin{pmatrix} i \\ j \end{pmatrix} = \boldsymbol{A} \begin{pmatrix} i \\ j \end{pmatrix}.$$

Then, again in a binary image, the first invariant moment in the rotated coordinate system is

$$\begin{aligned}
h_1 &= \frac{1}{n^2} \sum_{i',j' \in S} (i'^2 + j'^2) = \frac{1}{n^2} \sum_{i',j' \in S} (i', j') \begin{pmatrix} i' \\ j' \end{pmatrix} \\
&= \frac{1}{n^2} \sum_{i,j \in S} (i, j) \boldsymbol{A}^\top \boldsymbol{A} \begin{pmatrix} i \\ j \end{pmatrix} = \frac{1}{n^2} \sum_{i,j \in S} (i^2 + j^2),
\end{aligned}$$

since $\boldsymbol{A}^\top \boldsymbol{A} = \boldsymbol{I}$.

The Hu moments are strictly invariant only in the limit of continuous shapes. The discrete nature of image features introduces errors which become especially severe for higher-order moments. Liao and Pawlak (1996) give correction formulae for the geometric moments, Equation (5.10), which reduce the error. This forms the basis of the IDL function `HU_MOMENTS()` included in the `AUXIL` directory for calculating all seven invariant Hu moments. The program in Listing 5.5 generates the features shown in Figure 5.7 and then calls `HU_MOMENTS()` to calculate the logarithms of the invariant moments. The result is:

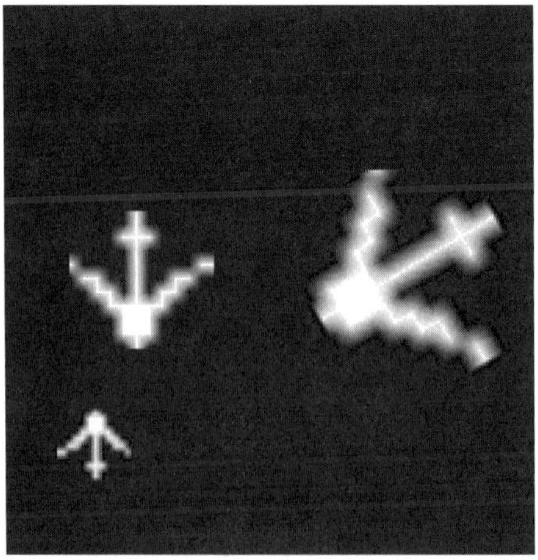

FIGURE 5.7
An "aircraft" feature translated, scaled and rotated; see Listing 5.5.

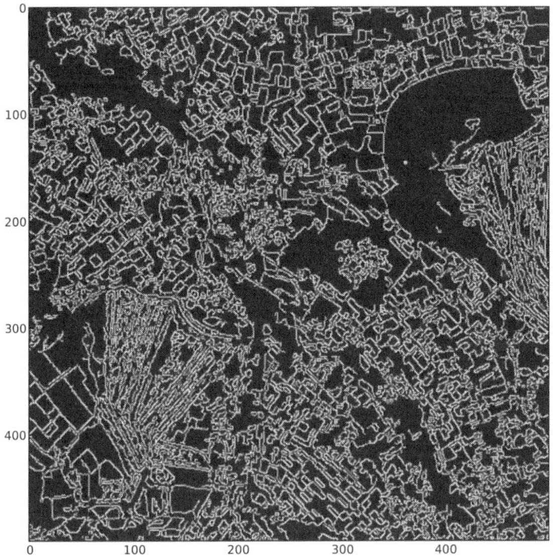

FIGURE 5.8
Edge contours for the first spectral band of the LANDSAT 7 ETM+ image of Figure 3.10.

```
ENVI> ex5_3
   -1.061   -7.363   -6.765   -6.241  -12.752   -9.934  -14.800
   -1.067   -6.819   -6.873   -6.211  -12.754   -9.621  -15.509
   -1.060   -7.355   -6.793   -6.246  -12.771   -9.939  -15.056
```

As a less trivial example, the Python script shown in Listing 5.6 extracts the "significant contours" from an image band, calculates their invariant moments and plots histograms of the moment logarithms, making extensive use of OpenCV functions. The `findContours()` function in line 24 extracts contours from the binary image returned from the Canny edge detector, line 23. After drawing all the contours onto an empty array (lines 26 and 27) and displaying them (line 28), the seven Hu moments of the individual contours are calculated and stored in an array (lines 32–36). This occurs in two steps: First the translational and scale invariants are determined with `moments()`, line 35, then these are passed to `HuMoments()`, line 36. For very simple contour shapes the higher moments will be zero. This condition is checked for in line 39 prior to taking logarithms and plotting the histogram. The edge contours together with one of the histograms generated by the script are depicted in Figures 5.8 and 5.9. See the OpenCV documentation for detailed explanation

Listing 5.6: Hu invariant moments of image contours.

```
#!/usr/bin/env python
#Name:  ex5_3.py
IMPORT auxil.auxil as auxil
FROM numpy IMPORT *
FROM osgeo IMPORT gdal
FROM osgeo.gdalconst IMPORT GA_ReadOnly
IMPORT cv2 as cv
IMPORT matplotlib.pyplot as plt

DEF main():
    gdal.AllRegister()
#  read first band of an MS image
    infile = auxil.select_infile()
    IF infile:
        inDataset = gdal.Open(infile,GA_ReadOnly)
        cols = inDataset.RasterXSize
        rows = inDataset.RasterYSize
    ELSE:
        RETURN
    rasterBand = inDataset.GetRasterBand(1)
    band = rasterBand.ReadAsArray(0,0,cols,rows)
#  find and display contours
    edges = cv.Canny(band, 20, 80)
    contours,hierarchy = cv.findContours(edges,\
             cv.RETR_LIST,cv.CHAIN_APPROX_NONE)
    arr = zeros((rows,cols),dtype=uint8)
    cv.drawContours(arr, contours, -1, 255)
    plt.imshow(arr,cmap='gray') ;  plt.show()
#  determine Hu moments
    num_contours = LEN(hierarchy[0])
    hus = zeros((num_contours,7),dtype=float32)
    FOR i IN RANGE(num_contours):
        arr = arr*0
        cv.drawContours(arr, contours, i, 1)
        m = cv.moments(arr)
        hus[i,:] = cv.HuMoments(m).ravel()
#  plot histograms of logarithms of the Hu moments
    FOR i IN RANGE(7):
        idx = where(hus[:,i]>0)
        hist,_ = histogram(log(hus[idx,i]),50)
        plt.plot(RANGE(50), hist, 'k-'); plt.show()

IF __name__ == '__main__':
    main()
```

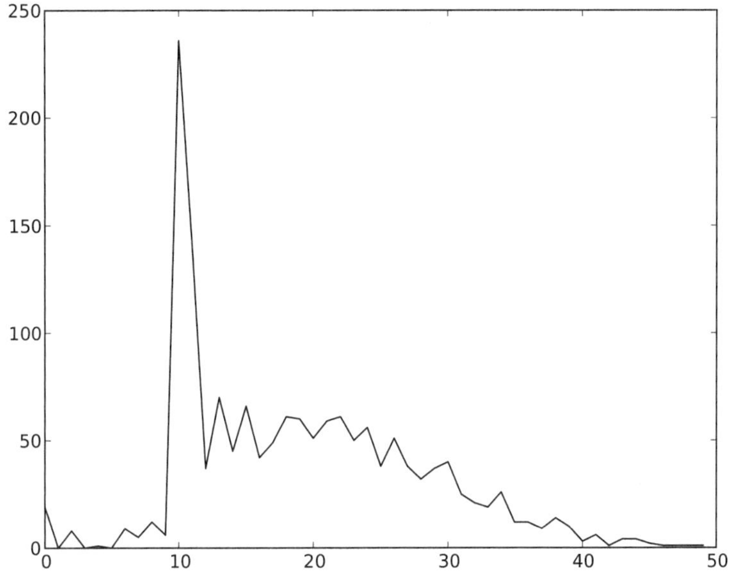

FIGURE 5.9
Histogram of the logarithms of the first Hu moments of the contours of Figure 5.8. The spike corresponds to short, simple line segments.

of the functions used and their parameters.

The geometric moments are projections of pixel intensities onto the monomials $f_{pq}(i,j) = i^p j^q$. The continuous equivalents $f_{pq}(\boldsymbol{x}) = x_1^p x_2^q$ are not orthogonal, i.e.,

$$\int f_{pq}(\boldsymbol{x}) f_{p'q'}(\boldsymbol{x}) d\boldsymbol{x} \neq 0, \quad p,q \neq p'q',$$

so that geometric moments and their invariant derivatives carry redundant information. Alternative moment systems can be defined in terms of orthogonal Legendre or Zernike polynomials, which are particularly relevant for image reconstruction; see Teague (1980).

5.3 Panchromatic sharpening

The change detection and classification algorithms that we will meet in the next chapters exploit both the spatial and the spectral information provided by remote sensing imagery. Many common satellite platforms (e.g., LANDSAT 8 OLI, SPOT, IKONOS, QuickBird, GeoEye) supply co-registered panchromatic images with considerably higher ground resolution than that of the multispectral bands. However, without additional processing, application of spectral change detection or classification methods is restricted in the first instance to the poorer spatial resolution of the multispectral data. Conventional image fusion techniques, such as the well-known HSV-transformation discussed below, can be used to sharpen the spectral components. However, the effect of mixing in the panchromatic image is often to "dilute" significantly the spectral information and, for instance in the case of classification, to reduce class separability in the multidimensional feature space. Another disadvantage of the HSV transformation is that one is restricted to using three of the available spectral channels. In the following, after outlining the HSV method, we consider a number of alternative (and better) fusion techniques. See Aanaes et al. (2008) for an excellent discussion of the issues involved in image fusion.

5.3.1 HSV fusion

In computers with 24-bit graphics (referred to as "true color" on Windows systems), three bands of a multispectral image can be displayed with 8 bits for

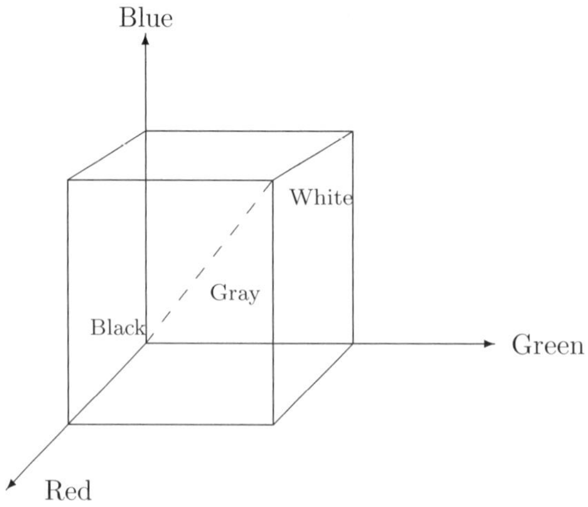

FIGURE 5.10
The RGB color cube.

Listing 5.7: HSV panchromatic sharpening with ENVI batch procedures and ENVI Classic.

```
PRO HSV, event
COMPILE_OPT IDL2
; get MS image
envi_select,title='Select␣3-band␣MS␣input␣file',$
                    fid=fid1, dims=dims1, pos=pos1
IF (fid1 EQ -1) OR (n_elements(pos1) NE 3) THEN RETURN
; get PAN image
envi_select, title='Select␣panchromatic␣image', $
             fid=fid2,pos=pos2,dims=dims2,/band_only
IF (fid2 EQ -1) THEN RETURN
; linear stretch the images and convert to byte format
envi_doit,'stretch_doit',fid=fid1,dims=dims1,pos=pos1,$
  method=1,r_fid=r_fid1,out_min=0,out_max=255, $
  range_by=0,i_min=0,i_max=100,out_dt=1,/in_memory
envi_doit,'stretch_doit',fid=fid2,dims=dims2,pos=pos2,$
  method=1,r_fid=r_fid2,out_min=0,out_max=255, $
  range_by=0, i_min=0, i_max=100, out_dt=1,/in_memory
envi_file_query,r_fid2,ns=f_ns,nl=f_nl
f_dims = [-1l,0,f_ns-1,0,f_nl-1]
; HSV pan-sharpening
envi_doit, 'sharpen_doit', $
  fid=[r_fid1,r_fid1,r_fid1],pos=[0,1,2],f_fid=r_fid2,$
  f_dims=f_dims,f_pos=[0],method=0,interp=0,/in_memory
; remove temporary files
envi_file_mng, id=r_fid1, /remove
envi_file_mng, id=r_fid2, /remove

END
```

each of the additive primary colors red, green, and blue. There are $2^{24} \approx 16$ million colors possible. The monitor displays the bands as an RGB color composite image which, depending on the choice of spectral bands and histogram functions, may or may not appear to be natural; see for instance Figure 1.1. The color of a given pixel in the image can be represented as a vector or point in a Cartesian coordinate system (the RGB color cube) as illustrated in Figure 5.10.

An alternative color representation is in terms of *hue*, *saturation* and *value* (HSV). Value (sometimes called *intensity*) can be thought of as the distance along an axis equidistant from the three orthogonal primary color axes in the RGB representation (the main diagonal of the RGB cube shown as the dashed line in Figure 5.10). Since all points on the axis have equal R, G, and B values, they appear as shades of gray. Hue refers to the actual color and is defined as an angle on a plane perpendicular to the value axis as measured from a reference line pointing in the direction the red vertex of the

Listing 5.8: HSV panchromatic sharpening with ENVI batch procedures and the new ENVI GUI.

```
PRO HSV5
COMPILE_OPT IDL2
e = envi()
; get MS image
envi_select, title='Select␣3-band␣MS␣input␣file', $
                     fid=fid1,dims=dims1,pos=pos1
IF (fid1 EQ -1) OR (n_elements(pos1) NE 3) THEN RETURN
; get PAN image
envi_select, title='Select␣panchromatic␣image', $
             fid=fid2,pos=pos2,dims=dims2,/band_only
IF (fid2 EQ -1) THEN RETURN
; linear stretch the images and convert to byte format
envi_doit,'stretch_doit',fid=fid1,dims=dims1,pos=pos1,$
  method=1, r_fid=r_fid1,out_min=0,out_max=255, $
  range_by=0,i_min=0,i_max=100,out_dt=1,/in_memory
envi_doit,'stretch_doit',fid=fid2,dims=dims2,pos=pos2,$
  method=1,r_fid=r_fid2,out_min=0,out_max=255, $
  range_by=0,i_min=0,i_max=100,out_dt=1,/in_memory
envi_file_query,r_fid2,ns=f_ns,nl=f_nl
f_dims = [-1l,0,f_ns-1,0,f_nl-1]
; HSV pan-sharpening
envi_doit, 'sharpen_doit', $
  fid=[r_fid1,r_fid1,r_fid1],pos=[0,1,2],f_fid=r_fid2,$
  f_dims=f_dims,f_pos=[0],method=0,interp=0,$
  r_fid=r_fid3,/in_memory
out_raster = ENVIFIDToRaster(r_fid3)
view = e.GetView()
layer = view.CreateLayer(out_raster)

END
```

RGB cube. Saturation is the "amount" of color present and is represented by the radius of a circle described in the hue plane. A commonly used method for fusion of three low-resolution multispectral bands with a higher-resolution panchromatic image is to resample the former to the panchromatic resolution, transform it from RGB to HSV coordinates, replace the V component with the gray-scale panchromatic image, eventually after performing some kind of histogram matching or normalization of the two, and then to transform back to RGB space.

This procedure is referred to as *HSV panchromatic image sharpening* and is illustrated in Listing 5.7. The code demonstrates the use of ENVI batch procedures in an IDL program to perform standard manipulations, in this case histogram stretching and HSV sharpening. The ENVI main menu command `Transform/Image Sharpening/HSV` offers similar functionality, however it does not work with 11-bit data such as IKONOS and QuickBird imagery. Apparently the routines which are invoked from the main menu command expect byte-encoded pixel intensities. This drawback is remedied in Listing 5.7. The `EVENT` parameter in the procedure declaration allows the program to be called from the ENVI Classic menu system. Listing 5.8 is an equivalent IDL script appropriate to the new ENVI user interface.

5.3.2 Brovey fusion

In *Brovey* or *color normalized* fusion (Vrabel, 1996), each resampled multispectral pixel is multiplied by the ratio of the corresponding panchromatic pixel intensity to the sum of all of the multispectral intensities. The corrected pixel intensities $\bar{g}_k(i,j)$ in the kth fused multispectral channel are given by

$$\bar{g}_k(i,j) = g_k(i,j) \cdot \frac{g_p(i,j)}{\sum_{k'} g_{k'}(i,j)}, \tag{5.15}$$

where $g_k(i,j)$ is the (resampled) pixel intensity in the kth spectral band and $g_p(i,j)$ is the corresponding pixel intensity in the panchromatic image. This technique assumes that the spectral range spanned by the panchromatic image is essentially the same as that covered by the multispectral bands. This is seldom the case. Moreover, to avoid bias, the intensities used should be the radiances at the satellite sensors, implying use of the sensors' calibration. The ENVI environment also offers Brovey fusion in its main menu.

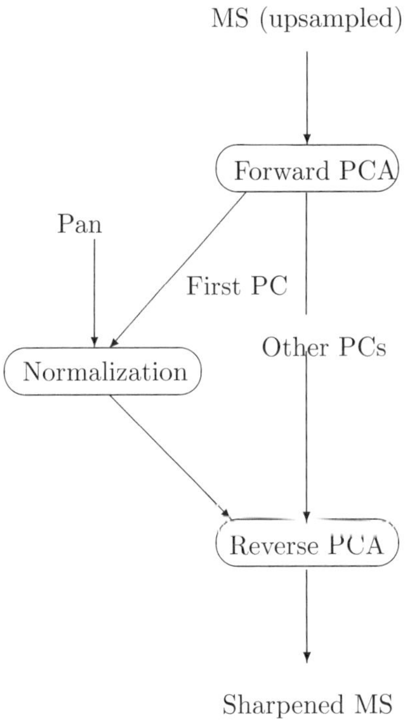

FIGURE 5.11
Panchromatic fusion with the principal components transformation.

5.3.3 PCA fusion

Panchromatic sharpening using principal components analysis (Welch and Ahlers, 1987) is similar to the HSV method. After the PCA transformation, the first principal component is replaced by the panchromatic image, again after some kind of normalization, and the transformation is inverted; see Figure 5.11. Image sharpening using PCA and the closely related Gram–Schmidt transformation are likewise available from the ENVI main menu.

5.3.4 DWT fusion

The discrete wavelet transform of a two-dimensional image was shown in Chapter 4 to be equivalent to an iterative application of the high-low-pass filter bank of Figure 4.10; see also Figures 4.11 and 4.12. The DWT filter bank can be used in an elegant way (Ranchin and Wald, 2000) to achieve panchromatic sharpening when the panchromatic and multispectral spatial resolutions differ by exactly a factor of 2^n for integer n. This is often the case, for example in SPOT, LANDSAT 7 ETM+, LANDSAT 8 OLI, IKONOS, QuickBird, GeoEye and ASTER imagery.

A DWT fusion procedure for IKONOS imagery for instance, in which the resolutions of panchromatic and the four multispectral bands differ exactly by a factor of 4, is as follows:

- The panchromatic image, represented by $f_n(i,j)$, is filtered twice to give a degraded image $f_{n-2}(i,j)$ plus wavelet coefficients at two scales (see Figures 4.10 and 4.12).

- Both the low-pass filtered portion of the panchromatic image as well as the four multispectral bands are filtered once again (the low-pass portions of all five bands now have 8 m pixel resolutions in the case of IKONOS data).

- The high-frequency components C^z_{n-3}, $z = H, V, D$, are sampled to estimate radiometric normalization coefficients a^z and b^z for each multispectral band:
$$\begin{aligned} a^z &= \hat\sigma^z_{ms}/\hat\sigma^z_{pan} \\ b^z &= \hat m^z_{ms} - a^z \hat m^z_{pan}, \end{aligned} \tag{5.16}$$
where $\hat m^z$ and $\hat\sigma^z$ denote estimated mean and standard deviation, respectively.

- These coefficients are then used to normalize the wavelet coefficients for the panchromatic image to those of each multispectral band at the $n-2$ and $n-1$ scales:
$$C^z_k(i,j) \to a^z C^z_k(i,j) + b^z, \quad z = H, V, D,\ k = n-2, n-1. \tag{5.17}$$

- Finally, the degraded portion of the panchromatic image $f_{n-2}(i,j)$ is replaced with the each of the four original multispectral bands in turn and the filter bank is inverted.

We thus obtain "what would be seen if the multispectral sensors had the resolution of the panchromatic sensor" (Ranchin and Wald, 2000). An ENVI/IDL

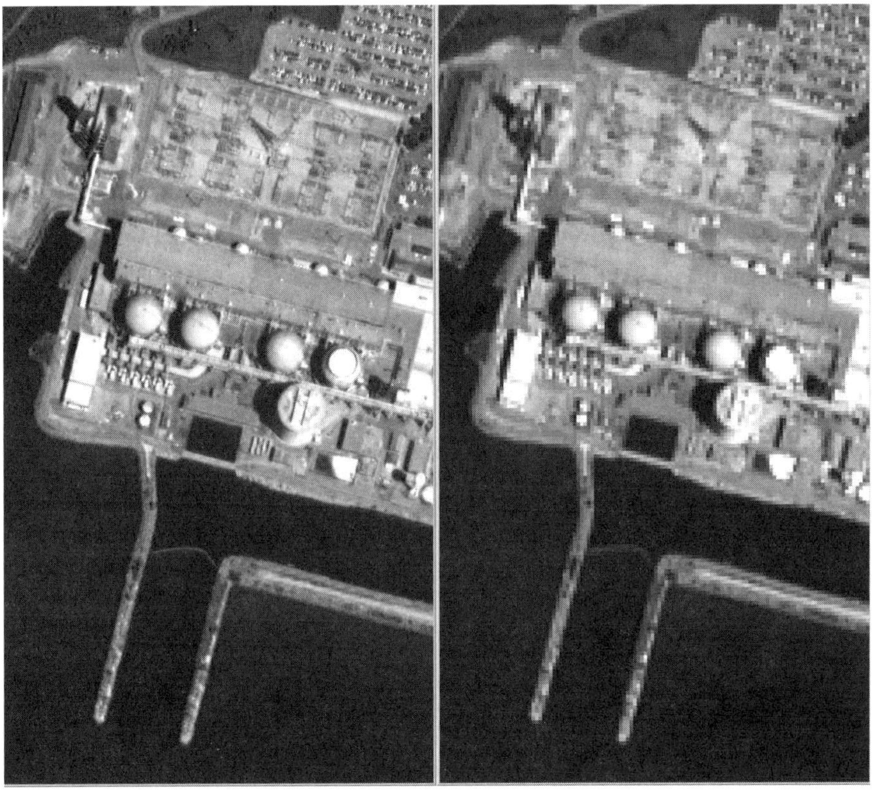

FIGURE 5.12
Panchromatic sharpening of an IKONOS multispectral image with the DWT filter bank. Left: panchromatically sharpened multispectral band of an image over a nuclear power reactor complex in Canada. Right: Original band upsampled by a factor of 4.

extension and a Python script for panchromatic sharpening with the DWT are described in Appendices C and D, respectively. Figure 5.12 shows an example of DWT pan-sharpening of IKONOS imagery.

5.3.5 *À trous* fusion

The spectral fidelity of the pan-sharpened images obtained with the discrete wavelet transform is excellent, as will be shown below. However the DWT is not *shift invariant* (Bradley, 2003), which means that small spatial displacements in the input array can cause major variations in the wavelet coefficients at their various scales. This has no effect on perfect reconstruction if one simply inverts the transformation. However a small misalignment can occur when the multispectral bands are "injected" into the panchromatic image pyramid

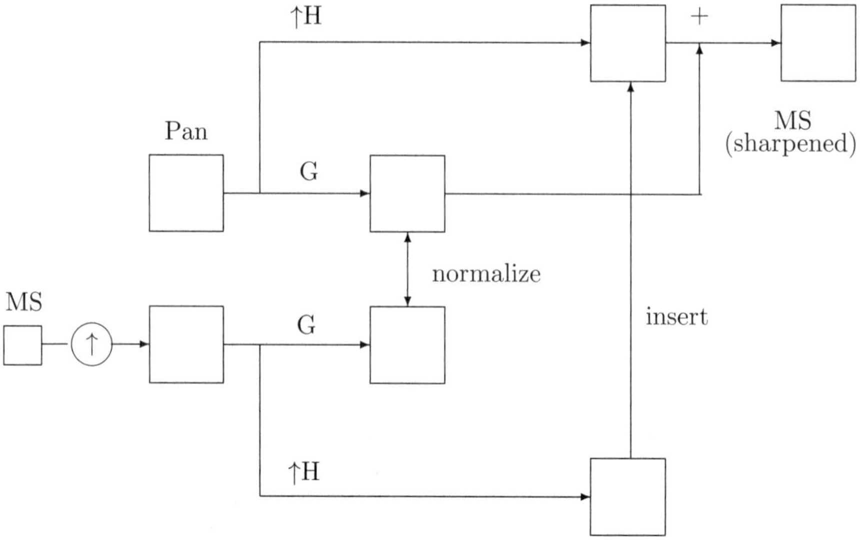

FIGURE 5.13
À trous image sharpening scheme for a multispectral to panchromatic resolution ratio of two. The symbol ↑H denotes the upsampled low-pass filter.

as described in the preceding section. This sometimes leads to spatial artifacts (blurring, shadowing, staircase effect) in the sharpened product (Yocky, 1996). These effects are visible in Figure 5.12 upon close inspection.

As one alternative to the DWT, the *à trous wavelet transform* (ATWT) has been proposed for image sharpening (Aiazzi et al., 2002). The ATWT is a multiresolution decomposition defined formally by a low-pass filter $H = \{h(0), h(1), \dots\}$ and a high-pass filter $G = \delta - H$, where δ denotes an all-pass filter. The high-frequency part is then just the difference between the original image and the low-pass filtered image. Not surprisingly, this transformation does not allow perfect reconstruction if the output is downsampled. Therefore downsampling is not performed at all. Rather, at the kth iteration of the low-pass filter, 2^{k-1} zeroes are inserted between the elements of H. This means that every other pixel is interpolated (averaged) on the first iteration:

$$H = \{h(0), 0, h(1), 0, \dots\},$$

while on the second iteration

$$H = \{h(0), 0, 0, h(1), 0, 0, \dots\},$$

etc. (hence the name *à trous* = with holes). The low-pass filter is usually chosen to be symmetric (unlike the Daubechies wavelet filters for example). A good choice turns out to be the cubic B-spline filter (Núnez et al., 1999), $H = \{1/16, 1/4, 3/8, 1/4, 1/16\}$; see Exercise 9 in Chapter 3.

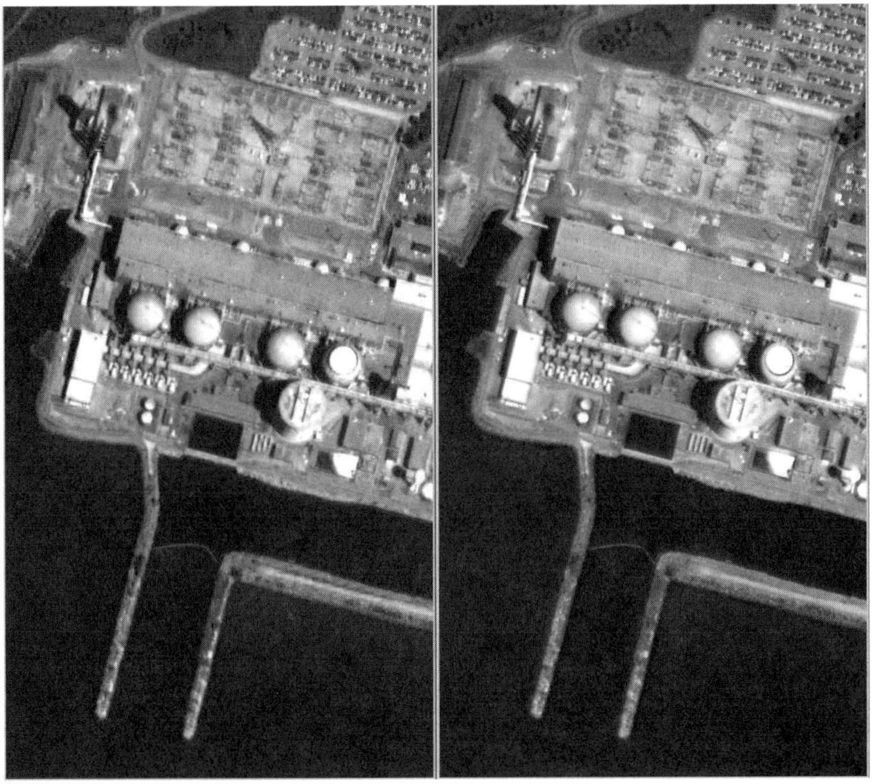

FIGURE 5.14
As Figure 5.12, except that the right-hand side now shows panchromatic sharpening with the *à trous* wavelet transform.

Figure 5.13 outlines the scheme implemented in the ENVI/IDL extension for ATWT panchromatic sharpening described in Appendix C, assuming a difference in spatial resolution of a factor of two: The MS band is resampled to match the dimensions of the high-resolution band. The *à trous* transformation is applied to both bands (columns and rows are filtered with the upsampled cubic spline filter, with the difference from the original image determining the high-pass result). The high-frequency component of the panchromatic image is normalized to that of the MS image in a similar way as for DWT sharpening, Equations (5.16) and (5.17). Then the smoothed panchromatic component is replaced by the filtered MS image and the transformation inverted. Appendix D documents the equivalent Python script.

The transformation is highly redundant and requires considerably more computer storage to implement. However, when used for image sharpening it is much less sensitive to misalignment between the multispectral and panchromatic images. A comparison with the DWT method is made in Figure 5.14.

5.3.6 A quality index

Several quantitative measures have been suggested for determining the degree to which the spectral properties of the multispectral bands have been degraded in the pan-sharpening process (Tsai, 1982). Wang and Bovik (2002) suggest a particularly simple and intuitive measure of spectral fidelity between two image bands with pixel intensities represented by f and g:

$$\mathrm{QI} = \frac{\sigma_{\mathrm{fg}}}{\sigma_{\mathrm{f}}\sigma_{\mathrm{g}}} \cdot \frac{2\bar{\mathrm{f}}\bar{\mathrm{g}}}{\bar{\mathrm{f}}^2 + \bar{\mathrm{g}}^2} \cdot \frac{2\sigma_{\mathrm{f}}\sigma_{\mathrm{g}}}{\sigma_{\mathrm{f}}^2 + \sigma_{\mathrm{g}}^2} = \frac{4\sigma_{\mathrm{fg}}\bar{\mathrm{f}}\bar{\mathrm{g}}}{(\bar{\mathrm{f}}^2 + \bar{\mathrm{g}}^2)(\sigma_{\mathrm{f}}^2 + \sigma_{\mathrm{g}}^2)}, \tag{5.18}$$

where $\bar{f}$ ($\bar{g}$) and σ_f^2 (σ_g^2) are mean, and variance of band f (g) and σ_{fg} is the covariance of the two bands; see Listing 5.9. The first factor in (5.18) is seen to be the correlation coefficient between the two images, with values in $[-1, 1]$; the second factor compares their average brightness, with values in $[0, 1]$; and the third factor compares their contrasts, also in $[0, 1]$. Thus, perfect spectral correspondence would give a value $Q = 1$.*

Listing 5.9: An IDL routine for the Wang-Bovik quality index.

```
FUNCTION QI, x, y
   Sxy  = correlate(x,y,/covariance)
   xm   = mean(x)
   ym   = mean(y)
   Sxx  = variance(x)
   Syy  = variance(y)
   RETURN, 4*Sxy*xm*ym/((Sxx+Syy)*(xm^2+ym^2))
END
```

TABLE 5.1
Quality indices for five panchromatic sharpening methods.

MS Band	HSV	PCA	G-S	DWT	ATWT
1	0.376	0.986	0.949	0.977	0.921
2	0.228	0.977	0.929	0.977	0.923
3	0.325	0.964	0.897	0.980	0.931
4	–	0.899	0.905	0.971	0.911
Mean	0.309	0.957	0.920	**0.976**	0.921

*It should be remarked that this index is "marginal" in the sense that the original and sharpened images are compared band-wise. A more stringent approach would be to test simultaneously for equal mean vectors and covariance matrices; see Anderson (2003).

Since image quality is normally not spatially invariant, it is usual to compute Q in M sliding windows and then average over all of the windows:

$$\mathrm{QI} = \frac{1}{\mathrm{M}} \sum_{\mathrm{j}=1}^{\mathrm{M}} \mathrm{QI_j}.$$

An ENVI/IDL extension for determining the quality index for pan-sharpened images is described in Appendix C. Table 5.1 gives the results for the HSV, PCA, Gram-Schmidt, DWT, and ATWT fusion methods for the image in Figure 5.12. The discrete wavelet transform performs best.

5.4 Radiometric correction of polarimetric SAR imagery

The most striking characteristic of SAR images, when compared to their visual/infrared counterparts, is the disconcerting "speckle" effect which makes visual interpretation very difficult. For the single-look data of Figure 1.2, for example, the effect is extreme. Speckle gives the appearance of random noise, but it is actually a deterministic consequence of the coherent nature of the radar signal. Although disadvantageous in some contexts, coherence effects in the amplitudes of the received radar signal constitute the basis of SAR interferometry and its many applications.

In our later treatment of classification and change detection, we will, however, be concerned primarily with SAR intensity (as opposed to amplitude) images, where speckle is indeed a nuisance. Further processing is necessary before the data can be used for the exacting tasks of thematic mapping or change detection. In this section we first examine the statistical properties of speckle in raw (single-look) polarimetric SAR images. We then consider the effect of averaging, or so-called *multi-looking*, and introduce the covariance matrix form for multi-look polarimetric imagery. Finally, we describe two adaptive filtering procedures to further reduce speckle in the data.

5.4.1 Speckle statistics

Goodman (1984) gives a definitive treatment of the statistical distributions of the noise-like speckle patterns in coherent and partially coherent radiation reflected from rough surfaces. We consider speckle statistics here in the context of the measured radiation amplitude and intensity in a SAR image.

For single polarization transmission and reception, e.g., horizontal-horizontal (hh), the backscattered amplitude signal is, from Equation (1.3),

$$E_h^b = s_{hh} E_h^i. \tag{5.19}$$

The effect of the scattering amplitude s_{hh} is to introduce a change in amplitude and phase of the incident signal E_h^i, which is characteristic of the (bio)physical properties of the reflecting area. The area involved is determined by the ground resolution. For example for TerraSAR-X in so-called *spotlight* acquisition mode, the pixel size may be as small as 1 m^2. The wavelength of the transmitted 9.65-GHz X-band signal is, on the other hand, only 31 mm. Therefore, even when the properties of the reflecting surface are uniform, random irregularities on the wavelength scale within the imaged pixel will act like a very large number of *incremental scatterers*, leading to a random superposition of small scattering amplitudes. These in turn combine coherently to form the backscattered signal and are responsible for speckle. A simple model for this effect is to write Equation (5.19) in the form

$$E_h^b = \frac{E}{\sqrt{n}} \sum_{k=1}^{n} e^{\mathbf{i}\phi_k}. \tag{5.20}$$

Here we assume that the only effect of the incremental scatterers is to introduce local random phase shifts ϕ_k, $k = 1 \dots n$, in an overall scattered signal amplitude E. The phase shifts may be considered to be uniformly distributed on the interval $[-\pi, \pi]$, so that we have

$$|E_h^b|^2 = \frac{|E|^2}{n} \sum_k \sum_\ell e^{\mathbf{i}(\phi_k - \phi_\ell)} = \frac{|E|^2}{n} \sum_k e^{\mathbf{i}(\phi_k - \phi_k)} = |E|^2.$$

If we neglect the phase of E*, which is a characteristic of the entire pixel area, then we can express the real and imaginary parts of Equation (5.20) in the form

$$\begin{aligned}\mathrm{Re}(E_h^b) &= \frac{E}{\sqrt{n}} \sum_k \cos\phi_k = \frac{E}{\sqrt{n}} X \\ \mathrm{Im}(E_h^b) &= \frac{E}{\sqrt{n}} \sum_k \sin\phi_k = \frac{E}{\sqrt{n}} Y,\end{aligned}$$

where $X = \sum_k \cos\phi_k$ and $Y = \sum_k \sin\phi_k$. Now the intensity of the backscattered signal can be written

$$|E_h^b|^2 = [\mathrm{Re}(E_h^b)]^2 + [\mathrm{Im}(E_h^b)]^2 = E^2 \frac{1}{n}(X^2 + Y^2). \tag{5.21}$$

We can think of X and Y as random variables. Since the sines and cosines contributing to X and Y are distributed symmetrically about zero, the means vanish: $\langle X \rangle = \langle Y \rangle = 0$. Furthermore,

$$\begin{aligned}\mathrm{var}(X) &= \langle X^2 \rangle - \langle X \rangle^2 \\ &= \sum_k \sum_\ell \langle \cos\phi_k \cos\phi_\ell \rangle \\ &= \sum_k \langle \cos^2\phi_k \rangle = \sum_k \langle \frac{1}{2}(\cos 2\phi_k + 1) \rangle = \frac{n}{2},\end{aligned}$$

*Or simply set it equal to zero.

with the same result for var(Y). Moreover, X and Y are uncorrelated:

$$\begin{aligned}\mathrm{cov}(X,Y) &= \langle XY\rangle - \langle X\rangle\langle Y\rangle \\ &= \sum_k\sum_\ell \langle \cos\phi_k \sin\phi_\ell\rangle = 0.\end{aligned}$$

Both X and Y are superpositions of a large number of identically distributed random variates ($\cos\phi_k$ and $\sin\phi_k$) and so the above results, together with the Central Limit Theorem 2.3, imply that X and Y are independently and normally distributed with zero mean and variance $\sigma^2 = n/2$. The joint probability density of X and Y is therefore

$$p(x,y) = \frac{1}{2\pi(n/2)}\exp\left(-\frac{1}{2(n/2)}(x^2+y^2)\right).$$

With $z = x + \mathbf{i}y$, this can be written in the form

$$p(z) = \frac{1}{\pi n}\exp\left(-z^*\left(\frac{1}{n}\right)z\right). \tag{5.22}$$

Comparison with Equation (2.46) shows that the complex random variable $Z = X + \mathbf{i}Y$, and hence the backscattered amplitude

$$E_h^b = \frac{E}{\sqrt{n}}(X + \mathbf{i}Y) = \frac{E}{\sqrt{n}}Z,$$

has a complex normal distribution. Furthermore, according to Theorem 2.5, the sum U of the squares of the standardized random variables X and Y,

$$U = \frac{X^2}{n/2} + \frac{Y^2}{n/2} = \frac{2}{n}(X^2+Y^2),$$

is chi-square distributed with 2 degrees of freedom, that is, with Equation (2.37),

$$p(u) = p_{\chi^2;2}(u) = \frac{1}{2}e^{-u/2}.$$

This is the exponential probability density, Equation (2.36), with parameter $\beta = 1/2$. With Equation (5.21), the backscattered signal intensity is then

$$|E_h^b|^2 = E^2U/2. \tag{5.23}$$

A measurement of the radar cross section $|s_{hh}|^2$ derived from the emitted and received polarized signal intensities (see Equation (1.2)) will of course have the same exponential distribution. In other words, if the random variable G represents the measured intensity and $x = |s_{hh}|^2$ is the underlying signal, then

$$G \sim xU/2.$$

A simple application of Theorem 2.1 (Exercise 10(a)) now shows that G has an exponential probability density with $\beta = x$,

$$p(g) = \frac{1}{x} e^{-g/x}.$$

Since $\mathrm{var}(G) = \langle G \rangle = x$, we can alternatively write

$$p(g) = \frac{1}{\langle G \rangle} e^{-g/\langle G \rangle}. \tag{5.24}$$

Thus, speckle behaves as an exponentially distributed "noise" with mean and variance equal to the underlying signal strength.

5.4.2 Multi-look data

The per-pixel polarimetric information in the scattering matrix $\boldsymbol{S}$ of Equation (1.3), under the assumption of reciprocity ($s_{hv} = s_{vh}$), can be expressed as a three-component complex vector

$$\boldsymbol{s} = \begin{pmatrix} s_{hh} \\ \sqrt{2} s_{hv} \\ s_{vv} \end{pmatrix}, \tag{5.25}$$

where the $\sqrt{2}$ ensures that the total intensity (received signal power) is consistent. The total intensity is referred to as the *span*,

$$\mathrm{span} = \boldsymbol{s}^\dagger \boldsymbol{s} = |s_{hh}|^2 + 2|s_{hv}|^2 + |s_{vv}|^2, \tag{5.26}$$

and the corresponding gray-scale image as the *span image.*

Multi-look processing essentially corresponds to the averaging of neighborhood pixels, with the objective of reducing speckle and compressing the data. In practice, the averaging is often not performed in the spatial domain, but rather in the frequency domain during range/azimuth compression of the received signal.* The speckle reduction can be understood as follows. We showed in the preceding subsection that a random variable G representing measurement of a physical quantity like $|s_{hh}|^2$ will have an exponential distribution with parameter $\beta = \langle G \rangle$. If these measurements are summed over m looks to give $\sum_{i=1}^m G_i$, then according to Theorem 2.4 the sums will be gamma distributed with $\beta = \langle G \rangle$ and $\alpha = m$, provided that the G_i are independent. The mean of the gamma distribution is $\alpha\beta = m\langle G \rangle$ and its variance is $\alpha\beta^2 = m\langle G \rangle^2$. Let us represent the look-averaged image by the random variable $\overline{G}$,

$$\overline{G} = \frac{1}{m} \sum_{i=1}^m G_i.$$

*Look averaging takes place at the cost of spatial resolution. See Richards (2009), Appendix D, for an excellent explanation of SAR image formation.

Then

$$\langle \overline{G} \rangle = \frac{1}{m} \cdot m \langle G \rangle = \langle G \rangle$$

and

$$\text{var}(\overline{G}) = \text{var}\left(\frac{1}{m}\sum_{i=1}^{m} G_i\right) = \frac{1}{m^2} \cdot m\langle G \rangle^2 = \frac{\langle G \rangle^2}{m} = \frac{\langle \overline{G} \rangle^2}{m}. \tag{5.27}$$

Hence, we see that the variance of the look-averaged image decreases inversely as the number of looks.

In real SAR imagery, the neighborhood pixel intensities contributing to the look average will be correlated to some extent. This is accounted for by defining an *equivalent number of looks* (ENL) whose definition is motivated by Equation (5.27), that is, by solving it for m:

$$\text{ENL} = m = \frac{\langle \overline{G} \rangle^2}{\text{var}(\overline{G})}. \tag{5.28}$$

The ENL can be estimated in homogeneous regions of an image, where the contribution of speckle to image statistics may be expected to dominate.* It is equivalent to the number of *independent* measurements averaged per pixel. An IDL program for ENL estimation is shown in Listing 5.10. This script also illustrates how to access ROI (region of interest) data programmatically.

Another representation of the polarimetric signal is the *complex covariance matrix* $\boldsymbol{c}$, obtained by taking the complex outer product of $\boldsymbol{s}$ with itself:

$$\boldsymbol{c} = \boldsymbol{s}\boldsymbol{s}^\dagger = \begin{pmatrix} |s_{hh}|^2 & \sqrt{2}\, s_{hh}s_{hv}^* & s_{hh}s_{vv}^* \\ \sqrt{2}\, s_{hv}s_{hh}^* & 2|s_{hv}|^2 & \sqrt{2}\, s_{hv}s_{vv}^* \\ s_{vv}s_{hh}^* & \sqrt{2}\, s_{vv}s_{hv}^* & |s_{vv}|^2 \end{pmatrix}. \tag{5.29}$$

The diagonal elements of $\boldsymbol{c}$ are real numbers, with span $= \text{tr}(\boldsymbol{c})$, and the off-diagonal elements are complex. The covariance matrix representation contains all of the information in the polarized signal. It can also be averaged over the number of looks to give

$$\begin{aligned} \bar{\boldsymbol{c}} &= \frac{1}{m}\sum_{\nu=1}^{m} \boldsymbol{s}(\nu)\boldsymbol{s}(\nu)^\dagger = \langle \boldsymbol{s}\boldsymbol{s}^\dagger \rangle \\ &= \begin{pmatrix} \langle |s_{hh}|^2 \rangle & \sqrt{2}\langle s_{hh}s_{hv}^* \rangle & \langle s_{hh}s_{vv}^* \rangle \\ \sqrt{2}\langle s_{hv}s_{hh}^* \rangle & \langle 2|s_{hv}|^2 \rangle & \sqrt{2}\langle s_{hv}s_{vv}^* \rangle \\ \langle s_{vv}s_{hh}^* \rangle & \sqrt{2}\langle s_{vv}s_{hv}^* \rangle & \langle |s_{vv}|^2 \rangle \end{pmatrix}. \end{aligned} \tag{5.30}$$

*For polarimetric data, one can simply average the values for each band, but see Exercise 11 and also Anfinsen et al. (2009a), who describe multivariate ENL estimators especially tuned to polarimetric SAR imagery. An ENVI/IDL extension and a Python script for one of their methods are described in Appendices C and D.

Listing 5.10: Estimation of the equivalent number of looks.

```
PRO ENL
; simple estimation of equivalent
; number of looks (ENL)
   compile_opt idl2

; select the polSAR image
   envi_select, title='Enter polSAR image band',$
               /band_only, fid=fid, dims=dims, pos=pos
   IF (fid EQ -1) THEN BEGIN
      PRINT,'cancelled'
      RETURN
   ENDIF
   envi_file_query, fid, fname=fname
   Gs = envi_get_data(/complex, dims=dims, $
                        fid=fid, pos=pos)

; get the ROI pixels
   roi_ids = envi_get_roi_ids(fid=fid)
   idx = envi_get_roi(roi_ids[0])

; calculate ENL
   mn = mean(abs(Gs[idx]))
   var = variance(abs(Gs[idx]))
   PRINT, 'Input file: ' + fname
   PRINT, 'ENL = ', real_part(mn^2/var)

END
```

Rewriting the first equation above,

$$m\bar{\boldsymbol{c}} = \sum_{\nu=1}^{m} \boldsymbol{s}(\nu)\boldsymbol{s}(\nu)^\dagger. \tag{5.31}$$

This is seen to be a realization of a complex sample matrix like Equation (2.57). According to Theorem 2.9, it will have a complex Wishart distribution with 3×3 covariance matrix $\boldsymbol{\Sigma}$ and m degrees of freedom, provided that the vectors $\boldsymbol{s}(\nu)$ are independent and drawn from a complex multivariate normal distribution. This is ideally the case, as we saw in the one-dimensional situation, Equation (5.22); see also Oliver and Quegan (2004), Chapter 11. However, as already mentioned, the $\boldsymbol{s}(\nu)$ will generally be correlated. In order to account for this, the complex Wishart distribution is often parameterized with ENL (rather than m) degrees of freedom. The complex Wishart distribution of $m\bar{\boldsymbol{c}}$ will be exploited in Chapters 6 and 9 in the contexts of SAR classification and change detection.

The scattering vector given in Equation (5.25) corresponds to so-called full, or *quad polarimetric* SAR. Satellite-based SAR sensors often operate in reduced polarization modes, emitting only one polarization and receiving two (dual polarization) or one (single polarization). The look-averaged covariance matrices are reduced in dimension correspondingly: for dual polarization and horizontal transmission,

$$\bar{\mathbf{c}} = \begin{pmatrix} \langle|s_{hh}|^2\rangle & \langle s_{hh}s_{hv}^*\rangle \\ \langle s_{hv}s_{hh}^*\rangle & \langle|s_{hv}|^2\rangle \end{pmatrix}, \tag{5.32}$$

and, for single polarization and horizontal transmission, simply the intensity image

$$\langle G\rangle = \langle|s_{hh}|^2\rangle. \tag{5.33}$$

In the former case, the observations are complex Wishart distributed with $N = 2$, in the latter case they are gamma distributed.

5.4.3 Speckle filtering

Let us represent an m look-averaged SAR intensity image by the random variable G (dropping for convenience the overbar) with mean $\langle G\rangle = x$, where x is again the underlying signal. From Equation (5.27), $\text{var}(G) = x^2/m$. Since G is gamma-distributed, its parameters α and β satisfy

$$\alpha\beta = x, \quad \alpha\beta^2 = x^2/m,$$

from which follows $\alpha = m$ and $\beta = x/m$. The density function of G for given value of x is therefore (see Equation (2.33))

$$p(g \mid x) = \frac{1}{(x/m)^m\Gamma(m)} g^{m-1}e^{-gm/x}. \tag{5.34}$$

Let $G = xV$. Then, again applying Theorem 2.1 (Exercise 10(b)), it follows that V has the density

$$p(v) = \frac{m^m}{\Gamma(m)} v^{m-1}e^{-vm}. \tag{5.35}$$

Therefore, in terms of the observed pixel intensities g (realizations of G), we can write

$$g = xv, \tag{5.36}$$

where v is a realization of a gamma-distributed random variable V with density given by Equation (5.35), and having mean 1 and variance $1/m$.

Because of this special *multiplicative noise* nature of speckle, conventional smoothing filters of the kind we met in Chapter 4 are not particularly suitable as an aid to SAR image interpretation. Moreover, since we will be especially concerned with polarimetric applications, a filtering algorithm which not only treats the statistical properties of the pixels correctly, but also preserves the polarization properties is to be preferred. In the following we will explain in some detail two speckle filters which try to meet these requirements, and also provide ENVI/IDL and Python scripts which implement them.

5.4.3.1 Minimum mean square error (MMSE) filter

Lee et al. (1999) proposed an adaptive SAR speckle filter which is performed in two steps:

1. Using the span image only, Equation (5.26), filter weights are computed across the image for local windows of size $n \times n$, where, typically, $n = 7$. The filter used is an adaptive one which takes into account the local speckle statistics. It is derived below.

2. The filter weights are then applied uniformly and separately to the elements of the look-averaged covariance matrix, Equation (5.30). By ensuring that all elements of the covariance matrix are filtered by the same amount, the polarimetric properties are preserved.

The speckle filter in step 1 calculates a filtered pixel intensity estimate $\hat{x}$ of the signal x at the window center according to the linear relation

$$\hat{x} = a\bar{g} + bg, \tag{5.37}$$

where $\bar{g}$ is the mean of the (locally) observed pixel intensities in the window. The local mean $\bar{g}$ serves an initial estimate of the signal in the absence of speckle noise. This estimate is adaptively corrected in Equation (5.37) by the parameters a and b and by the observed pixel intensity g to estimate the actual signal x at the window center. The filter parameters are chosen so as to minimize the mean square error

$$E = \langle(\hat{x} - x)^2\rangle = \langle(a\bar{g} + bg - x)^2\rangle,$$

where $\langle \cdot \rangle$ refers to mean value within the local window. Thus

$$\bar{g} = \langle g \rangle \approx \langle x \rangle = \bar{x}.$$

To obtain the parameters, we set the partial derivatives of E equal to zero:

$$\begin{aligned} \frac{\partial E}{\partial a} &= \langle 2\bar{g}(a\bar{g} + bg - x)\rangle = 0 \\ \frac{\partial E}{\partial b} &= \langle 2g(a\bar{g} + bg - x)\rangle = 0. \end{aligned} \tag{5.38}$$

The first of Equations (5.38) is equivalent to

$$a\bar{g} + b\bar{g} - \bar{g} = 0,$$

which implies

$$a = 1 - b. \tag{5.39}$$

Substituting this into the second of Equations (5.38) to eliminate a, we obtain

$$\langle g(x - \bar{g})\rangle + \langle b(\bar{g} - g)g\rangle = 0. \tag{5.40}$$

FIGURE 5.15
RGB composites (hhhh,hvhv,vvvv) of a quad polarimetric TerraSAR-X image acquired over the city of Mannheim, Germany at a ground resolution of 5 m ($\approx$ 7 looks), left: unfiltered; right: after adaptive filtering with the MMSE filter using a 7×7 window. **(See color insert.)**

From Equation (5.36), the first term on the left-hand side can be written

$$\langle g(x-\bar{g})\rangle = \langle xv(x-\bar{g})\rangle = \langle v\rangle\langle x(x-\bar{g})\rangle,$$

assuming that speckle noise v and signal intensity x are statistically independent. But $\langle v\rangle = 1$ and therefore

$$\langle g(x-\bar{g})\rangle = \langle x(x-\bar{g})\rangle.$$

Since $\langle \bar{g}(x-\bar{g})\rangle = \bar{g}\langle (x-\bar{g})\rangle = 0$, this can be rewritten in the form

$$\begin{aligned}\langle g(x-\bar{g})\rangle &= \langle x(x-\bar{g}) - \bar{g}(x-\bar{g})\rangle = \langle (x-\bar{g})^2\rangle \\ &= \langle (x-\bar{x})^2\rangle = \mathrm{var}(x)\end{aligned}$$

In a similar way we can write the second term in Equation (5.40) as

$$\langle b(\bar{g}-g)g\rangle = -b\langle (g-\bar{g})^2\rangle = -b\ \mathrm{var}(g)$$

and we get

$$b = \frac{\mathrm{var}(x)}{\mathrm{var}(g)}. \tag{5.41}$$

Combining Equations (5.37), (5.39) and (5.41), we obtain the adaptive filter

$$\hat{x} = \bar{g} + \frac{\mathrm{var}(x)}{\mathrm{var}(g)}(g-\bar{g}). \tag{5.42}$$

FIGURE 5.16
As Figure 5.15, except that the right-hand image is after adaptive filtering with the gamma MAP filter using a 7×7 window. **(See color insert.)**

In regions of the image where the signal variance $\mathrm{var}(x)$ is small compared to the overall variance (homogeneous regions), i.e., $\hat{x} \approx \bar{g}$, the pixels intensity is replaced by the window average. On the other hand, when the true signal intensity is varying strongly, $\mathrm{var}(x) \approx \mathrm{var}(g)$, then $\hat{x} \approx g$ and no filtering takes place at all.

While the overall variance within the window $\mathrm{var}(g)$ in Equation (5.42) can be determined at once, the signal variance $\mathrm{var}(x)$ is not directly available. However, we can again exploit the statistical independence of x and v and write, with Equation (5.36),

$$\langle g^2 \rangle = \langle x^2 \rangle \langle v^2 \rangle.$$

Equivalently,

$$\mathrm{var}(g) + \langle g \rangle^2 = (\mathrm{var}(x) + \langle x \rangle^2)(\mathrm{var}(v) + \langle v \rangle^2).$$

Solving for $\mathrm{var}(x)$ and using $\langle x \rangle = \langle g \rangle = \bar{g}$, $\mathrm{var}(v) = 1/m$, and $\langle v \rangle = 1$, we obtain

$$\mathrm{var}(x) = \frac{\mathrm{var}(g) - \frac{1}{m}\bar{g}^2}{1 + \frac{1}{m}}. \tag{5.43}$$

Appendices C and D describe, respectively, an ENVI/IDL extension and a Python script that implement the MMSE adaptive filter for polarimetric SAR imagery. Figure 5.15 shows an example of its application to quad polarimetric, multi-look TerraSAR-X data. The high degree of variability of the filter over the SAR image means that the global ENL which results from the initial multi-look processing will no longer be valid in the de-speckled image (Anfinsen et al., 2009b).

5.4.3.2 Gamma MAP filter

Oliver and Quegan (2004) discuss iterative versions of the MMSE filter and also other de-speckling algorithms that take into account explicit statistical models for the signal x. The simplest of these, referred to as *gamma maximum a posteriori* (gamma MAP) de-speckling, may be derived from Bayes' Theorem (Theorem 2.12). The *a posteriori* conditional probability for x, given intensity measurement g is (see Equation (2.66)),

$$\Pr(x \mid g) = \frac{p(g \mid x)\Pr(x)}{p(g)}, \tag{5.44}$$

where $p(g \mid x)$ is given by Equation (5.34), $\Pr(x)$ is the prior probability for x and $p(g)$ is the total probability density for g. This formulation allows us to include prior knowledge of the signal statistics (or texture) if available. An empirical statistical model for x is suggested by measurements of backscatter from ocean waves (Oliver and Quegan, 2004), namely

$$\Pr(x) \sim \left(\frac{\alpha}{\mu}\right)^{\alpha} \frac{x^{\alpha-1}}{\Gamma(\alpha)} e^{-\alpha x/\mu}. \tag{5.45}$$

This is just the gamma probability density with $\beta = \mu/\alpha$, and hence with mean $\alpha\beta = \mu$ and variance

$$\text{var}(x) = \alpha\beta^2 = \mu^2/\alpha. \tag{5.46}$$

The parameters μ and α can be estimated as follows. By passing an $n \times n$ window over the image we can obtain $\bar{g} = \langle g \rangle$ and $\text{var}(g)$. Then the estimates are

$$\hat{\mu} = \bar{g}, \tag{5.47}$$

and, with Equations (5.43) and (5.46),

$$\hat{\alpha} = \frac{\hat{\mu}^2}{\text{var}(x)} = \frac{\bar{g}^2}{\text{var}(x)} = \frac{1 + 1/m}{\text{var}(g)/\bar{g}^2 - 1/m}. \tag{5.48}$$

Now, combining Equations (5.34), (5.44) and (5.45)*, the posterior probability for x given measurement g is

$$\Pr(x \mid g) \sim \frac{1}{(x/m)^m \Gamma(m)} g^{m-1} e^{-gm/x} \left(\frac{\alpha}{\mu}\right)^{\alpha} \frac{x^{\alpha-1}}{\Gamma(\alpha)} e^{-\alpha x/\mu} =: L \tag{5.49}$$

Taking the logarithm,

$$\begin{aligned}\log L = &\ m \log m - m \log x + (m-1)\log g - \log\Gamma(m) - mg/x \\ &+ \alpha \log \alpha - \alpha \log \mu + (\alpha - 1)\log x - \log\Gamma(\alpha) - \alpha x/\mu.\end{aligned}$$

*And neglecting the total probability $p(g)$, which doesn't depend on x.

We get the maximum *a posteriori* (MAP) value for x given the observed pixel intensity g by maximizing $\log L$ with respect to x:

$$\frac{d}{dx}\log L = -m/x + mg/x^2 + (\alpha - 1)/x - \alpha/\mu = 0.$$

This leads to a quadratic equation for the most probable signal intensity x,

$$\frac{\alpha}{\mu}x^2 + (m + 1 - \alpha)x - mg = 0, \tag{5.50}$$

where the parameters μ and α are estimated locally with Equations (5.47) and (5.48). Note, from Equation (5.48), that in homogeneous regions where $m \approx \bar{g}^2/\mathrm{var}(g)$, $\hat{\alpha} \to \infty$. In that case, from Equation (5.50), $x \approx \hat{\mu} = \bar{g}$, as in the MMSE algorithm.*

The gamma MAP filter is not appropriate to the complex off-diagonal matrix elements in Equation (5.30) as their *a priori* statistics are not well understood. Appendices C and D provide ENVI/IDL and Python routines for gamma MAP filtering of the diagonal terms. Figure 5.16 shows an example. Comparison with Figure 5.15 indicates that the gamma MAP filter is more successful in preserving details than the MMSE filter.

5.5 Topographic correction

Satellite images are two-dimensional representations of the three-dimensional Earth surface. The inclusion of the third dimension — the elevation — is required for terrain modeling and accurate georeferencing.

5.5.1 Rotation, scaling and translation

Transformations of spatial coordinates in three dimensions, which involve only rotations, scaling and translations, can be conveniently represented by a 4×4 transformation matrix $\boldsymbol{A}$,

$$\boldsymbol{v}^* = \boldsymbol{A}\boldsymbol{v}, \tag{5.51}$$

where $\boldsymbol{v}$ is the column vector containing the original coordinates

$$\boldsymbol{v} = (X, Y, Z, 1)^\top$$

and $\boldsymbol{v}^*$ contains the transformed coordinates

$$\boldsymbol{v}^* = (X^*, Y^*, Z^*, 1)^\top.$$

*If the signal intensity x is integrated out of Equation (5.49), one obtains the probability density function $p(g)$ for the measured intensity in the presence of both speckle and texture, namely the so-called K-distribution; see Oliver and Quegan (2004), Chapter 5.

For example, the translation

$$
\begin{aligned}
X^* &= X + X_0 \\
Y^* &= Y + Y_0 \\
Z^* &= Z + Z_0
\end{aligned}
$$

corresponds to the transformation matrix

$$
\boldsymbol{T} = \begin{pmatrix} 1 & 0 & 0 & X_0 \\ 0 & 1 & 0 & Y_0 \\ 0 & 0 & 1 & Z_0 \\ 0 & 0 & 0 & 1 \end{pmatrix},
$$

a uniform scaling by 50% to

$$
\boldsymbol{S} = \begin{pmatrix} 1/2 & 0 & 0 & 0 \\ 0 & 1/2 & 0 & 0 \\ 0 & 0 & 1/2 & 0 \\ 0 & 0 & 0 & 1 \end{pmatrix},
$$

and a simple rotation θ about the Z-axis to

$$
\boldsymbol{R}_\theta = \begin{pmatrix} \cos\theta & -\sin\theta & 0 & 0 \\ \sin\theta & \cos\theta & 0 & 0 \\ 0 & 0 & 1 & 0 \\ 0 & 0 & 0 & 1 \end{pmatrix},
$$

etc. The combined rotation, scaling, translation (RST) transformation is in this case

$$
\boldsymbol{v}^* = (\boldsymbol{R}_\theta \boldsymbol{S} \boldsymbol{T})\boldsymbol{v} = \boldsymbol{A}\boldsymbol{v}.
$$

$\boldsymbol{A}$ is also referred to as a *similarity transformation* and is a special case of the more general *affine transformation*,

$$
\boldsymbol{A} = \begin{pmatrix} a_1 & a_2 & a_3 & X_0 \\ a_4 & a_5 & a_6 & Y_0 \\ a_7 & a_8 & a_9 & Z_0 \\ 0 & 0 & 0 & 1 \end{pmatrix} = \begin{pmatrix} \boldsymbol{a} & \boldsymbol{X} \\ \boldsymbol{0}^\top & 1 \end{pmatrix}, \tag{5.52}
$$

where the 3×3 matrix $\boldsymbol{a}$ is nonsingular. The two-dimensional counterpart of the similarity transformation will be met in Sections 5.6.1 and 5.6.3 when we discuss image–image registration.

5.5.2 Imaging transformations

An imaging (or perspective) transformation projects 3D points onto a plane. It is used to describe the formation of a camera image and, unlike the RST transformation, is nonlinear since it involves division by coordinate values.

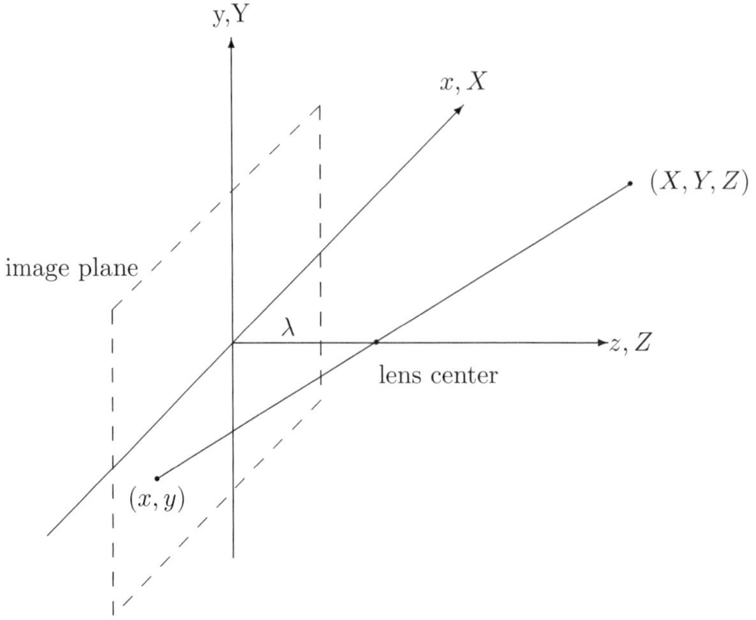

FIGURE 5.17
Basic imaging process. Camera coordinates are (x,y,z) and coincide with world coordinates (X,Y,Z). The focal length is λ.

In Figure 5.17, a camera coordinate system (x, y, x) is shown, aligned with a *world coordinate system* (X, Y, Z) describing the terrain to be imaged. The camera focal length is λ. From simple geometry (similar triangles) we obtain expressions for the image plane coordinates in terms of the world coordinates:

$$\begin{aligned} x &= \frac{\lambda X}{\lambda - Z} \\ y &= \frac{\lambda Y}{\lambda - Z}. \end{aligned} \tag{5.53}$$

Solving for the X and Y world coordinates:

$$\begin{aligned} X &= \frac{x}{\lambda}(\lambda - Z) \\ Y &= \frac{y}{\lambda}(\lambda - Z). \end{aligned} \tag{5.54}$$

Thus, in order to extract the geographical coordinates (X, Y) of a point on the Earth's surface from its image coordinates (x, y), we require knowledge of the elevation Z. Correcting for the elevation in this way constitutes the process of *orthorectification*, resulting in an image without perspective distortion — every point appears as if viewed from directly above.

5.5.3 Camera models and RFM approximations

Equations (5.54) are very simple because they assume that the world and image coordinates coincide. But in order to apply them, one has first to transform the world coordinate system to the satellite coordinate system. This can be done in a straightforward way with the rotation and translation transformations introduced above. However, it requires accurate knowledge of the altitude, orientation and detailed properties of the satellite imaging system at the time of the image acquisition (or, more precisely, *during* the acquisition, since the latter is never instantaneous). The resulting nonlinear equations that relate image and world coordinates, referred to as the *collinearity equations*, constitute the *camera model* for that particular image.

Direct use of a camera model for image processing is complicated as it requires extremely exact, sometimes proprietary information about the sensor system and its orbit. An alternative exists if the image provider also supplies a so-called *rational function model* (RFM) with the imagery. The RFM approximates the camera model for each acquisition in terms of ratios of polynomials, see, e.g., Tao and Hu (2001). The RFMs have the form

$$
\begin{aligned}
r' &= f(X', Y', Z') = \frac{a(X', Y', Z')}{b(X', Y', Z')} \\
c' &= g(X', Y', Z') = \frac{c(X', Y', Z')}{d(X', Y', Z')},
\end{aligned} \tag{5.55}
$$

where c' and r' are the column and row (x,y) coordinates in the image plane relative to an origin (c_0, r_0) and scaled by a factor c_s resp. r_s, i.e.,

$$
c' = \frac{c - c_0}{c_s}, \quad r' = \frac{r - r_0}{r_s}.
$$

Similarly X', Y', and Z' are relative, scaled world coordinates:

$$
X' = \frac{X - X_0}{X_s}, \quad Y' = \frac{Y - Y_0}{Y_s}, \quad Z' = \frac{Z - Z_0}{Z_s}.
$$

The polynomials a, b, c, and d in Equation (5.55) are typically to third order in the world coordinates, for example,

$$
\begin{aligned}
&a(X, Y, Z) = \\
&a_0 + a_1X + a_2Y + a_3Z + a_4XY + a_5XZ + a_6YZ + a_7X^2 + a_8Y^2 + a_9Z^2 \\
&+ a_{10}XYZ + a_{11}X^3 + a_{12}XY^2 + a_{13}XZ^2 + a_{14}X^2Y + a_{15}Y^3 + a_{16}YZ^2 \\
&+ a_{17}X^2Z + a_{18}Y^2Z + a_{19}Z^3.
\end{aligned}
$$

The advantage of using *ratios* of polynomials is that they are less subject to interpolation error than simple polynomials.

For a given acquisition, the provider will fit an RFM to his camera model with a least squares fitting procedure using a two- and three-dimensional grid

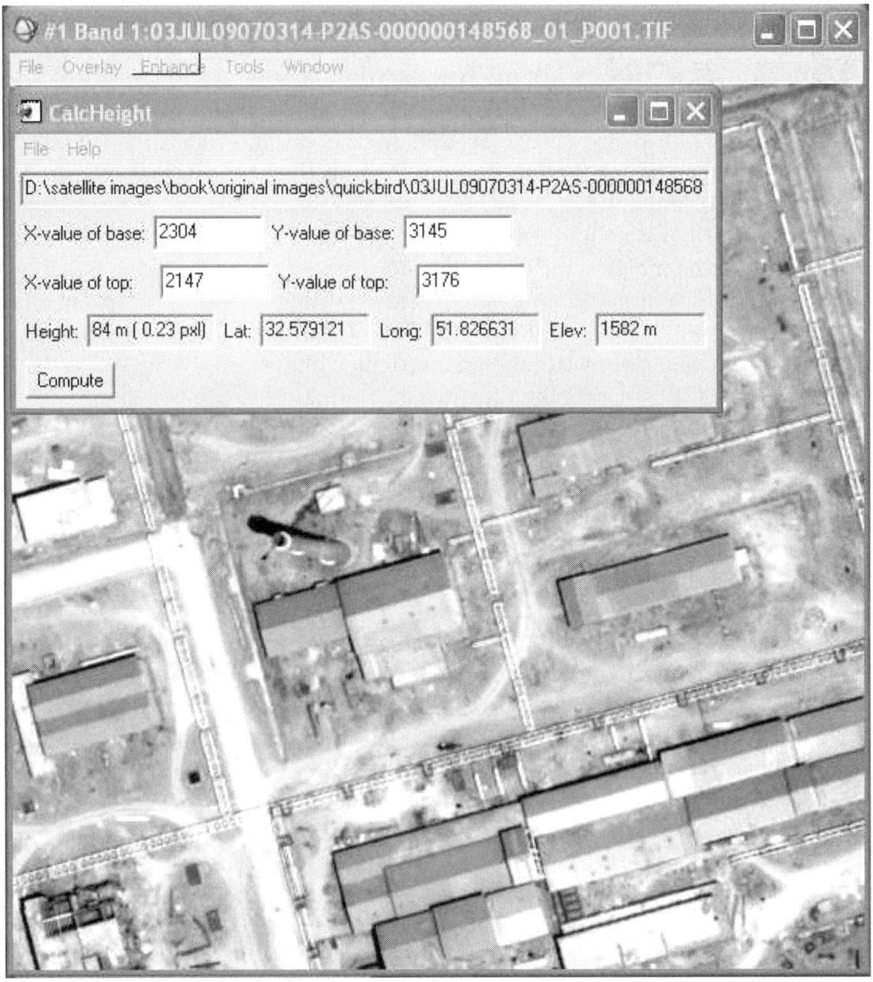

FIGURE 5.18
Determining the height of a vertical structure (the chimney, center left) in a QuickBird panchromatic image with the help of an RFM.

of points covering the image resp. world spaces. The RFM is capable of representing the camera model extremely well and can be used as a replacement for it. Both Space Imaging Corp. and DigitalGlobe Inc., for example, provide RFMs with their so-called "ortho-ready" high-resolution imagery (IKONOS, QuickBird, WorldView-1 platforms).

To illustrate a simple use of RFM data, consider a vertical structure in a high-resolution image, such as a chimney or building facade. Suppose we determine the image coordinates of the bottom and top of the structure to be

(r_b, c_b) and (r_t, c_t), respectively. Then, from Equations (5.55),

$$\begin{aligned} r'_b &= f(X', Y', Z'_b) \\ c'_b &= g(X', Y', Z'_b) \\ r'_t &= f(X', Y', Z'_t) \\ c'_t &= g(X', Y', Z'_t), \end{aligned} \tag{5.56}$$

since the (X', Y') coordinates must be the same. This would appear to constitute a set of four equations in four unknowns X', Y', Z'_b and Z'_t, however the solution is unstable because of the close similarity of Z'_t to Z'_b. Nevertheless, the object height $Z'_t - Z'_b$ can be obtained by the following procedure:

1. Get (r_b, c_b) and (r_t, c_t) from the image and convert to scaled values (r'_b, c'_b) and (r'_t, c'_t).
2. Solve the first two of Equations (5.56), for instance with Newton's method (Press et al., 2002), for X' and Y' with Z'_b set equal to the average elevation in the scene, i.e., Z_0/Z_s, if no digital elevation model (DEM) is available, otherwise to the true, properly scaled elevation.
3. For a range of Z'_t values increasing from Z'_b to some maximum value well exceeding the expected height, calculate (r'_t, c'_t) from the last two of Equations (5.56). Choose for Z'_t the value which gives closest agreement to the (r_t, c_t) values read in.

Quite generally, the RFM can be used as an alternative for providing end users with the necessary information to perform their own photogrammetric processing. An ENVI/IDL extension for object height determination from RFM data is given in Appendix C. It is implemented as a GUI (Graphics User Interface) making use of IDL widgets. A screenshot of its use with a QuickBird image is shown in Figure 5.18.

5.5.4 Stereo imaging and digital elevation models

The missing elevation information Z in Equations (5.53) or (5.54) can be obtained with stereoscopic imaging techniques. Figure 5.19 depicts two cameras viewing the same world point w from two positions. The separation B of the lens centers is the *baseline*. The objective is to find the coordinates of the point w if its image points have coordinates (x_1, y_1) and (x_2, y_2). We assume that the cameras are identical and that their image coordinate systems are perfectly aligned, differing only in the location of their origins. The Z coordinate of w is the same for both coordinate systems.

In the figure, the coordinate system of the first camera is shown as coinciding with the world coordinate system. Therefore, from Equation (5.54),

$$X_1 = \frac{x_1}{\lambda}(\lambda - Z).$$

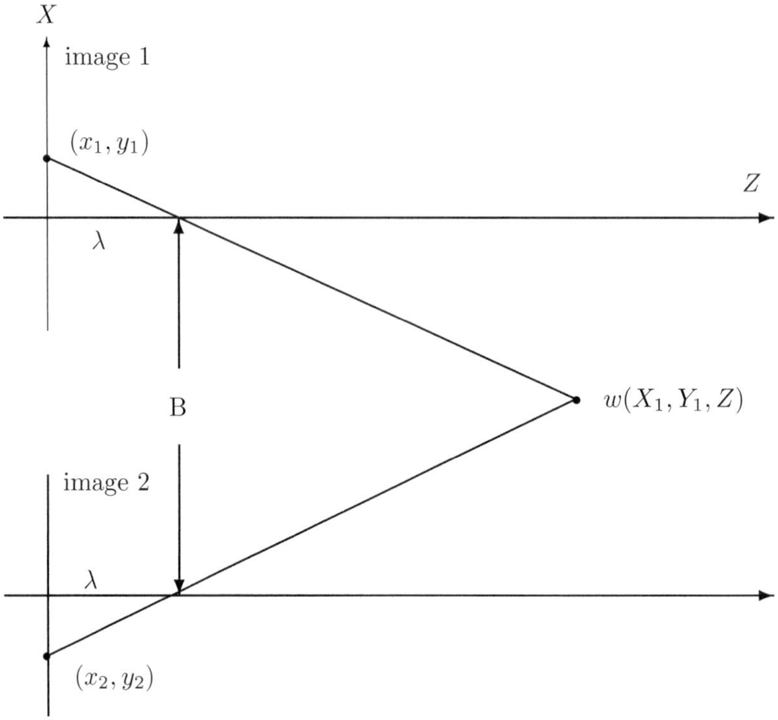

FIGURE 5.19
The stereo imaging process. The coordinates of the point w are relative to a world coordinate system coinciding with the upper camera.

Alternatively, if the second camera is brought to the origin of the world coordinate system,

$$X_2 = \frac{x_2}{\lambda}(\lambda - Z).$$

But, from the figure,

$$X_2 = X_1 + B,$$

where B is the baseline. We have from the above three equations:

$$Z = \lambda - \frac{\lambda B}{x_2 - x_1}. \tag{5.57}$$

Thus if the displacement of the image coordinates of the point w, namely $x_2 - x_1$, can be determined, then the Z coordinate can be calculated. The task is then to find two corresponding points in different images of the same scene. This is usually accomplished by spatial correlation techniques and is

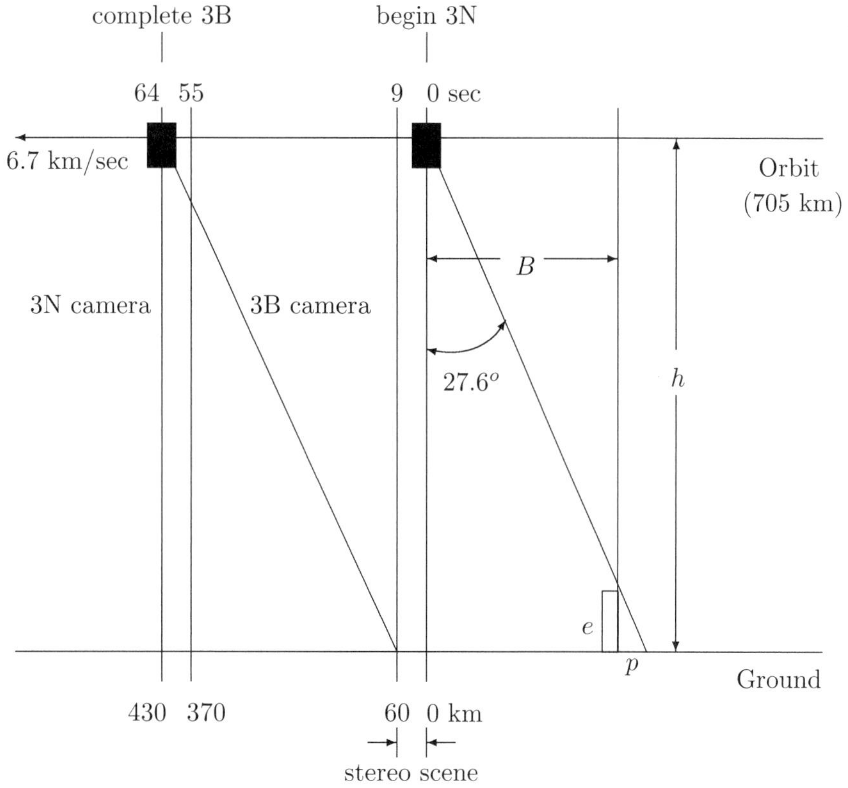

FIGURE 5.20
ASTER along-track stereo acquisition geometry, adapted from Lang and Welch (1999). Parallax p can be related to elevation e by similar triangles: $e/p = (h-e)/B \approx h/B$.

closely related to the problem of image–image registration discussed later in this chapter.

Because the stereo images must be correlated, best results are obtained if they are acquired within a very short time of each other, preferably "along track" if a single platform is used; see Figure 5.20. This figure shows the orientation and imaging geometry of the VNIR 3N and 3B cameras on the ASTER platform for acquiring a stereo full scene. The satellite travels at a speed of 6.7 km/sec at a height of 705 km. A 60×60 km^2 full scene is scanned in 9 seconds. Then, 55 seconds later, the same scene is scanned by the back-looking camera, corresponding to a baseline of $B = 370$ km. The along-track geometry means that the *epipolar lines* (Solem, 2012) of the stereo pair are parallel, i.e., the displacements due to viewing angle are only along a common direction, in this case the y axis, in the imaging planes of the two cameras. Therefore, the spatial correlation algorithm used to match points

FIGURE 5.21
ASTER 3N band (nadir camera) over a hilly region in North Korea.

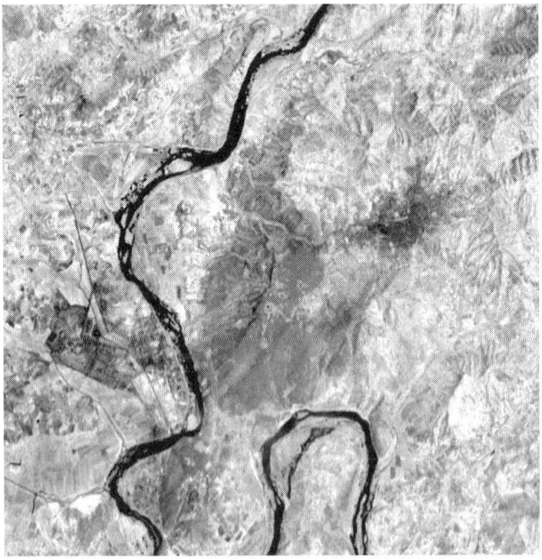

FIGURE 5.22
ASTER 3B band (back-looking camera) registered to Figure 5.21 by a first-order polynomial transformation (see Section 5.6.3).

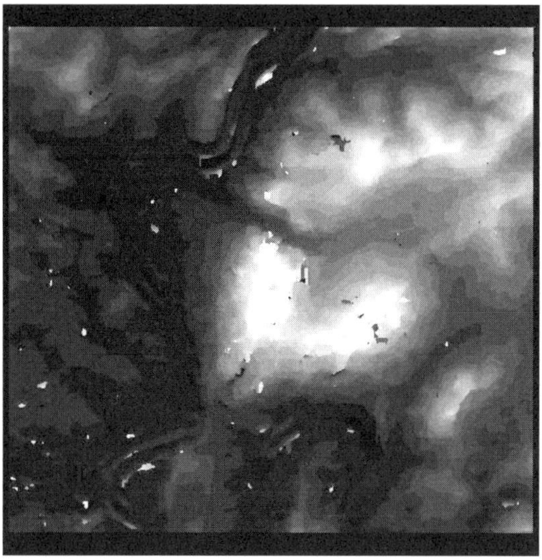

FIGURE 5.23
A rudimentary digital elevation model (DEM) from the ASTER stereo pair of Figures 5.21 and 5.22 using the program in Listing 5.11.

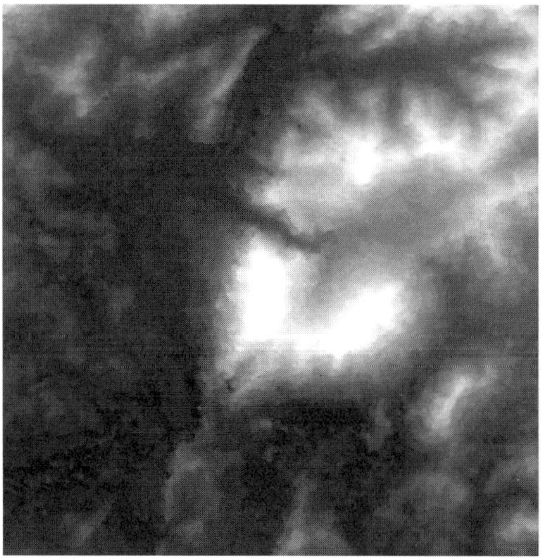

FIGURE 5.24
DEM generated with the commercial product AsterDTM (Sulsoft, 2003).

Listing 5.11: Calculation of a DEM with cross-correlation.

```
PRO ex5_4
COMPILE_OPT STRICTARR

height = 705.0
base = 370.0
pixel_size = 15.0
envi_select, title='Choose␣nadir␣image', $
   fid=fid1, dims=dims1, pos=pos1, /band_only
IF (fid1 EQ -1) THEN RETURN
envi_select, title='Choose␣back-looking␣image', $
   fid=fid2, dims=dims2, pos=pos2, /band_only
IF (fid2 EQ -1) THEN RETURN
im1 = envi_get_data(fid=fid1,dims=dims1,pos=pos1)
im2 = envi_get_data(fid=fid2,dims=dims2,pos=pos2)

n_cols = dims1[2]-dims1[1]+1
n_rows = dims1[4]-dims1[3]+1
parallax = fltarr(n_cols,n_rows)

progressbar = Obj_New('progressbar',Color='blue', $
 Text='0', title='Cross␣correlation,␣column␣...', $
                 xsize=250,ysize=20)
progressbar->start
FOR i=7L,n_cols-8 DO  BEGIN
   IF progressbar->CheckCancel() THEN BEGIN
      envi_enter_data,pixel_size*parallax*(height/base)
      progressbar->Destroy
      RETURN
   ENDIF
   progressbar->Update,(i*100)/n_cols,text=strtrim(i,2)
   FOR j=25L,n_rows-26 DO BEGIN
      cim = correl_images(im1[i-7:i+7,j-7:j+7], $
             im2[i-7:i+7,j-25:j+25],xoffset_b=0, $
             yoffset_b=-25,xshift=0,yshift=25)
      corrmat_analyze,cim,xoff,yoff,maxcorr,edge, $
             plateau,magnification=2
      IF (yoff LT -5) OR (yoff GT 18) THEN $
        parallax[i,j]=parallax[i,j-1] ELSE $
        parallax[i,j]=2*yoff
   ENDFOR
ENDFOR
progressbar->destroy
envi_enter_data, pixel_size*parallax*(height/base)

END
```

can be one-dimensional. If carried out on a pixel-for-pixel basis, one obtains a DEM having approximately the same resolution as that of the stereo imagery.

As an example, Figures 5.21 and 5.22 show an ASTER along-track stereo pair. The IDL program in Listing 5.11 calculates a very rudimentary DEM making use of the routines `CORREL_IMAGES()` and `CORRMAT_ANALYZE` from the IDL Astronomy User's Library* to calculate the local cross-correlations for the two images. For each pixel in the nadir image a 15×15 window is moved along a 15×51 window (called the *epipolar segment*) in the back-looking image centered at the corresponding position (lines 31 to 40), allowing for a maximum disparity of ± 18 pixels. The point of maximum correlation determines the parallax or disparity p. This is related to the elevation e of the pixel by

$$e = p \cdot \frac{h}{B} \times 15\text{m},$$

where h is the height of the sensor and B is the baseline; see Figure 5.20.

Figure 5.23 shows the result. Clearly there are problems due to correlation errors; however, the relative elevations are approximately correct when compared to the DEM determined with the ENVI commercial add-on `AsterDTM` (Sulsoft, 2003); see Figure 5.24. This much more sophisticated approach uses image pyramids to accumulate disparities at increasing scales; see also Quam (1987).

a	b	c
d	e	f
g	h	i

FIGURE 5.25
Pixel elevations in an 8-neighborhood. The letters represent elevations in meters.

In generating a DEM in this way, either the complete camera model or an RFM can be referred to, but usually neither is sufficient for determining absolute elevations. Most often, additional ground reference points within the image with known elevations are also required for absolute calibration. Orthorectification of the image (correction of displacement errors due to parallax and to off-nadir viewing angles) is carried out on the basis of the DEM and consists of referring the (X, Y, Z) coordinates of each pixel to the (X, Y) coordinates of a given map projection.

5.5.5 Slope and aspect

Topographic modeling, or terrain analysis, involves the processing of elevation data provided by a DEM. Specifically, we consider here the generation of *slope images*, which give the steepness of the terrain at each pixel, and *aspect*

*Available at www.astro.washington.edu/deutsch/idl/htmlhelp/slibrary01.html.

images, which give the direction relative to north of a vector normal to the landscape at each pixel.

A 3×3 pixel window can be used to determine both slope and aspect as follows; see Figure 5.25. Define

$$\begin{aligned} \Delta x_1 &= c - a \quad & \Delta y_1 &= a - g \\ \Delta x_2 &= f - d \quad & \Delta y_2 &= b - h \\ \Delta x_3 &= i - g \quad & \Delta y_3 &= c - i \end{aligned}$$

and

$$\begin{aligned} \Delta x &= (\Delta x_1 + \Delta x_2 + \Delta x_3)/3 \\ \Delta y &= (\Delta y_1 + \Delta y_2 + \Delta y_3)/3. \end{aligned}$$

Then the slope angle in radians at the central pixel position is given by (Exercise 12)

$$\theta_p = \tan^{-1}\left(\frac{\sqrt{(\Delta x)^2 + (\Delta y)^2}}{2 \cdot \mathrm{GSD}}\right), \tag{5.58}$$

whereas the aspect in radians measured clockwise from north is

$$\phi_o = \tan^{-1}\left(\frac{\Delta x}{\Delta y}\right). \tag{5.59}$$

Slope/aspect determinations from a DEM are available in the ENVI main menu under `Topographic/Topographic Modeling`; see Figures 5.26 and 5.27.

5.5.6 Illumination correction

Topographic modeling can be used to correct images for the effects of local solar illumination. The local illumination depends not only upon the sun's position (elevation and azimuth) but also upon the slope and aspect of the terrain being illuminated. Figure 5.28 shows the angles involved. The quantity to be determined is the local solar incidence angle γ_i, which determines the local irradiance. From trigonometry we can calculate the relation

$$\cos\gamma_i = \cos\theta_p\cos\theta_z + \sin\theta_p\sin\theta_z\cos(\phi_a - \phi_o). \tag{5.60}$$

An example of a $\cos\gamma_i$ image in hilly terrain is shown in Figure 5.29.

Image classification and change detection algorithms, the subject of the next chapters, will achieve better results if variable image properties extrinsic to the actual surface reflectance are first removed. For a Lambertian surface, the reflected radiance L_H from a horizontal surface toward a sensor (ignoring all atmospheric effects) is given by Equation (1.1), which we write in the simplified form

$$L_H = E \cdot \cos\theta_z \cdot R. \tag{5.61}$$

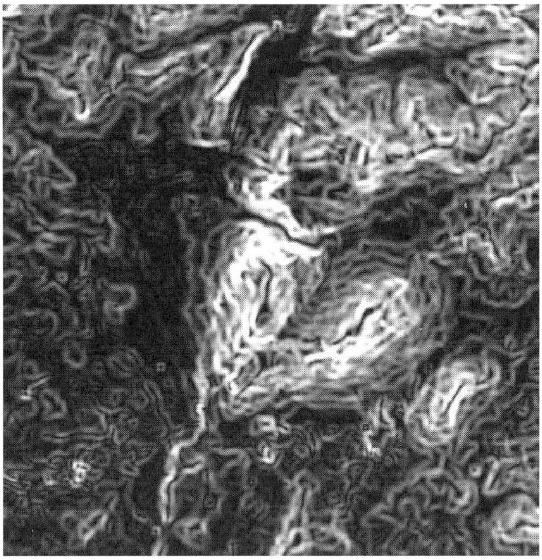

FIGURE 5.26
Slope image calculated with the DEM of Figure 5.24.

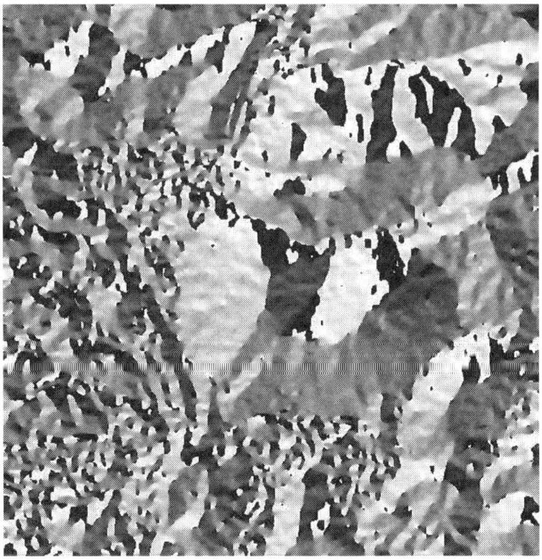

FIGURE 5.27
Aspect image calculated with the DEM of Figure 5.24.

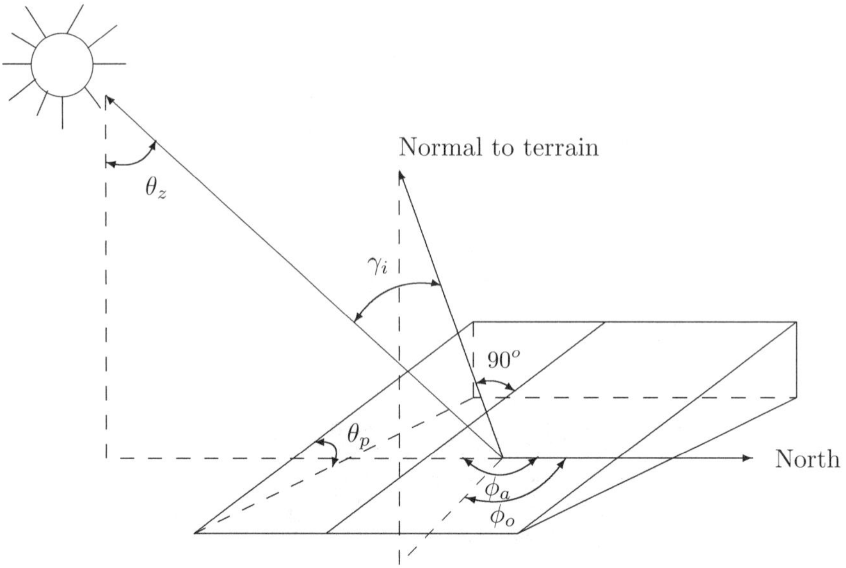

FIGURE 5.28
Angles involved in computation of local solar incidence (adapted from Riano et al. (2003)): θ_z = solar zenith angle, ϕ_a = solar azimuth, θ_p = slope, ϕ_o = aspect, γ_i = local solar incidence angle.

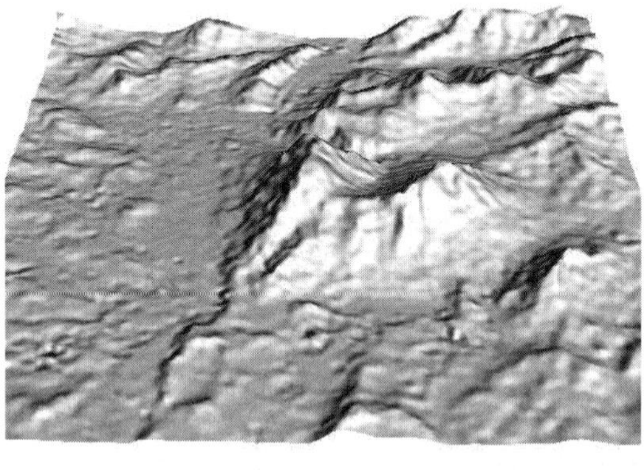

FIGURE 5.29
Cosine of solar incidence angle γ_i stretched across the DEM of Figure 5.24, solar zenith angle $\theta_z = 32.3^o$, solar azimuth $\phi_a = 151.3^o$. The 3D effect was generated from the ENVI main menu with the command `Topographic/3D SurfaceView`, elevation exaggeration factor = 2.

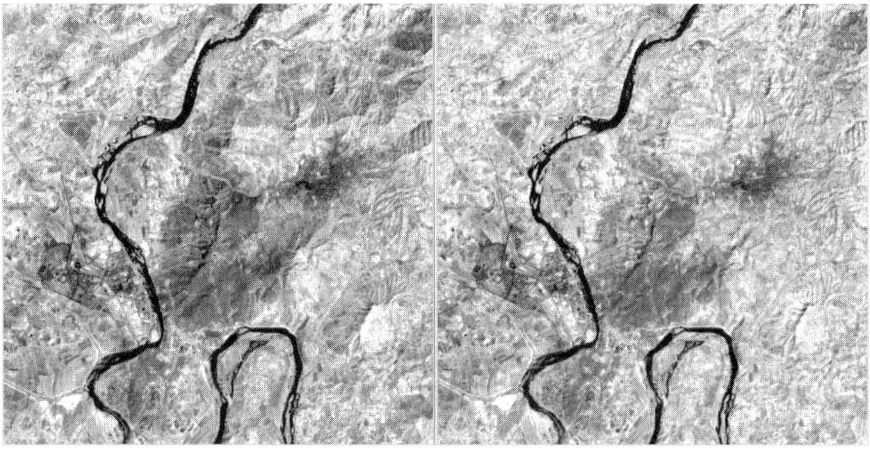

FIGURE 5.30
Original image (left) and after application of the C-correction method (right) for the scene of Figure 5.21. Hillsides sloped away from the sun (see Figure 5.29) are generally brighter after correction.

For a surface in rough terrain the reflected radiance L_T is similarly

$$L_T = E \cdot \cos \gamma_i \cdot R, \tag{5.62}$$

thus giving the standard *cosine correction* relating the observed radiance L_T to that which would have been observed had the terrain been flat, namely

$$L_H = L_T \frac{\cos \theta_z}{\cos \gamma_i}. \tag{5.63}$$

The Lambertian assumption is in general a poor approximation, the actual reflectance being governed by a *bidirectional reflectance distribution function* (BRDF), which describes the dependence of reflectance on both illumination and viewing angles as well as on wavelength (Beisl, 2001). Particularly for the large range of incident angles involved with rough terrain, the cosine correction will over- or underestimate the extremes and lead to unwanted artifacts in the corrected imagery.

An example of an approach which takes better account of BRDF effects is the semiempirical cosine correction (C-correction) method suggested by Teillet et al. (1982). We replace Equations (5.61) and (5.62) by

$$L_H = m \cdot \cos \theta_z + b, \quad L_T = m \cdot \cos \gamma_i + b.$$

The parameters m and b can be estimated from a linear regression of observed radiance L_T vs. $\cos \gamma_i$ for a particular image band.* Then, instead of Equation

*The regression may be carried out separately for different land cover categories in order to take into account the variation of BRDF effects with land cover.

(5.63), one uses

$$L_H = L_T \left(\frac{\cos\theta_z + b/m}{\cos\gamma_i + b/m} \right) \tag{5.64}$$

as a correction formula. An ENVI/IDL extension for illumination correction with the C-correction approximation, including land cover masking, is given in Appendix C. An example of its application is shown in Figure 5.30.

5.6 Image–image registration

Image registration, either to another image or to a map, is a fundamental task in remote sensing data processing. It is required for georeferencing, stereo imaging, accurate change detection, and indeed for any kind of multitemporal image analysis. A tedious task associated with manual coregistration in the past has been the setting of tie-points or, as they are often called, ground control points (GCPs), since in general it was necessary to resort to manual entry. Fortunately, there now exist many reliable automatic or semiautomatic procedures for locating tie-points. They can be divided roughly into four classes (Reddy and Chatterji, 1996):

1. Algorithms that use pixel intensity values directly, such as correlation methods or methods that maximize mutual information
2. Frequency- or wavelet-domain methods that use, e.g., the fast Fourier transform
3. Feature-based methods that use low-level features such as shapes, edges, or corners
4. Algorithms that use high-level features and the relations between them (object-based methods)

In ENVI Version 4.2, tie-point generation using both correlation methods (referred to in the ENVI documentation as "area-based") as well as feature-based methods was introduced. We will consider examples of frequency-domain and feature-based algorithms which can be used to complement or replace ENVI's built-in registration capability and which illustrate some of the principles involved.

5.6.1 Frequency domain registration

Consider two $c \times c$ gray-scale images $g_1(i,j)$ and $g_2(i,j)$, where g_2 is offset relative to g_1 by an integer number of pixels:

$$g_2(i,j) = g_1(i',j') = g_1(i-i_0, j-j_0).$$

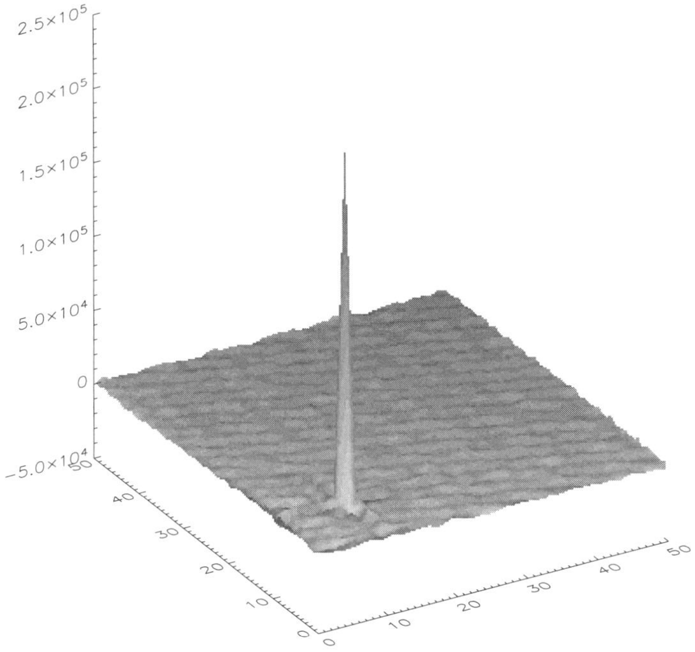

FIGURE 5.31
Phase correlation of two identical images shifted relative to one another by 10 pixels, image generated by the program shown in Listing 5.12.

Taking the Fourier transform we have,

$$\hat{g}_2(k,\ell) = \frac{1}{c^2}\sum_{ij} g_1(i-i_0, j-j_0)e^{-\mathbf{i}2\pi(ik+j\ell)/c},$$

or with a change of indices to $i'j'$,

$$\begin{aligned}\hat{g}_2(k,\ell) &= \frac{1}{c^2}\sum_{i'j'} g_1(i',j')e^{-\mathbf{i}2\pi(i'k+j'\ell)/c}e^{-\mathbf{i}2\pi(i_0k+j_0\ell)/c}\\ &= \hat{g}_1(k,\ell)e^{-\mathbf{i}2\pi(i_0k+j_0\ell)/c}.\end{aligned}$$

This the Fourier translation property that we met in Chapter 3; see Equation (3.11). Therefore, we can write

$$\hat{g}_2(k,\ell)\hat{g}_1^*(k,\ell) = |\hat{g}_1(k,\ell))|^2 e^{-\mathbf{i}2\pi(i_0k+j_0\ell)/c},$$

Listing 5.12: Image matching by phase correlation.

```
PRO ex5_5

; read a JPEG image, cut out two 512x512 pixel arrays
filename = Dialog_Pickfile(Filter='*.jpg',/READ)

IF filename EQ '' THEN RETURN ELSE BEGIN
   Read_JPeG,filename,image
   g1 = image[0,10:521,10:521]
   g2 = image[0,0:511,0:511]

; perform Fourier transforms
   f1 = fft(g1, /double)
   f2 = fft(g2, /double)

; Determine the offset
   g = fft(f2*conj(f1)/abs(f1*conj(f1)),$
        /inverse,/double)
   pos = where(g EQ max(g))
   PRINT,'Offset='+strtrim(pos MOD 512) $
        +strtrim(pos/512)

; output as EPS file
   thisDevice =!D.Name
   set_plot, 'PS'
   Device, Filename='c:\temp\fig5_25.eps', $
          xsize=4,ysize=4,/inches,/Encapsulated
   shade_surf,g[0,0:50,0:50]
   device,/close_file
   set_plot, thisDevice
ENDELSE

END
```

where $\hat{g}_1^*$ is the complex conjugate of $\hat{g}_1$, and hence

$$\frac{\hat{g}_2(k,\ell)\hat{g}_1^*(k,\ell)}{|\hat{g}_2(k,\ell)\hat{g}_1^*(k,\ell)|} = e^{-\mathrm{i}2\pi(i_0k+j_0\ell)/c}. \tag{5.65}$$

The inverse transform of the right-hand side of Equation (5.65) exhibits a delta function (spike) at the coordinates (i_0, j_0). Thus if two otherwise identical (or closely similar) images are offset by an integer number of pixels, the offset can be found by taking their Fourier transforms, computing the ratio on the left-hand side of Equation (5.65) (the so-called *cross-power spectrum*), and then taking the inverse transform of the result. The position of the maximum value in the inverse transform gives the offset values of i_0 and j_0. The IDL program in Listing 5.12 illustrates the procedure; the result is shown in Figure

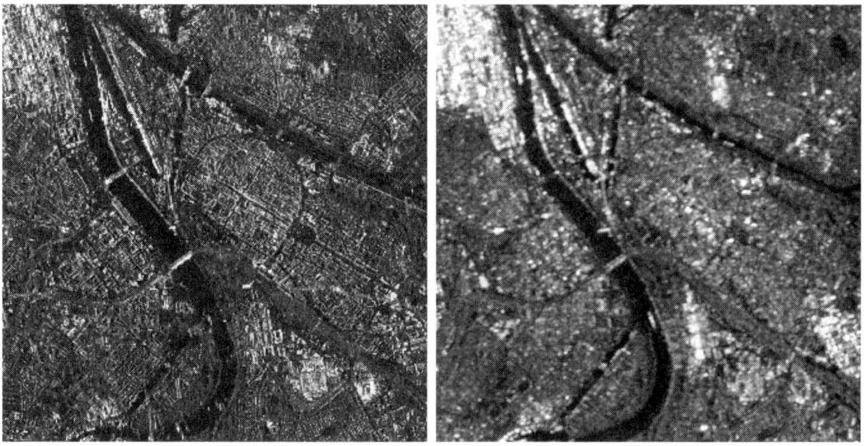

FIGURE 5.32
Frequency domain registration of quad polarimetric SAR imagery, RGB composites (HH,HV,VV), logarithmic intensity values in a linear 2% stretch. Left: TerraSAR-X, right: Radarsat-2. **(See color insert.)**

5.31. Subpixel registration is also possible with a refinement of the above method (Shekarforoush et al., 1995).

Images which differ not only by an offset but also by a rigid rotation and/or change of scale can be registered similarly (Reddy and Chatterji, 1996). Both ENVI/IDL (Xie et al., 2003) and Python scripts are available which calculate RST or similarity transformations in the frequency domain. The Python function `similarity()` included in the `auxil.py` module (Appendix D) estimates the similarity transformation parameters between two gray-scale images. It is a slight modification of code provided by C. Gohlke, see `http://www.lfd.uci.edu/~gohlke/code/imreg.py.html`.

A Python script for registration of polarimetric SAR images which makes use of `similarity()` is shown in Listings 5.13 and 5.14. The logarithms of the spans (see Equation (5.26)) of the reference and warp images are calculated (lines 23–29 in Listing 5.13, lines 3–5 in Listing 5.14) and then passed to the function `similarity()` (line 6 in Listing 5.14) in order to determine the scale, rotation angle, and linear shift parameters for warping. The warp is performed with the aid of the `interpolation` module imported from `scipy.ndimage` and aliased to `ndii`. Since this module does not accept complex arrays, the real and imaginary parts must be warped separately (lines 9–20 in Listing 5.14). An example is shown in Figure 5.32 in which a Radarsat-2 quad polarimetric image is registered to a TerraSAR-x quad polarimetric image. The latter was first re-sampled to the 15 m ground resolution of the Radarsat-2 image.

Listing 5.13: Polarimetric SAR image registration.

```
#!/usr/bin/env python
#Name:  ex5_4.py
IMPORT auxil.auxil as auxil
IMPORT os
FROM numpy IMPORT *
FROM osgeo IMPORT gdal
IMPORT scipy.ndimage.interpolation as ndii
FROM osgeo.gdalconst IMPORT GA_ReadOnly, GDT_CFloat32

DEF main():
    gdal.AllRegister()
    path = auxil.select_directory('Working␣directory')
    IF path:
        os.chdir(path)
    file0=auxil.select_infile(title='Base␣image')
    IF file0:
        inDataset0 = gdal.Open(file0,GA_ReadOnly)
        cols0 = inDataset0.RasterXSize
        rows0 = inDataset0.RasterYSize
        PRINT 'Base␣image:␣%s'%file0
    ELSE:
        RETURN
    rasterBand = inDataset0.GetRasterBand(1)
    span0 = rasterBand.ReadAsArray(0,0,cols0,rows0)
    rasterBand = inDataset0.GetRasterBand(4)
    span0 += 2*rasterBand.ReadAsArray(0,0,cols0,rows0)
    rasterBand = inDataset0.GetRasterBand(6)
    span0 += rasterBand.ReadAsArray(0,0,cols0,rows0)
    span0 = log(real(span0))
    inDataset0 = None
    file1=auxil.select_infile(title='Warp␣image')
    IF file1:
        inDataset1 = gdal.Open(file1,GA_ReadOnly)
        cols1 = inDataset1.RasterXSize
        rows1 = inDataset1.RasterYSize
        PRINT 'Warp␣image:␣%s'%file1
    ELSE:
        RETURN
    outfile,fmt = auxil.select_outfilefmt()
    IF NOT outfile:
        RETURN
    image1 = zeros((6,rows1,cols1),dtype=cfloat)
    FOR k IN RANGE(6):
        band = inDataset1.GetRasterBand(k+1)
        image1[k,:,:]=band\
```

Listing 5.14: Polarimetric SAR image registration (continued).

```
          .ReadAsArray(0,0,cols1,rows1).astype(cfloat)
    inDataset1 = None
    span1 = SUM(image1[[0,3,5] ,:,:],axis=0)\
                                             +image1[3,:,:]
    span1 = log(real(span1))
    scale,angle,shift = auxil.similarity(span0, span1)
    tmp_real = zeros((6,rows0,cols0))
    tmp_imag = zeros((6,rows0,cols0))
    FOR k IN RANGE(6):
        bn1 = real(image1[k,:,:])
        bn2 = ndii.zoom(bn1, 1.0/scale)
        bn2 = ndii.rotate(bn2, angle)
        bn2 = ndii.shift(bn2, shift)
        tmp_real[k,:,:] = bn2[0:rows0,0:cols0]
        bn1 = imag(image1[k,:,:])
        bn2 = ndii.zoom(bn1, 1.0/scale)
        bn2 = ndii.rotate(bn2, angle)
        bn2 = ndii.shift(bn2, shift)
        tmp_imag[k,:,:] = bn2[0:rows0,0:cols0]
    image2 = tmp_real + 1j*tmp_imag
    driver = gdal.GetDriverByName(fmt)
    outDataset = driver.Create(outfile,
                    cols0,rows0,6,GDT_CFloat32)
    FOR k IN RANGE(6):
        outBand = outDataset.GetRasterBand(k+1)
        outBand.WriteArray(image2[k,:,:],0,0)
        outBand.FlushCache()
    outDataset = None
    PRINT 'Warped␣image␣written␣to:␣%s'%outfile

IF __name__ == '__main__':
    main()
```

5.6.2 Feature matching

Various techniques for automatic determination of tie-points based on low-level features have been suggested in the literature. We next discuss and implement one such method, namely a contour matching procedure proposed by Li et al. (1995). It functions especially well in bitemporal scenes in which vegetation changes do not dominate, and can of course be augmented by other automatic feature matching methods or by manual selection. The required steps are shown in Figure 5.33 and are described below.

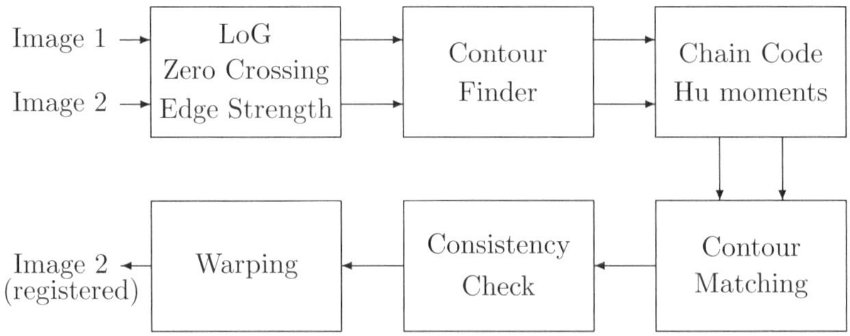

FIGURE 5.33
Image–image registration with contour matching.

5.6.2.1 High-pass filtering

The first step involves the application of a Laplacian-of-Gaussian filter to both images in the manner discussed in Section 5.2.2. After determining the contours by examining zero-crossings of the LoG-filtered image, the contour strengths are encoded in the pixel intensities. Strengths are taken to be proportional to the magnitude of the gradient at the zero-crossing determined by a Sobel filter as illustrated in the program shown in Listing 5.3.

5.6.2.2 Closed contours

In the next step, all closed contours with strengths above some given threshold are determined by tracing the contours. Pixels which have been visited during tracing are set to zero so that they will not be visited again. A typical result is shown in Figure 5.34.

5.6.2.3 Chain codes and moments

For subsequent matching purposes, all significant closed contours found in the preceding step are *chain encoded.* Any curve or contour can be represented by an integer sequence $\{a_1, a_2 \dots a_i \dots\}$, $a_i \in \{0, 1, 2, 3, 4, 5, 6, 7\}$, depending on the relative position of the current pixel with respect to the previous pixel in the curve. A shift to the east is coded as 0, to the northeast as 1 and so on. This simple code has the drawback that some contours produce wraparound. For example, the line in the direction -22.5^o has the chain code $\{707070\dots\}$. Li et al. (1995) suggest the smoothing operation:

$$\{a_1 a_2 \dots a_n\} \rightarrow \{b_1 b_2 \dots b_n\},$$

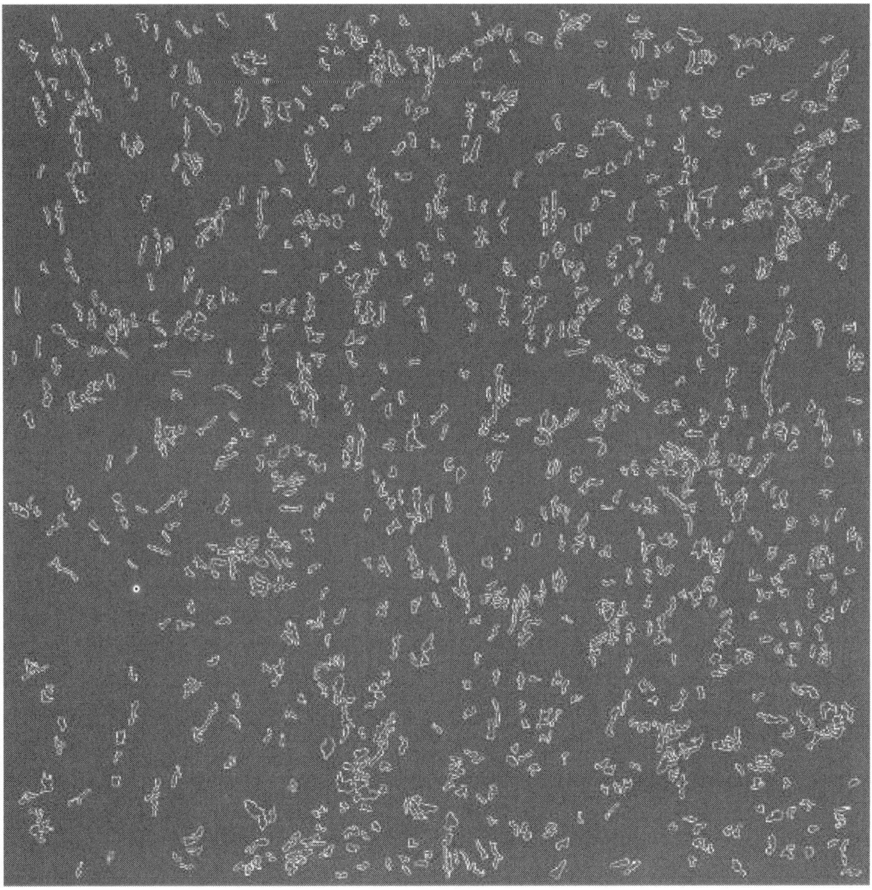

FIGURE 5.34
Closed contours derived from the 3N band of an ASTER image over Nevada acquired in July 2003.

where $b_1 = a_1$ and $b_i = q_i$. The integer q_i satisfies $(q_i - a_i) \bmod 8 = 0$ and $|q_i - b_{i-1}| \to \min$, $i = 2, 3 \ldots n$.* They also suggest applying the smoothing filter $\{0.1, 0.2, 0.4, 0.2, 0.1\}$ to the result. After both processing steps, two chain codes can be easily compared by "sliding" one over the other and determining their maximum correlation. The closed contours are further characterized by determining their first four Hu moments $h_1 \ldots h_4$, Equations (5.14).

*This is rather cryptic, so here is an example: For the wraparound sequence $\{707070\ldots\}$, we have $a_1 = b_1 = 7$ and $a_2 = 0$. Therefore, we must choose $q_2 = 8$, since this satisfies $(q_2 - a_2) \bmod 8 = 0$ and $|q_2 - b_1| = 1$. (For the alternatives $q_2 = 0, 16, 24 \ldots$ the difference $|q_2 - b_1|$ is larger.) Continuing the same argument leads to the new sequence $\{787878\ldots\}$ with no wraparound.

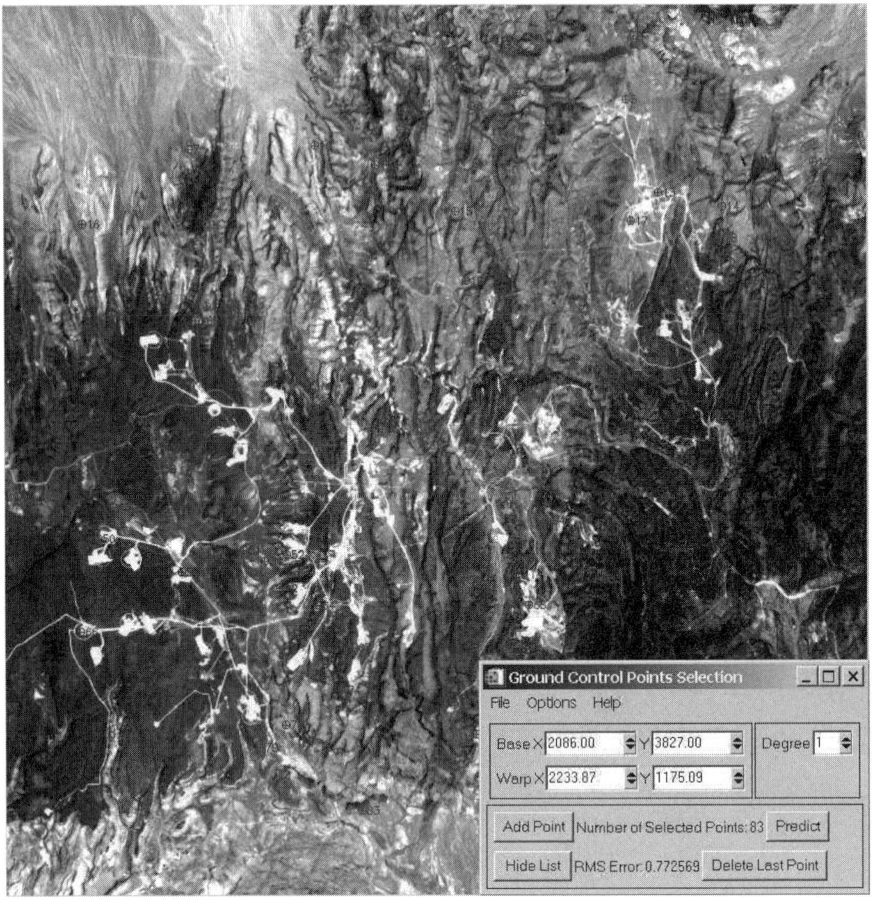

FIGURE 5.35
85 tie-points obtained by matching the contours of Figure 5.34 with those obtained from a similar image acquired in June 2001. The RMS error is 0.77 pixel for first-order polynomial warping; see text. **(See color insert.)**

5.6.2.4 Contour matching

Each significant contour in one image is first matched with contours in the second image according to their invariant moments. This is done by setting a threshold on the allowed differences, for instance, one standard deviation. If one or more matches are found, the best candidate for a tie-point is then chosen to be that matched contour in the second image for which the chain code correlation with the contour in the first image is maximum. If the maximum correlation is less than some threshold, e.g., 0.9, then the match is rejected. The tie-point coordinates are taken to be the centers of gravity $(\bar{x}_1, \bar{x}_2)$ of the matched contour pairs; see Equations (5.11).

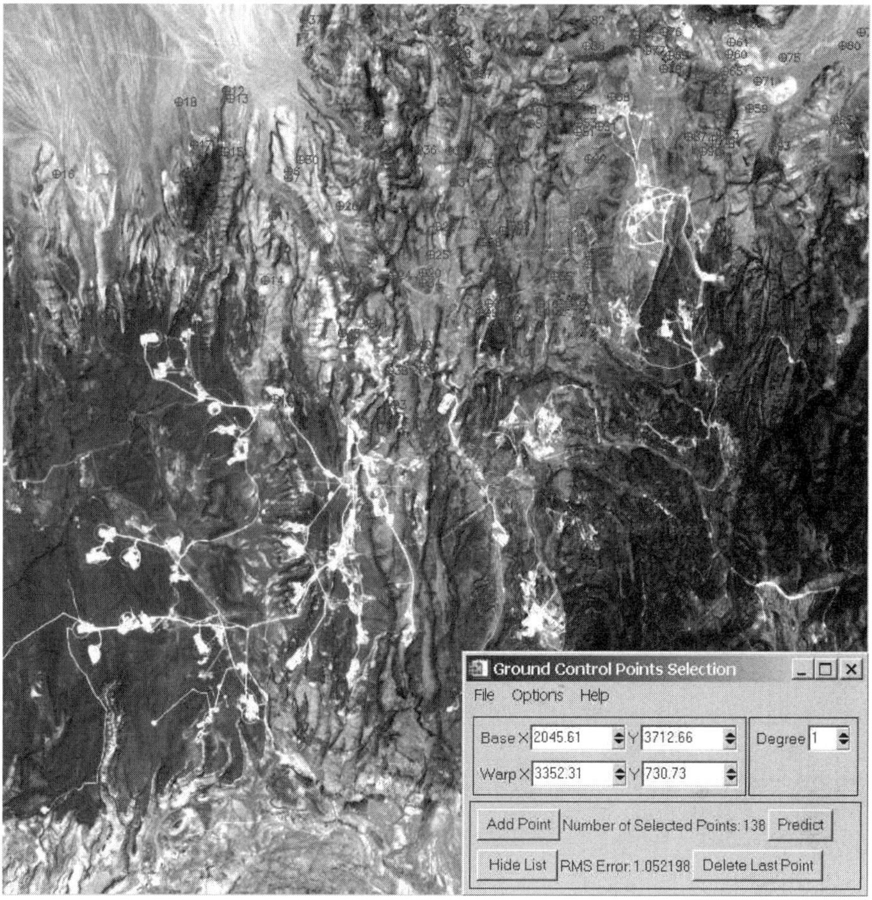

FIGURE 5.36
138 tie-points obtained with ENVI's feature-based matching procedure applied to the July 2003 and June 2001 ASTER scenes. The RMS error is 1.05 pixel for first-order polynomial warping. **(See color insert.)**

5.6.2.5 Consistency check

The contour matching procedure invariably generates some false tie-points, so a further processing step is required. In Li et al. (1995), use is made of the fact that distances are preserved under a rigid transformation. Let $\overline{A_1A_2}$ represent the distance between two points A_1 and A_2 in an image. For two sets of m matched contour centers $\{A_i \mid i = 1 \ldots m\}$ and $\{B_i \mid i = 1 \ldots m\}$ in image 1 and 2, the ratios

$$\overline{A_iA_j}/\overline{B_iB_j}, \quad i = 1 \ldots m,\ j = i + 1 \ldots m,$$

are calculated. These should form a cluster, so that indices associated with ratios scattered away from the cluster center can be rejected as false matches.

5.6.2.6 Implementation in IDL

An ENVI/IDL extension for tie-point determination using the above techniques is described in Appendix C. In this implementation, use is in fact not made of the scale and rotational invariance of the Hu moments because the chain codes as defined above are only translationally invariant.* The program's intended application is for accurate subpixel image–image registration of multitemporal scenes that are acquired with the same sensor (or that have been aligned and resampled to the same GSD), for instance, as a preliminary to change detection. Figures 5.34 and 5.35 illustrate the program's application to bitemporal ASTER data. For comparison, Figure 5.36 shows the tie-points generated by ENVI's automatic feature-based matching algorithm. Although ENVI finds considerably more reference points, they are less uniformly distributed than the matched contours.

5.6.3 Re-sampling with ground control points

Having determined a valid set of tie-points, transformation parameters which map the target image to the base image may be estimated. The ENVI environment provides several alternatives, including affine transformations and polynomial functions.

5.6.3.1 Similarity warping

If uniform scaling s, rotation θ and shift (x_0, y_0) of the target image are sufficient for registering it to the base image, then the two-dimensional equivalent of the RST transformation discussed in Section 5.5.1 can be applied:

$$\boldsymbol{A} = \begin{pmatrix} s\cos\theta & -s\sin\theta & x_0 \\ s\sin\theta & s\cos\theta & y_0 \\ 0 & 0 & 1 \end{pmatrix} = \begin{pmatrix} a & -b & x_0 \\ b & a & y_0 \\ 0 & 0 & 1 \end{pmatrix},$$

where $a^2 + b^2 = s^2(\cos^2\theta + \sin^2\theta) = s^2$. The base image points (u, v) and target image points (x, y) are related by

$$\begin{pmatrix} u \\ v \\ 1 \end{pmatrix} = \begin{pmatrix} a & -b & x_0 \\ b & a & y_0 \\ 0 & 0 & 1 \end{pmatrix} \begin{pmatrix} x \\ y \\ 1 \end{pmatrix}.$$

Equivalently,

$$\begin{pmatrix} u \\ v \end{pmatrix} = \begin{pmatrix} a & -b \\ b & a \end{pmatrix} \begin{pmatrix} x \\ y \end{pmatrix} + \begin{pmatrix} x_0 \\ y_0 \end{pmatrix},$$

*Rotational invariance can be achieved, for instance, by differencing the chain codes, scale invariance by resampling them; see Gonzalez and Woods (2002), Chapter 11.

which can easily be rewritten as

$$\begin{pmatrix} u \\ v \end{pmatrix} = \begin{pmatrix} x & -y & 1 & 0 \\ y & x & 0 & 1 \end{pmatrix} \begin{pmatrix} a \\ b \\ x_0 \\ y_0 \end{pmatrix}.$$

Thus, for n tie-points, we obtain a multiple linear regression problem of the form given in Section 2.6.3, Equation (2.91), namely,

$$\begin{pmatrix} u_1 \\ v_1 \\ u_2 \\ v_2 \\ \vdots \\ u_n \\ v_n \end{pmatrix} = \begin{pmatrix} x_1 & -y_1 & 1 & 0 \\ y_1 & x_1 & 0 & 1 \\ x_2 & -y_2 & 1 & 0 \\ y_2 & x_2 & 0 & 1 \\ \vdots & \vdots & \vdots & \vdots \\ x_n & -y_n & 1 & 0 \\ y_n & x_n & 0 & 1 \end{pmatrix} \begin{pmatrix} a \\ b \\ x_0 \\ y_0 \end{pmatrix},$$

from which the similarity transformation relating the target to the base image may be obtained. To illustrate the procedure, the Python script shown in Listings 5.15 and 5.16 reads a base and target image together with a tie-point file in ENVI format, and outputs the warped (registered) image clipped to the dimensions of the base image.

The transformation parameters are obtained in line 19, Listing 5.16, with the `linalg.lstsqr()` function exported from Numpy. The rotation part of the transformation is carried out in line 23 with the `ndimage.affine_transform()` function* provided by Scipy.

5.6.3.2 Polynomial warping

If the similarity transformation is not sufficient, then a polynomial map may be used: For instance, a second-order polynomial transformation of the target to the base image is given by

$$\begin{aligned} u &= a_0 + a_1 x + a_2 y + a_3 xy + a_4 x^2 + a_5 y^2 \\ v &= b_0 + b_1 x + b_2 y + b_3 xy + b_4 x^2 + b_5 y^2. \end{aligned}$$

Since there are 12 unknown coefficients, at least six tie-point pairs are needed to determine the map (each pair generates two equations). If more than six pairs are available, the coefficients can again be found by least squares fitting. Similar considerations apply for lower- or higher-order polynomial maps.

Having determined the transformation coefficients, the target image can be registered to the base by re-sampling. *Nearest neighbor re-sampling* simply

*At the time of writing, there is an issue in this function which complicates its use for performing the rotation and translation simultaneously.

Listing 5.15: Image–image registration with the similarity transform.

```
#!/usr/bin/env python
#Name:  ex5_5.py
IMPORT auxil.auxil as auxil
IMPORT os
FROM numpy IMPORT *
FROM osgeo IMPORT gdal
FROM scipy IMPORT ndimage
FROM osgeo.gdalconst IMPORT GA_ReadOnly, GDT_Byte

DEF parse_gcp(gcpfile):
    with OPEN(gcpfile) as f:
        pts = []
        FOR i IN RANGE(6):
            line =f.readline()
        WHILE line:
            pts.append(MAP(EVAL,line.split()))
            line = f.readline()
        f.close()
    pts = array(pts)
    RETURN (pts[:,:2],pts[:,2:])

DEF main():
    gdal.AllRegister()
    path = auxil.select_directory('Working␣directory')
    IF path:
        os.chdir(path)
    file1=auxil.select_infile(title='Base␣image')
    IF file1:
        inDataset1 = gdal.Open(file1,GA_ReadOnly)
        cols1 = inDataset1.RasterXSize
        rows1 = inDataset1.RasterYSize
        PRINT 'Base␣image:␣%s'%file1
    ELSE:
        RETURN
    file2=auxil.select_infile(title='Warp␣image')
    IF file2:
        inDataset2 = gdal.Open(file2,GA_ReadOnly)
        cols2 = inDataset2.RasterXSize
        rows2 = inDataset2.RasterYSize
        bands2 = inDataset2.RasterCount
        PRINT 'Warp␣image:␣%s'%file2
    ELSE:
        RETURN
    file3 = auxil.select_infile(title='GCP␣file',\
                                      filt='pts')
```

Listing 5.16: Image–image registration with the similarity transform (continued).

```
    IF file3:
        pts1,pts2 = parse_gcp(file3)
    ELSE:
        RETURN
    outfile,fmt = auxil.select_outfilefmt()
    IF NOT outfile:
        RETURN
    image2 = zeros((bands2,rows2,cols2))
    FOR k IN RANGE(bands2):
        band = inDataset2.GetRasterBand(k+1)
        image2[k,:,:]=band.ReadAsArray(0,0,cols2,rows2)
    inDataset2 = None
    n = LEN(pts1)
    y = pts1.ravel()
    A = zeros((2*n,4))
    FOR i IN RANGE(n):
        A[2*i,:] =   [pts2[i,0],-pts2[i,1],1,0]
        A[2*i+1,:] = [pts2[i,1], pts2[i,0],0,1]
    a,b,x0,y0 = linalg.lstsq(A,y)[0]
    R = array([[a,-b],[b,a]])
    warped = zeros((bands2,rows1,cols1),dtype=uint8)
    FOR k IN RANGE(bands2):
        tmp = ndimage.affine_transform(image2[k,:,:],R)
        warped[k,:,:]=tmp[-y0:-y0+rows1,-x0:-x0+cols1]
    driver = gdal.GetDriverByName(fmt)
    outDataset = driver.Create(outfile,
                    cols1,rows1,bands2,GDT_Byte)
    geotransform = inDataset1.GetGeoTransform()
    projection = inDataset1.GetProjection()
    IF geotransform IS NOT None:
        outDataset.SetGeoTransform(geotransform)
    IF projection IS NOT None:
        outDataset.SetProjection(projection)
    FOR k IN RANGE(bands2):
        outBand = outDataset.GetRasterBand(k+1)
        outBand.WriteArray(warped[k,:,:],0,0)
        outBand.FlushCache()
    outDataset = None
    inDataset1 = None
    PRINT 'Warped image written to: %s'%outfile

IF __name__ == '__main__':
    main()
```

chooses the pixel in the target image that has its transformed center nearest coordinates (i, j) in the warped image and transfers it to that location. This is often the preferred technique for classification or change detection, since the registered image consists of the original pixel intensities, simply rearranged in position to give a correct image geometry. Other commonly used re-sampling methods are *bilinear interpolation* and *cubic convolution interpolation*, see, e.g., Jensen (2005) for a good explanation. These methods interpolate, and therefore mix, the spectral intensities of neighboring pixels.

5.7 Exercises

1. Design a lookup table for byte-encoded data to perform 2% linear saturation (2% of the dark and bright pixels saturate to 0 and 255, respectively).

2. The *decorrelation stretch* generates a more color-intensive RGB composite image of highly correlated spectral bands than is obtained by simple linear stretching of the individual bands (Richards and Jia, 2006). Using ENVI batch procedures, write a routine to implement it:

   (a) Do a principal components transformation of three selected image bands.

   (b) Then do a linear histogram stretch of the principal components.

   (c) Finally, invert the transformation and place the result in ENVI's available bands list.

   (*Note:* This is merely an exercise in batch programming, as ENVI provides the decorrelation stretch transformation in its main menu.)

3. The *Roberts operator* or *Roberts filter* is available in the ENVI menu system and approximates intensity gradients in the diagonal directions:

$$
\begin{aligned}
\nabla_1(i,j) &= [g(i,j) - g(i+1,j+1)] \\
\nabla_2(i,j) &= [g(i+1,j) - g(i,j+1)]
\end{aligned}
$$

   Modify the program in Listing 5.2 to calculate its power spectrum, and write a Python equivalent.

4. An edge detector due to Smith and Brady (1997) called SUSAN (*Smallest Univalue Segment Assimilating Nucleus*) employs a circular mask, typically approximated by 37 pixels, i.e.,

```
  000
 00000
0000000
0000000
0000000
 00000
  000
```

Let r be any pixel under the mask, $g(r)$ its intensity and let r_0 be the central pixel. Define the function

$$c(r, r_0) = \begin{cases} 1 & \text{if } |g(r) - g(r_0)| \leq t \\ 0 & \text{if } |g(r) - g(r_0)| > t, \end{cases}$$

where t is a threshold. Associate with r_0 the sum

$$n(r_0) = \sum_r c(r, r_0).$$

If the mask covers a region of sufficiently low contrast, $n(r_0) = n_{max} = 37$. As the mask moves toward an intensity "edge" having any orientation in an image, the quantity $n(r_0)$ will decrease, reaching a minimum as the center crosses the edge. Accordingly, an edge strength can be defined as

$$E(r_0) = \begin{cases} g - n(r_0) & \text{if } n(r_0) < h \\ 0 & \text{otherwise.} \end{cases}$$

The parameter h is chosen (from experience) as $0.75 * n_{max}$.

(a) A convenient way to calculate $c(r, r_0)$ is to use the continuous approximation

$$c(r, r_0) = e^{-(g(r) - g(r_0))/t)^6}. \tag{5.66}$$

Write an IDL or Python procedure to plot this function for $g(r_0) = 127$ and for $g(r) = 0 \ldots 255$.

(b) Write an IDL or Python program to implement the SUSAN edge detector for arbitrary gray-scale images. *Hint:* Create a lookup table to evaluate the expression (5.66):

```
LUT  = fltarr(256)
FOR i=0,255 do LUT[i] = exp(-(i/t)^6)
```

5. One can approximate the centralized moments of a feature, Equation (5.12), by the integral

$$\mu_{pq} = \int\int (x - x_x)^p (y - y_c)^q f(x, y) dx dy,$$

where the integration is over the whole image and where $f(x, y) = 1$ if the point (x, y) lies on the feature and $f(x, y) = 0$ otherwise. Use this approximation to prove that the normalized centralized moments η_{pq} given in Equation (5.13) are invariant under scaling transformations of the form

$$\begin{pmatrix} x'_1 \\ x'_2 \end{pmatrix} = \begin{pmatrix} \alpha & 0 \\ 0 & \alpha \end{pmatrix} \begin{pmatrix} x_1 \\ x_2 \end{pmatrix}.$$

6. Wavelet noise reduction (Gonzalez and Woods, 2002).

(a) Apply the discrete wavelet transformation to reduce the noise in a multispectral image by modifying the example program in Listing 4.4 (or its Python equivalent) to perform the following steps:

- Select a multispectral image and determine the number of columns, rows, and spectral bands
- Create a band sequential (BSQ) array of the same dimensions for output
- For each band:
  - read the band into a new `DWT` (or `DWTarray`) object instance
  - filter once
  - for each of the three quadrants containing the detail wavelet coefficients:
    * extract the coefficients with the method `GET_QUADRANT()` (or `get_quadrant()`)
    * determine their mean and standard deviation
    * zero all coefficients with absolute value relative to the mean smaller than three standard deviations
    * inject them back into the transformed image with the class method `INJECT()` (or `put_quadrant()`)
  - expand back
  - store the modified band in the output array
  - destroy the object instance
- Return the output array to ENVI (or save to disk).

Note: The coefficients are extracted as one-dimensional arrays. When injecting them back, they must be reformed to two-dimensional arrays. The correct dimensions for doing this are returned from the `DWT` object with the methods `GET_NUM_COLS()` and `GET_NUM_ROWS()` after the compression. (See also the IDL function `WV_DENOISE` for similar functionality.) In the Python class `DWTArray`, use `<instance>.samples` and `<instance>.lines`.

(b) Test your program with a noisy 3-band image, for example, the last three components of the MNF transformation of a LANDSAT 7 TM+ image. Use the example program in Listing 3.8 or 3.9 to determine the noise covariance matrix before and after carrying through the above procedure.

7. Show that the means and standard deviations of the renormalized panchromatic wavelet coefficients C_k^z in Equation (5.17) are equal to those of the multispectral bands.

8. Write an ENVI/IDL or Python script to perform additive *à trous* fusion (see Núnez et al. (1999)).

9. Use the Wang-Bovik quality index routine (Appendix C) to assess the spectral fidelity of a pan-sharpened image using the complete range of methods offered by ENVI and by the extensions described in the present chapter: HSV, Brovey, Gram-Schmidt, PCA, DWT, and ATWT.

10. Show, with the help of Theorem 2.1, that

    (a) if the random variable U has density $(1/2)e^{-u/2}$, then $G = xU/2$ has density $e^{-g/x}/x$, and

    (b) if the random variable G has density

$$\frac{1}{(x/m)^m\Gamma(m)}g^{m-1}e^{-gm/x}$$

    and if $G = xV$, then V has density

$$\frac{m^m}{\Gamma(m)}v^{m-1}e^{-vm}.$$

11. Anfinsen et al. (2009b) suggest the following estimator, among others, for the ENL of a multi-look polarimetric SAR image which takes into account the full sample covariance matrix:

$$\mathrm{ENL} = \frac{\mathrm{tr}(\langle\bar{\boldsymbol{c}}\rangle\langle\bar{\boldsymbol{c}}\rangle)}{\langle\mathrm{tr}(\bar{\boldsymbol{c}})^2\rangle - \mathrm{tr}(\langle\bar{\boldsymbol{c}}\rangle)^2},$$

    where $\bar{\boldsymbol{c}}$ is the look-averaged complex covariance matrix given by Equation (5.30), and $\langle\cdot\rangle$ indicates local average over a homogeneous region.

    (a) Show that this expression reduces to Equation (5.28) for the single polarization case.

    (b) Write an IDL or Python script to calculate it.

    (c) (K. Condradsen (2013) private communication) Consider N-dimensional, complex-valued observations $\boldsymbol{z}(\nu) = \boldsymbol{x}(\nu) + \mathbf{i}\boldsymbol{y}(\nu)$, $\nu = 1\ldots m$, and organize them into real and imaginary parts in the $m \times N$ data matrices

$$\mathcal{X} = \begin{pmatrix} \boldsymbol{x}(1)^\top \\ \vdots \\ \boldsymbol{x}(m)^\top \end{pmatrix}, \quad \mathcal{Y} = \begin{pmatrix} \boldsymbol{y}(1)^\top \\ \vdots \\ \boldsymbol{y}(m)^\top \end{pmatrix}. \tag{5.67}$$

If the real and imaginary components $x(\nu)_1 \dots x(\nu)_N, y(\nu)_1 \dots x(\nu)_N$, $\nu = 1 \dots m$, are all standard normally distributed and independent, then

$$w = \frac{1}{2}(\mathcal{X}^\top \mathcal{X} + \mathcal{Y}^\top \mathcal{Y} - \mathrm{i}(\mathcal{X}^\top \mathcal{Y} - \mathcal{Y}^\top \mathcal{X})) \tag{5.68}$$

are realizations of a complex Wishart distributed random matrix $\boldsymbol{W} \sim \mathcal{W}_C(\boldsymbol{I}, N, m)$. Thus we may generate a complex Wishart distributed sample by generating $2Nm$ standardized Gaussian random samples, organizing them into the two data matrices, Equation (5.67), and then computing $\boldsymbol{w}$ as in Equation (5.68). Use this recipe to simulate a 500×500-pixel, quad polarimetric ($N = 3$) SAR image in covariance matrix format and verify the correctness of the script of part (b) above.

12. From the definition of the gradient, Equation (5.4), show that the terrain slope angle θ_p can be approximated from a DEM by Equation (5.58).

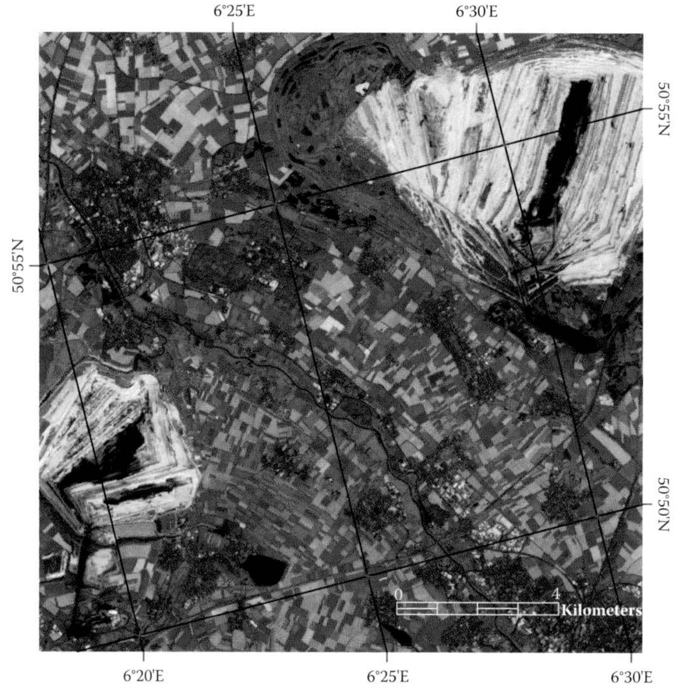

FIGURE 1.1
ASTER color composite image (1000 × 1000 pixels) of VNIR bands 1 (blue), 2 (green), and 3N (red) over the town of Jülich in Germany, acquired on May 1, 2007. The bright areas are open cast coal mines.

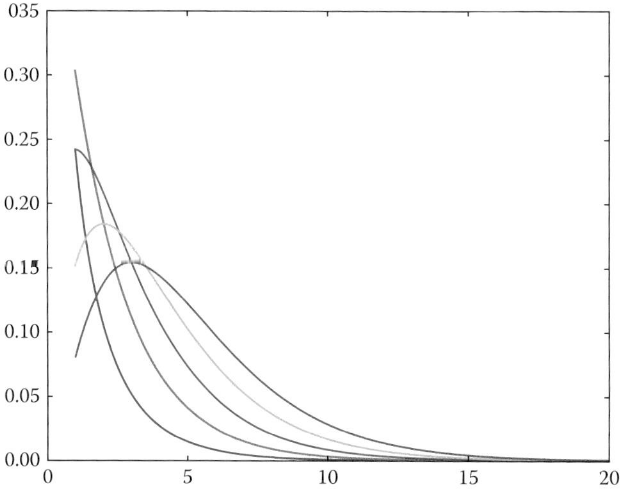

FIGURE 2.2
Plots of the chi-square probability density for $m = 1 \dots 5$ degrees of freedom.

FIGURE 3.10
RGB color composites (2% linear histogram stretch) of the first three (left) and last three (right) principal components of the six nonthermal bands of a LANDSAT 7 ETM+ image over Jülich, acquired August 29, 2001.

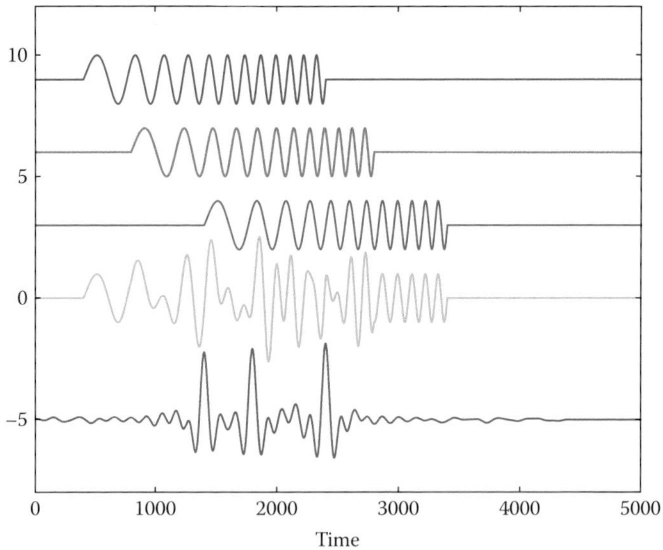

FIGURE 4.2
Radar ranging, see Listing 4.2. The upper three signals represent reflections of a frequency modulated radar pulse (chirp) from three ground points lying close to one another. Their separation in time is proportional to the distances separating the ground features along the direction of pulse emission, that is, transverse to the flight direction. The fourth signal is the superposition actually received. By convolving it with the emitted signal waveform the arrival times are resolved (bottom signal).

FIGURE 5.15
RGB composites (HHHH,HVHV,VVVV) of a quad polarimetric TerraSAR-X image acquired over the city of Mannheim, Germany at a ground resolution of 5 m ($\approx$ 7 looks), left: unfiltered; right: after adaptive filtering with the MMSE filter using a 7×7 window.

FIGURE 5.16
As Figure 5.15, except that the right hand image is after adaptive filtering with the gamma MAP filter using a 7×7 window.

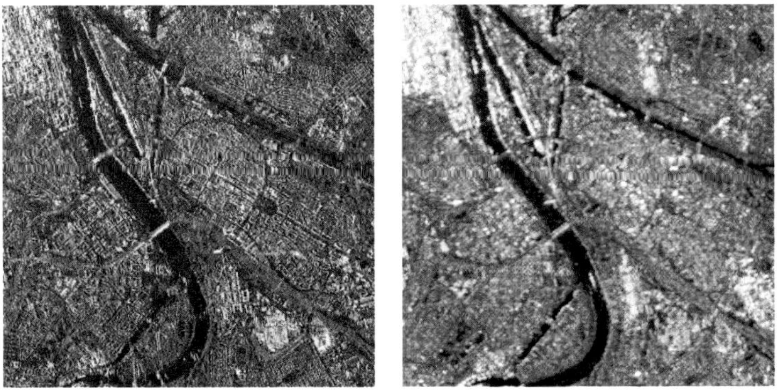

FIGURE 5.32
Frequency domain registration of fully polarimetric SAR imagery, RGB composites (HH,HV,VV), logarithmic intensity values in a linear 2% stretch. Left: TerraSAR-X, right: Radarsat-2.

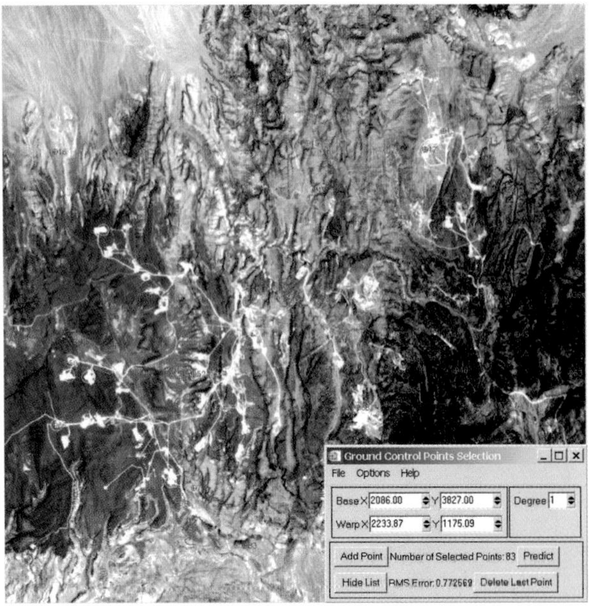

FIGURE 5.35
85 tie-points obtained by matching the contours of Figure 5.34 with those obtained from a similar image acquired in June, 2001. The RMS error is 0.77 pixel for first order polynomial warping, see text.

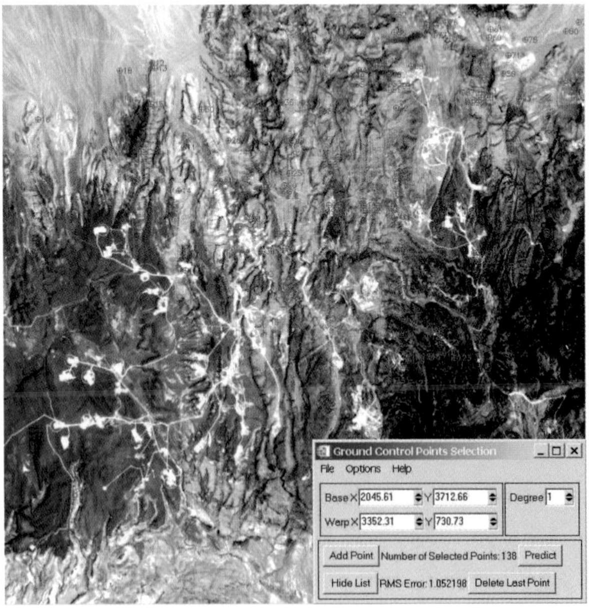

FIGURE 5.36
138 tie-points obtained with ENVI's feature-based matching procedure applied to the July 2003 and June 2001 ASTER scenes. The RMS error is 1.05 pixel for first order polynomial warping.

FIGURE 6.1
RGB color composite (1000 × 1000 pixels, linear 2% saturation stretch) of the first three principal components 1(red), 2(green), and 3(blue) of the nine nonthermal bands of the ASTER scene acquired over Jülich, Germany on May 1, 2007.

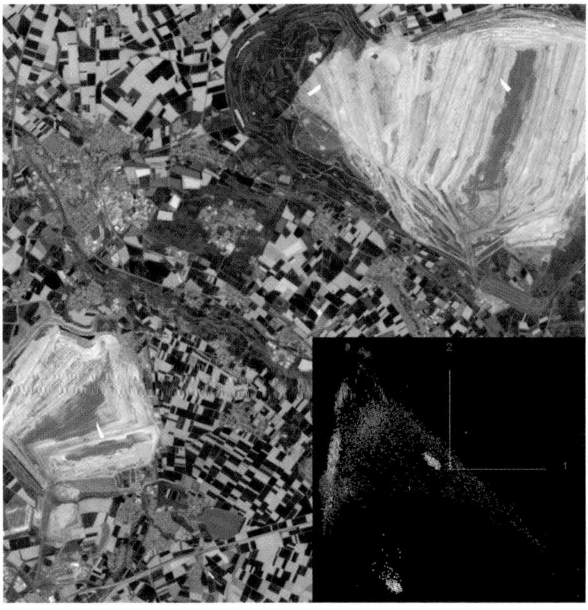

FIGURE 6.3
ROIs for supervised classification. The insert shows the training observations projected onto the plane of the first two principal axes.

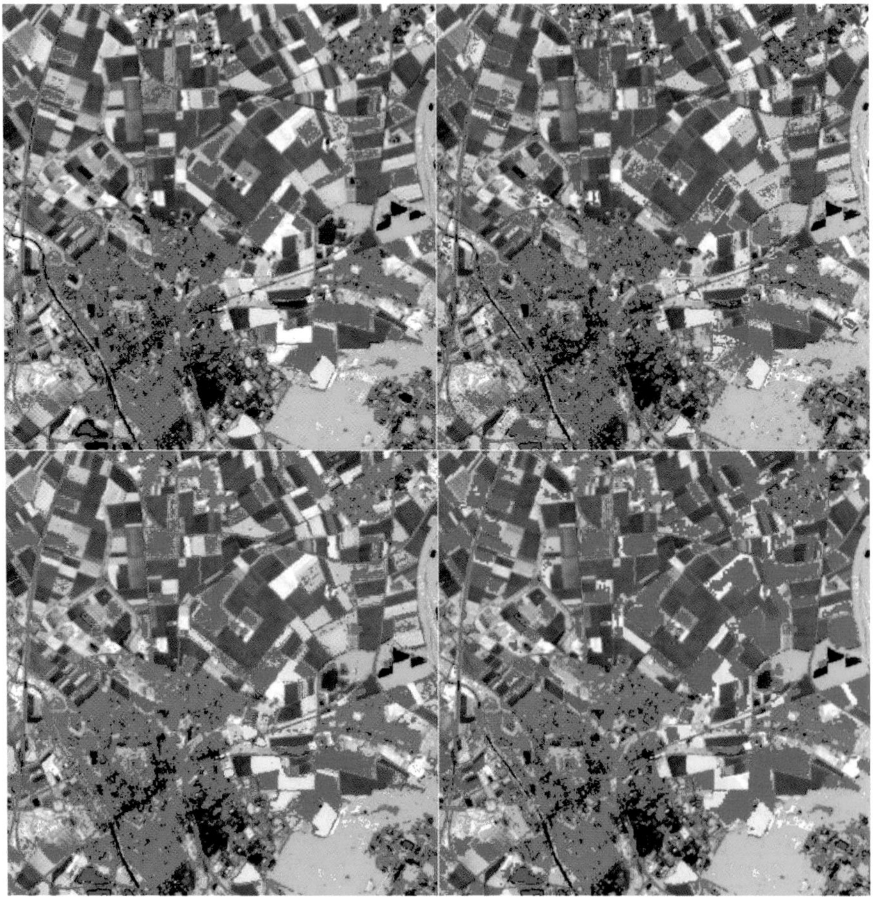

FIGURE 6.4
Supervised classification of a portion of the Jülich ASTER scene: top left: maximum likelihood, top right: Gaussian kernel, bottom left: neural network, bottom right: support vector machine. Five land cover categories (water: blue, settlement: magenta, herbifierous forest: green, rapeseed: yellow, cereal grain: red) are superimposed on VNIR band 3.

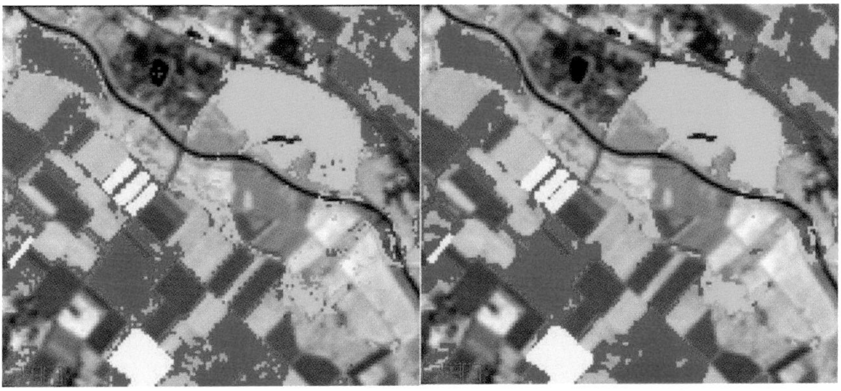

FIGURE 7.1
An example of post-classification processing. Left: original classification of a portion of the Jülich ASTER scene with a neural network. The classes shown are coniferous forest (green), rapeseed (yellow) and cereal grain (red). Right: after three iterations of PLR.

FIGURE 7.4
Maximum likelihood classification of an EMISAR L-band quad polarimetric SAR image acquired over a test agricultural area in Denmark, left: without prior adaptive filtering, right: with prior adaptive filtering. The land use categories are: winter wheat (red), rye (green), water (blue), spring barley (yellow), oats (cyan), beets (magenta), peas (purple), coniferous forest (coral).

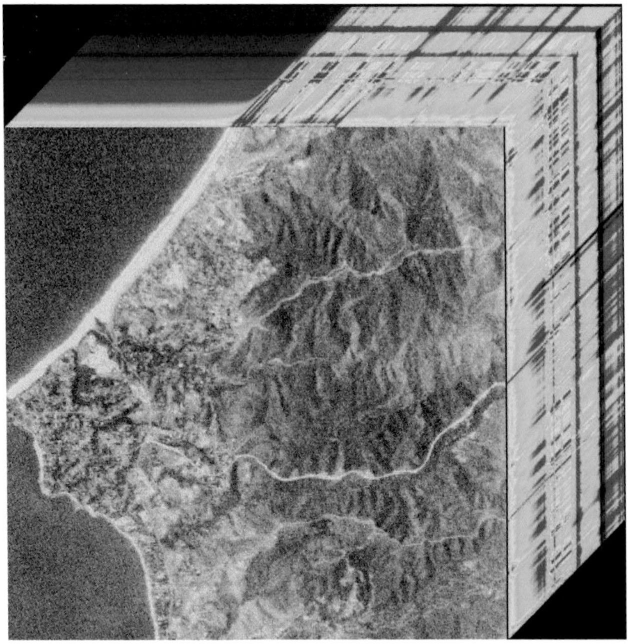

FIGURE 7.5
AVIRIS hyperspectral image cube over the Santa Monica Mountains acquired on April 7, 1997 at a GSD of 20m.

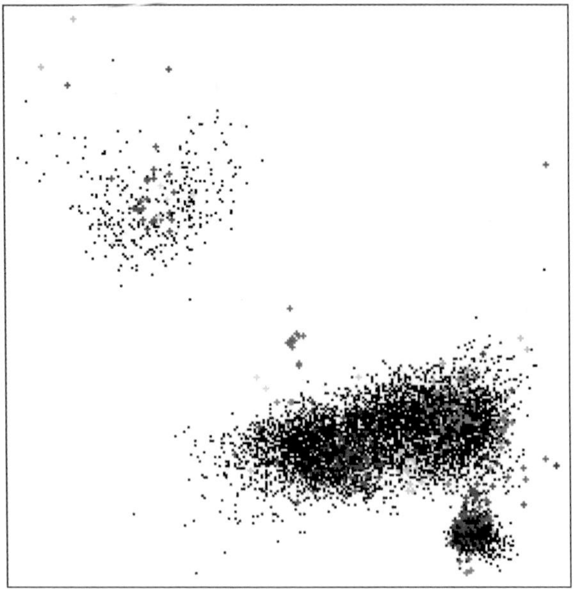

FIGURE 7.8
The n-D visualizer displaying pixel purity indices.

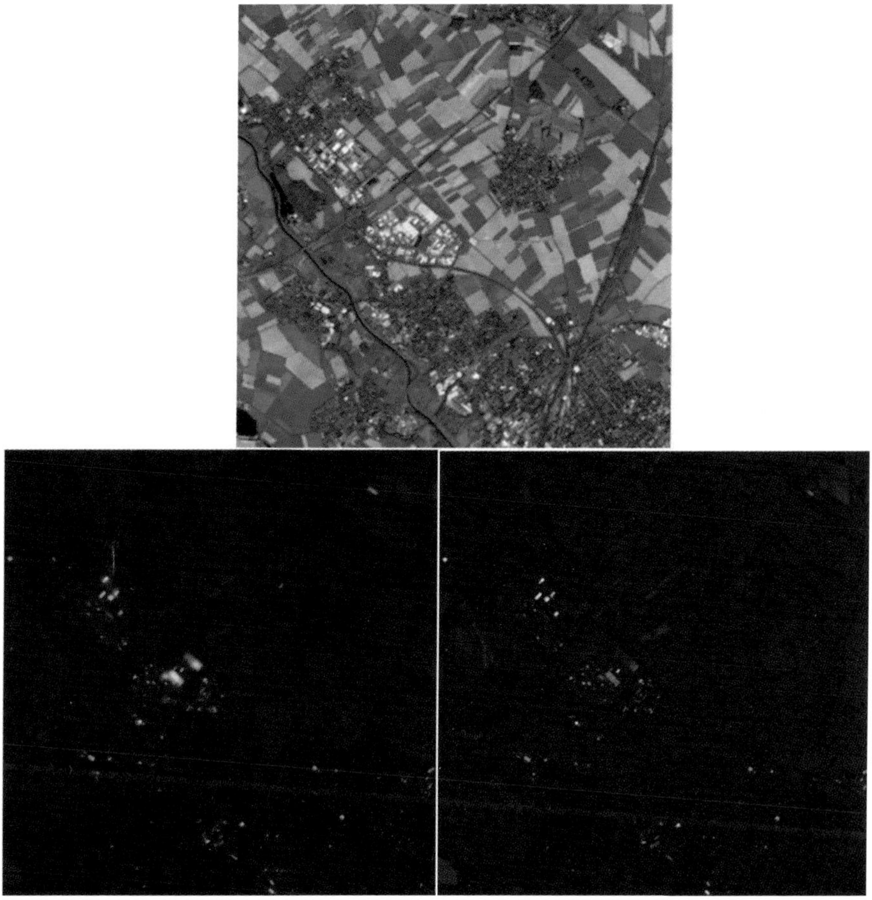

FIGURE 7.9

Anomaly detection: Top, RGB composite of bands 3, 2, and 1 of a spatial subset of the 9-band pan-sharpened ASTER image over Jülich. Bottom left, RX anomaly image calculated from the all 9 bands using the script in Listing 7.2. Bottom right, kernel RX anomaly image calculated from all 9 bands using the ENVI/IDL extension KRX_RUN.PRO.

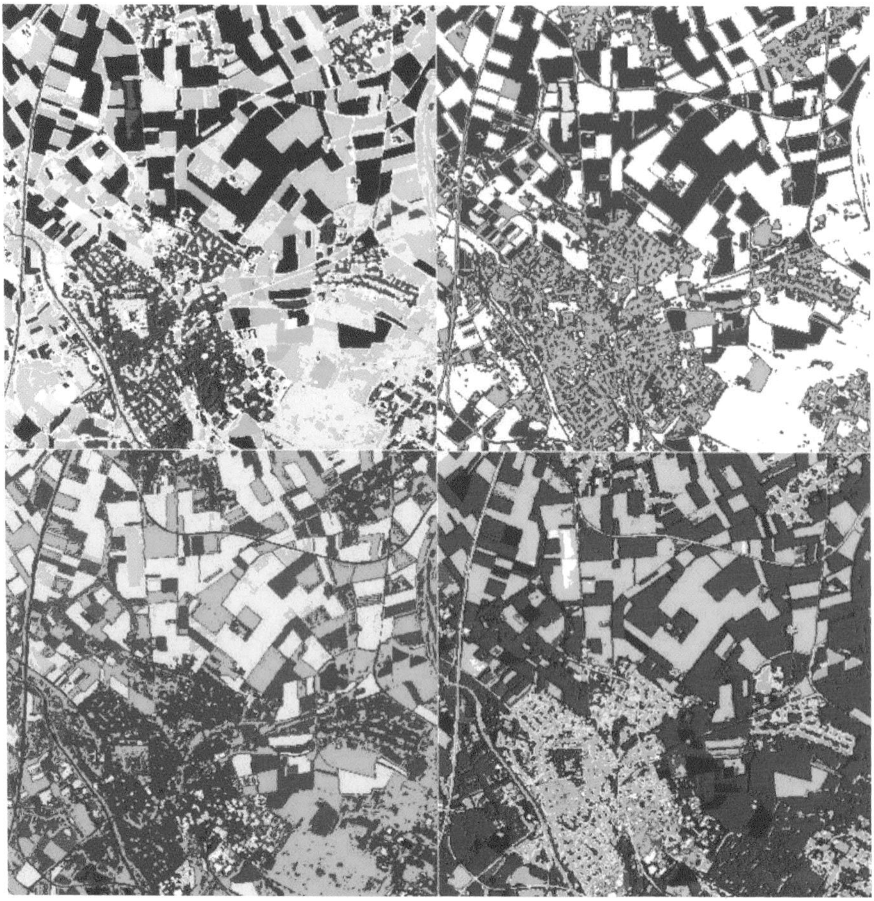

FIGURE 8.1
Unsupervised classification of a spatial subset of the principal components of the fused May 1, 2007 ASTER image over Jülich (Figure 6.1) using 8 classes. Upper left: kernel K-means clustering on the first four components; upper right: extended K-means clustering on the first component with $\tilde{K} = 8$, see Equation (8.22); lower left: agglomerative hierarchical clustering on the first four components; lower right: fuzzy K-means clustering on the first four components.

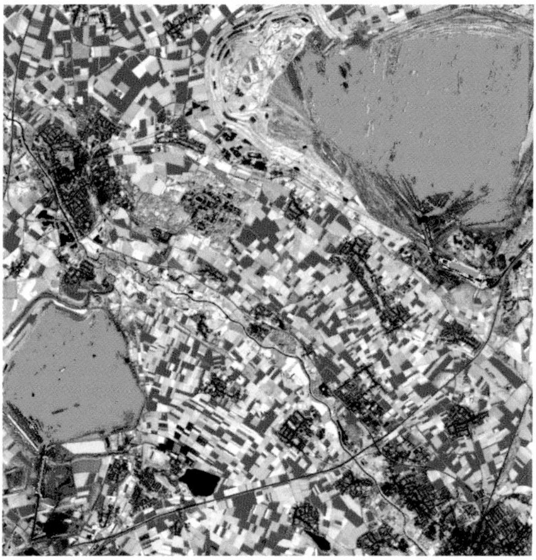

FIGURE 8.2
Gaussian mixture clustering of the first four principal components of the Jülich ASTER scene, eight clusters. Three clusters associated with new sugarbeet plantation (maroon), rapeseed (yellow), and open cast coal mining (coral) are shown superimposed on VNIR spectral band 3N (gray).

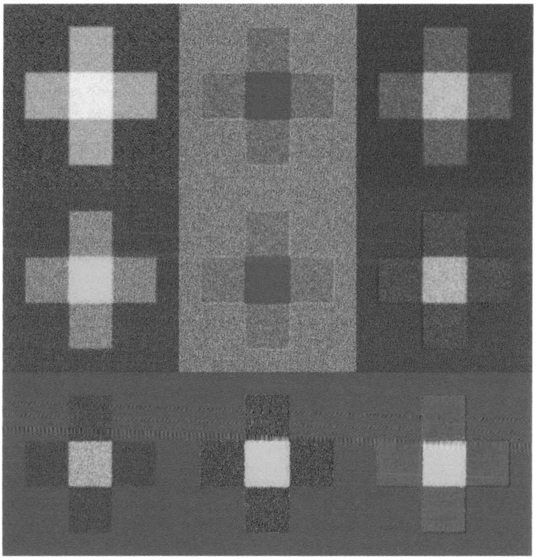

FIGURE 8.3
Unsupervised classification of a toy image. Row-wise, left to right, top to bottom: The toy image, K-means, kernel K-means, extended K-means (on first principal component), fuzzy K-means, agglomerative hierarchical, Gaussian mixture with depth 0 and $\beta = 1.0$, Gaussian mixture with depth 2 and $\beta = 0.0$, Gaussian mixture with depth 2 and $\beta = 1.0$.

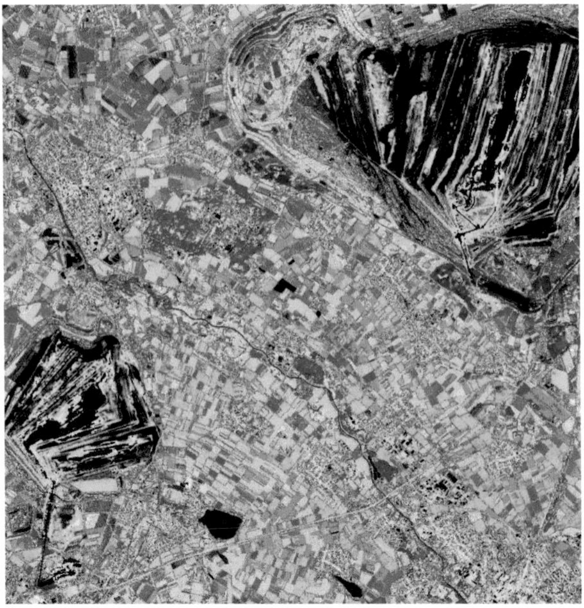

FIGURE 8.6
Kohonen self-organizing map of the nine VNIR and sharpened SWIR spectral bands of the Jülich ASTER scene. The network is a cube having dimensions $6 \times 6 \times 6$.

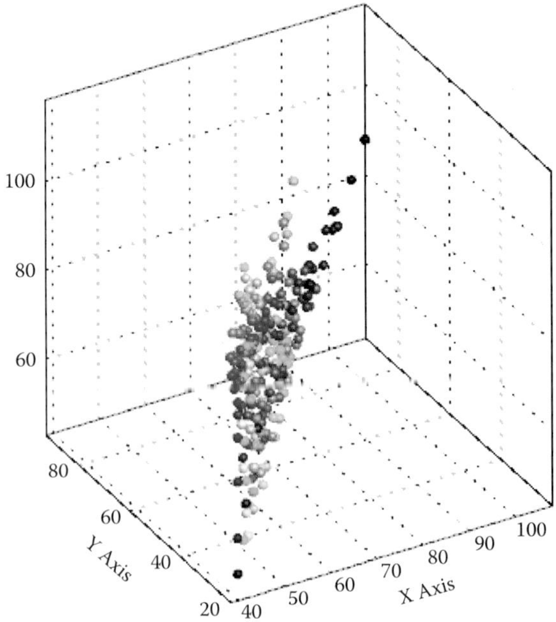

FIGURE 8.7
The $6^3 = 216$ SOM neurons in the feature space of the three VNIR spectral bands of the Jülich ASTER scene, see Figure 8.6.

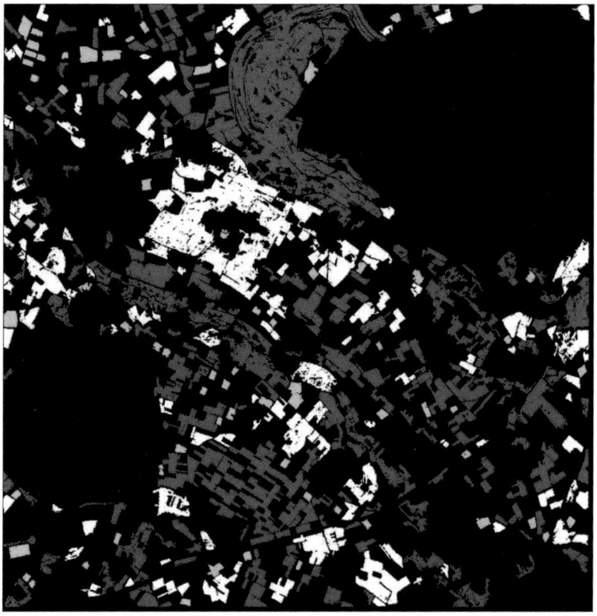

FIGURE 8.9
Classification of the segments in Figure 8.8 on the basis of their invariant moments.

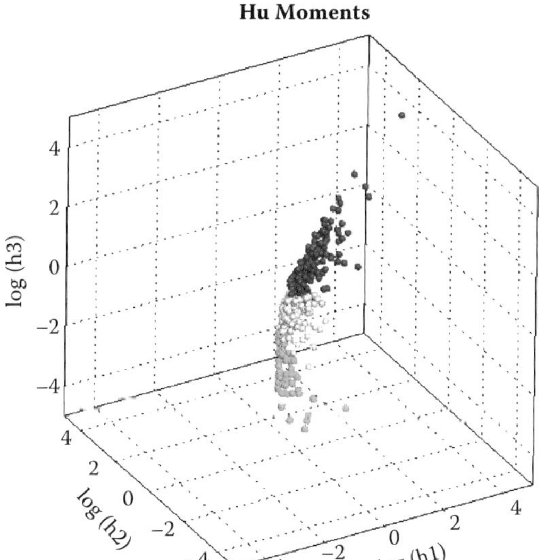

FIGURE 8.10
Standardized logarithms of the first three Hu moments of the classified segments of Figure 8.9.

FIGURE 8.11
Left, a spatial subset of the Jülich principal component image of Figure 6.1. Right, the mean shift filtered image and the segment boundaries.

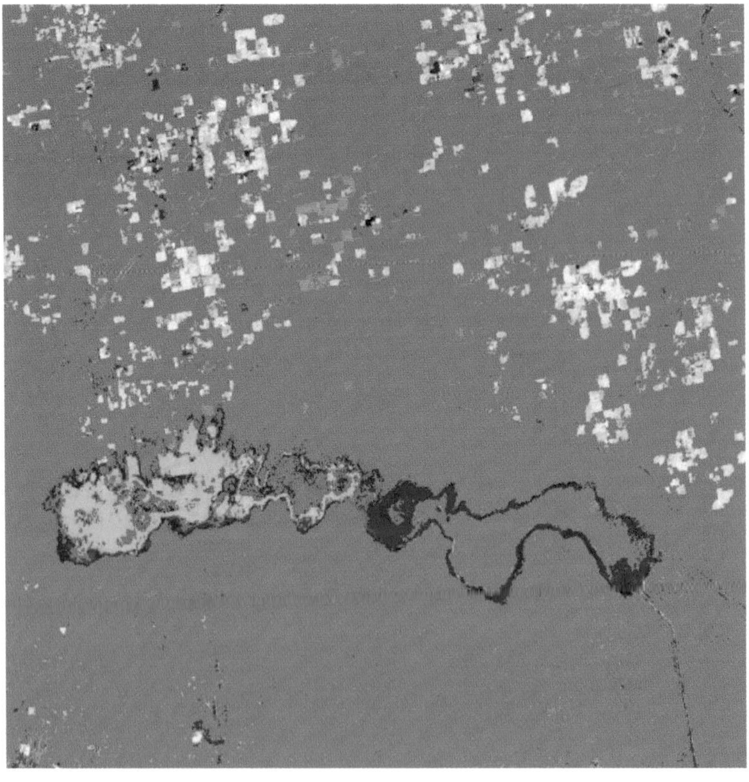

FIGURE 9.15
Color composite of MAD/MNF components 1 (red), 2 (green), and 3 (blue) for the bitemporal image of Figure 9.1. Image generated with `MAD_VIEW_RUN` using decision thresholds obtained from the Gaussian mixture model fit, linear stretch over ± 16 standard deviations of the no-change observations.

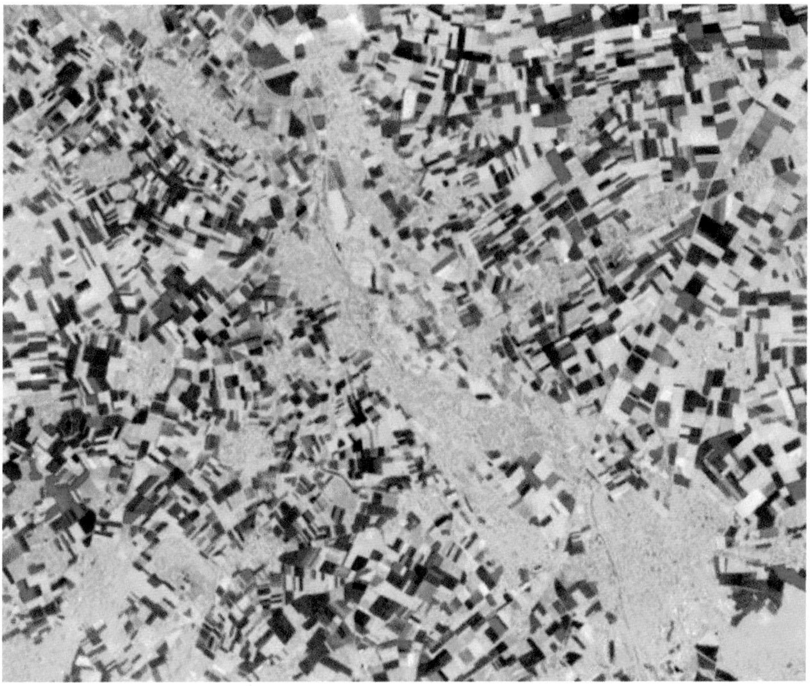

FIGURE 9.16
RGB composite of an iteratively re-weighted MAD image (MAD components 4,5,6 in a linear 2% stretch) obtained from LANDSAT ETM+ scenes acquired in June and August, 2001.

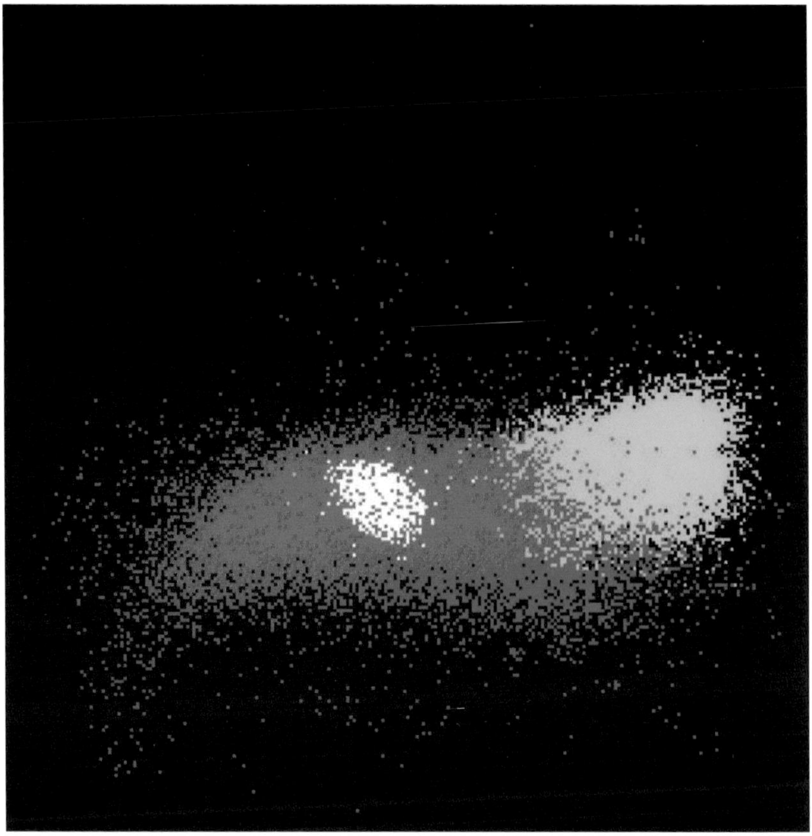

FIGURE 9.17
Four clusters in MAD feature space projected onto the plane of variates 5 and 6. The partially obscured white cluster is no change.

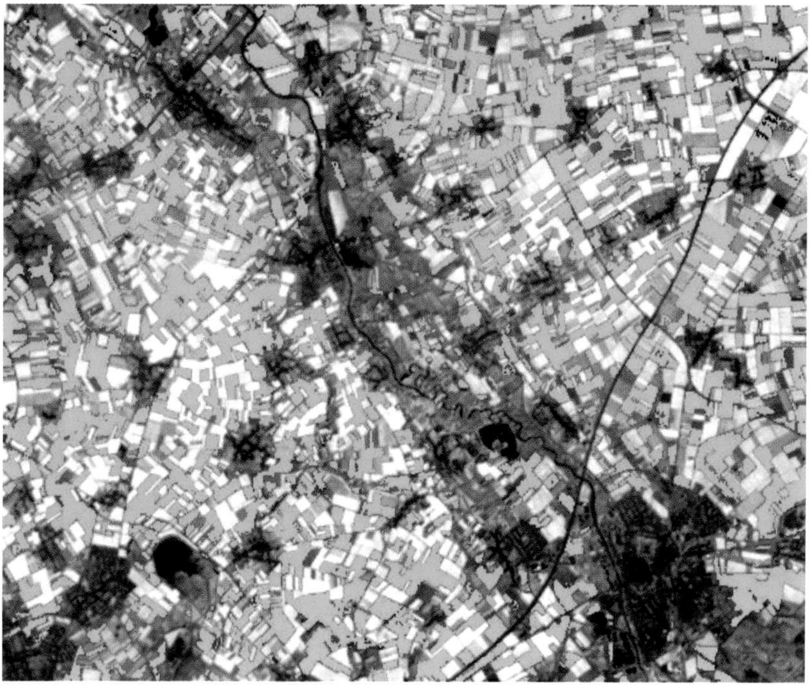

FIGURE 9.18
Overlay of the harvested grain cluster onto spectral band 4 of the August 2001 LANDSAT ETM+ image.

FIGURE 9.19
Wishart change detection. Upper left: RGB composite ($\langle|S_{hh}|^2\rangle, \langle|S_{hv}|^2\rangle, \langle|S_{vv}|^2\rangle$) of a TerraSAR-X strip map, quad polarimetric image over a region southwest of Bonn, Germany, acquired on April 21, 2010 (logarithmic intensity scale). Upper right: the same scene acquired on May 13, 2010. Lower left: rejection of the no-change hypothesis at the 1% level (red) overlayed onto the HH intensity band of the April image. Lower right: an aerial view of the sand quarry in the bottom right hand corner of the images, taken at an unknown date.

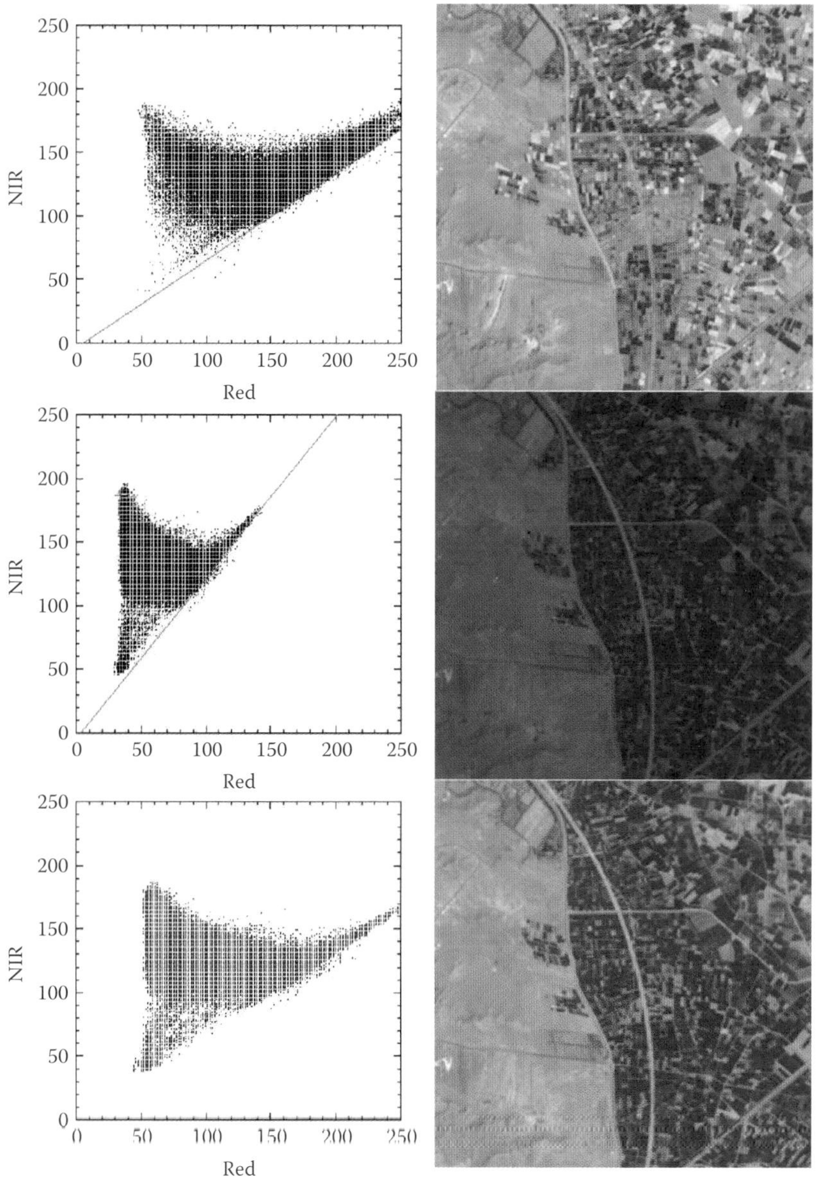

FIGURE 9.20
Scatterplot matching of two ASTER images. Top row: RGB composite of the July 2001 reference image (bands 2,3,2 in a 0-255 byte linear stretch) along with the NIR vs. red scatterplot showing the full canopy point (cross) and the bare soil line. Middle row: the September 2005 target image. Bottom row: the normalized target.

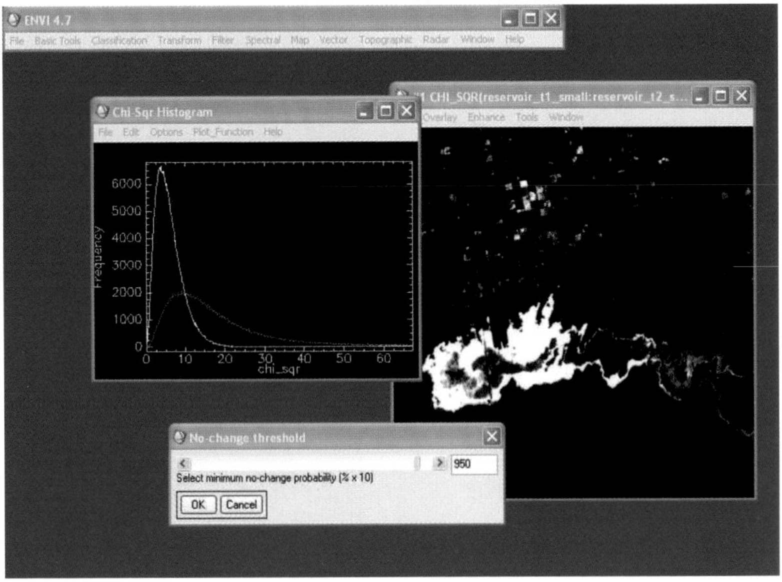

FIGURE C.2
Radiometric normalization.

6

Supervised Classification Part 1

Land cover classification of remote sensing imagery is a task which falls into the general category of *pattern recognition.* Pattern recognition problems, in turn, are usually approached by developing appropriate *machine learning* algorithms. Broadly speaking, machine learning involves tasks for which there is no known direct method to compute a desired output from a set of inputs. The strategy adopted is for the computer to "learn" from a set of representative examples.

In the case of supervised classification, the task can often be seen as one of modeling probability distributions. On the basis of representative data for K land cover classes presumed to be present in a scene, the *a posteriori* probabilities for class k conditional on observation $\boldsymbol{g}$, $\Pr(k \mid \boldsymbol{g})$, $k = 1 \dots K$, are "learned" or approximated. This is usually called the *training phase* of the classification procedure. Then these probabilities are used to classify all of the pixels in the image, a step referred to as the *generalization phase.*

In the present chapter we will consider three representative models for supervised classification which involve this sort of probability density estimation: a *parametric model* (the Bayes maximum likelihood classifier), a *nonparametric model* (Gaussian kernel classification), and a *semiparametric* or *mixture model* (the feed-forward neural network). For the nonparametric case, ENVI provides little functionality, and its neural network classifier uses a suboptimal training algorithm. Therefore, both here and in Appendix B, we shall make some considerable effort to develop our own classification routines. Finally, the *support vector machine* (SVM) classifier, available in ENVI since Version 4.3, will be discussed in detail. SVMs are also nonparametric in the sense that they make direct use of a subset of the labeled training data (the support vectors) to effect a partitioning of the feature space, however, unlike the aforementioned classifiers, without reference to the statistical distributions of the training data.

To illustrate the various algorithms developed here and in the following chapters on image classification, we will work with the ASTER scene shown in Figure 6.1 (see also Figure 1.1) on the next page. In the ASTER dataset, the six SWIR bands have been sharpened to the 15 m ground resolution of the three VNIR bands with the *à trous* wavelet fusion method of Section 5.3.5 and a principal components analysis of the stacked nine-band image has been performed. The classification examples in the present chapter will be carried out with subsets of the principal components of the ASTER image. Discussion of polarimetric SAR image classification will be postponed until Chapter 7.

FIGURE 6.1
RGB color composite (1000 × 1000 pixels, linear 2% saturation stretch) of the first three principal components 1(red), 2(green), and 3(blue) of the nine nonthermal bands of the ASTER scene acquired over Jülich, Germany, on May 1, 2007. **(See color insert.)**

6.1 Maximizing the *a posteriori* probability

The basis for most of the classifiers that we consider in this chapter is a decision rule based on the *a posteriori* probabilities $Pr(k \mid \boldsymbol{g})$, so this rule will be our starting point.

Let us begin by defining a *loss function* $L(k, \boldsymbol{g})$ which measures the cost of associating the observation $\boldsymbol{g}$ with the class k. Let λ_{kj} be the loss incurred if $\boldsymbol{g}$ in fact belongs to class k, but is classified as belonging to class j. It can reasonably be assumed that

$$\lambda_{kj} \begin{cases} = 0 & \text{if } k = j \\ > 0 & \text{otherwise,} \end{cases} \quad k, j = 1 \ldots K, \tag{6.1}$$

that is, correct classifications do not incur losses while misclassifications do. The loss function can then be expressed as a sum over the individual losses, weighted according to their probabilities of occurrence, $\Pr(j \mid \boldsymbol{g})$,

$$L(k, \boldsymbol{g}) = \sum_{j=1}^{K} \lambda_{kj} \Pr(j \mid \boldsymbol{g}). \tag{6.2}$$

Without further specifying λ_{kj}, a loss-minimizing decision rule for classification may be defined (ignoring the possibility of ties) as

$$\boldsymbol{g} \text{ is in class } k \text{ provided } L(k, \boldsymbol{g}) \le L(j, \boldsymbol{g}) \text{ for all } j = 1 \ldots K. \tag{6.3}$$

So far we have been quite general. Now suppose the losses are independent of the kind of misclassification that occurs (for instance, the classification of a "forest" pixel into the class "meadow" is just as costly as classifying it as "urban area," etc.). Then we can write

$$\lambda_{kj} = 1 - \delta_{kj}, \tag{6.4}$$

where $\delta_{kj} = 1$ for $k = j$ and 0 otherwise. Thus any given misclassification ($j \ne k$) has unit cost, and a correct classification ($j = k$) costs nothing, as before. We then obtain from Equation (6.2)

$$L(k, \boldsymbol{g}) = \sum_{j=1}^{K} \Pr(j \mid \boldsymbol{g}) - \Pr(k \mid \boldsymbol{g}) = 1 - \Pr(k \mid \boldsymbol{g}), \; k = 1 \ldots K, \tag{6.5}$$

and from Equation (6.3) the following decision rule:

$$\boldsymbol{g} \text{ is in class } k \text{ provided } \Pr(k \mid \boldsymbol{g}) \ge \Pr(j \mid \boldsymbol{g}) \text{ for all } j = 1 \ldots K; \tag{6.6}$$

in other words, assign each new observation to the class with the highest *a posteriori* probability. As indicated in the introduction, our main task will therefore be to determine the posterior probabilities $\Pr(k \mid \boldsymbol{g})$.

6.2 Training data and separability

The choice of training data is arguably the most difficult and critical part of the supervised classification process. The standard procedure is to select areas within a scene which are representative of each class of interest. In the ENVI Classic environment, the areas are entered as *regions of interest* (ROIs), from which the training observations are selected. Some fraction of the representative data may be retained for later accuracy assessment. These comprise the so-called *test data* and are withheld from the training phase in order not to bias the subsequent evaluation. We will refer to the set of labeled training data as *training pairs* or *training examples* and write it in the form

$$\mathcal{T} = \{\boldsymbol{g}(\nu), \ell(\nu)\}, \quad \nu = 1 \dots m, \tag{6.7}$$

where m is the number of observations and

$$\ell(\nu) \in \mathcal{K} = \{1 \dots K\} \tag{6.8}$$

is the class label of observation $\boldsymbol{g}(\nu)$.

Ground reference data were collected on the same day as the acquisition of the ASTER image in Figure 6.1. Figure 6.2 shows photographs for four of the ten land cover categories used for classification. The others were water, suburban settlements, urban areas/industrial parks, herbiferous forest, coniferous forest and open cast mining. In all, 30 regions of interest were identified in the scene as representative of the 10 classes, involving 7173 pixels. They are shown in Figure 6.3. Of these, 2/3 sampled uniformly across the ROIs were used for training and the remainder reserved for testing, i.e., estimating the generalization error on new observations.

The *degree of separability* of the training observations will give some indication of the prospects for success of the classification procedure and can help in deciding how the data should be processed prior to classification. A very commonly used separability measure may be derived by considering the *Bayes error*. Suppose that there are just two classes involved, $\mathcal{K} = \{1, 2\}$. If we apply the decision rule, Equation (6.6), for some pixel intensity vector $\boldsymbol{g}$, we must assign the class as that having maximum *a posteriori* probability. Therefore, the probability $r(\boldsymbol{g})$ of incorrectly classifying the pixel is given by

$$r(\boldsymbol{g}) = \min[\ \Pr(1 \mid \boldsymbol{g}), \Pr(2 \mid \boldsymbol{g})\ ].$$

The Bayes error ϵ is defined to be the average value of $r(\boldsymbol{g})$, which we can calculate as the integral of $r(\boldsymbol{g})$ times the probability density $p(\boldsymbol{g})$, taken over all of the observations $\boldsymbol{g}$:

$$\begin{aligned} \epsilon = \int r(\boldsymbol{g})p(\boldsymbol{g})d\boldsymbol{g} &= \int \min[\ \Pr(1 \mid \boldsymbol{g}), \Pr(2 \mid \boldsymbol{g})\ ]p(\boldsymbol{g})d\boldsymbol{g} \\ &= \int \min[\ p(\boldsymbol{g} \mid 1)\Pr(1), p(\boldsymbol{g} \mid 2)\Pr(2)\ ]d\boldsymbol{g}. \end{aligned} \tag{6.9}$$

Bayes' Theorem, Equation (2.66), was invoked in the last equality. The Bayes error may be used as a measure of the separability of the two classes: the smaller the error, the better the separability.

FIGURE 6.2
Ground reference data for four land cover categories, photographed on May 1st, 2007. Top left: cereal grain, top right: grassland, bottom left: rapeseed, bottom right: sugar beets.

Calculating the Bayes error is in general difficult, but we can at least get an approximate upper bound on it as follows (Fukunaga, 1990). First note that, for any $a, b \ge 0$,

$$\min[\, a, b \,] \le a^s b^{1-s}, \quad 0 \le s \le 1.$$

For example, if $a < b$, then the inequality can be written

$$a < a\left(\frac{b}{a}\right)^{1-s},$$

which is clearly true. Applying this inequality to Equation (6.9), we get the *Chernoff bound* ϵ_u on the Bayes error,

$$\epsilon \le \epsilon_u = \Pr(1)^s \Pr(2)^{1-s} \int p(\boldsymbol{g} \mid 1)^s p(\boldsymbol{g} \mid 2)^{1-s} d\boldsymbol{g}. \tag{6.10}$$

The least upper bound is then determined by minimizing ϵ_u with respect to s. If $p(\boldsymbol{g} \mid 1)$ and $p(\boldsymbol{g} \mid 2)$ are multivariate normal distributions with equal

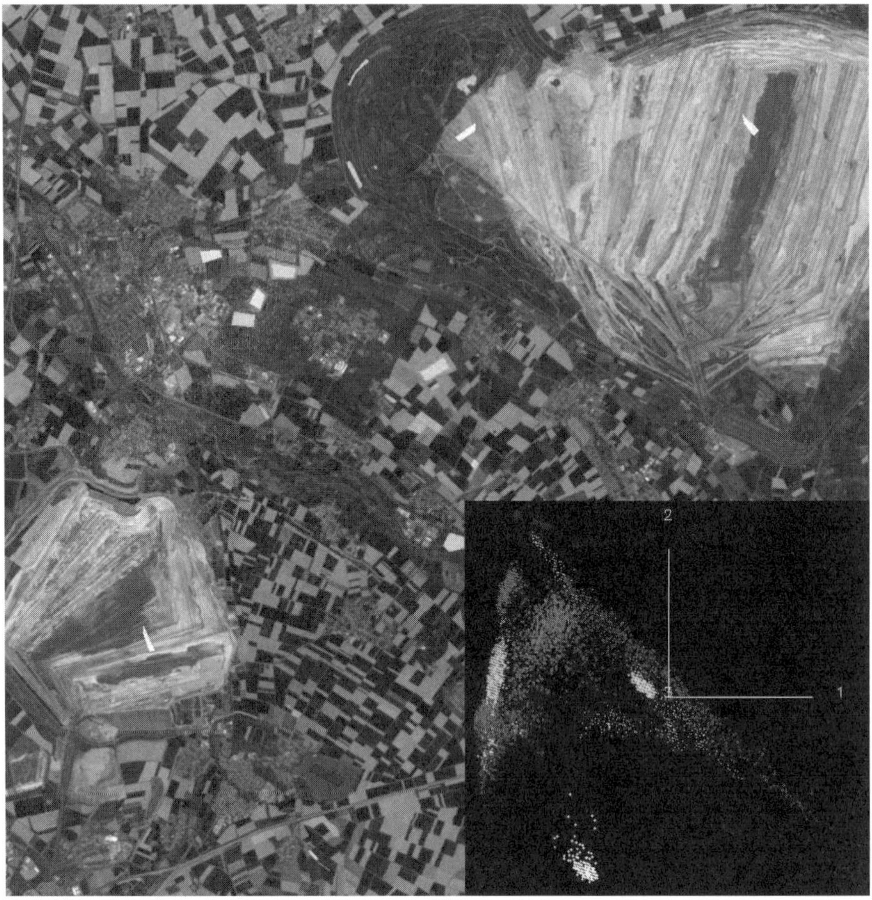

FIGURE 6.3
ROIs for supervised classification. The insert shows the training observations projected onto the plane of the first two principal axes. **(See color insert.)**

covariance matrices $\boldsymbol{\Sigma}_1 = \boldsymbol{\Sigma}_2$, then it can be shown that the minimum in fact occurs at $s = 1/2$. Approximating the minimum as $s = 1/2$ also for the case where $\boldsymbol{\Sigma}_1 \neq \boldsymbol{\Sigma}_2$ leads to the (somewhat less tight) *Bhattacharyya bound* ϵ_B,

$$\epsilon \leq \epsilon_B = \sqrt{\Pr(1)\Pr(2)} \int \sqrt{p(\boldsymbol{g} \mid 1)p(\boldsymbol{g} \mid 2)}\, d\boldsymbol{g}. \tag{6.11}$$

This integral can be evaluated explicitly (Exercise 1). The result is

$$\epsilon_B = \sqrt{\Pr(1)\Pr(2)}\, e^{-B},$$

where B is the *Bhattacharyya distance*, given by

$$B = \frac{1}{8}(\boldsymbol{\mu}_2 - \boldsymbol{\mu}_1)^\top \left[\frac{\boldsymbol{\Sigma}_1 + \boldsymbol{\Sigma}_2}{2}\right]^{-1} (\boldsymbol{\mu}_2 - \boldsymbol{\mu}_1) + \frac{1}{2}\log\left(\frac{|\boldsymbol{\Sigma}_1 + \boldsymbol{\Sigma}_2|/2}{\sqrt{|\boldsymbol{\Sigma}_1||\boldsymbol{\Sigma}_2|}}\right). \tag{6.12}$$

Large values of B imply small upper limits on the Bayes error and hence good separability. The first term in B is a squared average *Mahalanobis distance* (see Section 6.3) and expresses the class separability due to the dissimilarity of the class means.* The second term measures the difference between the covariance matrices of the two classes. It vanishes when $\boldsymbol{\Sigma}_1 = \boldsymbol{\Sigma}_2$.

TABLE 6.1
The lowest 10 paired class separabilities for the first four principal components of the ASTER scene.

Class 1	Class 2	J-M Distance
Grain [Red]	Grassland [Red2]	1.42
Settlement [Magenta]	Industry [Maroon]	1.51
Grassland [Red2]	Herbiferous [Green]	1.68
Settlement [Magenta]	Herbiferous [Green]	1.91
Settlement [Magenta]	Grassland [Red2]	1.94
Industry [Maroon]	Coniferous [Sea Green]	1.94
Coniferous [Sea Green]	Herbiferous [Green]	1.95
Sugar beet [Cyan]	Mining [White]	1.98
Grain [Red]	Herbiferous [Green]	1.99
Industry[Maroon]	Herbiferous [Green]	1.99

The Bhattacharyya distance as a measure of separability has the disadvantage that it continues to grow even after the classes have become so well separated that any classification procedure could distinguish them perfectly. The *Jeffries–Matusita distance* measures separability of two classes on a more convenient scale [0 – 2] in terms of B:

$$J = 2(1 - e^{-B}). \tag{6.13}$$

As B continues to grow, the measure saturates at the value 2. The factor 2 comes from the fact that the Jeffries–Matusita distance can be derived independently as the average distance between two density functions; see Richards and Jia (2006) and Exercise 1. The ENVI menu command

```
Basic Tools/Region of Interest/Compute ROI Separability
```

calculates Jeffries–Matusita distances between all pairs of classes defined in a given set of ROIs by estimating the class means and covariance matrices from the pixels contained within them and then using Equations (6.12) and (6.13). Some examples are shown in Table 6.1 for the training data of Figure 6.3.

*This term is proportional to the maximum value of the Fisher linear discriminant, as can be seen by substituting Equation (3.82) into Equation (3.81); see Exercise 13, Chapter 3.

6.3 Maximum likelihood classification

Consider once again Bayes' Theorem, expressed in the form of Equation (2.66),

$$\Pr(k \mid \boldsymbol{g}) = \frac{p(\boldsymbol{g} \mid k)\Pr(k)}{p(\boldsymbol{g})}, \tag{6.14}$$

where $\Pr(k)$, $k = 1 \dots K$, are prior probabilities, $p(\boldsymbol{g} \mid k)$ is a class-specific probability density function, and where $p(\boldsymbol{g})$ is given by

$$p(\boldsymbol{g}) = \sum_{j=1}^{K} p(\boldsymbol{g} \mid j)\Pr(j).$$

Since $p(\boldsymbol{g})$ is independent of k, we can write the decision rule, Equation (6.6), as

$$\boldsymbol{g} \text{ is in class } k \text{ provided } p(\boldsymbol{g} \mid k)\Pr(k) \geq p(\boldsymbol{g} \mid j)\Pr(j) \text{ for all } j = 1 \dots K. \tag{6.15}$$

Now suppose that the observations from class k are sampled from a multivariate normal distribution. Then the density functions are given by

$$p(\boldsymbol{g} \mid k) = \frac{1}{(2\pi)^{N/2}|\boldsymbol{\Sigma}_k|^{1/2}} \exp\left(-\frac{1}{2}(\boldsymbol{g} - \boldsymbol{\mu}_k)^\top \boldsymbol{\Sigma}_k^{-1}(\boldsymbol{g} - \boldsymbol{\mu}_k)\right). \tag{6.16}$$

Taking the logarithm of Equation (6.16) gives

$$\log\big(p(\boldsymbol{g} \mid k)\big) = -\frac{N}{2}\log(2\pi) - \frac{1}{2}\log|\boldsymbol{\Sigma}_k| - \frac{1}{2}(\boldsymbol{g} - \boldsymbol{\mu}_k)^\top \boldsymbol{\Sigma}_k^{-1}(\boldsymbol{g} - \boldsymbol{\mu}_k).$$

The first term may be ignored, as it too is independent of k. Together with Equation (6.15) and the definition of the *discriminant function*

$$d_k(\boldsymbol{g}) = \log(\Pr(k)) - \frac{1}{2}\log|\boldsymbol{\Sigma}_k| - \frac{1}{2}(\boldsymbol{g} - \boldsymbol{\mu}_k)^\top \boldsymbol{\Sigma}_k^{-1}(\boldsymbol{g} - \boldsymbol{\mu}_k), \tag{6.17}$$

we obtain the *maximum-likelihood classifier*:

$$\boldsymbol{g} \text{ is in class } k \text{ provided } d_k(\boldsymbol{g}) \geq d_j(\boldsymbol{g}) \text{ for all } j = 1 \dots K. \tag{6.18}$$

There may be no information about the prior class probabilities $\Pr(k)$, in which case they can be set equal and ignored in the classification. Then the factor 1/2 in Equation (6.17) can be dropped as well and the discriminant becomes

$$d_k(\boldsymbol{g}) = -\log|\boldsymbol{\Sigma}_k| - (\boldsymbol{g} - \boldsymbol{\mu}_k)^\top \boldsymbol{\Sigma}_k^{-1}(\boldsymbol{g} - \boldsymbol{\mu}_k). \tag{6.19}$$

The second term in Equation (6.19) is the square of the so-called *Mahalanobis distance*

$$\sqrt{(\boldsymbol{g} - \boldsymbol{\mu}_k)^\top \boldsymbol{\Sigma}_k^{-1}(\boldsymbol{g} - \boldsymbol{\mu}_k)}.$$

The contours of constant multivariate probability density in Equation (6.16) are hyper-ellipsoids of constant Mahalanobis distance to the mean $\boldsymbol{\mu}_k$.

The moments $\boldsymbol{\mu}_k$ and $\boldsymbol{\Sigma}_k$, which appear in the discriminant functions, may be estimated from the training data using the maximum likelihood parameter estimates (see Section 2.4)

$$\begin{aligned}\hat{\boldsymbol{\mu}}_k &= \frac{1}{m_k}\sum_{\{\nu|\ell(\nu)=k\}} \boldsymbol{g}(\nu)\\ \hat{\boldsymbol{\Sigma}}_k &= \frac{1}{m_k}\sum_{\{\nu|\ell(\nu)=k\}} (\boldsymbol{g}(\nu)-\boldsymbol{\mu}_k)(\boldsymbol{g}(\nu)-\boldsymbol{\mu}_k)^\top,\end{aligned} \tag{6.20}$$

where m_k is the number of training pixels with class label k.

Having estimated the parameters from the training data, the generalization phase consists simply of applying the rule (6.18) to all of the pixels in the image. Because of the small number of parameters to be estimated, maximum likelihood classification is extremely fast. Its weakness lies in the restrictiveness of the assumption that all observations are drawn from multivariate normal probability distributions. Computational efficiency eventually achieved at the cost of generality is a characteristic of *parametric classification models*, to which category the maximum likelihood classifier belongs (Bishop, 1995).

6.3.1 ENVI's maximum likelihood classifier

Note that applying the rule of Equation (6.18) will place any observation into one of the K classes no matter how small its maximum discriminant function turns out to be. If it is thought that some classes may have been overlooked, or if no training data were available for one or two known classes, then it might be reasonable to assume that observations with small maximum discriminant functions belong to one of these inaccessible classes. Then they should perhaps not be classified at all. If this is desired, it may be achieved simply by setting a threshold on the maximum discriminant $d_k(\boldsymbol{g})$, marking the observations lying below the threshold as "unclassified." This possibility is provided with ENVI's maximum likelihood classification algorithm. The algorithm is called from the ENVI main menu with

`Classification/Supervised/Maximum Likelihood`

To facilitate setting thresholds for unclassified observations, the ENVI maximum likelihood classifier optionally generates a *rule image* consisting of the discriminant functions, Equation (6.17), for each class and training observation, whereby the prior probabilities $\Pr(k)$ are set equal to one another. If $\boldsymbol{g}$ is normally distributed, the squared Mahalanobis distance term in Equation (6.17) is chi-square distributed with N degrees of freedom, where N is the dimensionality of the observations. The histogram of a rule image band for class k will therefore resemble a chi-square distribution, reflected about zero

Listing 6.1: Calculating class membership probabilities from rule images.

```
PRO rule_convert

envi_select, title='Choose␣rule␣image', $
     fid=fid, dims=dims,pos=pos,/no_dims,/no_spec
IF (fid EQ -1) THEN BEGIN
   PRINT, 'cancelled'
   RETURN
ENDIF

map_info = envi_get_map_info(fid=fid)

num_cols = dims[2]-dims[1]+1
num_rows = dims[4]-dims[3]+1
num_classes = n_elements(pos)
num_pixels = num_cols*num_rows

; get MaxLike rule image
rule_image = dblarr(num_pixels, num_classes)
FOR i=0,num_classes-1 DO rule_image[*,i] = $
     envi_get_data(fid=fid,dims=dims,pos=pos[i])

; exponentiate and normalize
prob_image = exp(rule_image - alog(1.0/num_classes))
den = total(prob_image,2,/double)
FOR i=0,num_classes-1 DO  prob_image[*,i]= $
     prob_image[*,i]/den

; write to memory
envi_enter_data, $
  reform(bytscl(prob_image,min=0.0,max=1.0),num_cols, $
   num_rows,num_classes),map_info=map_info

END
```

due to the minus sign, and shifted due to the $\log(\Pr(k))$ and $\frac{1}{2}\log|\mathbf{\Sigma}_k|$ terms. Thresholds can be set in a postclassification session with the *rule classifier*, accessible from the ENVI main menu under

`Classification/Post Classification/Rule Classifier`

and a new classified image generated.

For further postprocessing of classification results we shall later require the posterior class membership probabilities $\Pr(k \mid \boldsymbol{g})$. The IDL routine in Listing 6.1 converts ENVI maximum likelihood rule images to membership probabilities (byte-scaled to save storage space) under the assumption that there is no "unclassified" class (the membership probabilities sum to unity).

6.3.2 A modified classifier for ENVI and a Python script

In order to carry out an unbiased assessment of the accuracy of supervised classification methods, ENVI's built-in evaluation procedures allow comparison with so-called "ground truth" ROIs or images containing areas of labeled data not used during the training phase. We will prefer a somewhat different evaluation philosophy, arguing that, if other representative training areas are indeed available for evaluation, then they should also be used to train the classifier. For evaluation purposes, some portion of the pixels in *all* of the training areas can be held back, but such test data should be selected from the pool of available labeled observations. This point of view assumes that all training/test areas are equally representative of their respective classes, but if that were not the case, then there would be no justification to use them at all.

A wrapper for the ENVI maximum likelihood classifier called MAXLIKE_RUN, which separates ROI training area pixels into a training dataset and a test

Listing 6.2: Excerpt from the program module MAXLIKE_RUN.PRO.

```
; split into training and test data
  seed = 12345L
  num_test = m/3

; sampling with replacement
  test_indices = randomu(seed,num_test,/long) MOD m
  test_indices =  test_indices[sort(test_indices)]
  train_indices=difference(lindgen(m),test_indices)
  Gs_test = Gs[*,test_indices]
  Ls_test = Ls[*,test_indices]
  m = n_elements(train_indices)
  Gs = Gs[*,train_indices]
  Ls = Ls[*,train_indices]

; train the classifier
  mn =  fltarr(num_bands,K)
  cov = fltarr(num_bands,num_bands,K)
  class_names = strarr(K+1)
  class_names[0]='unclassified'
  void = max(transpose(Ls),labels,dimension=2)
  labels = byte((labels/m))
  FOR i=0,K-1 DO BEGIN
     class_names[i+1]='class'+string(i+1)
     indices = where(labels EQ i,count)
     IF count GT 1 THEN BEGIN
        GGs = Gs[*,indices]
        FOR j=0,num_bands-1 DO mn[j,i] = mean(GGs[j,*])
        cov[*,*,i] = correlate(GGs,/covariance)
     ENDIF
  ENDFOR
```

Listing 6.3: Excerpt from the Python module `supervisedclass.py`.

```
IMPORT numpy as np
FROM mlpy IMPORT MaximumLikelihoodC , LibSvm

CLASS Maxlike(MaximumLikelihoodC):

    DEF __init__(self,Gs,ls):
        MaximumLikelihoodC.__init__(self)
        self._K = ls.shape[1]
        self._Gs = Gs
        self._N = Gs.shape[1]
        self._ls = ls

    DEF train(self):
        TRY:
            labels = np.argmax(self._ls,axis=1)
            idx = np.where(labels == 0)[0]
            ls = np.ones(LEN(idx),dtype=np.INT)
            Gs = self._Gs[idx,:]
            FOR k IN RANGE(1,self._K):
                idx = np.where(labels == k)[0]
                ls = np.concatenate((ls, \
                (k+1)*np.ones(LEN(idx),dtype=np.INT)))
                Gs = np.concatenate((Gs,\
                               self._Gs[idx,:]),axis=0)
            self.learn(Gs,ls)
            RETURN True
        EXCEPT Exception as e:
            PRINT 'Error:␣%s'%e
            RETURN False

    DEF classify(self,Gs):
        classes = self.pred(Gs)
        RETURN (classes, None)
```

dataset in the ratio 2:1, is described in Appendix C. It makes use of the ENVI batch procedure `CLASS_DOIT` to perform the actual classification of both the test data and the full image and can serve as an ENVI/IDL extension. Test results can optionally be saved to a file in a format consistent with that used by the other classification routines to be described in the remainder of this chapter. Classification evaluation using the test results will be discussed in Chapter 7.

An excerpt from `MAXLIKE_RUN` is shown in Listing 6.2. The labeled observations are stored in data matrix format in the $N \times m$ array variable `Gs` and the corresponding labels in the $K \times m$ array variable `Ls`. Here, m is the number of ground reference observations, N is their dimensionality and K is the number of classes. (The individual labels are in fact K-element arrays with zeroes everywhere except at the position of the class. For example, if there are 5 classes, class 2 corresponds to the label array $(0, 1, 0, 0, 0)$. This convention will turn out to be convenient when we come to consider neural network classifiers.) In lines 6 to 13 the training set is split, with a random sample (with replacement) of 1/3 of the observations held back in the variables `GS_TEST, LS_TEST`. In lines 27 and 28 the means and covariance matrices of the individual classes are estimated from the remaining training pixels. They serve as input to the `CLASS_DOIT` batch procedure (not shown).

All of the Python scripts for supervised classification discussed in this Chapter are bundled for convenience into the module `supervisedclass.py`; see Appendix D. In particular, the Python script for maximum likelihood classification takes advantage of `mlpy`, an open source package for machine learning described by Albanese et al. (2012). Among a large number of other algorithms and utilities, this package exports the Python object class `MaximumLikelihoodC`. Listing 6.3 shows the corresponding excerpt from `supervisedvlass.py`. The class `Maxlike` specializes (inherits from) `MaximumLikelihoodC`, primarily in order to convert from the class labeling convention discussed above to the one required by `mlpy`. This occurs in lines 15–24, following which the inherited `learn()` method is invoked on the training data. New observations are classified in line 32 with the `pred()` (predict) method inherited from `MaximumLikelihoodC` and wrapped in the method `Maxlike.classify()`. Since `pred()` does not calculate class membership probabilities, `None` is returned as a place-holder. The front-end routine `classify.py`, also described in Appendix D, reads the image and training data, creates an instance of a classifier (in this case `Maxlike`) and then generates both a thematic map and test results file.

An example of a maximum likelihood classification is shown in Figure 6.4, top left. The input data consisted of the first four principal components of Jülich ASTER scene of Figure 6.1 together with the training data of Figure 6.3.

6.4 Gaussian kernel classification

Nonparametric classification models may estimate the class-specific probability densities $p(\boldsymbol{g} \mid k)$, as in the preceding section, from a set of training data. However, unlike the maximum likelihood classifier, no strong prior assumptions about the nature of the densities are made. In the *Parzen window* approach to nonparametric classification (Duda et al., 2001), each training observation $\boldsymbol{g}(\nu)$, $\nu = 1 \dots m$, is used as the center of a local kernel function. The probability density for class k at a point $\boldsymbol{g}$ is taken to be the average of

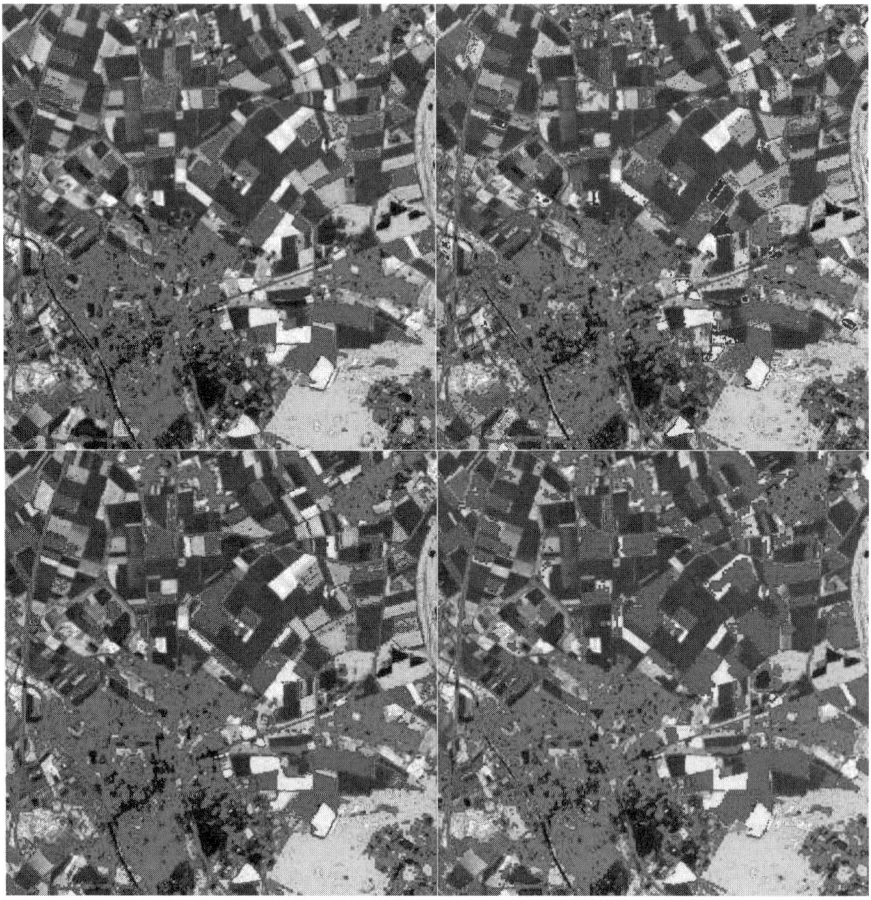

FIGURE 6.4
Supervised classification of a portion of the Jülich ASTER scene: top left: maximum likelihood, top right: Gaussian kernel, bottom left: neural network, bottom right: support vector machine. Five land cover categories (water: blue, settlement: magenta, herbiferous forest: green, rapeseed: yellow, cereal grain: red) are superimposed on VNIR band 3. **(See color insert.)**

the kernel functions for the training data in that class, evaluated at $\boldsymbol{g}$. For example, using a Gaussian kernel, the probability density for the kth class is estimated as

$$p(\boldsymbol{g} \mid k) \approx \frac{1}{m_k} \sum_{\{\nu \mid \ell(\nu)=k\}} \frac{1}{\sqrt{2\pi}\sigma} \exp\left(-\frac{\|\boldsymbol{g}-\boldsymbol{g}(\nu)\|^2}{2\sigma^2}\right). \tag{6.21}$$

The quantity σ is a smoothing parameter, which has been chosen in this case to be class-independent. Since the Gaussian functions are normalized, we have

$$\int_{-\infty}^{\infty} p(\boldsymbol{g} \mid k) d\boldsymbol{g} = \frac{1}{m_k} \sum_{\{\nu \mid \ell(\nu)=k\}} 1 = 1,$$

as required of a probability density. Under fairly general conditions, the right-hand side of Equation (6.21) can be shown to converge to $p(\boldsymbol{g} \mid k)$ as the number of training observations tends to infinity. If we set the prior probabilities in the decision rule, Equation (6.6), equal as before, then an observation $\boldsymbol{g}$ will be assigned to class k when

$$p(\boldsymbol{g} \mid k) \geq p(\boldsymbol{g} \mid j), \quad j = 1 \dots K.$$

Training the Gaussian kernel classifier involves searching for an optimal value of the smoothing parameter σ. Too large a value will wash out the class dependency, and too small a value will lead to poor generalization on new data. Training can be effected very conveniently by minimizing the misclassification rate with respect to σ. When presenting an observation vector $\boldsymbol{g}(\nu)$ to the classifier during the training phase, the contribution to the probability density at the point $\boldsymbol{g}(\nu)$ from class k is, from Equation (6.21) and apart from a constant factor, given by

$$\begin{aligned} p(\boldsymbol{g}(\nu) \mid k) &= \frac{1}{m_k} \sum_{\{\nu' \mid \ell(\nu')=k\}} \exp\left(-\frac{\|\boldsymbol{g}(\nu) - \boldsymbol{g}(\nu')\|^2}{2\sigma^2}\right) \\ &= \frac{1}{m_k} \sum_{\{\nu' \mid \ell(\nu')=k\}} (\boldsymbol{\mathcal{K}})_{\nu\nu'}, \end{aligned} \tag{6.22}$$

where $\boldsymbol{\mathcal{K}}$ is an $m \times m$ Gaussian kernel matrix. It is advisable to delete the contribution of $\boldsymbol{g}(\nu)$ itself to the sum in the above equation in order to avoid biasing the classification in favor of the training observation's own label, a bias which would otherwise arise due to the appearance of a zero in the exponent and a dominating contribution to $p(\boldsymbol{g}(\nu) \mid k)$; see Masters (1995). This amounts to zeroing the diagonal of $\boldsymbol{\mathcal{K}}$ before performing the sum in Equation (6.22).

In Appendix C an ENVI/IDL extension `KERNEL_RUN` is described which implements these ideas. Listings 6.4 and 6.5 show excerpts. The function `OUTPUT(SIGMA,HS)` in Listing 6.4 calculates the arrays of class probability densities for all pixel vectors in the data matrix `HS` using the current value of the smoothing parameter `SIGMA`. The common block variable `GS` contains the training pixels, also in data matrix format. Their labels are stored in the m-dimensional array `ELLS`. In the training phase, the keyword `SYMM` is set, indicating that `HS` and `GS` are in fact identical. The diagonal of the symmetric kernel matrix `KAPPA` is then set to zero (line 19). The sums in Equation (6.22) are calculated in the `FOR`-loop in lines 20 to 25. As was the case for the kernel

Listing 6.4: Excerpt from the program module KERNEL_RUN.PRO.

```
FUNCTION Output, sigma, Hs, symm=symm
COMPILE_OPT STRICTARR
   COMMON examples, Gs, Gs_gpu, ells, K, m, cuda
   IF n_elements(symm) EQ 0 THEN symm = 0
   result = fltarr(n_elements(Hs[0,*]),K)
   IF cuda THEN BEGIN
      IF symm THEN Kappa_gpu = $
        gpukernel_matrix(Gs_gpu,gma=0.5/sigma^2) $
      ELSE BEGIN
         Hs_gpu = gpuputarr(Hs)
         Kappa_gpu = gpukernel_matrix(Gs_gpu, $
                              Hs_gpu,gma=0.5/sigma^2)
         gpufree,Hs_gpu
      ENDELSE
      Kappa = gpugetarr(Kappa_gpu)
      gpufree,Kappa_gpu
   END $
   ELSE Kappa=gausskernel_matrix(Gs,Hs,gma=0.5/sigma^2)
   IF symm THEN Kappa[indgen(m),indgen(m)] = 0.0
   FOR j=0,K-1 DO BEGIN
      Kpa = Kappa
      idx = where(ells NE j, ncomplement=nj)
      Kpa[*,idx] = 0
      result[*,j] = total(Kpa,2)/nj
   ENDFOR
   RETURN, result
END

FUNCTION Theta, sigma
COMPILE_OPT STRICTARR
   COMMON examples, Gs, Gs_gpu, ells, K, m, cuda
   _ = max(output(sigma,Gs,/symm),labels,dimension=2)
   _ = where(labels/m NE ells, count)
   result = float(count)/m
   oplot,[alog10(sigma)],[result],psym=5,color=0
   RETURN, result
```

PCA program (Section 4.4.2), this ENVI extension makes use of the GPULib interface to CUDA, if available, in order to accelerate calculation.

The minimization of the misclassification rate with respect to σ takes place in two steps (Listing 6.5): First the minimum is bracketed by a call to procedure MINF_BRACKET with initial estimates $0.1 \le \sigma \le 10$, line 28, then it is approximated iteratively using Brent's parabolic interpolation method, procedure MINF_PARABOLIC, line 39. The two procedures are described in Press et al. (2002), and the code was taken from the IDL Astronomy User's Library (see

Listing 6.5: Excerpt from the program module KERNEL_RUN.PRO.

```
; set training class labels in common block
   _ = max(transpose(Ls),ells,dimension=2)
   ells = ells/m

; set GPU array for training data if CUDA available
   cuda = 0
   IF gpu_detect() THEN BEGIN
      PRINT,'Running␣CUDA...'
      cuda = 1
      IF m GT 2000 THEN BEGIN
;        too many training data
         PRINT, 'Subsampling␣to␣2000␣training␣examples'
         idx = randomu(seed,2000,/long) MOD m
         ells = ells[idx]
         Gs = Gs[*,idx]
         m = 2000
      ENDIF
      Gs_gpu = gpuputarr(Gs)
   END ELSE PRINT,'CUDA␣not␣available␣...'

   window,12,xsize=600,ysize=400, $
     title='Bracketing,␣please␣wait␣...'
   wset, 12
   widget_control, /hourglass

; bracket the minimum
   s1=0.1  & s2=10.0
   minF_bracket, s1,s2,s3, FUNC_NAME="Theta"

   window,12,xsize=600,ysize=400, $
     title='Iteration␣...'
   PLOT,[alog10(s1),alog10(s3)],[theta(s1),theta(s3)],$
      psym=5,color=0,background='FFFFFF'XL, $
      xtitle = 'log(sigma)', $
      ytitle = 'theta'

; hunt it down
   minF_parabolic, s1,s2,s3, sigma, theta_min, $
                 FUNC_NAME='Theta'
```

Section 5.5.4). The function THETA(SIGMA), Listing 6.4, passed to both procedures, calculates the misclassification rate with a call to OUTPUT(SIGMA,GS,/SYMM). Figure 6.5 shows a plot of the misclassification rate vs. σ during the training phase. Classification of the entire image proceeds row-by-row using ENVI's tiling facility with repeated calls to OUTPUT(SIGMA,HS), where SIGMA is fixed to

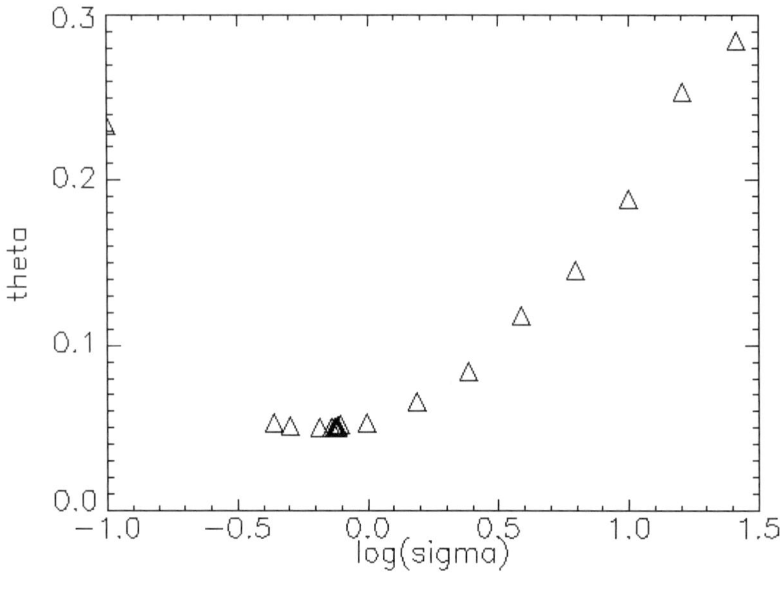

FIGURE 6.5
Training the Gaussian kernel classifier with parabolic interpolation.

its optimum value and where HS is a row of image pixel vectors in data matrix format. A classification result, again for the Jülich ASTER scene, is shown in Figure 6.4, top right. As with the modified maximum likelihood classifier, the labeled observations can be split into training and test pixels with the latter held back for later accuracy evaluation.

The Gaussian kernel classifier, like many other nonparametric methods, suffers from the drawback of requiring that all training data points be used in the generalization phase (a so-called *memory-based* classifier). Evaluation is very slow if the number of training points is large, which, on the other hand, it should be if a reasonable approximation of the class-specific densities is to be achieved. The routine KERNEL_RUN is in fact unacceptably slow for training datasets exceeding a few thousand pixels, even with CUDA enabled.* Moreover, the required number of training samples grows exponentially with the dimensionality N of the data — the so-called *curse of dimensionality* (Bellman, 1961). Quite generally, the complexity of the calculation is determined by the amount of training data, not by the difficulty of the classification problem itself — an undesirable state of affairs.

*For this reason we won't bother with a Python implementation, but see the Parzen classifier described in the mlpy documentation (Albanese et al., 2012).

6.5 Neural networks

Neural networks belong to the category of semiparametric models for probability density estimation, a category which lies somewhere between the parametric and nonparametric extremes (Bishop, 1995). They make no strong assumptions about the form of the probability distributions and can be adjusted flexibly to the complexity of the system that they are being used to model. They therefore provide an attractive compromise.

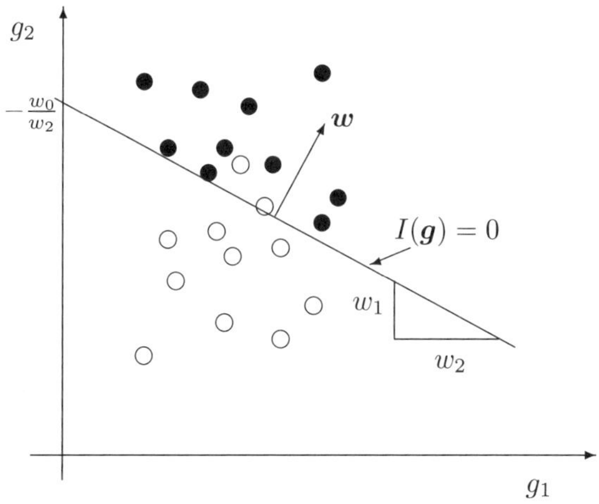

FIGURE 6.6
A linear discriminant for two classes. The vector $\boldsymbol{w} = (w_1, w_2)^\top$ is normal to the separating line in the direction of class $k = 1$, shown as black dots.

To motivate their use for classification, let us consider two classes $k = 1$ and $k = 2$ for two-dimensional observations $\boldsymbol{g} = (g_1, g_2)^\top$. We can write the maximum likelihood decision rule, Equation (6.18) with Equation (6.17), in terms of a new discriminant function

$$I(\boldsymbol{g}) = d_1(\boldsymbol{g}) - d_2(\boldsymbol{g})$$

and say that

$$\boldsymbol{g} \text{ is class } \begin{cases} 1 & \text{if } I(\boldsymbol{g}) \geq 0 \\ 2 & \text{if } I(\boldsymbol{g}) < 0. \end{cases}$$

The discriminant $I(\boldsymbol{g})$ is a rather complicated quadratic function of $\boldsymbol{g}$. The *simplest* discriminant that could conceivably decide between the two classes is a linear function of the form*

$$I(\boldsymbol{g}) = w_0 + w_1 g_1 + w_2 g_2, \tag{6.23}$$

where w_0, w_1 and w_2 are parameters. The decision boundary occurs for $I(\boldsymbol{g}) = 0$, i.e., for

$$g_2 = -\frac{w_1}{w_2} g_1 - \frac{w_0}{w_2},$$

as depicted in Figure 6.6.

Extending discussion now to N-dimensional observations, we can work with the discriminant

$$I(\boldsymbol{g}) = w_0 + w_1 g_1 + \ldots + w_N g_N = \boldsymbol{w}^\top \boldsymbol{g} + w_0. \tag{6.24}$$

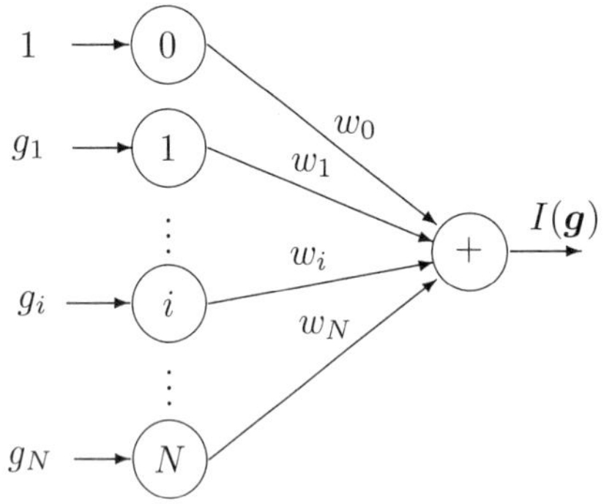

FIGURE 6.7
An artificial neuron representing Equation (6.24). The first input is always unity and is called the bias.

In this higher dimensional feature space, the decision boundary $I(\boldsymbol{g}) = 0$ generalizes to an *oriented hyperplane*. Equation (6.24) can be represented schematically as an *artificial neuron* or *perceptron*, as shown Figure 6.7, along with some additional jargon. Thus the "input signals" $g_1 \ldots g_N$ are multiplied with "synaptic weights" $w_1 \ldots w_N$ and the results arc summed in a "neuron" to produce the "output signal" $I(\boldsymbol{g})$. The w_0 term is treated by introducing a "bias" input of unity, which is multiplied by w_0 and included in the summation.

In keeping with the biological analogy, the output $I(\boldsymbol{g})$ may be modified by a so-called *sigmoid* (= S-shaped) "activation function", for example by the *logistic* function

$$f(\boldsymbol{g}) = \frac{1}{1 + e^{-I(\boldsymbol{g})}}.$$

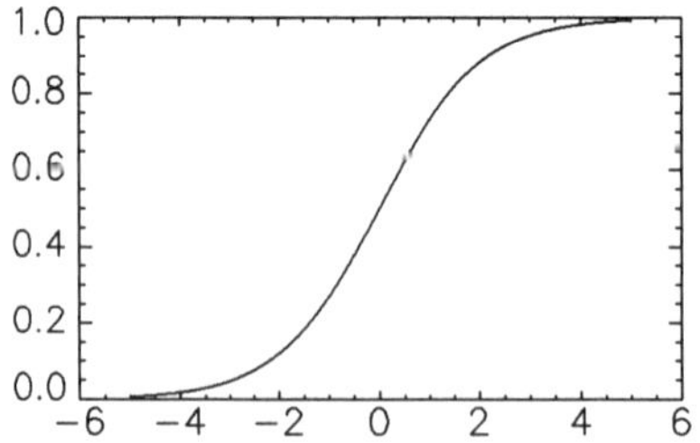

FIGURE 6.8
The logistic activation function.

$I(\boldsymbol{g})$ is then referred to as the *activation* of the neuron inducing the output signal $f(\boldsymbol{g})$. The IDL command `PLOT, x,1/(1+exp(-x))` generates the plot of the logistic function shown in

*A linear decision boundary will arise in a maximum likelihood classifier if the covariance matrices for the two classes are identical, see Exercise 2.

Figure 6.8. This modification of the discriminant has the advantage that the output signal saturates at values zero or one for large negative or positive inputs, respectively. However, Bishop (1995) suggests that there is also a good statistical justification for using it. Suppose the two classes are normally distributed with $\boldsymbol{\Sigma}_1 = \boldsymbol{\Sigma}_2 = \boldsymbol{I}$. Then

$$p(\boldsymbol{g} \mid k) = \frac{1}{2\pi} \exp\Big(\frac{-\|\boldsymbol{g} - \boldsymbol{\mu}_k\|^2}{2}\Big),$$

for $k = 1, 2$, and we have with Bayes' Theorem

$$\begin{aligned}
\Pr(1 \mid \boldsymbol{g}) &= \frac{p(\boldsymbol{g} \mid 1)\Pr(1)}{p(\boldsymbol{g} \mid 1)\Pr(1) + p(\boldsymbol{g} \mid 2)\Pr(2)} \\
&= \frac{1}{1 + p(\boldsymbol{g} \mid 2)\Pr(2)/(p(\boldsymbol{g} \mid 1)\Pr(1))} \\
&= \frac{1}{1 + \exp(-\frac{1}{2}[\|\boldsymbol{g} - \boldsymbol{\mu}_2\|^2 - \|\boldsymbol{g} - \boldsymbol{\mu}_1\|^2])(\Pr(2)/\Pr(1))}.
\end{aligned}$$

With the substitution

$$e^{-a} = \Pr(2)/\Pr(1)$$

we get

$$\begin{aligned}
\Pr(1 \mid \boldsymbol{g}) &= \frac{1}{1 + \exp(-\frac{1}{2}[\|\boldsymbol{g} - \boldsymbol{\mu}_2\|^2 - \|\boldsymbol{g} - \boldsymbol{\mu}_1\|^2] - a)} \\
&= \frac{1}{1 + \exp(-\boldsymbol{w}^\top \boldsymbol{g} - w_0)} \\
&= \frac{1}{1 + e^{-I(\boldsymbol{g})}} = f(\boldsymbol{g}).
\end{aligned}$$

In the second equality above we have made the additional substitutions

$$\begin{aligned}
\boldsymbol{w} &= \boldsymbol{\mu}_1 - \boldsymbol{\mu}_2 \\
w_0 &= -\frac{1}{2}\|\boldsymbol{\mu}_1\|^2 + \frac{1}{2}\|\boldsymbol{\mu}_2\|^2 + a.
\end{aligned}$$

Thus we expect that the output signal $f(\boldsymbol{g})$ of the neuron will not only discriminate between the two classes, *but also that it will approximate the posterior class membership probability* $\Pr(1 \mid \boldsymbol{g})$.

The extension of linear discriminants from two to K classes is straightforward, and leads to the *single-layer neural network* of Figure 6.9. There are K neurons (the circles on the right), each of which calculates its own discriminant

$$n_j(\boldsymbol{g}) = f(I_j(\boldsymbol{g})), \ j = 1 \dots K.$$

The observation $\boldsymbol{g}$ is assigned to the class whose neuron produces the maximum output signal, i.e.,

$$k = \arg\max_j n_j(\boldsymbol{g}).$$

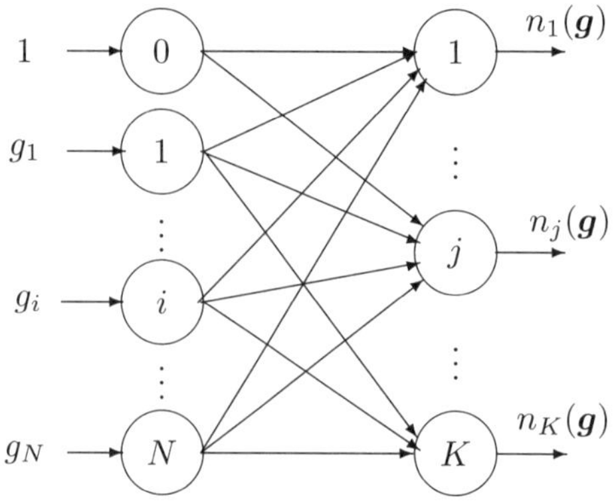

FIGURE 6.9
A single-layer neural network.

Each neuron is associated with a synaptic weight vector $\boldsymbol{w}_j$, which we from now on will understand to include the bias weight w_0. Thus, for the jth neuron,

$$\boldsymbol{w}_j = (w_{0j}, w_{1j} \dots w_{Nj})^\top,$$

and, for the whole network,

$$\boldsymbol{W} = (\boldsymbol{w}_1, \boldsymbol{w}_2 \dots \boldsymbol{w}_K) = \begin{pmatrix} w_{01} & w_{02} & \cdots & w_{0K} \\ w_{11} & w_{12} & \cdots & w_{1K} \\ \vdots & \vdots & \ddots & \vdots \\ w_{N1} & w_{N2} & \cdots & w_{NK} \end{pmatrix}, \tag{6.25}$$

which we shall call the *synaptic weight matrix* for the neuron layer.

6.5.1 The neural network classifier

Single-layer networks turn out to be rather limited in the kind of classification tasks that they can handle. In fact, only so-called *linearly separable* problems, problems in which classes of training observations can be separated by hyperplanes, are fully solvable; see Exercise 3. On the other hand, networks with just one additional layer of processing neurons can approximate any given decision boundary arbitrarily closely (Bishop, 1995) provided that the first, or *hidden*, layer outputs are nonlinear. The input data are transformed by the hidden layer into a higher dimensional function space in which the problem becomes linearly separable (Müller et al., 2001). This is also the strategy used for designing *support vector machines* (Belousov et al., 2002), as we will see later.

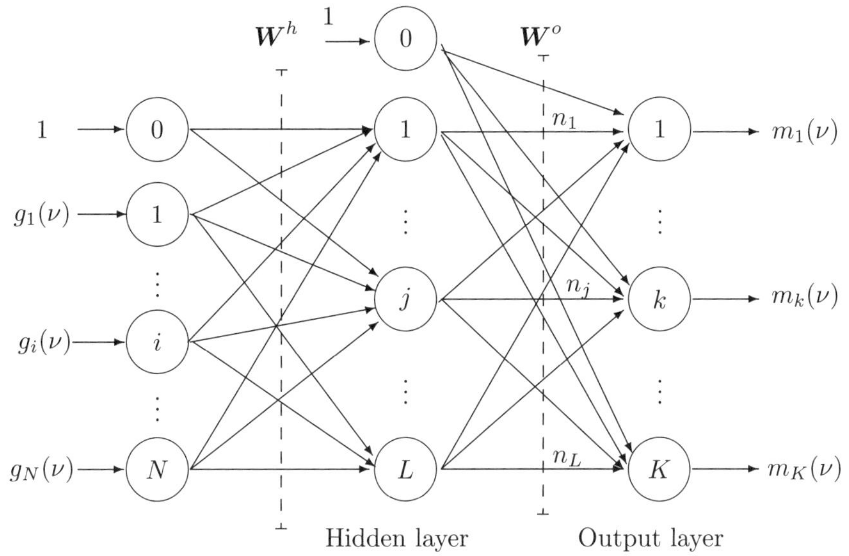

FIGURE 6.10
A two-layer feed-forward neural network with L hidden neurons for classification of N-dimensional data into K classes. The argument ν identifies a training example.

Accordingly, we shall develop a classifier based on the two-layer, feed-forward architecture* shown in Figure 6.10. For the νth training pixel, the input to the network is the $(N+1)$-component (biased) observation vector

$$\boldsymbol{g}(\nu) = (1, g_1(\nu) \dots g_N(\nu))^\top.$$

This input is distributed simultaneously to all of the L neurons in the hidden layer of neurons. These in turn determine an $(L+1)$-component vector of intermediate outputs (adding in the bias input for the next layer)

$$\boldsymbol{n}(\nu) = (1, n_1(\nu) \dots n_L(\nu))^\top$$

in which $n_j(\nu)$ is shorthand for

$$f(I_j^h(\boldsymbol{g}(\nu))), \quad j = 1 \dots L.$$

In this expression, the activation I_j^h of the hidden neurons is given by

$$I_j^h(\boldsymbol{g}(\nu)) = \boldsymbol{w}_j^{h\top} \boldsymbol{g}(\nu),$$

*The adjective *feed-forward* merely serves to differentiate this network structure from other networks having *feedback* connections.

where the vector $\boldsymbol{w}_j^h$ is the weight vector for the jth neuron in the hidden layer,

$$\boldsymbol{w}_j^h = (w_{0j}^h, w_{1j}^h \dots w_{Nj}^h)^\top.$$

In terms of a hidden weight matrix $\boldsymbol{W}^h$ having the form of Equation (6.25), namely

$$\boldsymbol{W}^h = (\boldsymbol{w}_1^h, \boldsymbol{w}_2^h \dots \boldsymbol{w}_L^h),$$

we can write all of this more compactly in vector notation as

$$\boldsymbol{n}(\nu) = \begin{pmatrix} 1 \\ f(\boldsymbol{W}^{h\top} \boldsymbol{g}(\nu)) \end{pmatrix}. \tag{6.26}$$

Here we just have to interpret the logistic function of a vector $\boldsymbol{v}$, $f(\boldsymbol{v})$, as a vector of logistic functions of the components of $\boldsymbol{v}$.

The vector $\boldsymbol{n}(\nu)$ is then fed in the same manner to the *output layer* with its associated output weight matrix

$$\boldsymbol{W}^o = (\boldsymbol{w}_1^o, \boldsymbol{w}_2^o \dots \boldsymbol{w}_K^o),$$

and the output signal $\boldsymbol{m}(\nu)$ is calculated, in a manner similar to Equation (6.26), as

$$\boldsymbol{m}(\nu) = f(\boldsymbol{W}^{o\top} \boldsymbol{n}(\nu)). \tag{6.27}$$

However, this last equation is not quite satisfactory. According to our previous considerations, we would like to interpret the network outputs as class membership probabilities. That means that we must ensure that

$$0 \le m_k(\nu) \le 1, \quad k = 1 \dots K,$$

and, furthermore, that

$$\sum_{k=1}^{K} m_k(\nu) = 1.$$

The logistic function f satisfies the first condition, but there is no reason why the second condition should be met. It can be enforced, however, by using a modified logistic activation function for the output neurons, called *softmax* (Bridle, 1990). The softmax function is defined as

$$m_k(\nu) = \frac{e^{I_k^o(\boldsymbol{n}(\nu))}}{e^{I_1^o(\boldsymbol{n}(\nu))} + e^{I_2^o(\boldsymbol{n}(\nu))} + \dots + e^{I_M^o(\boldsymbol{n}(\nu))}}, \tag{6.28}$$

where

$$I_k^o(\boldsymbol{n}(\nu)) = \boldsymbol{w}_k^{o\top} \boldsymbol{n}(\nu), \quad k = 1 \dots K, \tag{6.29}$$

and clearly guarantees that the output signals sum to unity.

The Equations (6.26), (6.28) and (6.29) now provide a complete mathematical representation of the neural network classifier shown in Figure 6.10. It turns out to be a very useful classifier indeed. To quote Bishop (1995):

Listing 6.6: Excerpt from the object class FFN.

```
FUNCTION FFN::Init, Gs, Ls, L
; network architecture
   self.LL = L
   self.np = n_elements(Gs[*,0])
   self.NN = n_elements(Gs[0,*])
   self.KK = n_elements(Ls[0,*])
; biased output vector from hidden layer
   self.N= ptr_new(dblarr(L+1))
; biased exemplars (column vectors)
   self.Gs = ptr_new([[dblarr(self.np)+1],[Gs]])
   self.Ls = ptr_new(Ls)
; weight matrices
   self.Wh = ptr_new(randomu(seed,L,self.NN+1,/double)$
                                                -0.5)
   self.Wo = ptr_new(randomu(seed,self.KK,L+1,/double)$
                                                -0.5)
   RETURN,1
END

FUNCTION FFN::forwardPass, G
; forward pass through network
   expnt = transpose(*self.Wh)##G
   *self.N = [[1.0],[1/(1+exp(-expnt))]]
; softmax activation for output neurons
   I = transpose(*self.Wo)##*self.N
   A = exp(I-max(I))
   RETURN, A/total(A)
END

FUNCTION FFN::classify, Gs, Probs
; vectorized class membership probabilities
   nx = n_elements(Gs[*,0])
   Ones = dblarr(nx) + 1.0
   expnt = transpose(*self.Wh)##[[Ones],[Gs]]
   N = [[Ones],[1/(1+exp(-expnt))]]
   Io = transpose(*self.Wo)##N
   maxIo = max(Io,dimension=2)
   FOR k=0,self.KK-1 DO Io[*,k]=Io[*,k]-maxIo
   A = exp(Io)
   sum = total(A,2)
   Probs = fltarr(nx,self.KK)
   FOR k=0,self.KK-1 DO Probs[*,k] = A[*,k]/sum
   void = max(probs,labels,dimension=2)
   RETURN, byte(labels/nx+1)
END
```

Listing 6.7: Excerpt from the Python module supervisedclass.py.

```
    DEF __init__(self,Gs,ls,L):
#       setup the network architecture
        self._L=L
        self._m,self._N = Gs.shape
        self._K = ls.shape[1]
#       biased input as column vectors
        Gs = np.mat(Gs).T
        self._Gs = np.vstack((np.ones(self._m),Gs))
#       biased output vector from hidden layer
        self._n = np.mat(np.zeros(L+1))
#       labels as column vectors
        self._ls = np.mat(ls).T
#       weight matrices
        self._Wh=np.mat(np.random. \
                        random((self._N+1,L)))-0.5
        self._Wo=np.mat(np.random. \
                        random((L+1,self._K)))-0.5
    DEF forwardpass(self,G):
#       forward pass through the network
        expnt = self._Wh.T*G
        self._n = np.vstack((np.ones(1),1.0/ \
                                  (1+np.exp(-expnt))))
#       softwmax activation
        I = self._Wo.T*self._n
        A = np.exp(I-MAX(I))
        RETURN A/np.SUM(A)

    DEF classify(self,Gs):
#       vectorized classes and membership probabilities
        Gs = np.mat(Gs).T
        m = Gs.shape[1]
        Gs = np.vstack((np.ones(m),Gs))
        expnt = self._Wh.T*Gs
        expnt[np.where(expnt<-100.0)] = -100.0
        expnt[np.where(expnt>100.0)] = 100.0
        n=np.vstack((np.ones(m),1/(1+np.exp(-expnt))))
        Io = self._Wo.T*n
        maxIo = np.MAX(Io,axis=0)
        FOR k IN RANGE(self._K):
            Io[k,:] -= maxIo
        A = np.exp(Io)
        sm = np.SUM(A,axis=0)
        Ms = np.zeros((self._K,m))
        FOR k IN RANGE(self._K):
            Ms[k,:] = A[k,:]/sm
        classes = np.argmax(Ms,axis=0)+1
        RETURN (classes, Ms)
```

> ... [two-layer, feed-forward] networks can approximate arbitrarily well any functional continuous mapping from one finite dimensional space to another, provided the number $[L]$ of hidden units is sufficiently large. ... An important corollary of this result is, that in the context of a classification problem, networks with sigmoidal non-linearities and two layers of weights can approximate any decision boundary to arbitrary accuracy. ... More generally, the capability of such networks to approximate general smooth functions allows them to model posterior probabilities of class membership.

The IDL and Python object classes `FFN` (Appendix C) and `supervisedclass.Ffn` (Appendix D), excerpts of which are given in Listings 6.6 and 6.7, mirror the network architecture of Figure 6.10. They will form the basis for the implementation of the backpropagation training algorithm developed below and also for the more efficient training algorithms described in Appendix B.

6.5.2 Cost functions

We have not yet considered the correct choice of synaptic weights, that is, how to go about training the neural network classifier. As mentioned in Section 6.3.2, the training data are most conveniently represented as the set of labeled pairs

$$\mathcal{T} = \{(\boldsymbol{g}(\nu), \boldsymbol{\ell}(\nu)) \mid \nu = 1 \dots m\},$$

where the label

$$\boldsymbol{\ell}(\nu) = (0 \dots 0, 1, 0 \dots 0)^\top$$

is a K-dimensional column vector of zeroes, except with the "1" at the kth position to indicate that $\boldsymbol{g}(\nu)$ belongs to class k.

Under certain assumptions about the distribution of the training data, the *quadratic cost function*

$$E(\boldsymbol{W}^h, \boldsymbol{W}^o) = \frac{1}{2}\sum_{\nu=1}^{m} \|\boldsymbol{\ell}(\nu) - \boldsymbol{m}(\nu)\|^2 \tag{6.30}$$

can be justified as a training criterion for feed-forward networks (Exercise 4). The network weights $\boldsymbol{W}^h$ and $\boldsymbol{W}^o$ must be adjusted so as to minimize E. This minimization will clearly tend to make the network produce the output signal

$$\boldsymbol{m} = (1, 0 \dots 0 \dots 0)^\top$$

whenever it is presented with a training observation $\boldsymbol{g}(\nu)$ from class $k = 1$, and similarly for the other classes.

This, of course, is what we wish it to do. However a more appropriate cost function for classification problems can be obtained with the maximum likelihood criterion: Choose the synaptic weights so as to *maximize the probability of observing the training data.* The joint probability for observing the training example $(\boldsymbol{g}(\nu), \boldsymbol{\ell}(\nu))$ is

$$\Pr(\boldsymbol{g}(\nu), \boldsymbol{\ell}(\nu)) = \Pr(\boldsymbol{\ell}(\nu) \mid \boldsymbol{g}(\nu))\Pr(\boldsymbol{g}(\nu)), \tag{6.31}$$

where we have used Equation (2.59). The neural network, as was argued, approximates the posterior class membership probability $\Pr(\boldsymbol{\ell}(\nu) \mid \boldsymbol{g}(\nu))$. In fact, this probability can be expressed directly in terms of the network output signal in the form

$$\Pr(\boldsymbol{\ell}(\nu) \mid \boldsymbol{g}(\nu)) = \prod_{k=1}^{K} [\ m_k(\boldsymbol{g}(\nu))\ ]^{\ell_k(\nu)}. \tag{6.32}$$

In order to see this, consider the case $\boldsymbol{\ell} = (1, 0 \ldots 0)^\top$. Then, according to Equation (6.32),

$$\Pr((1, 0 \ldots 0)^\top \mid \boldsymbol{g}) = m_1(\boldsymbol{g})^1 \cdot m_2(\boldsymbol{g})^0 \cdots m_K(\boldsymbol{g})^0 = m_1(\boldsymbol{g}),$$

which is the probability that $\boldsymbol{g}$ is in class 1, as desired. Now, substituting Equation (6.32) into Equation (6.31), we therefore wish to maximize

$$\Pr(\boldsymbol{g}(\nu), \boldsymbol{\ell}(\nu)) = \prod_{k=1}^{K} [\ m_k(\boldsymbol{g}(\nu))\ ]^{\ell_k(\nu)} \Pr(\boldsymbol{g}(\nu)).$$

Taking logarithms, dropping terms which are independent of the synaptic weights and summing over all of the training data, we see that this is equivalent to minimizing the *cross-entropy* cost function

$$E(\boldsymbol{W}^h, \boldsymbol{W}^o) = -\sum_{\nu=1}^{m} \sum_{k=1}^{K} \ell_k(\nu) \log[m_k(\boldsymbol{g}(\nu))] \tag{6.33}$$

with respect to the synaptic weight parameters.

6.5.3 Backpropagation

A minimum of the cost function, Equation (6.33), can be found with various search algorithms. *Backpropagation* is the most well-known and extensively used method and is described below. It is implemented in the standard ENVI neural network for supervised classification under the main menu command

```
Classification/Supervised/Neural Net
```

Two considerably faster algorithms, *scaled conjugate gradient* and the *Kalman filter*, are discussed in detail in Appendix B. An ENVI/IDL extension and Python script for supervised classification with a feed-forward neural network trained with these algorithms are described in Appendices C and D.

The discussion in Appendix B builds on the following development and coding of the backpropagation algorithm. Our starting point is the so-called *local version* of the cost function of Equation (6.33),

$$E(\boldsymbol{W}^h, \boldsymbol{W}^o, \nu) = -\sum_{k=1}^{K} \ell_k(\nu) \log[m_k(\boldsymbol{g}(\nu))], \quad \nu = 1 \ldots m.$$

This is just the cost function for a single training example. If we manage to make it smaller at each step of the calculation and cycle, either sequentially or randomly, through the available training pairs, then we are obviously minimizing the overall cost function as well. With the abbreviation $m_k(\nu) = m_k(\boldsymbol{g}(\nu))$, the local cost function can be written a little more compactly as

$$E(\nu) = -\sum_{k=1}^{K} \ell_k(\nu) \log[m_k(\nu)].$$

Here the dependence of the cost function on the synaptic weights is also implicit. More compactly still, it can be represented in vector form as an inner product:

$$E(\nu) = -\boldsymbol{\ell}(\nu)^\top \log[\boldsymbol{m}(\nu)]. \tag{6.34}$$

Our problem then is to minimize Equation (6.34) with respect to the synaptic weights, which are the $(N+1) \times L$ elements of the matrix $\boldsymbol{W}^h$ and the $(L+1) \times K$ elements of $\boldsymbol{W}^o$. Let us consider the following algorithm:

Algorithm (Backpropagation)

1. Initialize the synaptic weights with random numbers and set ν equal to a random integer in the interval $[1, m]$.

2. Choose training pair $(\boldsymbol{g}(\nu), \boldsymbol{\ell}(\nu))$ and determine the output response $\boldsymbol{m}(\nu)$ of the network.

3. For $k = 1 \ldots K$ and $j = 0 \ldots L$, replace w^o_{jk} with $w^o_{jk} - \eta \frac{\partial E(\nu)}{\partial w^o_{jk}}$.

4. For $j = 1 \ldots L$ and $i = 0 \ldots N$, replace w^h_{ij} with $w^h_{ij} - \eta \frac{\partial E(\nu)}{\partial w^h_{ij}}$.

5. If $\sum_\nu E(\nu)$ ceases to change significantly, stop, otherwise set ν equal to a new random integer in $[1, m]$ and go to step 2.

The algorithm jumps randomly through the training data, reducing the local cost function at each step. The reduction is accomplished by changing each synaptic weight w by an amount proportional to the negative slope $-\partial E(\nu)/\partial w$ of the local cost function with respect to that weight parameter, stopping when the overall cost function, Equation (6.33), can no longer be reduced. The constant of proportionality η is referred to as the *learning rate* for the network. This algorithm only makes use of the first derivatives of the cost function with respect to the synaptic weight parameters and belongs to the class of *gradient descent* methods.

To implement the algorithm, the partial derivatives of $E(\nu)$ with respect to the synaptic weights are required. Let us begin with the output neurons, which generate the softmax output signals

$$m_k(\nu) = \frac{e^{I_k^o(\nu)}}{e^{I_1^o(\nu)} + e^{I_2^o(\nu)} + \ldots + e^{I_K^o(\nu)}}, \tag{6.35}$$

where

$$I_k^o(\nu) = \boldsymbol{w}_k^{o\top}\boldsymbol{n}(\nu).$$

We wish to determine (step 3 of the backpropagation algorithm)

$$\frac{\partial E(\nu)}{\partial w_{jk}^o}, \quad j = 0 \ldots L, \; k = 1 \ldots K.$$

Recalling the rules for vector differentiation in Chapter 1 and applying the chain rule we get

$$\frac{\partial E(\nu)}{\partial \boldsymbol{w}_k^o} = \frac{\partial E(\nu)}{\partial I_k^o(\nu)}\frac{\partial I_k^o(\nu)}{\partial \boldsymbol{w}_k^o} = -\delta_k^o(\nu)\boldsymbol{n}(\nu), \quad k = 1 \ldots K, \tag{6.36}$$

where we have introduced the quantity $\delta_k^o(\nu)$ given by

$$\delta_k^o(\nu) = -\frac{\partial E(\nu)}{\partial I_k^o(\nu)}. \tag{6.37}$$

This is the negative rate of change of the local cost function with respect to the activation of the kth output neuron.

Again, applying the chain rule and using Equations (6.34) and (6.35),

$$\begin{aligned} -\delta_k^o(\nu) = \frac{\partial E(\nu)}{\partial I_k^o(\nu)} &= \sum_{k'=1}^K \frac{\partial E(\nu)}{\partial m_{k'}(\nu)}\frac{\partial m_{k'}(\nu)}{\partial I_k^o(\nu)} \\ &= \sum_{k'=1}^K -\frac{\ell_{k'}(\nu)}{m_{k'}(\nu)}\left(\frac{e^{I_k^o(\nu)}\delta_{kk'}}{\sum_{k''=1}^K e^{I_{k''}^o(\nu)}} - \frac{e^{I_{k'}^o(\nu)}e^{I_k^o(\nu)}}{(\sum_{k''=1}^K e^{I_{k''}^o(\nu)})^2}\right). \end{aligned}$$

Here, $\delta_{kk'}$ is given by

$$\delta_{kk'} = \begin{cases} 0 & \text{if } k \neq k' \\ 1 & \text{if } k = k'. \end{cases}$$

Continuing, making use of Equation (6.35),

$$\begin{aligned} -\delta_k^o(\nu) &= \sum_{k'=1}^K -\frac{\ell_{k'}(\nu)}{m_{k'}(\nu)} m_k(\nu)(\delta_{kk'} - m_{k'}(\nu)) \\ &= -\ell_k(\nu) + m_k(\nu)\sum_{k'=1}^K \ell_{k'}(\nu). \end{aligned}$$

But this last sum over the K components of the label $\boldsymbol{\ell}(\nu)$ is just unity, and therefore we have

$$-\delta_k^o(\nu) = -\ell_k(\nu) + m_k(\nu), \quad k = 1 \dots K,$$

which may be written as the K-component vector

$$\boldsymbol{\delta}^o(\nu) = \boldsymbol{\ell}(\nu) - \boldsymbol{m}(\nu). \tag{6.38}$$

From Equation (6.36), we can therefore express the third step in the back-propagation algorithm in the form of the matrix equation (see Exercise 6)

$$\boldsymbol{W}^o(\nu+1) = \boldsymbol{W}^o(\nu) + \eta\, \boldsymbol{n}(\nu)\boldsymbol{\delta}^o(\nu)^\top. \tag{6.39}$$

Here $\boldsymbol{W}^o(\nu+1)$ indicates the synaptic weight matrix *after* the update for νth training pair. Note that the second term on the right-hand side of Equation (6.39) is an outer product, yielding a matrix of dimension $(L+1) \times K$ and so matching the dimension of $\boldsymbol{W}^o(\nu)$.

For the hidden weights, step 4 of the algorithm, we proceed similarly:

$$\frac{\partial E(\nu)}{\partial \boldsymbol{w}_j^h} = \frac{\partial E(\nu)}{\partial I_j^h(\nu)} \frac{\partial I_j^h(\nu)}{\partial \boldsymbol{w}_j^h} = -\delta_j^h(\nu)\boldsymbol{g}(\nu), \quad j = 1 \dots L, \tag{6.40}$$

where $\delta_j^h(\nu)$ is the negative rate of change of the local cost function with respect to the activation of the jth hidden neuron:

$$\delta_j^h(\nu) = -\frac{\partial E(\nu)}{\partial I_j^h(\nu)}.$$

Applying the chain rule again:

$$\begin{aligned}
-\delta_j^h(\nu) &= \sum_{k=1}^{K} \frac{\partial E(\nu)}{\partial I_k^0(\nu)} \frac{\partial I_k^o(\nu)}{\partial I_j^h(\nu)} = -\sum_{k=1}^{K} \delta_k^o(\nu) \frac{\partial I_k^o(\nu)}{\partial I_j^h(\nu)} \\
&= -\sum_{k=1}^{K} \delta_k^o(\nu) \frac{\partial \boldsymbol{w}_k^{o\top} \boldsymbol{n}(\nu)}{\partial I_j^h(\nu)} = -\sum_{k=1}^{K} \delta_k^o(\nu) \boldsymbol{w}_k^{o\top} \frac{\partial \boldsymbol{n}(\nu)}{\partial I_j^h(\nu)}.
\end{aligned}$$

In the last partial derivative, since $I_j^h(\nu) = \boldsymbol{w}_j^{h\top}\boldsymbol{g}(\nu)$, only the output of the jth hidden neuron is a function of $I_j^h(\nu)$. Therefore

$$\delta_j^h(\nu) = \sum_{k=1}^{K} \delta_k^o(\nu) w_{jk}^o \frac{\partial n_j(\nu)}{\partial I_j^h(\nu)}. \tag{6.41}$$

Recall that the hidden units use the logistic activation function

$$n_j(I_j^h) = f(I_j^h) = \frac{1}{1 + e^{-I_j^h}}.$$

This function has a very simple derivative:

$$\frac{\partial n_j(x)}{\partial x} = n_j(x)(1 - n_j(x)).$$

Therefore we can write Equation (6.41) as

$$\delta_j^h(\nu) = \sum_{k=1}^{K} \delta_k^o(\nu) w_{jk}^o n_j(\nu)(1 - n_j(\nu)), \quad j = 1 \dots L,$$

or more compactly as the matrix equation

$$\begin{pmatrix} 0 \\ \boldsymbol{\delta}^h(\nu) \end{pmatrix} = \boldsymbol{n}(\nu) \cdot (\mathbf{1} - \boldsymbol{n}(\nu)) \cdot \big(\boldsymbol{W}^o \boldsymbol{\delta}^o(\nu)\big). \tag{6.42}$$

The dot is intended to denote simple component-by-component (so-called *Hadamard*) multiplication. The equation must be written in this rather awkward way because the expression on the right-hand side has $L+1$ components. This also makes the fact that $1 - n_0(\nu) = 0$ explicit. Equation (6.42) is the origin of the term "backpropagation," since it propagates the negative rate of change of the cost function with respect to the output activations $\boldsymbol{\delta}^o(\nu)$ backwards through the network to determine the negative rate of change with respect to the hidden activations $\boldsymbol{\delta}^h(\nu)$.

Finally, with Equation (6.40) we obtain the update rule for step 4 of the backpropagation algorithm:

$$\boldsymbol{W}^h(\nu + 1) = \boldsymbol{W}^h(\nu) + \eta\, \boldsymbol{g}(\nu) \boldsymbol{\delta}^h(\nu)^\top. \tag{6.43}$$

The choice of an appropriate learning rate η is problematic: small values imply slow convergence and large values produce oscillation. Some improvement can be achieved with an additional, purely heuristic parameter called *momentum*, which maintains a portion of the preceding weight increments in the current iteration. Equation (6.39) is replaced with

$$\boldsymbol{W}^o(\nu + 1) = \boldsymbol{W}^o(\nu) + \Delta^o(\nu) + \alpha \Delta^o(\nu - 1), \tag{6.44}$$

where $\Delta^o(\nu) = \eta\, \boldsymbol{n}(\nu) \boldsymbol{\delta}^{o\top}(\nu)$ and α is the momentum parameter. A similar expression replaces Equation (6.43). Typical choices for the backpropagation parameters are $\eta = 0.01$ and $\alpha = 0.5$.

Listing 6.8 gives an excerpt from the object class `FFNBP` (Appendix C) extending (i.e., inheriting) the class `FFN` to implement the backpropagation algorithm. It shows the code for the procedure method `TRAIN`, which closely parallels the equations developed above. Listing 6.9 shows the Python equivalent (Appendix D).

Listing 6.8: Excerpt from the object class FFNBP.

```
PRO FFNBP::Train
   iter = 0L & epoch = 0L
   iterations = 100*self.np
   eta = 0.01 &  alpha = 0.5 ; learn rate & momentum
   progressbar=Obj_New('cgprogressbar', $
    /cancel,Text='0',$
    title='Training:␣example␣No...',xsize=250,ysize=20)
   progressbar->start
   window,12,xsize=600,ysize=400,title='Cost␣Function'
   wset,12
   inc_o1 = 0 & inc_h1 = 0
   REPEAT BEGIN
      IF progressbar->CheckCancel() THEN BEGIN
         PRINT,'Training␣interrupted'
         progressbar->Destroy & RETURN
      ENDIF
; select example pair at random
      nu = long(self.np*randomu(seed))
      x=(*self.Gs)[nu,*]
      ell=(*self.Ls)[nu,*]
; send it through the network
      m=self->forwardPass(x)
; determine the deltas
      d_o = ell - m
      d_h = (*self.N*(1-*self.N)* $
            (*self.Wo##d_o))[1:self.LL]
; update the synaptic weights
      inc_o = eta*(*self.N##transpose(d_o))
      inc_h = eta*(x##d_h)
      *self.Wo = *self.Wo + inc_o + alpha*inc_o1
      *self.Wh = *self.Wh + inc_h + alpha*inc_h1
      inc_o1 = inc_o & inc_h1 = inc_h
; record cost history
      IF iter MOD self.np EQ 0 THEN BEGIN
         (*self.cost_array)[epoch] = $
                    alog10(self->cost(0))
         epoch++
         progressbar->Update,iter*100/iterations
         PLOT,*self.cost_array,xrange=[0,epoch],$
             color=0,background='FFFFFF'XL,$
            xtitle='Epoch)',$
            ytitle='log(cross␣entropy)'
      END
      iter=iter+1
   ENDREP UNTIL iter EQ iterations
   progressbar->destroy
END
```

Listing 6.9: Excerpt from the Python module `supervisedclass.py`.

```
    DEF train(self):
        eta = 0.01
        alpha = 0.5
        maxitr = epochs*self._m
        inc_o1 = 0.0
        inc_h1 = 0.0
        epoch = 0
        cost = []
        itr = 0
        TRY:
            WHILE itr<maxitr:
#               select example pair at random
                nu = np.random.randint(0,self._m)
                x = self._Gs[:,nu]
                ell = self._ls[:,nu]
#               send it through the network
                m = self.forwardpass(x)
#               determine the deltas
                d_o = ell - m
                d_h = np.multiply(np.multiply(self._n,\
                     (1-self._n)),(self._Wo*d_o))[1::]
#               update synaptic weights
                inc_o = eta*(self._n*d_o.T)
                inc_h = eta*(x*d_h.T)
                self._Wo += inc_o + alpha*inc_o1
                self._Wh += inc_h + alpha*inc_h1
                inc_o1 = inc_o
                inc_h1 = inc_h
#               record cost function
                IF itr % self._m == 0:
                    cost.append(self.cost())
                    epoch += 1
                itr += 1
        EXCEPT Exception as e:
            PRINT 'Error:␣%s'%e
            RETURN None
        RETURN np.array(cost)
```

6.5.4 Overfitting and generalization

A fundamental and much-discussed dilemma in the application of neural networks (and other learning algorithms) is that of *overfitting*. Reduced to its essentials, the question is: "How many hidden neurons are enough?" The number of neurons in the output layer of the network in Figure 6.10 is determined by the number of training classes. The number in the hidden layer is fully undetermined. If "too few" hidden neurons are chosen (and thus too few

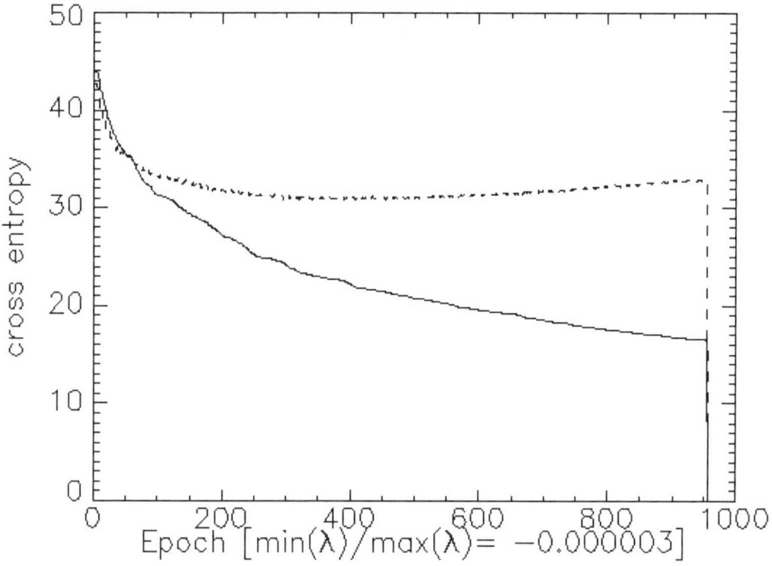

FIGURE 6.11
An example of overfitting during neural network training with the scaled conjugate gradient algorithm. The lower curve shows the cost function calculated with the training data, the upper curve with the validation data. One "epoch" corresponds to a cycle through all of the training examples. The quantity λ is an array of eigenvalues of the Hessian matrix; see Appendix B.

synaptic weights), there is danger that the classification will be suboptimal: there will be an insufficient number of adjustable parameters to resolve the class structure of the training data. If, on the other hand, "too many" hidden neurons are selected, and if the training data have a large noise variance, there will be a danger that the network will fit the data all too well, including their detailed random structure. Such detailed structure is a characteristic of the particular training sample chosen and not of the underlying class distributions that the network is supposed to learn. It is here that one speaks of overfitting. In either case, the capability of the network to generalize to unknown inputs will be impaired. One can find excellent discussions of this subject in Hertz et al. (1991), Chapter 6, and in Bishop (1995), Chapter 9, where regularization techniques are introduced to penalize overfitting. Alternatively, so-called *growth* and *pruning* algorithms can be applied in which the network architecture is optimized during the training procedure. A popular growth algorithm is the *cascade correlation neural network* of Fahlman and LeBiere (1990).

We shall restrict ourselves here to a solution which presupposes an overdimensioned network, that is, one with too many hidden weights, as well as the availability of a second dataset, which is usually referred to as the *validation* dataset. An option in the ENVI/IDL neural network program described in Appendix C allows the training data to be split in two, so that half the data are reserved for validation purposes. (If test data were held back, the training dataset is then one third of its original size.) Both halves are still representative of the class distributions and are statistically independent. During the training phase, which is carried out only with the training data, cost functions calculated both with the training data as well as with the validation data are displayed. Overfitting is indicated by a continued decrease in the training cost function accompanied by a gradual increase in the validation cost function. Figure 6.11 shows a fairly typical example, achieved with 12 hidden neurons and 7 classes. The gradual, albeit very slight, increase in the validation cost function is indicative of overfitting: the network is learning the detailed structure of the training data at the cost of its ability to generalize. The algorithm should thus be stopped when the upper curve starts to rise (so-called *early stopping*).

Figure 6.4, bottom left, shows a classification of the Jülich ASTER scene as obtained with the neural network classifier. In the next chapter we shall see how to compare the generalization capability of neural networks with that of both the maximum likelihood and Gaussian kernel classifiers which we developed previously.

6.6 Support vector machines

Let us return to the simple linear discriminant function $I(\boldsymbol{g})$ for a two-class problem given by Equation (6.24), with the convention that the weight vector does not include the bias term, i.e.,

$$I(\boldsymbol{g}) = \boldsymbol{w}^\top \boldsymbol{g}(\nu) + w_0,$$

where

$$\boldsymbol{w} = (w_1 \ldots w_N)^\top$$

and where the training observations are

$$\boldsymbol{g}(\nu) = (g_1(\nu) \ldots g_N(\nu))^\top,$$

with corresponding (this time scalar) labels

$$\ell(\nu) \in \{0, 1\}, \quad \nu = 1 \ldots m.$$

A quadratic cost function for training this discriminant on two classes would then be

$$E(\boldsymbol{w}) = \frac{1}{2}\sum_{\nu=1}^{m}(\boldsymbol{w}^\top \boldsymbol{g}(\nu) + w_0 - \ell(\nu))^2. \tag{6.45}$$

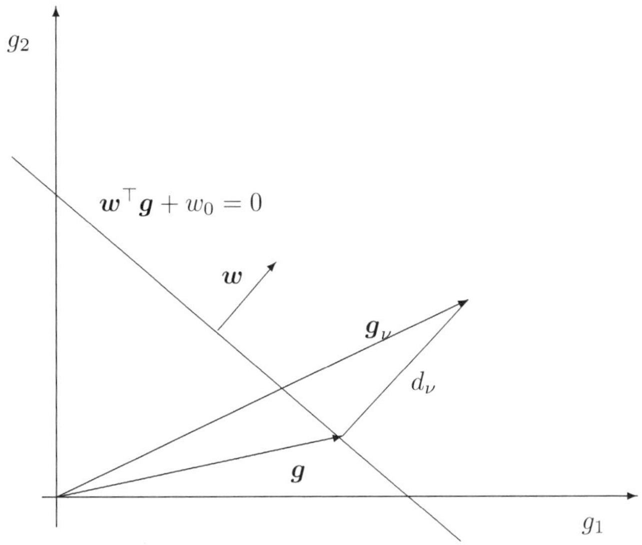

FIGURE 6.12
Distance d to the separating hyperplane.

This expression is essentially the same as Equation (6.30), when it is written for the case of a single neuron. Training the neuron of Figure 6.7 to discriminate the two classes means finding the weight parameters which minimize the above cost function. This can be done, for example, by using a gradient descent method, or more efficiently with the *perceptron algorithm* as explained in Exercise 3. We shall consider in the following an alternative to such a cost function approach. The method we describe is reminiscent of the Gaussian kernel method of Section 6.4 in that the training observations are also used at the classification phase, but as we shall see, not all of them.

6.6.1 Linearly separable classes

It is convenient first of all to relabel the training observations as $\ell(\nu) \in \{+1, -1\}$, rather than $\ell(\nu) \in \{0, 1\}$. With this convention, the product

$$\ell(\nu)(\boldsymbol{w}^\top \boldsymbol{g}(\nu) + w_0), \quad \nu = 1 \dots m,$$

is called *margin* of the νth training pair $(\boldsymbol{g}(\nu), \ell(\nu))$ relative to the hyperplane

$$I(\boldsymbol{g}) = \boldsymbol{w}^\top \boldsymbol{g} + w_0 = 0.$$

The perpendicular distance d_ν of a point $\boldsymbol{g}(\nu)$ to the hyperplane is, see Figure 6.12, given by

$$d_\nu = \frac{1}{\|\boldsymbol{w}\|}(\boldsymbol{w}^\top \boldsymbol{g}(\nu) + w_0). \tag{6.46}$$

That is, from the figure,

$$d_\nu = \|\boldsymbol{g}(\nu) - \boldsymbol{g}\|,$$

and, since $\boldsymbol{w}$ is perpendicular to the hyperplane,

$$\boldsymbol{w}^\top(\boldsymbol{g}(\nu) - \boldsymbol{g}) = \|\boldsymbol{w}\|\|\boldsymbol{g}(\nu) - \boldsymbol{g}\| = \|\boldsymbol{w}\|d_\nu.$$

But the left-hand side of the above equation is just

$$\boldsymbol{w}^\top(\boldsymbol{g}(\nu) - \boldsymbol{g}) = \boldsymbol{w}^\top\boldsymbol{g}(\nu) - \boldsymbol{w}^\top\boldsymbol{g} = \boldsymbol{w}^\top\boldsymbol{g}(\nu) + w_0,$$

from which Equation (6.46) follows. The distance d_ν is understood to be positive for points above the hyperplane and negative for points below. The quantity

$$\gamma_\nu = \ell(\nu)d_\nu = \frac{1}{\|\boldsymbol{w}\|}\ell(\nu)(\boldsymbol{w}^\top\boldsymbol{g}(\nu) + w_0) \tag{6.47}$$

is called the *geometric margin* for the observation. Observations lying on the hyperplane have zero geometric margins, incorrectly classified observations have negative geometric margins, and correctly classified observations have positive geometric margins.

The *geometric margin of a hyperplane* (relative to a given training set), is defined as the smallest (geometric) margin over the observations in that set, and the *maximal margin hyperplane* is the hyperplane which maximizes the smallest margin, i.e., the hyperplane with parameters $\boldsymbol{w}, w_0$ given by

$$\arg\max_{\boldsymbol{w},w_0}\left(\frac{1}{\|\boldsymbol{w}\|}\min_\nu\left(\ell(\nu)(\boldsymbol{w}^\top\boldsymbol{g}(\nu) + w_0)\right)\right). \tag{6.48}$$

If the training data are linearly separable, then the resulting smallest margin will be positive since all observations are correctly classified. A maximal margin hyperplane is illustrated in Figure 6.13.

6.6.1.1 Primal formulation

The maxmin problem (6.48) can be reformulated as follows (Cristianini and Shawe–Taylor, 2000; Bishop, 2006). If we transform the parameters according to

$$\boldsymbol{w} \to \kappa\boldsymbol{w}, \quad w_0 \to \kappa w_0$$

for some constant κ, then the distance d_ν will remain unchanged, as is clear from Equation (6.46). Now let us choose κ such that

$$\ell(\nu)(\boldsymbol{w}^\top\boldsymbol{g}(\nu') + w_0) = 1 \tag{6.49}$$

for whichever training observation $\boldsymbol{g}(\nu')$ happens to be closest to the hyperplane. This implies that the following constraints are met:

$$\ell(\nu)I(\boldsymbol{g}(\nu)) = \ell(\nu)(\boldsymbol{w}^\top\boldsymbol{g}(\nu) + w_0) \ge 1, \quad \nu = 1 \ldots m, \tag{6.50}$$

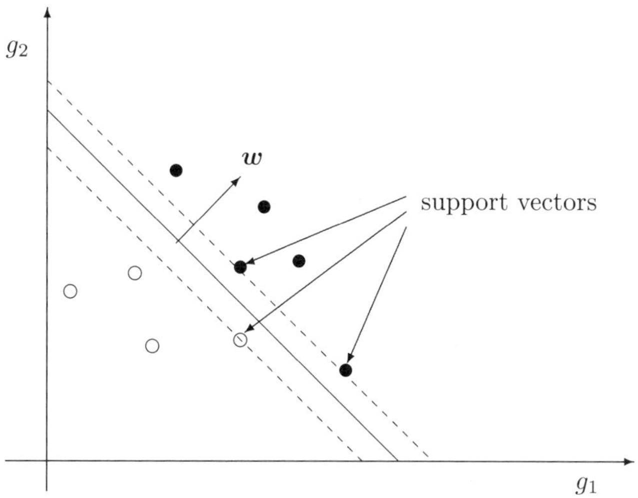

FIGURE 6.13
A maximal margin hyperplane for linearly separable training data and its support vectors (see text).

and the geometric margin of the hyperplane (the smallest geometric margin over the observations) is, with Equation (6.47), simply $\|\boldsymbol{w}\|^{-1}$. For observations for which equality holds in Equation (6.50), the constraints are called *active*, otherwise *inactive*. Clearly, for any choice of $\boldsymbol{w}, w_0$, there will always be at least one active constraint. So the problem expressed in Equation (6.48) is equivalent to maximizing $\|\boldsymbol{w}\|^{-1}$ or, expressed more conveniently, to solving

$$\arg\min_{\boldsymbol{w}} \frac{1}{2}\|\boldsymbol{w}\|^2 \tag{6.51}$$

subject to the constraints of Equation (6.50). These constraints define the *feasible region* for the minimization problem. Taken together, Equations (6.50) and (6.51) constitute the primal formulation of the original maxmin problem, Equation (6.48). The bias parameter w_0 is determined implicitly by the constraints, as we will see later.

6.6.1.2 Dual formulation

To solve Equation (6.51) we will apply the Lagrangian formalism for *inequality constraints*, a generalization of the method which we have used extensively up till now and first introduced in Section 1.6. Bishop (2006), Appendix E and Cristianini and Shawe–Taylor (2000), Chapter 5, provide good discussions. We introduce a Lagrange multiplier α_ν for each of the inequality constraints, to obtain in the Lagrange function

$$L(\boldsymbol{w}, w_0, \boldsymbol{\alpha}) = \frac{1}{2}\|\boldsymbol{w}\|^2 - \sum_{\nu=1}^{m} \alpha_\nu \big(\ell(\nu)(\boldsymbol{w}^\top \boldsymbol{g}(\nu) + w_0) - 1\big), \tag{6.52}$$

where $\boldsymbol{\alpha} = (\alpha_1 \dots \alpha_m)^\top \geq \boldsymbol{0}$. Minimization over $\boldsymbol{w}$ and w_0 requires that the respective derivatives be set equal to zero:

$$\frac{\partial L}{\partial \boldsymbol{w}} = \boldsymbol{w} - \sum_\nu \ell(\nu)\alpha_\nu \boldsymbol{g}(\nu) = \boldsymbol{0}, \tag{6.53}$$

$$\frac{\partial L}{\partial w_0} = \sum_\nu \ell(\nu)\alpha_\nu = 0. \tag{6.54}$$

Therefore, from Equation (6.53),

$$\boldsymbol{w} = \sum_\nu \ell(\nu)\alpha_\nu \boldsymbol{g}(\nu).$$

Substituting this back into Equation (6.52) and using Equation (6.54) gives

$$\begin{aligned} L(\boldsymbol{w}, w_0, \boldsymbol{\alpha}) &= \frac{1}{2}\sum_{\nu\nu'} \ell(\nu)\ell(\nu')\alpha_\nu\alpha_{\nu'}(\boldsymbol{g}(\nu)^\top \boldsymbol{g}(\nu')) \\ &\quad - \sum_{\nu\nu'} \ell(\nu)\ell(\nu')\alpha_\nu\alpha_{\nu'}(\boldsymbol{g}(\nu)^\top \boldsymbol{g}(\nu')) + \sum_\nu \alpha_\nu \\ &= \sum_\nu \alpha_\nu - \frac{1}{2}\sum_{\nu\nu'} \ell(\nu)\ell(\nu')\alpha_\nu\alpha_{\nu'}(\boldsymbol{g}(\nu)^\top \boldsymbol{g}(\nu')), \end{aligned} \tag{6.55}$$

in which $\boldsymbol{w}$ and w_0 no longer appear. We thus obtain the *dual formulation* of the minimization problem, Equations (6.50) and (6.51), namely maximize

$$\tilde{L}(\boldsymbol{\alpha}) = \sum_\nu \alpha_\nu - \frac{1}{2}\sum_{\nu\nu'} \ell(\nu)\ell(\nu')\alpha_\nu\alpha_{\nu'}(\boldsymbol{g}(\nu)^\top \boldsymbol{g}(\nu')) \tag{6.56}$$

with respect to the *dual variables* $\boldsymbol{\alpha} = (\alpha_1 \dots \alpha_m)^\top$ subject to the constraints

$$\begin{aligned} \boldsymbol{\alpha} &\geq \boldsymbol{0}, \\ \sum_\nu \ell(\nu)\alpha_\nu &= 0. \end{aligned} \tag{6.57}$$

We must maximize because, according to *duality theory*, $\tilde{L}(\boldsymbol{\alpha})$ is a lower limit for $L(\boldsymbol{w}, w_0, \boldsymbol{\alpha})$ and, for our problem,

$$\max_{\boldsymbol{\alpha}} \tilde{L}(\boldsymbol{\alpha}) = \min_{\boldsymbol{w}, w_0} L(\boldsymbol{w}, w_0, \boldsymbol{\alpha}),$$

see, e.g., Cristianini and Shawe–Taylor (2000), Chapter 5. We shall see how to do this in the sequel, however let us for now suppose that we have found the solution $\boldsymbol{\alpha}^*$ which maximizes $\tilde{L}(\boldsymbol{\alpha})$. Then

$$\boldsymbol{w}^* = \sum_{\nu} \ell(\nu)\alpha^*_\nu \boldsymbol{g}(\nu)$$

determines the maximal margin hyperplane and it has geometric margin $\|\boldsymbol{w}^*\|^{-1}$. In order to classify a new observation $\boldsymbol{g}$, we simply evaluate the sign of

$$I^*(\boldsymbol{g}) = \boldsymbol{w}^{*\top}\boldsymbol{g} + w_0^* = \sum_{\nu} \ell(\nu)\alpha^*_\nu (\boldsymbol{g}(\nu)^\top \boldsymbol{g}) + w_0^*, \tag{6.58}$$

that is, we ascertain on which side of the hyperplane the observation lies. (We still need an expression for w_0^*. This is described below.) Note that both the training phase, i.e., the solution of the dual problem Equations (6.56) and (6.57), as well as the generalization phase, Equation (6.58), involve only inner products of the observations $\boldsymbol{g}$.

6.6.1.3 Quadratic programming and support vectors

The optimization problem represented by Equations (6.56) and (6.57) is a *quadratic programming problem*, the objective function $\tilde{L}(\boldsymbol{\alpha})$ being quadratic in the dual variables α_υ. According to the *Karush–Kuhn–Tucker* (KKT) conditions for quadratic programs,* in addition to the constraints of Equations (6.57), the *complementarity condition*

$$\alpha_\nu\big(\ell(\nu)I(\boldsymbol{g}(\nu)) - 1\big) = 0, \quad \nu = 1 \dots m, \tag{6.59}$$

must be satisfied. Taken together, Equations (6.50), (6.57) and (6.59) are *necessary and sufficient conditions* for a solution $\boldsymbol{\alpha} = \boldsymbol{\alpha}^*$. The complementarity condition says that each of the constraints in Equation (6.50) is either active, that is, $\ell(\nu)I(\boldsymbol{g}(\nu)) = 1$, or it is inactive, $\ell(\nu)I(\boldsymbol{g}(\nu)) > 1$, in which case, from Equation (6.59), $\alpha_\nu = 0$.

When classifying a new observation with Equation (6.58), either the training observation $\boldsymbol{g}(\nu)$ satisfies $\ell(\nu)I^*(\boldsymbol{g}(\nu)) = 1$, which is to say it has minimum margin (Equation (6.49)), or $\alpha_\nu = 0$ by virtue of the complementarity condition, meaning that it *plays no role in the classification*. The labeled training observations with minimum margin are called *support vectors* (see Figure 6.13). After solution of the quadratic programming problem they are the only observations which can contribute to the classification of new data.

Let us call SV the set of support vectors. Then from the complementarity condition Equation (6.59) we have, for $\nu \in SV$,

$$\ell(\nu)\left(\sum_{\nu' \in SV} \ell(\nu')\alpha^*_{\nu'}(\boldsymbol{g}(\nu')^\top \boldsymbol{g}(\nu)) + w_0^*\right) = 1. \tag{6.60}$$

*See again Cristianini and Shawe–Taylor (2000), Chapter 5.

We can therefore write

$$\|\boldsymbol{w}^*\|^2 = {\boldsymbol{w}^*}^\top \boldsymbol{w}^* = \sum_{\nu,\nu'} \ell(\nu)\ell(\nu')\alpha_\nu\alpha_{\nu'}(\boldsymbol{g}(\nu)^\top \boldsymbol{g}(\nu'))$$

$$= \sum_{\nu\in SV} \alpha_\nu^* \left(\ell(\nu) \sum_{\nu'\in SV} \ell(\nu')\alpha_{\nu'}^*(\boldsymbol{g}(\nu)^\top \boldsymbol{g}(\nu'))\right),$$

and, from Equation (6.60),

$$\|\boldsymbol{w}^*\|^2 = \sum_{\nu\in SV} \alpha_\nu^*(1-\ell(\nu)w_0^*) = \sum_{\nu\in SV} \alpha_\nu^*, \tag{6.61}$$

where in the second equality we have made use of the second constraint in Equation (6.57). Thus the geometric margin of the maximal hyperplane is given in terms of the dual variables by

$$\|\boldsymbol{w}^*\|^{-1} = \left(\sum_{\nu\in SV} \alpha_\nu^*\right)^{-1/2}. \tag{6.62}$$

Equation (6.60) can be used to determine w_0^* once the quadratic program has been solved, simply by choosing an arbitrary $\nu \in SV$. Bishop (2006) suggests a numerically more stable procedure: Multiply Equation (6.60) by $\ell(\nu)$ and make use of the fact that $\ell(\nu)^2 = 1$. Then take the average of the equations for all support vectors and solve for w_0^* to get

$$w_0^* = \frac{1}{|SV|} \sum_{\nu\in SV} \left(\ell(\nu) - \sum_{\nu'\in SV} \alpha_{\nu'}\ell(\nu')(\boldsymbol{g}(\nu)^\top \boldsymbol{g}(\nu'))\right). \tag{6.63}$$

6.6.2 Overlapping classes

With the SVM formalism introduced so far, one can solve two-class classification problems which are linearly separable. Using kernel substitution, which will be treated shortly, nonlinearly separable problems can also be solved. In general, though, one is faced with labeled training data that overlap considerably, so that their complete separation would imply gross overfitting.

To overcome this problem, we introduce so-called *slack variables* ξ_ν associated with each training vector, such that

$$\xi_\nu = \begin{cases} |\ell(\nu) - I(\boldsymbol{g}(\nu))| & \text{if } \boldsymbol{g}(\nu) \text{ is on the wrong side of the margin boundary} \\ 0 & \text{otherwise.} \end{cases}$$

The situation is illustrated in Figure 6.14. The constraints of Equation (6.50) now become

$$\ell(\nu)I(\boldsymbol{g}(\nu)) \ge 1 - \xi_\nu, \quad \nu = 1\ldots m, \tag{6.64}$$

and are often referred to as *soft margin* constraints.

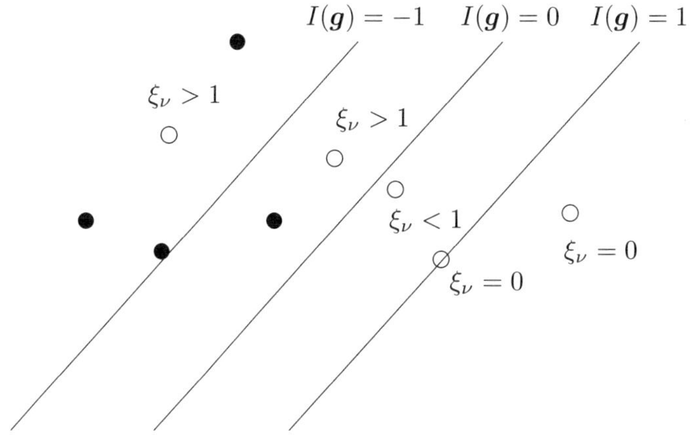

FIGURE 6.14
Slack variables. Observations with values less than 1 are correctly classified.

Our objective is again to maximize the margin, but to penalize points with large values of ξ. We therefore modify the objective function of Equation (6.51) by introducing a regularization term:

$$C \sum_{\nu=1}^{m} \xi_\nu + \frac{1}{2}\|\boldsymbol{w}\|^2. \tag{6.65}$$

The parameter C determines the degree of penalization. When minimizing this objective function, in addition to the above inequality constraints we require that

$$\xi_\nu \geq 0, \quad \nu = 1 \ldots m.$$

The Lagrange function is now given by

$$L(\boldsymbol{w}, w_0, \boldsymbol{\alpha}) = \frac{1}{2}\|\boldsymbol{w}\|^2 + C\sum_{\nu=1}^{m} \xi_\nu - \sum_{\nu} \alpha_\nu(\ell(\nu)(\boldsymbol{w}^\top \boldsymbol{g}(\nu) + w_0) - 1 + \xi_\nu) - \sum_{\nu=1}^{m} \mu_\nu \xi_\nu, \tag{6.66}$$

where $\mu_\nu \geq 0$, $\nu = 1 \ldots m$, are additional Lagrange multipliers. Setting the derivatives of the Lagrange function with respect to $\boldsymbol{w}$, w_0 and ξ_ν equal to zero, we get

$$\boldsymbol{w} = \sum_{\nu=1}^{m} \alpha_\nu \ell(\nu) \boldsymbol{g}(\nu) \tag{6.67}$$

$$\sum_{\nu=1}^{m} \alpha_\nu \ell(\nu) = 0 \tag{6.68}$$

$$\alpha_\nu = C - \mu_\nu, \tag{6.69}$$

which leads to the dual form

$$\tilde{L}(\boldsymbol{\alpha}) = \sum_{\nu} \alpha_\nu - \frac{1}{2}\sum_{\nu\nu'} \ell(\nu)\ell(\nu')\alpha_\nu\alpha_{\nu'}(\boldsymbol{g}(\nu)^\top \boldsymbol{g}(\nu')). \tag{6.70}$$

This is the same as for the separable case, Equation (6.56), except that the constraints are slightly different. From Equation (6.69), $\alpha_\nu = C - \mu_\nu \le C$, so that

$$0 \le \alpha_\nu \le C, \quad \nu = 1 \dots m. \tag{6.71}$$

Furthermore, the complementarity conditions now read

$$\begin{aligned} \alpha_\nu\big(\ell(\nu)I(\boldsymbol{g}(\nu)) - 1 + \xi_\nu\big) &= 0 \\ \mu_\nu\xi_\nu &= 0, \quad \nu = 1\dots m. \end{aligned} \tag{6.72}$$

Equations (6.64), (6.68), (6.71) and (6.72) are again necessary and sufficient conditions for a solution $\boldsymbol{\alpha} = \boldsymbol{\alpha}^*$.

As before, when classifying new data with Equation (6.58), only the support vectors, i.e., the training observations for which $\alpha_\nu > 0$, will play a role. If $\alpha_\nu > 0$, there are now two possibilities to be distinguished:

- $0 \le \alpha_\nu < C$. Then we must have $\mu_\nu > 0$ (Equation (6.69)), implying from the second condition in Equation (6.72) that $\xi_\nu = 0$. The training vector thus lies exactly on the margin and corresponds to the support vector for the separable case.

- $\alpha_\nu = C$. The training vector lies inside the margin and is correctly classified ($\xi \le 1$) or incorrectly classified ($\xi > 1$).

As before we can use Equation (6.63) to determine $\boldsymbol{w}_0^*$ after solving the quadratic program for $\boldsymbol{w}^*$, except that the set SV must be restricted to those observations for which $0 < \alpha_\nu < C$, i.e., to the support vectors lying on the margin.

6.6.3 Solution with sequential minimal optimization

To train the classifier, we must maximize Equation (6.70) subject to the boundary conditions given by Equations (6.68) and (6.71), which define the feasible region for the problem. We outline here a popular method for doing this called *sequential minimal optimization* (SMO); see Cristianini and Shawe–Taylor (2000). In this algorithm, pairs of Lagrange multipliers $(\alpha_\nu, \alpha_{\nu'})$ are chosen and varied so as to increase the objective function while still satisfying the constraints. (At least 2 multipliers must be considered in order to guarantee fulfillment of the equality constraint, Equation (6.68)). The SMO method has the advantage that the maximum can be found analytically at each step, thus leading to fast computation.

Algorithm (Sequential minimal optimization)

1. Set $\alpha_\nu^{old} = 0,\ \nu = 1 \dots m$ (clearly $\boldsymbol{\alpha}^{old}$ is in the feasible region).
2. Choose a pair of Lagrange multipliers, calling them without loss of generality α_1 and α_2. Let all of the other multipliers be fixed.
3. Maximize $\tilde{L}(\boldsymbol{\alpha})$ with respect to α_1 and α_2.
4. If the complementarity conditions are satisfied (within some tolerance) stop, else go to step 2.

In in step 2, heuristic rules must be used to choose appropriate pairs. Step 3 can be carried out analytically by maximizing the quadratic function

$$\begin{aligned}\tilde{L}(\alpha_1, \alpha_2) = \alpha_1 + \alpha_2 &- \frac{1}{2}(\boldsymbol{g}(1)^\top \boldsymbol{g}(1))\alpha_1^2 - \frac{1}{2}(\boldsymbol{g}(2)^\top \boldsymbol{g}(2))\alpha_2^2 - \ell_1\ell_2\alpha_1\alpha_2 \\ &- \ell_1\alpha_1 v_1 - \ell_2\alpha_2 v_2 + \text{Const},\end{aligned}$$

where

$$v_i = \sum_{\nu=3}^{n} \ell(\nu)\alpha_\nu^{old}(\boldsymbol{g}(\nu)^\top \boldsymbol{g}(i)), \quad i = 1, 2.$$

The maximization is carried out in the feasible region for α_1, α_2. The equality constraint, Equation (6.68), requires

$$\ell_1\alpha_1 + \ell_2\alpha_2 = \ell_1\alpha_1^{old} + \ell_2\alpha_2^{old},$$

or equivalently

$$\alpha_2 = -\ell_1\ell_2\alpha_1 + \gamma,$$

where $\gamma = \ell_1\ell_2\alpha_1^{old} + \alpha_2^{old}$, so that (α_1, α_2) lies on a line with slope ± 1, depending on the value of $\ell_1\ell_2$. The inequality constraint, Equation (6.71), defines the endpoints of the line segment that need to be considered, namely $(\alpha_1 = 0, \alpha_2 = \max(0, \gamma))$ and $(\alpha_1 = C, \alpha_2 = \min(C, -\ell_1\ell_2 C + \gamma))$. Regarding step 4, the complementarity conditions are as follows; see Equations (6.72):

For all ν,

$$\begin{aligned}&\text{if } \alpha_\nu = 0, \text{ then } \ell(\nu)I(\boldsymbol{g}(\nu)) - 1 \geq 0 \\ &\text{if } \alpha_\nu > 0, \text{ then } \ell(\nu)I(\boldsymbol{g}(\nu)) - 1 = 0 \\ &\text{if } \alpha_\nu = C, \text{ then } \ell(\nu)I(\boldsymbol{g}(\nu)) - 1 \leq 0.\end{aligned}$$

6.6.4 Multiclass SVMs

A distinct advantage of the SVM for remote sensing image classification over neural networks is the unambiguity of the solution. Since one maximizes a quadratic function, the maximum is global and will always be found. There is no possibility of becoming trapped in a local optimum. However, SVM classifiers have two disadvantages. They are designed for two-class problems and their outputs, unlike feed-forward neural networks, do not model posterior class membership probabilities in a natural way.

A common way to overcome the two-class restriction is to determine all possible two-class results and then use a voting scheme to decide on the class label (Wu et al., 2004). That is, for K classes, we train $K(K-1)/2$ SVM's on each of the possible pairs $(i,j) \in \mathcal{K} \otimes \mathcal{K}$. For a new observation $\boldsymbol{g}$ and the SVM for (i,j), let

$$\mu_{ij}(\boldsymbol{g}) = \Pr(\ell = i \mid \ell = i \text{ or } j,\ \boldsymbol{g}) = \frac{\Pr(\ell = i \mid \boldsymbol{g})}{\Pr(\ell = i \text{ or } j \mid \boldsymbol{g})}. \tag{6.73}$$

The last equality follows from the definition of conditional probability, Equation (2.59), since the joint probability (given $\boldsymbol{g}$) for i or j and i is just the probability for i. Now suppose that r_{ij} is some rough estimator for μ_{ij}, perhaps simply $r_{ij} = 1$ if $\mu_{ij} > 0.5$ and 0 otherwise. The voting rule is then

$$k = \arg\max_i \left(\sum_{j \neq i}^{K} [[r_{ij} > r_{ji}]] \right), \tag{6.74}$$

where $[[\cdots]]$ is the *indicator function*

$$[[x]] = \begin{cases} 1 & \text{if } x \text{ is true} \\ 0 & \text{if } x \text{ is false.} \end{cases}$$

With regard to the second restriction, a very simple estimate of the posterior class membership probability is the ratio of the number of votes to the total number of classifications,

$$\Pr(k \mid \boldsymbol{g}) \approx \frac{2}{K(K-1)} \sum_{j \neq i}^{K} [[r_{kj} > r_{jk}]]. \tag{6.75}$$

If r_{ij} is a more realistic estimate, we can proceed as follows (Price et al., 1995):

$$\begin{aligned}
\sum_{j \neq i} \Pr(\ell = i \text{ or } j \mid \boldsymbol{g}) &= \sum_{j \neq i} \left(\Pr(\ell = i \mid \boldsymbol{g}) + \Pr(\ell = j \mid \boldsymbol{g}) \right) \\
&= (K-1)\Pr(\ell = i \mid \boldsymbol{g}) + \sum_{j \neq i} \Pr(\ell = j \mid \boldsymbol{g}) \\
&= (K-2)\Pr(\ell = i \mid \boldsymbol{g}) + \sum_{j} \Pr(\ell = j \mid \boldsymbol{g}) \\
&= (K-2)\Pr(\ell = i \mid \boldsymbol{g}) + 1.
\end{aligned} \tag{6.76}$$

Combining this with Equation (6.73) gives, see Exercise 8,

$$\Pr(\ell = i \mid \boldsymbol{g}) = \frac{1}{\sum_{j \neq i} \mu_{ij}^{-1} - (K-2)} \tag{6.77}$$

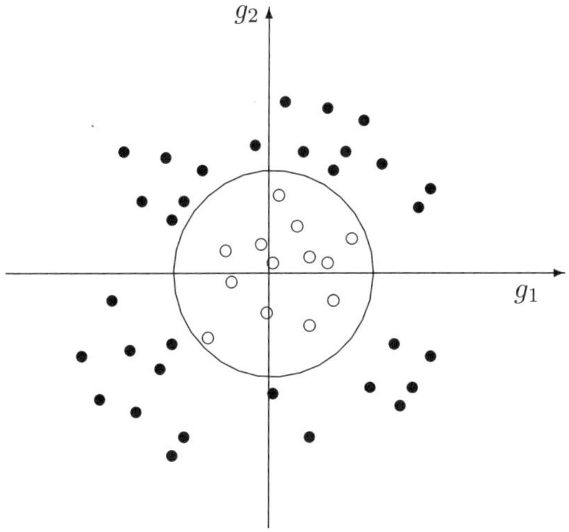

FIGURE 6.15
Two classes which are not linearly separable in the two-dimensional space of observations.

or, substituting the estimate r_{ij}, we estimate the posterior class membership probabilities as

$$\Pr(k \mid \boldsymbol{g}) \approx \frac{1}{\sum_{j \neq k} r_{kj}^{-1} - (K-2)}. \tag{6.78}$$

Because of the substitution, the probabilities will not sum exactly to one, and must be normalized.

6.6.5 Kernel substitution

Of course, the reason why we have replaced the relatively simple optimization problem, Equation (6.51), with the more involved dual formulation is due to the fact that the labeled training observations only enter into the dual formulation in the form of scalar products $\boldsymbol{g}(\nu)^\top \boldsymbol{g}(\nu')$. This allows us once again to use the elegant method of kernelization introduced in Chapter 4. Then we can apply support vector machines to situations in which the classes are not linearly separable.

To give a simple example (Müller et al., 2001), consider the classification problem illustrated in Figure 6.15. While the classes are clearly separable, they cannot be separated with a hyperplane. Now introduce the transformation

$$\phi : \mathbb{R}^2 \mapsto \mathbb{R}^3,$$

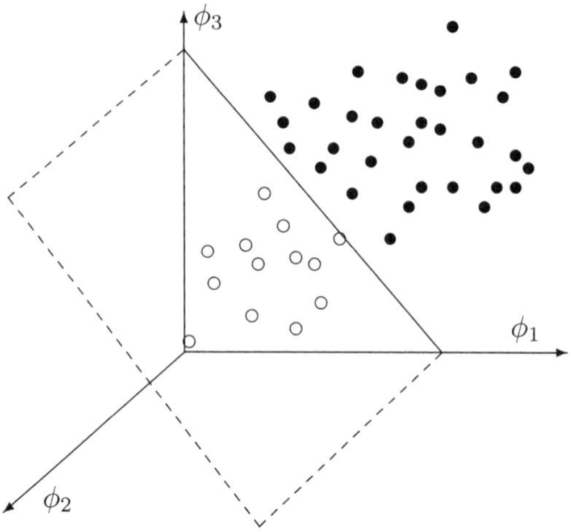

FIGURE 6.16
The classes of Figure 6.15 become linearly separable in a three-dimensional, nonlinear feature space.

such that

$$\phi(\boldsymbol{g}) = \begin{pmatrix} g_1^2 \\ \sqrt{2}g_1g_2 \\ g_2^2 \end{pmatrix}.$$

In this new feature space, the observations are transformed as shown in Figure 6.16, and can be separated by a two-dimensional hyperplane. We can thus apply the support vector formalism unchanged, simply replacing $\boldsymbol{g}_v^\top \boldsymbol{g}_{v'}$ by $\boldsymbol{\phi}(\boldsymbol{g}(\nu))^\top \boldsymbol{\phi}(\boldsymbol{g}(\nu'))$. In this case, the kernel function is given by

$$k(\boldsymbol{g}(\nu), \boldsymbol{g}(\nu')) = (g_1^2, \sqrt{2}g_1g_2, g_2^2)_\nu \begin{pmatrix} g_1^2 \\ \sqrt{2}g_1g_2 \\ g_2^2 \end{pmatrix}_{\nu'} = (\boldsymbol{g}(\nu)^\top \boldsymbol{g}(\nu'))^2,$$

called the quadratic kernel.

6.6.6 A modified SVM classifier

Supervised classification with the SVM can be called from the ENVI Classic menu with

`Classification/Supervised/Support Vector Machine`

The kernels available in the ENVI SVM are:

$$\begin{aligned} k_{\text{lin}}(\boldsymbol{g}_i, \boldsymbol{g}_j) &= \boldsymbol{g}_i^\top \boldsymbol{g}_j \\ k_{\text{poly}}(\boldsymbol{g}_i, \boldsymbol{g}_j) &= (\gamma \boldsymbol{g}_i^\top \boldsymbol{g}_j + r)^d \\ k_{\text{rbf}}(\boldsymbol{g}_i, \boldsymbol{g}_j) &= \exp(-\gamma \|\boldsymbol{g}_i - \boldsymbol{g}_j\|^2) \\ k_{\text{sig}}(\boldsymbol{g}_i, \boldsymbol{g}_j) &= \tanh(\gamma \boldsymbol{g}_i^\top \boldsymbol{g}_j + r). \end{aligned}$$

Listing 6.10: Excerpt from the Python module `supervisedclass.py`.

```
CLASS Svm(OBJECT):

    DEF __init__(self,Gs,ls):
        self._K = ls.shape[1]
        self._Gs = Gs
        self._N = Gs.shape[1]
        self._ls = ls
        self._svm = LibSvm('c_svc','rbf',\
            gamma=1.0/self._N,C=100,probability=True)

    DEF train(self):
        TRY:
            labels = np.argmax(self._ls,axis=1)
            idx = np.where(labels == 0)[0]
            ls = np.ones(LEN(idx),dtype=np.INT)
            Gs = self._Gs[idx,:]
            FOR k IN RANGE(1,self._K):
                idx = np.where(labels == k)[0]
                ls = np.concatenate((ls, \
                (k+1)*np.ones(LEN(idx),dtype=np.INT)))
                Gs = np.concatenate((Gs,\
                               self._Gs[idx,:]),axis=0)
            self._svm.learn(Gs,ls)
            RETURN True
        EXCEPT Exception as e:
            PRINT 'Error: %s'%e
            RETURN False

    DEF classify(self,Gs):
        probs = np.transpose(self._svm. \
                             pred_probability(Gs))
        classes = np.argmax(probs,axis=0)+1
        RETURN (classes, probs)
```

The most popular kernel is the Gaussian kernel (k_{rbf}) and this is the default for ENVI. The parameter γ essentially determines the training/generalization tradeoff, with large values leading to overfitting (Shawe–Taylor and Cristianini, 2004). Whereas both the maximum likelihood and Gaussian kernel training procedures have essentially no externally adjustable parameters,* this is not the case for support vector machines. First, one must decide upon a kernel, and then choose associated kernel parameters as well as the soft margin penalization constant C. In addition, the ENVI implementation provides for

*Neural network training with backpropagation has three (learn rate, momentum and the number of hidden neurons).

multi-resolution classification with a choice of different pyramid depths. This feature is mainly offered to speed up classification, however it can influence classification accuracy as well.

The ENVI/IDL extension `SVM_RUN` described in Appendix C, like the extension `MAXLIKE_RUN`, provides a wrapper for ENVI's built-in SVM algorithm, separating ROI training area pixels into a training dataset and a test dataset in the ratio 2:1. It makes use of the ENVI batch procedure `ENVI_SVM_DOIT`. The extension only generates classification and rule images in memory. Again, test results are saved to a file in a format consistent with that used by the other classification routines described earlier, see Appendix C. A similar wrapper for the `mlpy.LibSvm` Python class (Albanese et al., 2012; Chang and Lin, 2011) is shown in Listing 6.10.* It is exported by the module `supervisedclass.py` and is described in Appendix D. A SVM classification is shown in Figure 6.4, bottom right.

6.7 Exercises

1. (a) Perform the integration in Equation (6.11) for the one-dimensional case:

$$p(g \mid k) = \frac{1}{\sqrt{2\pi}\sigma_k}\exp\left(-\frac{1}{2\sigma_k^2}(g-\mu_k)^2\right), \quad k = 1, 2. \tag{6.79}$$

(*Hint:* Use the definite integral $\int_{-\infty}^{\infty}\exp(ag - bg^2)dg = \sqrt{\frac{\pi}{b}}\exp(a^2/4b)$.)
(b) The Jeffries–Matusita distance between two probability densities $p(\boldsymbol{g} \mid 1)$ and $p(\boldsymbol{g} \mid 2)$ is defined as

$$J = \int_{-\infty}^{\infty}\left(p(\boldsymbol{g} \mid 1)^{1/2} - p(\boldsymbol{g} \mid 2)^{1/2}\right)^2 d\boldsymbol{g}.$$

Show that this is equivalent to the definition in Equation (6.13).

(c) A measure of the separability of $p(\boldsymbol{g} \mid 1)$ and $p(\boldsymbol{g} \mid 2)$ can be written in terms of the Kullback–Leibler divergence, see Equation (2.92), as follows (Richards and Jia, 2006):

$$d_{12} = \mathrm{KL}\left(p(\boldsymbol{g} \mid 1), p(\boldsymbol{g} \mid 2)\right) + \mathrm{KL}\left(p(\boldsymbol{g} \mid 2), p(\boldsymbol{g} \mid 1)\right).$$

Explain why this is a satisfactory separability measure.

*The wrapper is somewhat less elegant here, because the C-library `LibSvm` does not permit sub-classing.

(d) Show that, for the one-dimensional distributions of Equation (6.79),

$$d_{12} = \frac{1}{2}\left(\frac{1}{\sigma_1^2} - \frac{1}{\sigma_2^2}\right)(\sigma_1^2 - \sigma_2^2) + \frac{1}{2}\left(\frac{1}{\sigma_1^2} + \frac{1}{\sigma_2^2}\right)(\mu_1 - \mu_2)^2.$$

2. (Ripley, 1996) Assuming that all K land cover classes have identical covariance matrices $\boldsymbol{\Sigma}_k = \boldsymbol{\Sigma}$:

(a) Show that the discriminant in Equation (6.17) can be replaced by the linear discriminant

$$d_k(\boldsymbol{g}) = \log(\Pr(k)) - \boldsymbol{\mu}_k^\top \boldsymbol{\Sigma}^{-1}\boldsymbol{g} + \frac{1}{2}\boldsymbol{\mu}_k^\top \boldsymbol{\Sigma}^{-1}\boldsymbol{\mu}_k.$$

(b) Suppose that there are just two classes $k = 1$ and $k = 2$. Show that the maximum likelihood classifier will choose $k = 1$ if

$$h = (\boldsymbol{\mu}_1 - \boldsymbol{\mu}_2)^\top \boldsymbol{\Sigma}^{-1}\left(\boldsymbol{g} - \frac{\boldsymbol{\mu}_1 + \boldsymbol{\mu}_2}{2}\right) > \log\left(\frac{\Pr(2)}{\Pr(1)}\right).$$

(c) The quantity $d = \sqrt{(\boldsymbol{\mu}_1 - \boldsymbol{\mu}_2)^\top \boldsymbol{\Sigma}^{-1}(\boldsymbol{\mu}_1 - \boldsymbol{\mu}_2)}$ is the Mahalanobis distance between the class means. Demonstrate that, if $\boldsymbol{g}$ belongs to class 1, then h is the realization of a normally distributed random variable H_1 with mean $d^2/2$ and variance d^2. What is the corresponding distribution if $\boldsymbol{g}$ belongs to class 2?

(d) Prove from the above considerations that the probability of misclassification is given by

$$\Pr(1)\cdot\Phi\left(-\frac{1}{2}d + \frac{1}{d}\log\left(\frac{\Pr(2)}{\Pr(1)}\right)\right) + \Pr(2)\cdot\Phi\left(-\frac{1}{2}d - \frac{1}{d}\log\left(\frac{\Pr(2)}{\Pr(1)}\right)\right),$$

where Φ is the standard normal distribution function.

(e)What is the minimum possible probability of misclassification?

3. (Linear separability) With reference to Figures 6.6 and 6.7:

(a) Show that the vector $\boldsymbol{w}$ is perpendicular to the hyperplane

$$I(\boldsymbol{g}) = \boldsymbol{w}^\top \boldsymbol{g} + w_0 = 0.$$

(b) Suppose that there are just three training observations or points $\boldsymbol{g}_i$, $i = 1, 2, 3$, in a two-dimensional feature space and that these do not lie on a straight line. Since there are only two possible classes, the three points can be labeled in $2^3 = 8$ possible ways. Each possibility is obviously *linearly separable*. That is to say, one can find an oriented hyperplane which will correctly classify the three points, i.e., all class 1 points (if any) lie on one side and all class 2 points (if any) lie on the other side, with the vector $\boldsymbol{w}$ pointing to the class 1 side. The three points are said to be *shattered* by the set of hyperplanes. The maximum number of points that can be shattered is called the *Vapnik-Chervonenskis* (VC) dimension of the hyperplane classifier. What is the VC dimension in this case?

(c) A training set is linearly separable if a hyperplane can be found for which the smallest margin is positive. The following *perceptron algorithm* is guaranteed to find such a hyperplane, that is, to train the neuron of Figure 6.7 to classify any linearly separable training set of n observations (Cristianini and Shawe–Taylor, 2000):

i. Set $\boldsymbol{w} = \boldsymbol{0}$, $b = 0$, and $R = \max_\nu \|\boldsymbol{g}(\nu)\|$.

ii. Set $m = 0$.

iii. For $i = 1$ to n do: if $\gamma_\nu \leq 0$, then set $\boldsymbol{w} = \boldsymbol{w} + \ell(\nu)\boldsymbol{g}(\nu)$, $b = b + \ell(\nu)R^2$ and $m = 1$.

iv. If $m = 1$ go to ii, else stop.

The algorithm stops with a separating hyperplane $\boldsymbol{w}^\top \boldsymbol{g} + b = 0$. Implement this algorithm in IDL or Python and test it with linearly separable training data, see, e.g., Listing 3.10 and Figure 3.12.

4. (Bishop, 1995) Neural networks can also be used to approximate continuous vector functions $\boldsymbol{h}(\boldsymbol{g})$ of their inputs $\boldsymbol{g}$. Suppose that, for a given observation $\boldsymbol{g}(\nu)$, the corresponding training value $\boldsymbol{\ell}(\nu)$ is not known exactly, but that its components $\ell_k(\nu)$ are normally and independently distributed about the (unknown) functions $h_k(\boldsymbol{g}(\nu))$, i.e.,

$$p(\ell_k(\nu) \mid \boldsymbol{g}(\nu)) = \frac{1}{\sqrt{2\pi}\sigma} \exp\left(-\frac{(h_k(\boldsymbol{g}(\nu)) - \ell_k(\nu))^2}{2\sigma^2}\right), \quad k = 1 \ldots K. \tag{6.80}$$

Show that the appropriate cost function to train the synaptic weights so as to best approximate $\boldsymbol{h}$ with the network outputs $\boldsymbol{m}$ is the quadratic cost function, Equation (6.30). *Hint:* The probability density for a particular training pair $(\boldsymbol{g}(\nu), \boldsymbol{\ell}(\nu))$ can be written as (see Equation (2.61))

$$p(\boldsymbol{g}(\nu), \boldsymbol{\ell}(\nu)) = p(\boldsymbol{\ell}(\nu) \mid \boldsymbol{g}(\nu))p(\boldsymbol{g}(\nu)).$$

The likelihood function for n training examples chosen independently from the same distribution is, accordingly,

$$\prod_{\nu=1}^{n} p(\boldsymbol{\ell}(\nu) \mid \boldsymbol{g}(\nu))p(\boldsymbol{g}(\nu)).$$

Argue that maximizing this likelihood with respect to the synaptic weights is equivalent to minimizing the cost function

$$E = -\sum_{\nu=1}^{n} \log p(\boldsymbol{\ell}(\nu) \mid \boldsymbol{g}(\nu))$$

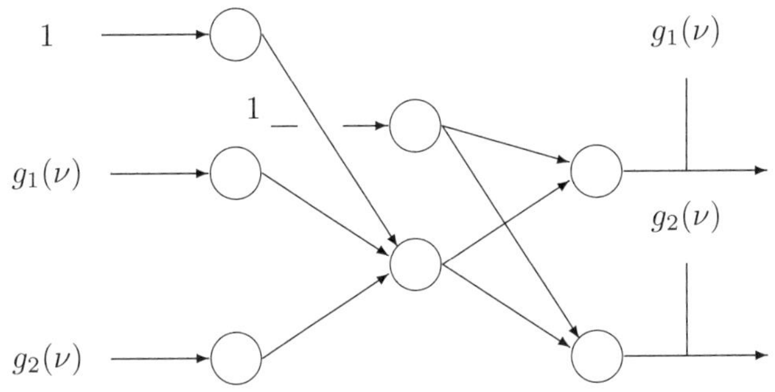

FIGURE 6.17
Principal components analysis with a neural network.

and then show that, with Equation (6.80), this reduces to Equation (6.30).

5. As an example of continuous function approximation, the neural network of Figure 6.10 can be used to perform principal components analysis on sequential data. The method applied involves so-called *self-supervised* classification. The idea is illustrated in Figure 6.17 for two-dimensional observations. The training data are presented to a network with a single hidden neuron. The output is constrained to be identical to the input (self-supervision). Since the output from the hidden layer is one-dimensional, the constraint requires that as much information as possible about the input signal be coded by the hidden neuron. This is the case if the data are projected along the first principal axis. As more and more data are presented, the synaptic weight vector (w_1^h, w_2^h) will therefore point more and more in the direction of that axis. Use the object class `FFNBP` to implement the network in Figure 6.17 in IDL and test it with simulated data. (According to the previous exercise, the cross-entropy cost function used in `FFNBP` is not fully appropriate, but suffices nevertheless.) Alternatively, use the Python class `Ffnbp` defined in the `supervisedclass.py` module to implement the network in Python.

6. Demonstrate Equation (6.39).

7. (Network symmetries) The hyperbolic tangent is defined as

$$\tanh(x) = \frac{e^x - e^{-x}}{e^x + e^{-x}}.$$

(a) Show that the logistic function $f(x)$ can be expressed in the form

$$f(x) = \frac{1}{2}\tanh\left(\frac{x}{2}\right) + \frac{1}{2}.$$

(b) Noting that $\tanh(x)$ is an odd function, i.e., $\tanh(x) = -\tanh(-x)$, argue that, for the feed-forward network of Figure 6.10, there are (at least) 2^L identical local minima in the cost function of Equation (6.33).

8. Show that Equation (6.77) follows from Equations (6.73) and (6.76).

7

Supervised Classification Part 2

Continuing in the present chapter on the subject of supervised classification, we will begin with a discussion of postclassification processing methods to improve classification results on the basis of contextual information. Then we turn our attention to statistical procedures for evaluating classification accuracy and for making quantitative comparisons between different classifiers. In this context, the computationally expensive n-fold cross-validation procedure will provide a good excuse to illustrate how to take advantage of modern cloud-computing services and perform several tasks in parallel. As an example of so-called *ensembles* or *committees* of classifiers, we then examine the *adaptive boosting* technique, applying it in particular to improve the generalization accuracy of neural network classifiers. This is followed by the derivation and implementation of a maximum likelihood classifier for polarimetric SAR imagery. The chapter concludes with a discussion of the problems posed by the classification of images with high spectral resolution, including an introduction to linear spectral un-mixing and the derivation of algorithms for linear and kernel anomaly detection.

7.1 Postprocessing

Intermediate resolution remote sensing satellite platforms used for land cover/land use classification, e.g., LANDSAT 7 TM+, SPOT, RapidEye, ASTER, have ground sample distances (GSDs), ranging between a few to a few tens of meters. These are typically smaller than the landscape objects being classified (agricultural fields, forests, urban areas, etc.). The imagery that they generate is therefore characterized by a high degree of spatial correlation. In Chapter 8 we will see examples of how spatial or contextual information might be incorporated into unsupervised classification. The supervised classification case is somewhat different, since reference is always being made to a — generally quite small — subset of labeled training data. Two approaches can be distinguished for inclusion of contextual information into supervised classification: moving window (or filtering) methods and segmentation (or region growing) methods. Both approaches can be applied either during classification or as a postprocessing step. We will restrict ourselves here to postclassification filtering, in particular mentioning briefly the majority filtering function offered in the standard ENVI environment, and then discussing in some detail a modification of a probabilistic technique described in Richards and Jia (2006). For an overview (and an example of) the use of segmentation for contextual classification, see Stuckens et al. (2000) and references therein.

7.1.1 Majority filtering

Majority postclassification filtering employs a moving window, with each central pixel assigned to the majority class of the pixels within the window. This clearly will have the effect of reducing the "salt-and-pepper" appearance typical of the thematic maps generated by pixel-oriented classifiers, and, to quote Stuckens et al. (2000), "it also results in larger classification units that might adhere more to the human perception of land cover." The ENVI implementation of the method is accessible from

`Classification/Post Classification/Majority/Minority Analysis`

Majority filtering merely examines the labels of neighborhood pixels. The classifiers we have discussed in the last chapter generate, in addition to class labels, class membership probability vectors for each observation. Therefore, neighboring pixels offer considerably more information than that exploited in majority filtering, information which can also be included in the relabeling process.

7.1.2 Probabilistic label relaxation

Recalling Figure 4.14, the 4-neighborhood $\mathcal{N}_i$ of an image pixel with intensity vector $\boldsymbol{g}_i$ consists of the four pixels above, below, to the left, and to the right of the pixel. The *a posteriori* class membership probabilities of the central pixel, as approximated by one of the classifiers of Chapter 6, are given by

$$\Pr(k \mid \boldsymbol{g}_i), \quad k = 1 \dots K, \quad \text{where} \sum_{k=1}^{K} \Pr(k \mid \boldsymbol{g}_i) = 1,$$

which, for notational convenience, we represent in the following as the K-component column vector $\boldsymbol{P}_i$ having components

$$P_i(k) = \Pr(k \mid \boldsymbol{g}_i), \quad k = 1 \dots K. \tag{7.1}$$

According to the standard decision rule, Equation (6.6), the maximum component of $\boldsymbol{P}_i$ determines the class membership of the ith pixel.

In analogy to majority filtering, we might expect that a possible misclassification of the pixel can be corrected by examining the membership probabilities in its neighborhood. If that is so, the neighboring pixels will have in some way to modify $\boldsymbol{P}_i$ such that its maximum component is more likely to correspond to the true class. We now describe a purely heuristic but nevertheless intuitively satisfying procedure to do just that, the so-called *probabilistic label relaxation* (PLR) method (Richards and Jia, 2006).

Let us postulate a multiplicative *neighborhood function* $Q_i(k)$ for the ith pixel which corrects $P_i(k)$ in the above sense, that is,

$$P_i'(k) = P_i(k)\frac{Q_i(k)}{\sum_j P_i(j)Q_i(j)}, \quad k = 1 \dots K. \tag{7.2}$$

The denominator ensures that the corrected values sum to unity and so still constitute a probability vector. In an obvious vector notation, we can write

$$\boldsymbol{P}_i' = \boldsymbol{P}_i \cdot \frac{\boldsymbol{Q}_i}{\boldsymbol{P}_i^\top \boldsymbol{Q}_i}, \tag{7.3}$$

where the dot signifies ordinary component-by-component multiplication.

The vector $\boldsymbol{Q}_i$ must somehow reflect the contextual information of the neighborhood. In order to define it, a *compatibility measure*

$$P_{ij}(k \mid m), \quad j \in \mathcal{N}_i$$

is introduced, namely, the conditional probability that pixel i has class label k, given that a neighboring pixel $j \in \mathcal{N}_i$ belongs to class m. A "small piece of evidence" (Richards and Jia, 2006) that i should be classified to k would then be

$$P_{ij}(k \mid m)P_j(m), \quad j \in \mathcal{N}_i.$$

This is the conditional probability that pixel i is in class k if neighboring pixel j is in class m multiplied by the probability that pixel j actually is in class m. We obtain the component $Q_i(k)$ of the neighborhood function by summing over all pieces of evidence and then averaging over the neighborhood:

$$\begin{aligned} Q_i(k) &= \frac{1}{4}\sum_{j\in\mathcal{N}_i}\sum_{m=1}^{K} P_{ij}(k \mid m)P_j(m) \\ &= \sum_{m=1}^{K} P_{i\mathcal{N}_i}(k \mid m)P_{\mathcal{N}_i}(m). \end{aligned} \tag{7.4}$$

Here $P_{\mathcal{N}_i}(m)$ is an average over all four neighborhood pixels:

$$P_{\mathcal{N}_i}(m) = \frac{1}{4}\sum_{j\in\mathcal{N}_i} P_j(m),$$

and $P_{i\mathcal{N}_i}(k \mid m)$ also corresponds to the *average compatibility* of pixel i with its entire neighborhood. We can write Equation (7.4) in matrix notation in the form

$$\boldsymbol{Q}_i = \boldsymbol{P}_{i\mathcal{N}_i}\boldsymbol{P}_{\mathcal{N}_i}$$

and Equation (7.3) finally as

$$\boldsymbol{P}_i' = \boldsymbol{P}_i \cdot \frac{\boldsymbol{P}_{i\mathcal{N}_i}\boldsymbol{P}_{\mathcal{N}_i}}{\boldsymbol{P}_i^\top \boldsymbol{P}_{i\mathcal{N}_i}\boldsymbol{P}_{\mathcal{N}_i}}. \tag{7.5}$$

For supervised classification, the matrix of average compatibilities $\boldsymbol{P}_{i\mathcal{N}_i}$ is not *a priori* available. However, it may easily be estimated directly from the initially classified image by assuming that it is independent of pixel location. First, a random central pixel i is chosen and its class label $\ell_i = k$ determined. Then, again randomly, a pixel $j \in \mathcal{N}_i$ is chosen and its class label $\ell_j = m$ is also determined. Thereupon the matrix element $P_{i\mathcal{N}_i}(k \mid m)$ (which was initialized to 0) is incremented by 1. This is repeated many times and finally the rows of the matrix are normalized. All of which constitutes the first step of the following algorithm:

Algorithm (Probabilistic Label Relaxation)

1. Carry out a supervised classification and determine the $K \times K$ compatibility matrix $\boldsymbol{P}_{i\mathcal{N}_i}$.
2. For each pixel i, determine the average neighborhood vector $\boldsymbol{P}_{\mathcal{N}_i}$ and replace $\boldsymbol{P}_i$ with $\boldsymbol{P}_i'$ as in Equation (7.5). Reclassify pixel i according to $\ell_i = \arg\max_k P_i'(k)$.
3. If only a few reclassifications took place, stop; otherwise go to step 2.

The stopping condition in the algorithm is obviously rather vague. Experience shows that the best results are obtained after 3 to 4 iterations; see Richards and Jia (2006). Too many iterations lead to a widening of the effective neighborhood of a pixel to such an extent that fully irrelevant spatial information falsifies the final product.

The PLR method can also be applied to any unsupervised classification algorithm that generates posterior class membership probabilities.* ENVI/IDL extensions and Python scripts for probabilistic label relaxation are given in Appendices C and D. Figure 7.1 shows a neural network classification result before and after PLR. The spatial coherence of the classes is improved.

7.2 Evaluation and comparison of classification accuracy

Assuming that sufficient labeled data are available for some to be set aside for test purposes, test data can be used to make an unbiased estimate of the

*An example is the Gaussian mixture clustering algorithm that will be met in Chapter 8.

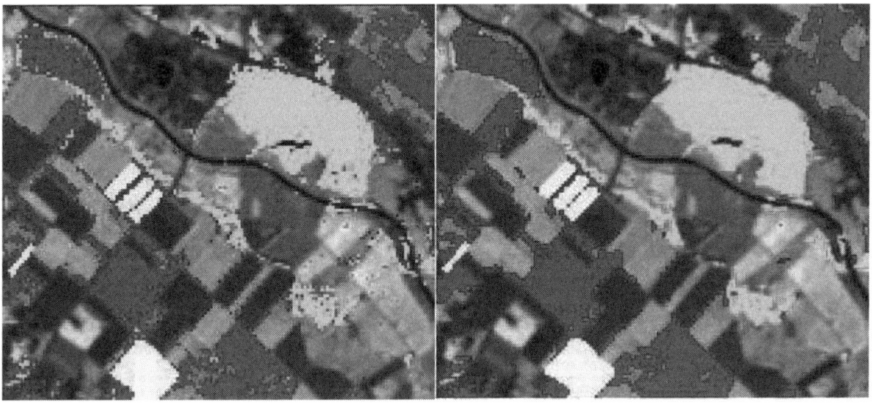

FIGURE 7.1
An example of postclassification processing. Left: original classification of a portion of the Jülich ASTER scene with a neural network. The classes shown are coniferous forest (green), rapeseed (yellow), and cereal grain (red). Right: after three iterations of PLR. **(See color insert.)**

misclassification rate of a trained classifier, i.e., the fraction of new data that will be incorrectly classified. This quantity provides a reasonable yardstick not only for evaluating the overall accuracy of supervised classifiers, but also for comparison of alternatives, for example, to compare the performance of a neural network with a maximum likelihood classifier on the same set of data.

7.2.1 Accuracy assessment

The classification of a single test datum is a random experiment, the possible outcomes of which constitute the sample space $\{\bar{A}, A\}$, where $\bar{A}$ = *misclassified*, A = *correctly classified*. Let us define a real-valued function X on this set, i.e., a random variable

$$X(\bar{A}) = 1, \quad X(A) = 0, \tag{7.6}$$

with mass function

$$\Pr(X = 1) = \theta, \quad \Pr(X = 0) = 1 - \theta.$$

The mean value of X is then

$$\langle X \rangle = 1\theta + 0(1 - \theta) = \theta \tag{7.7}$$

and its variance is (see Equation (2.13))

$$\mathrm{var}(X) = \langle X^2 \rangle - \langle X \rangle^2 = 1^2\theta + 0^2(1 - \theta) - \theta^2 = \theta(1 - \theta). \tag{7.8}$$

For the classification of n test data, which are represented by the i.i.d. sample $X_1 \dots X_n$, the random variable

$$Y = X_1 + X_2 + \dots X_n$$

corresponds to the total number of misclassifications. The random variable describing the misclassification rate is therefore Y/n having mean value

$$\langle \frac{1}{n} Y \rangle = \frac{1}{n}(\langle X_1 \rangle + \dots + \langle X_n \rangle) = \frac{1}{n} \cdot n\theta = \theta. \tag{7.9}$$

From the independence of the X_i, $i = 1 \dots n$, the variance of Y is given by

$$\mathrm{var}(Y) = \mathrm{var}(X_1) + \dots + \mathrm{var}(X_n) = n\theta(1-\theta), \tag{7.10}$$

so the variance of the misclassification rate is

$$\sigma^2 = \mathrm{var}\left(\frac{Y}{n}\right) = \frac{1}{n^2}\mathrm{var}(Y) = \frac{\theta(1-\theta)}{n}. \tag{7.11}$$

For y observed misclassifications we estimate θ as $\hat{\theta} = y/n$. Then the estimated variance is given by

$$\hat{\sigma}^2 = \frac{\hat{\theta}(1-\hat{\theta})}{n} = \frac{\frac{y}{n}\left(1-\frac{y}{n}\right)}{n} = \frac{y(n-y)}{n^3},$$

and the estimated standard deviation by

$$\hat{\sigma} = \sqrt{\frac{y(n-y)}{n^3}}. \tag{7.12}$$

As was pointed out in Section 2.1.1, Y is (n, θ)-binomially distributed. However, for a sufficiently large number n of test data, the binomial distribution is well approximated by the normal distribution. This is illustrated by the following IDL code, which generates the mass function for the binomial distribution, Equation (2.3). The result is shown in Figure 7.2 for $n = 2000$.

```
n = 2000
theta = 0.1
x = lindgen(n)
f = fltarr(n)
FOR i=1,n-1 do $
   f[i]=binomial(i,n,theta)-binomial(i+1,n,theta)
thisDevice =!D.Name
set_plot, 'PS'
Device, Filename='c:\temp\fig7_2.eps', $
   xsize=15,ysize=10,/Encapsulated
plot, x[100:300],f[100:300]
device,/close_file
set_plot,thisDevice
```

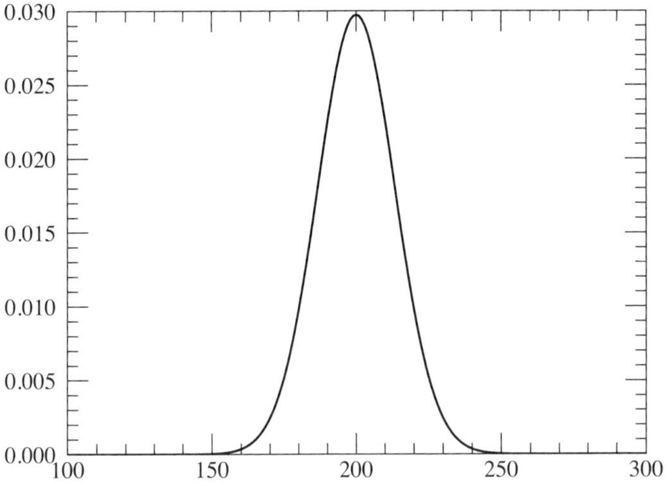

FIGURE 7.2
The binomial distribution for $n = 2000$ and $\theta = 0.1$ closely approximates a normal distribution.

Mean and standard deviation are thus generally sufficient to characterize the distribution of misclassification rates completely. To obtain an interval estimation for θ, recalling the discussion in Section 2.2.2, we make use of the fact that the random variable $(Y/n - \theta)/\sigma$ is approximately standard normally distributed. Then

$$\Pr\left(-s < \frac{Y/n - \theta}{\sigma} \le s\right) = 2\Phi(s) - 1, \tag{7.13}$$

so that the random interval $(Y/n - s\sigma, Y/n + s\sigma)$ covers the unknown misclassification rate θ with probability $2\Phi(s) - 1$. However, note that from Equation (7.11), σ is itself a function of θ. It is easy to show (Exercise 1) that, for $1 - \alpha = 0.95 = 2\Phi(s) - 1$ (which gives $s = 1.96$ from the normal distribution table), a 95% confidence interval for θ is

$$\left(\frac{y + 1.92 - 1.96 \cdot \sqrt{0.96 + \frac{y(n-y)}{n}}}{3.84 + n}, \frac{y + 1.92 + 1.96 \cdot \sqrt{0.96 + \frac{y(n-y)}{n}}}{3.84 + n}\right). \tag{7.14}$$

This interval should be routinely stated for any supervised classification of land use/land cover.

Detailed test results for supervised classification are usually presented in the form of a *contingency table*, or *confusion matrix*, which for K classes is defined as

$$\boldsymbol{C} = \begin{pmatrix} c_{11} & c_{12} & \cdots & c_{1K} \\ c_{21} & c_{22} & \cdots & c_{2K} \\ \vdots & \vdots & \ddots & \vdots \\ c_{K1} & c_{K2} & \cdots & c_{KK} \end{pmatrix}. \tag{7.15}$$

The matrix element c_{ij} is the number of test pixels with class label j which are classified as i. Note that the estimated misclassification rate is

$$\hat{\theta} = \frac{y}{n} = \frac{n - \sum_{i=1}^K c_{ii}}{n} = \frac{n - \mathrm{tr}(\boldsymbol{C})}{n}$$

and only takes into account the diagonal elements of the confusion matrix. The so-called *Kappa-coefficient*, on the other hand, makes use of all the matrix elements. It corrects the classification rate for the possibility of chance correct classifications and is defined as follows (Cohen, 1960):

$$\kappa = \frac{\mathrm{Pr(correct\ classification) - Pr(chance\ classification)}}{1 - \mathrm{Pr(chance\ classification)}}.$$

An expression for κ can be obtained in terms of the row and column sums in the matrix $\boldsymbol{C}$, which we write as

$$c_{i\cdot} = \sum_{j=1}^K c_{ij} \quad \text{and} \quad c_{\cdot i} = \sum_{j=1}^K c_{ji},$$

respectively. For n randomly labeled test pixels, the proportion of entries in the ith row is $c_{i\cdot}/n$ and in the ith column $c_{\cdot i}/n$. The probability of a chance correct classification (chance coincidence of row and column index) is therefore approximately given by the sum over i of the product of these two proportions:

$$\sum_{i=1}^K \frac{c_{i\cdot}\ c_{\cdot i}}{n^2}.$$

Hence an estimate for the Kappa coefficient is

$$\hat{\kappa} = \frac{\sum_i c_{ii}/n - \sum_i c_{i\cdot}c_{\cdot i}/n^2}{1 - \sum_i c_{i\cdot}c_{\cdot i}/n^2}. \tag{7.16}$$

Again, the Kappa coefficient alone tells us little about the quality of the classifier. We require its uncertainty. This can be calculated in the large sample limit $n \to \infty$ to be (Bishop et al., 1975)

$$\hat{\sigma}_\kappa^2 = \frac{1}{n}\left(\frac{\theta_1(1-\theta_1)}{(1-\theta_2)^2} + \frac{2(1-\theta_1)(2\theta_1\theta_2 - \theta_3)}{(1-\theta_2)^3} + \frac{(1-\theta_1)^2(\theta_4 - 4\theta_2^2)}{(1-\theta_2)^4}\right), \tag{7.17}$$

where

$$\theta_1 = \frac{1}{n}\sum_{i=1}^{K} c_{ii}, \quad \theta_2 = \frac{1}{n^2}\sum_{i=1}^{K} c_{i.}c_{.i}, \quad \theta_3 = \frac{1}{n^2}\sum_{i=1}^{K} c_{ii}(c_{i.} + c_{.i}),$$

$$\theta_4 = \frac{1}{n^3}\sum_{i,j=1}^{K} c_{ij}(c_{j.} + c_{.i})^2.$$

The ENVI/IDL extension `CT_RUN.PRO` described in Appendix C* prints out misclassification rate $\hat{\theta}$, standard deviation $\hat{\sigma}$, the confidence interval of Equation (7.14), Kappa coefficient $\hat{\kappa}$, standard deviation $\hat{\sigma}_\kappa$ and a contingency table $\boldsymbol{C}$ from the test result files generated by the maximum likelihood classifier, the Gaussian kernel classifier, the neural network and the support vector machine described in Sections 6.3 to 6.6. Here is a sample output for $K = 7$ classes:

```
1 Test observations: 4505
2 Classes: 7
3  Misclassification rate   0.0249
4      Standard deviation   0.0023
5    Conf. interval (95%)   0.0207   0.0298
6       Kappa coefficient   0.9705
7      Standard deviation   0.0028
8
9 Contingency Table
10    600    12     2     4     0     0    13   631  0.951
11     14   516     0     2     0     0     4   536  0.963
12      0     0   668     0     2     0     0   670  0.997
13      0     0     0   363     0     0     0   363  1.000
14      1     0     5     0   447     0    31   484  0.924
15      0     4     0     0     0   755     0   759  0.995
16      4    11     2     0     1     0  1044  1062  0.983
17    619   543   677   369   450   755  1092  4505  0.000
18 0.969 0.950 0.987 0.984 0.993 1.000 0.956 0.000  0.000
```

The matrix $\boldsymbol{C}$, Equation (7.15), is in the upper left 7×7 block in the table. Row 8 (line 17 in the above listing) gives the column sums $c_{.i}$ and column 8 the row sums $c_{i.}$. Row 9 (line 18) contains the ratios $c_{ii}/c_{.i}$, $i = 1 \ldots 7$, which are referred to as the *producer accuracies*. Column 9 contains the *user accuracies* $c_{ii}/c_{i.}$, $i = 1 \ldots 7$. The producer accuracy is the probability that an observation with label i will be classified as such. The user accuracy is the probability that the true class of an observation is i given that the classifier has labeled it as i. The rate of *correct classification*, which is more commonly quoted in the literature, is of course one minus the misclassification rate. A detailed discussion of confusion matrices for assessment of classification accuracy is provided in Congalton and Green (1999).

*Resp. its Python equivalent `ct.py`, Appendix D.

7.2.2 Cross-validation on the cloud

Representative ground reference data at or sufficiently near the time of image acquisition are generally difficult and/or expensive to come by; see Section 6.2. In this regard, the simple 2:1 train:test split is rather wasteful of the available labeled training pixels. Moreover, the variability due to the training data is not taken properly into account, since the data are sampled just once from their underlying distributions. In the case of neural networks, we have also so far ignored the variability of the training procedure itself with respect to the random initialization of the synaptic weights. Different initializations may lead to different local minima in the cost function and correspondingly different misclassification rates. Only if these aspects are considered to be negligible should the simple procedures discussed above be applied. Including them properly may constitute a very computationally intensive task (Ripley, 1996).

An alternative approach, one which at least makes more efficient use of the training data, is to apply *n-fold cross-validation*: A small fraction (one nth of the labeled pixels) is held back for testing, and the remaining data are used to train the classifier. This is repeated n times for n complementary test data subsets and then the results, e.g., misclassification rates, are averaged. In this way a larger fraction of the labeled data, namely $(n-1)/n$, is used for training. Moreover, all of the data are used for both training/testing and each observation is used for testing exactly once. For neural network classifiers, the effect of synaptic weight initialization is also reflected in the variance of the test results.

The drawback here, of course, is the necessity to repeat the train/test procedure n times rather than carrying it through only once. This is a problem especially for classifiers like neural networks or support vector machines with computationally expensive training algorithms. The cross-validation steps can, however, be performed in parallel given appropriate computer resources. Fortunately these are becoming increasingly available, not only in the form of multi-core processors, GPU hardware, etc., but also as so-called "cloud services". Since we have programmed our neural network classifiers in Python as well as in IDL, we shall illustrate n-fold cross-validation on a particularly easy-to-use, Python-based cloud platform called PiCloud.* Python functions coded on a local machine can be executed on PiCloud servers with slight, almost trivial, modifications in the code. And they can be run in parallel if so desired.

Appendix D includes a description of the Python program `ffncg.py`, which trains a feed-forward neural network with the efficient scaled conjugate gra-

*At the time of writing, the commercial PiCloud platform is in the process of transition to a compatible open-source service called Multyvac; see http://www.multyvac.com/. The code examples in this section are written against the original PiCloud API and will require slight modification.

Listing 7.1: Cross-validation on the cloud.

```
1  IMPORT numpy as np
2  IMPORT cloud, gdal, ogr, osr, os, time
3
4  DEF crossvalidate((Gstrn,lstrn,Gstst,lstst,L)):
5      affn = Ffncg(Gstrn,lstrn,L)
6      IF affn.train(epochs=1000):
7          RETURN affn.test(Gstst,lstst)
8      ELSE:
9          RETURN None
10
11 DEF traintst(Gs,ls,L):
12     m = np.shape(Gs)[0]
13     traintest = []
14     FOR i IN RANGE(10):
15         sl = SLICE(i*m//10,(i+1)*m//10)
16         traintest.append( (np.delete(Gs,sl,0), \
17         np.delete(ls,sl,0),Gs[sl,:],ls[sl,:],L) )
18     jids=cloud.MAP(crossvalidate,traintest,_type='c1')
19     RETURN cloud.result(jids)
```

dient algorithm derived in Appendix B and then performs 10-fold cross-validation "on the cloud" to estimate the misclassification rate. An excerpt is shown in Listing 7.1. The imported `cloud` module (line 2) makes the service available (assuming the user has registered for it). The function `crossvalidate`, defined in lines 4–9, takes labeled train and test pairs as its input, instantiates a network with L hidden neurons (line 5), trains it with 1000 epochs (line 6) and returns a misclassification rate on success (line 7). The function `traintst()`, lines 11–19, called from the main program, breaks the training data matrix `Gs` and class labels `ls` into a list of 10 disjunct train/test pairs in the ratio 9:1. In line 18, the function `crossvalidate` is mapped (applied) to this list with `cloud.MAP()`, the effect of which is to run the `crossvalidation` function on each train/test pair in parallel on as many computing units (cores) as are available at the time, up to 10. Each process returns a job identification number (`jid`) and these are collected into the list `jids` returned from `cloud.MAP()`. Finally, the list of job results, i.e., the actual misclassification rates returned by `crossvalidate()`, are fetched with `cloud.result()` and returned (line 19). The corresponding part of the `main()` Python script module is shown below:

```
    start = time.time()
    jid = cloud.call(traintst,Gs,ls,L)
    PRINT 'submission␣time:␣%s' %STR(time.time()-start)
    start = time.time()
    result = cloud.result(jid)
    PRINT 'cloud␣execution␣time:␣%s'%STR(time.time()-start)
```

```
7      PRINT 'misclassification␣rate:␣%f' %np.mean(result)
8      PRINT 'standard␣deviation:␣␣␣␣␣%f' %np.std(result)
```

In line 2, the function `cloud.call()` causes the Python code to be uploaded to PiCloud and the `traintst()` function to be executed there. When the upload is completed, a job identification number is returned, which is queried in line 5 with `cloud.result()`. The script waits at this point until `traintst()` has completed, upon which the result, the list of misclassification rates, is returned. The script then prints their mean and standard deviation, lines 7 and 8. Here is the console printout for classification of the Jülich ASTER scene of Chapter 6:

```
=========================
       ffncg
=========================
Mon Feb 04 14:14:34 2013
image:    D:/imagery/CRC/Chapters6-7/may0107pca.tif
training: D:/imagery/CRC/Chapters6-7/train.shp
reading training data...
7173 training pixel vectors were read in
training on 6455 pixel vectors...
elapsed time 199.502000093
classifying...
thematic map written to: D:/imagery/CRC/Chapters6-7/ffn.tif
please close the cross entropy plot to continue
submitting cross-validation to cloud
submission time: 16.2350001335
cloud execution time: 152.707999945
misclassification rate: 0.048513
standard deviation:     0.007567
--------done---------------------
```

The initial training of the network on the local machine (1.7 GHz Intel i5 CPU, running 64-bit Windows 7) took 200 seconds, while the 10-fold cross-validation required only 153 seconds* on the cloud.

7.2.3 Model comparison

A good value for a misclassification rate is $\theta \approx 0.05$. In order to claim that two rates produced by two different classifiers differ from one another significantly, a rule of thumb is that they should lie at least two standard deviations apart. A commonly used heuristic (van Niel et al., 2005) is to choose a minimum of $n \approx 30 \cdot N \cdot K$ training samples in all, where, as always, N is the data

*This will vary, depending on server availability. The charges incurred were about 0.7 core hours, or US$ 0.04.

dimensionality and K is the number of classes. For the Jülich classification examples using $N = 4$ principal components and $K = 10$ classes, this gives $n \approx 30 \cdot 4 \cdot 10 = 1200$. If one third are to be reserved for testing, the number should be increased to $1200 \cdot 3/2 = 1800$. But suppose we wish to claim, for instance, that misclassification rates 0.05 and 0.06 are significantly different. Then, according to our thumb rule, their standard deviations should be no greater than 0.005. From Equation (7.11), this means $0.05(1 - 0.05)/n \approx 0.005^2$, or $n \approx 2000$ observations are needed for testing alone. Since we are dealing with pixel data, this number of test observations (assuming sufficient training areas are available) is still realistic.

In order to decide whether classifier A is better than classifier B in a more precise manner, a hypothesis test must be formulated. The individual misclassifications Y_A and Y_B are, as we have seen, approximately normally distributed. If they were also independent, then the test statistic

$$S = \frac{Y_A - Y_B}{\sqrt{\text{var}(Y_A) + \text{var}(Y_B)}}$$

would be standard normally distributed under the null hypothesis $\theta_A = \theta_B$. In fact, the independence of the misclassification rates is not given, since they are determined with the same set of test data.

There exist computationally expensive alternatives. The buzzwords here are cross-validation, as discussed in the preceding section, and *bootstrapping*; see Weiss and Kulikowski (1991), Chapter 2, for an excellent introduction. As an example, suppose that a series of p trials is carried out. In the ith trial, the training data are split randomly into training and test sets in the ratio 2:1 as before. Both classifiers are then trained and tested on these sets to give misclassifications represented by the random variables Y_A^i and Y_B^i. If the differences $Y_i = Y_A^i - Y_B^i$, $i = 1 \dots p$, are independent and normally distributed, then a Student-t statistic, Equation (2.75), can be constructed to test the null hypothesis that the mean numbers of misclassifications are equal:

$$T = \frac{\bar{Y}}{\sqrt{S/p}},$$

where

$$\bar{Y} = \sum_{i=1}^{p} Y_i, \quad S = \frac{1}{p-1}\sum_{i=1}^{p}(Y_i - \bar{Y})^2.$$

T is Student-t distributed with $p - 1$ degrees of freedom.

There are some objections to this approach, the most obvious again being the need to repeat the training/test cycle many times (typically $p \approx 30$). Moreover, as Dietterich (1998) points out in a comparative investigation of several such test procedures, Y_i is not normally distributed since Y_A^i and Y_B^i are not independent. He recommends a *nonparametric* hypothesis test which avoids these problems and which we shall adopt here; see also Ripley (1996).

After training of the two classifiers which are to be compared, the following events for classification of the test data can be distinguished:

$$\bar{A}B,\ A\bar{B},\ \bar{A}\bar{B},\ \text{and}\ AB.$$

The event $\bar{A}B$ is *test observation is misclassified by A and correctly classified by B*, while $A\bar{B}$ is the event *test observation is correctly classified by A and misclassified by B* and so on. As before, we define random variables:

$$X_{\bar{A}B},\ X_{A\bar{B}},\ X_{\bar{A}\bar{B}}\ \text{and}\ X_{AB}$$

where

$$X_{\bar{A}B}(\bar{A}B) = 1, \quad X_{\bar{A}B}(A\bar{B}) = X_{\bar{A}B}(\bar{A}\bar{B}) = X_{\bar{A}B}(AB) = 0,$$

with mass function

$$\Pr(X_{\bar{A}B} = 1) = \theta_{\bar{A}B}, \quad \Pr(X_{\bar{A}B} = 0) = 1 - \theta_{\bar{A}B}.$$

Corresponding definitions are made for $X_{A\bar{B}}$, $X_{\bar{A}\bar{B}}$ and X_{AB}.

Now, in comparing the two classifiers, we are interested in the events $\bar{A}B$ and $A\bar{B}$. If the number of the former is significantly smaller than the number of the latter, then A is better than B and vice versa. Events $\bar{A}\bar{B}$ in which *both* methods perform poorly are excluded.

For n test observations, the random variables

$$\begin{aligned} Y_{\bar{A}B} &= X_{\bar{A}B_1} + \dots X_{\bar{A}B_n} \quad \text{and} \\ Y_{A\bar{B}} &= X_{A\bar{B}_1} + \dots X_{A\bar{B}_n} \end{aligned}$$

are the frequencies of the respective events. We then have

$$\begin{aligned} \langle Y_{\bar{A}B}\rangle &= n\theta_{\bar{A}B}, \quad \text{var}(Y_{\bar{A}B}) = n\theta_{\bar{A}B}(1-\theta_{\bar{A}B}) \\ \langle Y_{A\bar{B}}\rangle &= n\theta_{A\bar{B}}, \quad \text{var}(Y_{A\bar{B}}) = n\theta_{A\bar{B}}(1-\theta_{A\bar{B}}). \end{aligned}$$

We expect that $\theta_{\bar{A}B} \ll 1$, that is, $\text{var}(Y_{\bar{A}B}) \approx n\theta_{\bar{A}B} = \langle Y_{\bar{A}B}\rangle$. The same holds for $Y_{A\bar{B}}$. It follows that the random variables

$$\frac{Y_{\bar{A}B} - \langle Y_{\bar{A}B}\rangle}{\sqrt{\langle Y_{\bar{A}B}\rangle}} \quad \text{and} \quad \frac{Y_{A\bar{B}} - \langle Y_{A\bar{B}}\rangle}{\sqrt{\langle Y_{A\bar{B}}\rangle}}$$

are approximately standard normally distributed.

Under the null hypothesis (equivalence of the two classifiers), the expectation values of $Y_{\bar{A}B}$ and $Y_{A\bar{B}}$ satisfy

$$\langle Y_{\bar{A}B}\rangle = \langle Y_{A\bar{B}}\rangle =: \langle Y\rangle.$$

We form the *McNemar test statistic*

$$S = \frac{(Y_{\bar{A}B} - \langle Y\rangle)^2}{\langle Y\rangle} + \frac{(Y_{A\bar{B}} - \langle Y\rangle)^2}{\langle Y\rangle}, \tag{7.18}$$

which is chi-square distributed with one degree of freedom; see Section 2.1.5 and, e.g., Siegel (1965). Let $y_{\bar{A}B}$ and $y_{A\bar{B}}$ be the number of events actually measured. Then the mean $\langle Y \rangle$ is estimated as

$$\langle \hat{Y} \rangle = \frac{y_{\bar{A}B} + y_{A\bar{B}}}{2}$$

and a realization of the test statistic S is

$$s = \frac{(y_{\bar{A}B} - \frac{y_{\bar{A}B}+y_{A\bar{B}}}{2})^2}{\frac{y_{\bar{A}B}+y_{A\bar{B}}}{2}} + \frac{(y_{A\bar{B}} - \frac{y_{\bar{A}B}+y_{A\bar{B}}}{2})^2}{\frac{y_{\bar{A}B}+y_{A\bar{B}}}{2}}.$$

With a little algebra, this expression can be simplified to

$$s = \frac{(y_{\bar{A}B} - y_{A\bar{B}})^2}{y_{\bar{A}B} + y_{A\bar{B}}}. \tag{7.19}$$

A correction is usually made to Equation (7.19), writing it in the form

$$s = \frac{(|y_{\bar{A}B} - y_{A\bar{B}}| - 1)^2}{y_{\bar{A}B} + y_{A\bar{B}}}, \tag{7.20}$$

which takes into approximate account the fact that the statistic is discrete, while the chi-square distribution is continuous. From the percentiles of the chi-square distribution, the critical region for rejection of the null hypothesis of equal misclassification rates at the 5% significance level is $s \geq 3.841$. The ENVI/IDL extension `McNEMAR_RUN` (Appendix C) or the Python script `mcnemar.py` (Appendix D) compares two classifiers on the basis of their test result files, printing out $y_{\bar{A}B}$, $y_{A\bar{B}}$, s and the P-value $1 - P_{\chi^2;1}(s)$. Here is an example comparing the maximum likelihood and neural network classifiers for the Jülich ASTER scene:

```
ENVI> mcnemar_run
--------------------------
Classification Comparison
Fri Jan 09 12:53:50 2009
--------------------------
First classifier
; FFN test results FOR juelich aster\may0107_pca
Second classifier
; MaxLike test results FOR juelich aster\may0107_pca
Test observations: 2391
Classes: 10
     First classifier:        21
    Second classifier:        54
    McNemar statistic: 13.6533
              P-value:  0.0002
```

In this case the null hypothesis can certainly be rejected in favor of the neural network. Comparing the neural network with ENVI's SVM classifier:

```
--------------------------
Classification Comparison
Fri Jan 09 12:57:45 2009
--------------------------
First classifier
; FFN test results FOR juelich aster\may0107_pca
Second classifier
; SVM test results FOR juelich aster\may0107_pca
Test observations: 2391
Classes: 10
      First classifier:        64
     Second classifier:        88
     McNemar statistic:   3.4803
               P-value:   0.0621
```

where now the neural network is "better", but not at the 5% significance level.

7.3 Adaptive boosting

Further enhancement of classifier accuracy is sometimes possible by combining several classifiers into an *ensemble* or *committee* and then applying some kind of "voting scheme" to generalize to new data. An excellent introduction to ensemble-based systems is given by Polikar (2006). The basic idea is to generate several classifiers and pool them in such a way as to improve on the performance of any single one. This implies that the pooled classifiers make errors on *different* observations, implying further that each classifier be as unique as possible, particularly with respect to misclassified instances (Polikar, 2006). One way of achieving this uniqueness is to use different training sets for each classifier, for example, by re-sampling the training data with replacement, a procedure referred to as "bootstrap aggregation" or *bagging* (Breiman, 1996).

Representative of ensemble methods, we consider here a powerful technique called *adaptive boosting* or *AdaBoost* for short (Freund and Shapire, 1996). It involves training a sequence of classifiers, placing increasing emphasis on hard-to-classify data, and then combining the sequence so as to reduce the overall training error. AdaBoost was originally suggested for combining binary classifiers, i.e., for two-class problems. However, Freund and Shapire (1997) proposed two multi-class extensions, the more commonly used of which is *AdaBoost.M1*. In the following we shall apply AdaBoost.M1 to an ensemble of neural network classifiers. For other examples of adaptive boosting of neural networks, see Schwenk and Bengio (2000) and Murphey et al. (2001).

In order to motivate the adaptive boosting idea, consider the training of the feed-forward neural network classifier of Chapter 6 when there are just two classes to choose between. Making use as before of stochastic training, we train the network by minimizing the local cost function, Equation (6.34), on randomly selected labeled examples. To begin with, the training data are sampled uniformly, as is done, for example, in the backpropagation training algorithm of Listing 6.6 or 6.9. We can represent such a sampling scheme with the uniform, discrete probability distribution

$$p_1(\nu) = 1/m, \quad \nu = 1 \dots m,$$

over the m training examples. Let U_1 be the set of incorrectly classified examples after completion of the training procedure. Then the classification error is given by

$$\epsilon_1 = \sum_{\nu \in U_1} p_1(\nu).$$

Let us now find a new sampling distribution $p_2(\nu)$ such that the trained classifier would achieve an error of 50% if trained with respect to that distribution. In other words, it would perform as well as uninformed random guessing. The intention is, through the new distribution, to achieve a new classifier–training set combination which is as different as possible from the one just used. We obtain the new distribution $p_2(\nu)$ by reducing the probability for correctly classified examples by a factor $\beta_1 < 1$ so that the accuracy obtained is 1/2, that is,

$$\frac{1}{2} = \sum_{\nu \in U_1} p_2(\nu) = \sum_{\nu \notin U_1} p_2(\nu) = \frac{1}{Z} \sum_{\nu \notin U_1} \beta_1 p_1(\nu) = \frac{1}{Z} \beta_1 (1 - \epsilon_1). \tag{7.21}$$

The denominator Z is a normalization which ensures that $\sum_\nu p_2(\nu) = 1$ so that $p_2(\nu)$ is indeed a probability distribution,

$$Z = \sum_{\nu \in U_1} p_1(\nu) + \beta_1 \sum_{\nu \notin U_1} p_1(\nu) = \epsilon_1 + \beta_1 (1 - \epsilon_1). \tag{7.22}$$

Combining Equations (7.21) and (7.22) gives

$$\beta_1 = \frac{\epsilon_1}{1 - \epsilon_1}. \tag{7.23}$$

If we now train the network with respect to $p_2(\nu)$, we will get (it is to be hoped) a different set U_2 of incorrectly classified training examples and, correspondingly, a different classification error

$$\epsilon_2 = \sum_{\nu \in U_2} p_2(\nu).$$

This leads to a new reduction factor β_2 and the procedure is repeated. The sequence must, of course, terminate at i classifiers when $\epsilon_{i+1} > 1/2$, as then the incorrectly classified examples can no longer be emphasized since $\beta_{i+1} > 1$.

At the generalization phase, the "importance" of each classifier is set to some function of β_i, the smaller β_i, the more important the classifier. As we shall see below, an appropriate weight is $\log(1/\beta_i)$. Thus, if C_k is the set of networks which classify feature vector $\boldsymbol{g}$ as k, then that class receives the "vote"

$$V_k = \sum_{i \in C_k} \log(1/\beta_i), \quad k = 1, 2,$$

after which $\boldsymbol{g}$ is assigned to the class with the maximum vote.

To place things on a more precise footing, we will define the *hypothesis* generated by the neural network classifier for input observation $\boldsymbol{g}(\nu)$ as

$$h(\boldsymbol{g}(\nu)) = h(\nu) = \arg\max_k (m_k(\boldsymbol{g}(\nu))), \quad \nu = 1 \dots m. \tag{7.24}$$

This is just the index of the output neuron whose signal is largest. For a two-class problem, $h(\nu) \in \{1, 2\}$. Suppose that $k(\nu)$ is the label of observation $\boldsymbol{g}(\nu)$. Following Freund and Shapire (1997), define the indicator

$$[[h(\nu) \neq k(\nu)]] = \begin{cases} 1 & \text{if the hypothesis } h(\nu) \text{ is incorrect} \\ 0 & \text{if it is correct.} \end{cases} \tag{7.25}$$

With this notation, we can give an exact formulation of the adaptive boosting algorithm for a sequence of neural networks applied to two-class problems. Then we can prove a theorem on the upper bound of the overall training error for that sequence. Here, first of all, is the algorithm:

Algorithm (AdaBoost)

1. Define an initial uniform probability distribution $p_1(\nu) = 1/m$, $\nu = 1 \dots m$, and the number N_c of classifiers in the sequence. Define initial weights $w_1(\nu) = p_1(\nu)$, $\nu = 1 \dots m$.

2. For $i = 1 \dots N_c$ do the following:

   (a) Set $p_i(\nu) = w_i(\nu) / \sum_{\nu'=1}^{m} w_i(\nu')$, $\nu = 1 \dots m$.
   (b) Train a network with the sampling distribution $p_i(\nu)$ to get back the hypotheses $h_i(\nu)$, $\nu = 1 \dots m$.
   (c) Calculate the error $\epsilon_i = \sum_{\nu=1}^{m} p_i(\nu)[[h_i(\nu) \neq k(\nu)]]$.
   (d) Set $\beta_i = \epsilon_i/(1 - \epsilon_i)$.
   (e) Determine a new weights w_{i+1} according to

$$w_{i+1}(\nu) = w_i(\nu)\beta_i^{1-[[h_i(\nu) \neq k(\nu)]]}, \quad \nu = 1 \dots m.$$

3. Given an unlabeled observation $\boldsymbol{g}$, obtain the total vote received by each class,

$$V_k = \sum_{\{i | h_i(\boldsymbol{g}) = k\}} \log(1/\beta_i), \quad k = 1, 2,$$

   and assign $\boldsymbol{g}$ to the class with maximum vote.

The training error in the AdaBoost algorithm is the fraction of training examples that will be incorrectly classified when put into the voting procedure in step 3 above. We have the following theorem (the proof is given in Appendix A):

THEOREM 7.1
The training error ϵ for the algorithm AdaBoost is bounded above according to

$$\epsilon \leq 2^{N_c} \prod_{i=1}^{N_c} \sqrt{\epsilon_i(1-\epsilon_i)}. \tag{7.26}$$

This theorem tells us that, provided each classifier in the sequence can return an error $\epsilon_i < 1/2$, the training error will approach zero exponentially. It can be shown (Freund and Shapire, 1997) that the result is also valid for the multiclass case $K > 2$. The boosting algorithm is then referred to as AdaBoost.M1.

An ENVI/IDL extension `FFN3AB_RUN` for boosting neural networks (trained with the fast Kalman filter algorithm of Appendix B) is described in Appendix C and in Canty (2009). A sequence of neural networks, $i = 1, 2, \ldots$, is trained on samples chosen with respect to distributions $p_1(\nu), p_2(\nu) \ldots$, the sequence terminating at i when $\epsilon_{i+1} \geq 1/2$ or when a maximum sequence length is reached. In order to take into account the fact that a network may become trapped in a local minimum of the cost function, training is restarted with a new random synaptic weight configuration if the current training error ϵ_i exceeds 1/2. The maximum number of restarts for a given network is five, after which the boosting terminates. The classifiers in the sequence are implemented in an object-oriented framework as instances of a neural network object class; see Chapter 6. The algorithm is as follows:

Algorithm (Adaptive boosting of a sequence of neural network classifiers)

1. Set $p_1(\nu) = 1/m$, $\nu = 1 \ldots m$, where m is the number of observations in the set of labeled training data. Choose maximum sequence length N_{max}. Set $i = 1$.

2. Set $r = 0$.

3. Create a new neural network instance FFN(i) with random synaptic weights. Train FFN(i) with sampling distribution $p_i(\nu)$. Let U_i be the set of incorrectly classified training observations after completion of the training procedure.

4. Calculate $\epsilon_i = \sum_{\nu \in U_i} p_i(\nu)$. If $\epsilon_i < 1/2$, then continue, else if $r < 5$, then set $r = r + 1$, destroy the instance FFN(i), and go to 3, else stop.

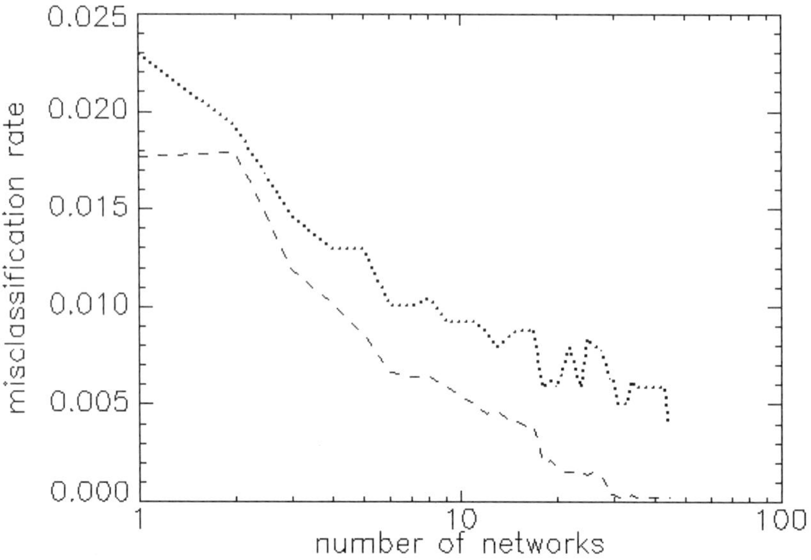

FIGURE 7.3
Adaptive boost training of an ensemble of neural networks. The dashed (dotted) line is the error on the training (test) observations.

5. Set $\beta_i = \epsilon_i/(1-\epsilon_i)$ and update the distribution:

$$p_{i+1}(\nu) = \frac{p_i(\nu)}{Z_i} \times \begin{cases} \beta_i & \text{if } \nu \notin U_i \\ 1 & \text{otherwise} \end{cases}, \quad \nu = 1 \ldots m,$$

   where $Z_i = \sum_{\nu \in U_i} p_i(\nu) + \beta_i \sum_{\nu \notin U_i} p_i(\nu)$.

6. Set $i = i+1$. If $i > N_{\max}$, then stop, else go to 2.

During the training phase, the program displays the cost function of the currently trained network and the training and generalization errors of the boosted sequence. Boosting can be interrupted at any time, upon which the sequence generated so far will be used to classify the input data. An example is shown in Figure 7.3 for the training/test datasets of the Jülich ASTER scene. For illustration purposes, six principal components were used for training, as the boosting effect is most evident for higher-dimensional input spaces. In the figure, the training error is actually boosted to zero, while the generalization error is reduced by a factor of around 2.5. For a detailed comparison with other classifiers, see Canty (2009).

7.4 Classification of polarimetric SAR imagery

We saw in Chapter 6, Section 6.3, how to derive a Bayes maximum likelihood classifier for normally distributed optical/infrared pixels. In the case of the fully polarimetric m-look SAR data discussed in Chapter 5, the image observations were expressed in complex covariance matrix form

$$\bar{\boldsymbol{c}} = \frac{1}{m}\boldsymbol{x},$$

where

$$\boldsymbol{x} = \sum_{\nu=1}^{m} \boldsymbol{s}(\nu)\boldsymbol{s}(\nu)^{\dagger}, \quad \boldsymbol{s} = (s_{hh}, \sqrt{2}s_{hv}, s_{vv})^{T}.$$

The corresponding random matrix $\boldsymbol{X}$, as was pointed out, will follow a complex Wishart distribution, Equation (2.58),

$$p_{W_c}(\boldsymbol{x}) = \frac{|\boldsymbol{x}|^{(m-N)}\exp(-\mathrm{tr}(\boldsymbol{\Sigma}^{-1}\boldsymbol{x}))}{\pi^{N(N-1)/2}|\boldsymbol{\Sigma}|^{m}\prod_{i=1}^{N}\Gamma(m+1-i)}, \tag{7.27}$$

where N is the dimension of the covariance matrix: $N = 3$ for quad, $N = 2$ for dual and $N = 1$ for single polarimetric SAR images.* Following exactly the same argument as in Chapter 6 (see also Exercise 3 and Lee et al. (1994)) we can derive a maximum likelihood discriminant function for observations $\bar{\boldsymbol{c}}$, namely,

$$d_k(\bar{\boldsymbol{c}}) = \log(\Pr(k)) - m\left(\log|\boldsymbol{\Sigma}_k| + \mathrm{tr}(\boldsymbol{\Sigma}_k^{-1}\bar{\boldsymbol{c}})\right). \tag{7.28}$$

In the training phase, the class-specific complex covariance matrices $\boldsymbol{\Sigma}_k$, $k = 1 \dots K$, are estimated using pixels within selected training areas of the SAR image.

From Equation (7.28) it is evident that the prior class membership probabilities, $\Pr(k)$, will play a smaller role in the classification when the number of looks is large. In fact, if we set all prior probabilities equal, then the discriminant is simply

$$d_k(\bar{\boldsymbol{c}}) = -\log|\boldsymbol{\Sigma}_k| - \mathrm{tr}(\boldsymbol{\Sigma}_k^{-1}\bar{\boldsymbol{c}}), \tag{7.29}$$

independent of the number of looks m. This independence holds provided that m is a global parameter for the entire image, which for look-averaged imagery is certainly the case. However, if an adaptive filter, such as the MMSE filter of Chapter 5, is applied prior to classification, then m may vary somewhat from one land cover class to the next.

*For $N = 1$, Equation (7.27) reduces to the gamma distribution.

FIGURE 7.4
Maximum likelihood classification of an EMISAR L-band quad polarimetric SAR image acquired over a test agricultural area in Denmark, left: without prior adaptive filtering, right: with prior adaptive filtering. The land use categories are: winter wheat (red), rye (green), water (blue), spring barley (yellow), oats (cyan), beets (magenta), peas (purple), coniferous forest (coral). **(See color insert.)**

AN ENVI/IDL extension `MAXLIKESAR_RUN` for supervised classification of polarimetric SAR images is presented in Appendix C. An example is shown in Figure 7.4, where a quad polarimetric SAR image obtained with the EMISAR airborne sensor (Conradsen et al., 2003) is classified with and without prior adaptive filtering using the MMSE filter of Chapter 5.

TABLE 7.1
Kappa values for SAR classification.

Polarization	Filter	Kappa	Sigma
Dual	None	0.468	0.006
Dual	MMSE	0.558	0.006
Quad	None	0.619	0.005
Quad	MMSE	0.666	0.005

Qualitatively, the classification is seen to be improved by prior filtering. This is confirmed quantitatively in Table 7.1, which lists the Kappa coefficients calculated with the `CT_RUN.PRO` script described in Section 7.2.1. The table also shows reduced Kappa values for dual polarimetry,* confirming that classification accuracy improves significantly with the number of polarimetric channels.

7.5 Hyperspectral image analysis

Hyperspectral — as opposed to multispectral — images combine both high or moderate spatial resolution with high spectral resolution. Typical remote sensing imaging spectrometers generate in excess of two hundred spectral channels. Figure 7.5 shows part of a so-called *image cube* for the Airborne Visible/Infrared Imaging Spectrometer (AVIRIS) sensor taken over a region of the California coast. Figure 7.6 displays the spectrum of a single pixel in the image. Sensors of this kind produce much more complex data and provide correspondingly much more information about the reflecting surfaces examined than their multispectral counterparts.

The classification methods discussed in Chapter 6 and in the present chapter must in general be modified considerably in order to cope with the volume of data provided in a hyperspectral image. For example, the covariance matrix of an AVIRIS scene has dimension 224×224, so that the modeling of image or class probability distributions is much more difficult. Here one speaks of "ill-posed" classification problems, meaning that the available training data may be insufficient to estimate the model parameters adequately, and some sort of dimensionality reduction is needed (Richards and Jia, 2006). We will restrict discussion in the following to the concept of spectral unmixing, a method often used in lieu of "conventional" classification, and one of the most common techniques for hyperspectral image analysis.

7.5.1 Spectral mixture modeling

In multispectral image classification the fact that, at the scale of observation, a pixel often contains a mixture of land cover categories is generally treated as a second-order effect and more or less ignored. When working with hyperspectral imaging spectrometers, it is possible to treat the problem of the "mixed pixel" quantitatively. While one can still speak of classification of land surfaces, the training data now consist of external spectral libraries or, in some instances, reference spectra derived from within the images themselves. Comparison with external spectral libraries requires detailed attention to atmospheric correction. The end product is not the discrete labeling of the pixels that we have become familiar with, but consists rather of image planes

*In which case the observations are 2×2 matrices; see Equation (5.32).

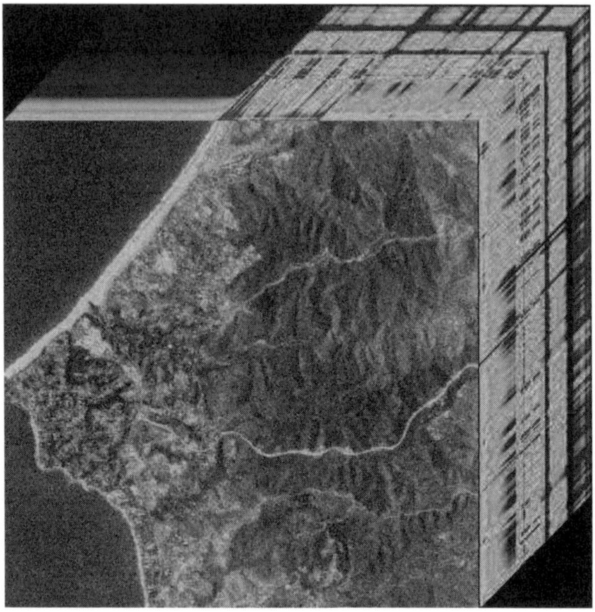

FIGURE 7.5
AVIRIS hyperspectral image cube over the Santa Monica Mountains acquired on April 7, 1997 at a GSD of 20 m. **(See color insert.)**

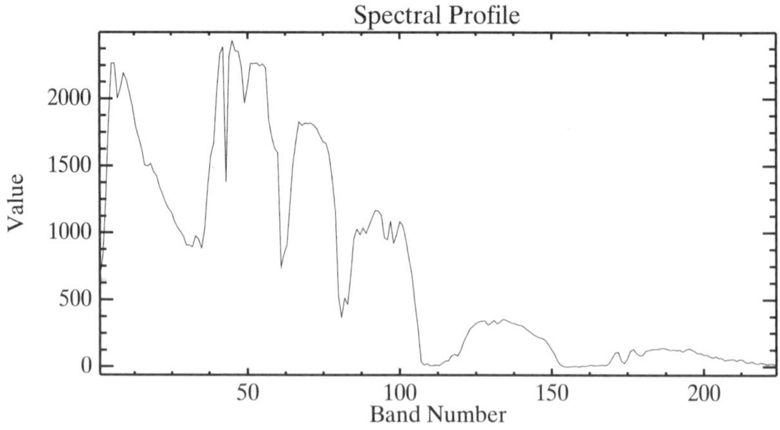

FIGURE 7.6
AVIRIS spectrum at one pixel location in Figure 7.5. There are 224 spectral bands covering the 0.4–1.5 μm wavelength interval.

or maps showing, at each pixel location, the proportion of surface material contributing to the observed reflectance.

The basic premise of mixture modeling is that, within a given scene, the surface is dominated by a small number of common materials that have characteristic spectral properties. These are referred to as the *end-members* and it is assumed that the spectral variability captured by the remote sensing system can be modeled by mixtures of end-members.

Suppose that there are K end-members and N spectral bands. Denote the spectrum of the ith end-member by the column vector

$$\boldsymbol{m}^i = (m_1^i, m_2 \dots m_N^i)^\top$$

and the matrix of end-member spectra $\boldsymbol{M}$ by

$$\boldsymbol{M} = (\boldsymbol{m}^1 \dots \boldsymbol{m}^K) = \begin{pmatrix} m_1^1 & \dots & m_1^K \\ \vdots & \ddots & \vdots \\ m_N^1 & \dots & m_N^K \end{pmatrix},$$

with one column for each end-member. For hyperspectral imagery we always have $K \ll N$, unlike the situation for multispectral data where $K \approx N$.

The measured spectrum, represented by random vector $\boldsymbol{G}$, may be modeled as a linear combination of end-members plus a residual term $\boldsymbol{R}$ which is understood to be the variation in $\boldsymbol{G}$ not explained by the mixture model:

$$\boldsymbol{G} = \alpha_1 \boldsymbol{m}^1 + \dots + \alpha_K \boldsymbol{m}^K + \boldsymbol{R} = \boldsymbol{M}\boldsymbol{\alpha} + \boldsymbol{R}. \tag{7.30}$$

The vector $\boldsymbol{\alpha} = (\alpha_1 \dots \alpha_K)^\top$ contains nonnegative mixing coefficients which are to be determined. Let us assume that the residual $\boldsymbol{R}$ is a normally distributed, zero mean random vector with covariance matrix

$$\boldsymbol{\Sigma}_R = \begin{pmatrix} \sigma_1^2 & 0 & \dots & 0 \\ 0 & \sigma_2^2 & \dots & 0 \\ \vdots & \vdots & \ddots & \vdots \\ 0 & 0 & \dots & \sigma_N^2 \end{pmatrix}.$$

The standardized residual is $\boldsymbol{\Sigma}_R^{-1/2}\boldsymbol{R}$ (why?) and the square of the standardized residual is

$$(\boldsymbol{\Sigma}_R^{-1/2}\boldsymbol{R})^\top(\boldsymbol{\Sigma}_R^{-1/2}\boldsymbol{R}) = \boldsymbol{R}^\top\boldsymbol{\Sigma}_R^{-1}\boldsymbol{R}. \tag{7.31}$$

The mixing coefficients $\boldsymbol{\alpha}$ may then be determined by minimizing this quantity with respect to the α_i under the condition that they sum to unity,

$$\sum_{i=1}^{K} \alpha_i = 1, \tag{7.32}$$

and are all nonnegative,

$$\alpha_i \ge 0, \quad i = 1 \dots K. \tag{7.33}$$

Ignoring the requirement of Equation (7.33) for the time being, a Lagrange function for minimization of Equation (7.31) under the constraint of Equation (7.32) is

$$
\begin{aligned}
L &= \boldsymbol{R}^\top \boldsymbol{\Sigma}_R^{-1} \boldsymbol{R} + 2\lambda(\sum_{i=1}^{K} \alpha_i - 1) \\
&= (\boldsymbol{G} - \boldsymbol{M}\boldsymbol{\alpha})^\top \boldsymbol{\Sigma}_R^{-1} (\boldsymbol{G} - \boldsymbol{M}\boldsymbol{\alpha}) + 2\lambda(\sum_{i=1}^{K} \alpha_i - 1),
\end{aligned}
$$

the last equality following from Equation (7.30). Solving the set of equations

$$
\frac{\partial L}{\partial \boldsymbol{\alpha}} = 0, \quad \frac{\partial L}{\partial \lambda} = 0,
$$

and replacing $\boldsymbol{G}$ by its realization $\boldsymbol{g}$, we obtain the estimates for the mixing coefficients (Exercise 4)

$$
\begin{aligned}
\hat{\boldsymbol{\alpha}} &= (\boldsymbol{M}^\top \boldsymbol{\Sigma}_R^{-1} \boldsymbol{M})^{-1} (\boldsymbol{M}^\top \boldsymbol{\Sigma}_R^{-1} \boldsymbol{g} - \lambda \boldsymbol{1}_K) \\
\hat{\boldsymbol{\alpha}}^\top \boldsymbol{1}_K &= 1,
\end{aligned} \tag{7.34}
$$

where $\boldsymbol{1}_K$ is a column vector of K ones. The first equation determines the mixing coefficients in terms of known quantities and λ. The second equation can be used to eliminate λ. Neglecting the constraint in Equation (7.33) is common practice. It can, however, be dealt with using appropriate numerical methods (Nielsen, 2001).

7.5.2 Unconstrained linear unmixing

If we work, for example, with MNF-transformed data (see Section 3.4), then we can assume that $\boldsymbol{\Sigma}_R = \boldsymbol{I}$.* If furthermore we ignore both of the constraints on $\boldsymbol{\alpha}$, Equations (7.32) and (7.33), which amounts to the assumption that the end-member spectra $\boldsymbol{M}$ are capable of explaining the observations completely apart from random noise, then Equation (7.34) reduces to the ordinary least squares estimate for $\boldsymbol{\alpha}$,

$$
\hat{\boldsymbol{\alpha}} = [(\boldsymbol{M}^\top \boldsymbol{M})^{-1} \boldsymbol{M}^\top] \boldsymbol{g}. \tag{7.35}
$$

The expression in square brackets is the pseudoinverse of the matrix $\boldsymbol{M}$ and the covariance matrix for $\boldsymbol{\alpha}$ is $\sigma^2(\boldsymbol{M}^\top \boldsymbol{M})^{-1}$, as is explained in Section 2.6.3. This calculation can be invoked from the ENVI menu under

```
Spectral/Mapping Methods/Linear Spectral Unmixing
```

*This is at least the case for the MNF transformation of Section 3.4.1. If ENVI's MNF algorithm is used, Section 3.4.2, then the components of $\boldsymbol{R}$ must first be divided by the corresponding eigenvalues of the transformation.

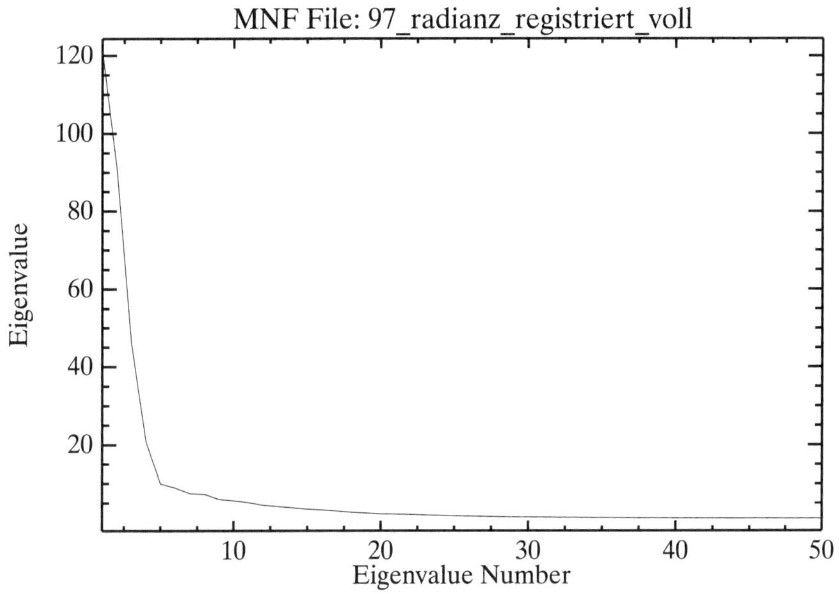

FIGURE 7.7
The first 50 eigenvalues of the MNF transformation of the image in Figure 7.5.

which optionally can also include the constraint (7.32). More sophisticated approaches, which are applicable when not all of the end-members are known, are discussed in the Exercises.

7.5.3 Intrinsic end-members and pixel purity

When a spectral library for all of the K end-members in $\boldsymbol{M}$ is available, the mixture coefficients can be calculated directly using the above methods. The primary product of the spectral mixture analysis consists of fraction images which show the spatial distribution and abundance of the end-member components in the scene. If such external data are unavailable, there are various strategies for determining end-members from the hyperspectral image itself. We describe briefly the method recommended and implemented in ENVI.

The first step is to reduce the dimensionality of the data. This may be accomplished in ENVI with the minimum noise fraction transformation described in Section 3.4.2. By examining the eigenvalues of the transformation and retaining only the components with eigenvalues exceeding one (non-noise components), the number of dimensions can be reduced substantially; see Figure 7.7.

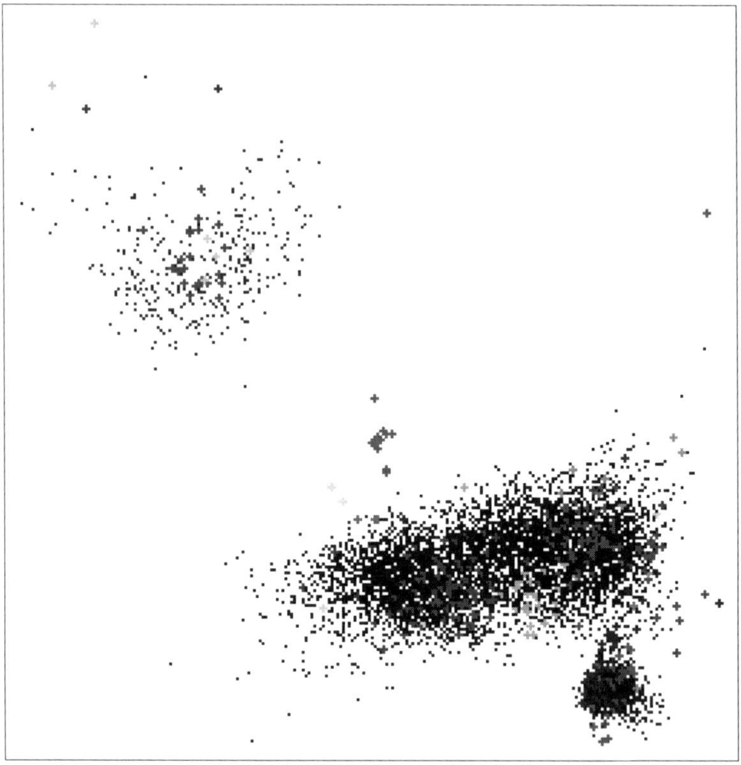

FIGURE 7.8
The n-D visualizer displaying pixel purity indices. **(See color insert.)**

The so-called *pixel purity index* (PPI) is then used to find the most spectrally pure, or extreme, pixels in the reduced feature space. The most spectrally pure pixels typically correspond to end-members. These pixels must be on the corners, edges or faces of the data cloud. The PPI is computed by repeatedly projecting n-dimensional scatterplots onto a random unit vector. The extreme pixels in each projection are noted and the number of times each pixel is marked as extreme is recorded. A threshold value is used to define how many pixels are marked as extreme at the ends of the projected vector. This value should be 2 to 3 times the variance in the data, which is 1 when using the MNF-transformed bands. A minimum of about 5000 iterations is usually required to produce useful results.

When the iterations are completed, a PPI image is created in which the intensity of each pixel corresponds to the number of times that pixel was recorded as extreme; see Figure 7.8. Therefore, bright pixels are generally end-members. (The end-members, projected back onto image coordinates, also hint at locations and sites that could be visited for ground truth measurements, should that be feasible.)

ENVI's n-dimensional visualizer can be used interactively to define classes of pixels corresponding to end-members and to plot their spectra. These may be saved along with their pixel locations as ROIs for later use in spectral unmixing or related procedures such as spectral angle mapping (Kruse et al., 1993). This sort of "data-driven" analysis has both advantages and disadvantages. To quote Mustard and Sunshine (1999):

> This method is repeatable and has distinct advantages for objective analysis of a data set to assess the general dimensionality and to define end-members. The primary disadvantage of this method is that it is fundamentally a statistical approach dependent on the specific spectral variance of the scene and its components. Thus the resulting end-members are mathematical constructs and may not be physically realistic.

7.5.4 Anomaly detection: The RX algorithm

The highly resolved spectral information provided by hyperspectral imagery has led to its frequent use in so-called *target detection*, the discovery of small-scale features of interest, most often of military or law enforcement relevance. Target detection typically involves two steps (Kwon and Nasrabadi, 2005): First, localized spectral anomalies are pinpointed by an unsupervised filtering procedure. Second, the significance of each identified anomaly, i.e., whether or not it is a target, is ascertained. The latter step usually involves comparison with known spectral signatures. In the following we outline a well-known procedure for carrying out the first step, the *RX anomaly detector* proposed by Reed and Yu (1990). It has been extensively used in the context of target detection and is equally applicable to multispectral imagery. Theiler and Matsekh (2009) discuss the algorithm in the context of anomalous change detection.

By referring to the likelihood ratio test introduced in Section 2.5, a derivation of the RX algorithm is straightforward. Consider a local neighborhood, e.g., a rectangular window within a multi- or hyperspectral image, and distinguish the central pixel as a possible anomaly. Let $\boldsymbol{g}(\nu)$, $\nu = 1 \ldots m$, denote the observed pixel vectors in the background neighborhood and $\boldsymbol{g}(m+1)$ the central pixel. We construct the following null and alternative hypotheses:

$$
\begin{aligned}
H_0 &: \boldsymbol{g}(\nu) \sim \mathcal{N}(\boldsymbol{\mu}_b, \boldsymbol{\Sigma}_b),\ \nu = 1 \ldots m+1, \\
H_1 &: \boldsymbol{g}(\nu) \sim \mathcal{N}(\boldsymbol{\mu}_b, \boldsymbol{\Sigma}_b),\ \nu = 1 \ldots m, \quad \boldsymbol{g}(m+1) \sim \mathcal{N}(\boldsymbol{\mu}, \boldsymbol{\Sigma}_b).
\end{aligned}
\tag{7.36}
$$

Thus the null, or no-anomaly, hypothesis is that all of the observations in the window are uniformly sampled from a multivariate normal distribution with mean $\boldsymbol{\mu}_b$ and covariance matrix $\boldsymbol{\Sigma}_b$. The alternative hypothesis states that the central pixel is characterized by the *same* covariance matrix but by

Listing 7.2: Excerpt from the script `rx.py`

```
FROM spectral.algorithms.detectors IMPORT RX
FROM spectral.algorithms.algorithms IMPORT calc_stats
IMPORT spectral.io.envi as envi

DEF main():
    gdal.AllRegister()
    path = auxil.select_directory('Input␣directory')
    IF path:
        os.chdir(path)
#   input image, convert to ENVI format
    infile = auxil.select_infile(title='Image␣file')
    IF infile:
        inDataset = gdal.Open(infile,GA_ReadOnly)
        cols = inDataset.RasterXSize
        rows = inDataset.RasterYSize
        projection = inDataset.GetProjection()
        geotransform = inDataset.GetGeoTransform()
        driver = gdal.GetDriverByName('ENVI')
        enviDataset=driver\
           .CreateCopy('entmp',inDataset)
        inDataset = None
        enviDataset = None
    ELSE:
        RETURN
    outfile, outfmt= \
           auxil.select_outfilefmt(title='Output␣file')
#   RX-algorithm
    img = envi.OPEN('entmp.hdr')
    arr = img.load()
    rx = RX(background=calc_stats(arr))
    res = rx(arr)
#   output
    driver = gdal.GetDriverByName(outfmt)
    outDataset = driver.Create(outfile,cols,rows,1,\
                                     GDT_Float32)
    IF geotransform IS NOT None:
        outDataset.SetGeoTransform(geotransform)
    IF projection IS NOT None:
        outDataset.SetProjection(projection)
    outBand = outDataset.GetRasterBand(1)
    outBand.WriteArray(np.asarray(res,np.float32),0,0)
    outBand.FlushCache()
    outDataset = None
    PRINT 'Result␣written␣to␣%s'%outfile
```

a different mean vector $\boldsymbol{\mu}$. In the notation of Definition 2.7, the maximized likelihoods are given by

$$\max_{\theta\in\omega_0} L(\theta) = \prod_{\nu=1}^{m+1} \exp\left(-\frac{1}{2}(\boldsymbol{g}(\nu)-\hat{\boldsymbol{\mu}}_b)^\top \hat{\boldsymbol{\Sigma}}_b^{-1}(\boldsymbol{g}(\nu)-\hat{\boldsymbol{\mu}}_b)\right)$$

and by

$$\max_{\theta\in\omega} L(\theta) = \prod_{\nu=1}^{m} \exp\left(-\frac{1}{2}(\boldsymbol{g}(\nu)-\hat{\boldsymbol{\mu}}_b)^\top \hat{\boldsymbol{\Sigma}}_b^{-1}(\boldsymbol{g}(\nu)-\hat{\boldsymbol{\mu}}_b)\right) \cdot \exp\left(-\frac{1}{2}(\boldsymbol{g}(m+1)-\hat{\boldsymbol{\mu}})^\top \hat{\boldsymbol{\Sigma}}_b^{-1}(\boldsymbol{g}(m+1)-\hat{\boldsymbol{\mu}})\right),$$

where $\hat{\boldsymbol{\mu}}_b$ and $\hat{\boldsymbol{\Sigma}}_b$ are the maximum likelihood estimates of the mean and covariance matrix of the background. But $\hat{\boldsymbol{\mu}} = \boldsymbol{g}(m+1)$ (there is only one observation), so the last exponential factor in the above equation is unity. Therefore the likelihood ratio test has the critical region

$$Q = \frac{\max_{\theta\in\omega_0} L(\theta)}{\max_{\theta\in\omega} L(\theta)} = \exp\left(-\frac{1}{2}(\boldsymbol{g}(m+1)-\hat{\boldsymbol{\mu}}_b)^\top \hat{\boldsymbol{\Sigma}}_b^{-1}(\boldsymbol{g}(m+1)-\hat{\boldsymbol{\mu}}_b)\right) \le k.$$

Thus we reject the null hypothesis (that no anomaly is present) for central observation $\boldsymbol{g} = \boldsymbol{g}(m+1)$ when the squared Mahalanobis distance

$$d = (\boldsymbol{g}-\hat{\boldsymbol{\mu}}_b)^\top \hat{\boldsymbol{\Sigma}}_b^{-1}(\boldsymbol{g}-\hat{\boldsymbol{\mu}}_b) \tag{7.37}$$

exceeds some threshold. This distance serves as the RX anomaly detector. In practice, quite good results can be obtained with a global, rather than local, estimate of the background statistical parameters $\hat{\boldsymbol{\mu}}_b$ and $\hat{\boldsymbol{\Sigma}}_b$.

The RX-algorithm is available in both the ENVI Classic and new ENVI graphical interfaces with global and local background options. A simple Python implementation using the `Spy` package* and allowing only for global background statistics is shown in Listing 7.2; see also Appendix D. In lines 13–22 the image filename is entered and the image copied to ENVI standard format in order to be read into a `Spy ImageFile` object. This occurs in lines 28 and 29. In line 30 the global background statistics are set, and the result (the image of squared Mahalanobis distances) is returned in line 31. An example is shown in Figure 7.9.

7.5.5 Anomaly detection: The kernel RX algorithm

An improvement in anomaly detection might be expected if nonlinearities in the data are included in the model. A kernelized variant of the RX algorithm was suggested by Kwon and Nasrabadi (2005). To quote from their introduction:

---

*`http://spectralpython.sourceforge.net/index.html`.

> The conventional RX distance measure does not take into account the higher order relationships between the spectral bands at different wavelengths. The nonlinear relationships between different spectral bands within the target or clutter spectral signature need to be exploited in order to better distinguish between the two hypotheses. Furthermore the Gaussian assumption in the RX-algorithm for the distributions [under] the two hypotheses H_0 and H_1 in general is not valid.

Their derivation of the kernel RX algorithm is reproduced in the following, merely adapting it to our notation.

To begin with, we write the mapping of the terms in Equation (7.37) from the linear input space to a nonlinear feature space in the form

$$\begin{aligned} \boldsymbol{g} &\to \boldsymbol{\phi}(\boldsymbol{g}) \equiv \boldsymbol{\phi} \\ \hat{\boldsymbol{\mu}}_b &\to (\hat{\boldsymbol{\mu}}_\phi)_b \equiv \boldsymbol{\mu}_\phi \\ \hat{\boldsymbol{\Sigma}}_b &\to (\hat{\boldsymbol{\Sigma}}_\phi)_b \equiv \boldsymbol{\Sigma}_\phi. \end{aligned}$$

The definitions on the right serve to simplify the notation. The centered data matrix of m observations in the feature space is, see Equation (4.30),

$$\tilde{\boldsymbol{\Phi}} = \begin{pmatrix} \tilde{\boldsymbol{\phi}}(1)^\top \\ \vdots \\ \tilde{\boldsymbol{\phi}}(m)^\top \end{pmatrix},$$

where

$$\tilde{\boldsymbol{\phi}}(\nu) = \boldsymbol{\phi}(\nu) - \boldsymbol{\mu}_\phi,\ \nu = 1 \dots m, \quad \boldsymbol{\mu}_\phi = \frac{1}{m}\sum_{\nu=1}^m \boldsymbol{\phi}(\nu), \tag{7.38}$$

and the $m \times m$ centered kernel matrix for the m observations is

$$\tilde{\mathcal{K}} = \tilde{\boldsymbol{\Phi}}\tilde{\boldsymbol{\Phi}}^\top,$$

which can be calculated with Equation (4.31). The Mahalanobis distance measure, Equation (7.37), expressed in the nonlinear feature space, is given by

$$d_\phi(\boldsymbol{g}) = (\boldsymbol{\phi} - \boldsymbol{\mu}_\phi)^\top \boldsymbol{\Sigma}_\phi^{-1}(\boldsymbol{\phi} - \boldsymbol{\mu}_\phi), \tag{7.39}$$

and we wish to express it purely in terms of kernel functions. Let's begin with the covariance matrix $\boldsymbol{\Sigma}_\phi$. It is given by the outer product

$$\boldsymbol{\Sigma}_\phi = \frac{1}{m}\sum_{\nu=1}^m (\boldsymbol{\phi}(\nu) - \boldsymbol{\mu}_\phi)(\boldsymbol{\phi}(\nu) - \boldsymbol{\mu}_\phi)^\top = \frac{1}{m}\tilde{\boldsymbol{\Phi}}^\top\tilde{\boldsymbol{\Phi}}. \tag{7.40}$$

Let $\boldsymbol{w}_\phi^i$, $i = 1 \dots r$, be its first r eigenvectors in order of decreasing eigenvalue, and define

$$\boldsymbol{W}_\phi = (\boldsymbol{w}_\phi^1, \dots \boldsymbol{w}_\phi^r).$$

In general, we must assume that the eigenvalues of $\mathbf{\Sigma}_\phi$ beyond the rth one are effectively zero. Then the eigendecomposition of $\mathbf{\Sigma}_\phi$ is given by, see Equation (1.49),

$$\mathbf{\Sigma}_\phi = \boldsymbol{W}_\phi \mathbf{\Lambda} \boldsymbol{W}_\phi{}^\top,$$

where $\mathbf{\Lambda}$ is a diagonal matrix of the first r eigenvalues of $\mathbf{\Sigma}_\phi$. We replace $\mathbf{\Sigma}_\phi^{-1}$ in Equation (7.39) by the pseudoinverse, Equation (1.50),

$$\mathbf{\Sigma}_\phi^+ = \boldsymbol{W}_\phi \mathbf{\Lambda}^{-1} \boldsymbol{W}_\phi{}^\top. \tag{7.41}$$

Recall that Equation (3.49) in Chapter 3 expressed the eigenvectors $\boldsymbol{w}_i$ of the covariance matrix in terms of the dual vectors $\boldsymbol{\alpha}_i$, which are eigenvectors of the centered Gram matrix. In the nonlinear space we do the same, writing Equation (3.49) in the form (see also Equation (1.13))

$$\boldsymbol{w}_\phi^i = \sum_{\nu=1}^{m} (\boldsymbol{\alpha}_i)_\nu \tilde{\boldsymbol{\phi}}(\nu) = (\tilde{\boldsymbol{\phi}}(1), \dots \tilde{\boldsymbol{\phi}}(m))\boldsymbol{\alpha}_i = \tilde{\mathbf{\Phi}}^\top \boldsymbol{\alpha}_i, \quad i = 1 \dots r, \tag{7.42}$$

where the $\boldsymbol{\alpha}_i$ are now eigenvectors of the centered kernel matrix, that is,

$$\tilde{\mathcal{K}} \boldsymbol{\alpha}_i = \lambda_i \boldsymbol{\alpha}_i.$$

Now let us write Equation (7.42) in matrix form,

$$\boldsymbol{W}_\phi = \tilde{\mathbf{\Phi}}^\top \boldsymbol{\alpha}, \tag{7.43}$$

where

$$\boldsymbol{\alpha} = (\boldsymbol{\alpha}_1, \dots \boldsymbol{\alpha}_r).$$

Substituting Equation (7.43) into Equation (7.41), we obtain

$$\mathbf{\Sigma}_\phi^+ = \tilde{\mathbf{\Phi}}^\top \boldsymbol{\alpha} \mathbf{\Lambda}^{-1} \boldsymbol{\alpha}^\top \tilde{\mathbf{\Phi}}$$

and accordingly, with Equation (7.39), the distance measure

$$d_\phi(\boldsymbol{g}) = (\boldsymbol{\phi} - \boldsymbol{\mu}_\phi)^\top \tilde{\mathbf{\Phi}}^\top \boldsymbol{\alpha} \mathbf{\Lambda}^{-1} \boldsymbol{\alpha}^\top \tilde{\mathbf{\Phi}} (\boldsymbol{\phi} - \boldsymbol{\mu}_\phi). \tag{7.44}$$

The eigenvalues of $\mathbf{\Sigma}_\phi$, which appear along the diagonal of $\mathbf{\Lambda}$, are in fact the same as those of the centered kernel matrix $\tilde{\mathcal{K}}$ except for a factor $1/m$. That is,

$$\mathbf{\Sigma}_\phi \boldsymbol{w}_i = \mathbf{\Sigma}_\phi \tilde{\mathbf{\Phi}}^\top \boldsymbol{\alpha}_i = \frac{1}{m} \tilde{\mathbf{\Phi}}^\top \tilde{\mathbf{\Phi}} \tilde{\mathbf{\Phi}}^\top \boldsymbol{\alpha}_1 = \frac{1}{m} \tilde{\mathbf{\Phi}}^\top \tilde{\mathcal{K}} \boldsymbol{\alpha}_1 = \frac{1}{m} \tilde{\mathbf{\Phi}}^\top \lambda_i \boldsymbol{\alpha}_1 = \frac{\lambda_i}{m} \boldsymbol{w}_i.$$

Hence the eigendecomposition of $\tilde{\mathcal{K}}$ is

$$\tilde{\mathcal{K}} = m \boldsymbol{\alpha} \mathbf{\Lambda} \boldsymbol{\alpha}^\top$$

and its pseudoinverse is

$$\tilde{\mathcal{K}}^+ = \frac{1}{m} \boldsymbol{\alpha} \mathbf{\Lambda}^{-1} \boldsymbol{\alpha}^\top. \tag{7.45}$$

Listing 7.3: Excerpt from the ENVI/IDL extension KRX_RUN.PRO.

```
K = kernel_matrix(G,gma=gma,nscale=nscale)
Kc = center(K)
PRINT,'GMA:␣' + strtrim(gma,2)

PRINT, 'Pseudoinverse␣of␣centered␣kernel␣matrix␣...'
; pseudoinvert centered kernel matrix
lambda = la_eigenql(Kc,/double,eigenvectors=alpha)
alpha = transpose(alpha)
idx = reverse(sort(lambda))
lambda = lambda[idx]
alpha = alpha[idx,*]
tol = m*eps(max(lambda))
_ = where(lambda GT tol, r)
alpha = alpha[0:r,*]
lambda = lambda[0:r]
Kci = alpha##diag_matrix(1.0/lambda)##transpose(alpha)
```

Listing 7.4: Excerpt from the ENVI/IDL extension KRX_RUN.PRO.

```
   outimage = fltarr(num_cols,num_rows)
   Ku = total(K,1)/m-total(K)/m^2
   Ku = (dblarr(num_cols)+1.0)##Ku
   i = 0L
   WHILE i LT num_rows DO BEGIN
      pct=i*100/num_rows
      IF progressbar->CheckCancel() THEN BEGIN
         PRINT,'RX␣aborted'
         i = num_rows-1
      ENDIF
      progressbar->update,fix(pct)
      GGi = GG[*,i*num_cols:(i+1)*num_cols-1]
      Kg = kernel_matrix(GGi,G,gma=gma,nscale=nscale)
      a = total(Kg,1,/double)
      a = a##(dblarr(m)+1.0)
      Kg = Kg - a/m
      Kgu = Kg - Ku
      d = total(Kgu*(Kgu##Kci),1,/double)
      outimage[*,i] = float(d)
      i++ ; next row
   ENDWHILE
```

We can therefore write the nonlinear anomaly detector, Equation (7.44), in the form

$$d_\phi(\boldsymbol{g}) = m(\boldsymbol{\phi} - \boldsymbol{\mu}_\phi)^\top \tilde{\boldsymbol{\Phi}}^\top \tilde{\mathcal{K}}^+ \tilde{\boldsymbol{\Phi}}(\boldsymbol{\phi} - \boldsymbol{\mu}_\phi).$$

The factor m is irrelevant and can be dropped. To complete the kernelization of this expression, consider the inner product

$$\begin{aligned}
\boldsymbol{\phi}^\top \tilde{\boldsymbol{\Phi}}^\top &= \boldsymbol{\phi}^\top (\tilde{\boldsymbol{\phi}}(1) \dots \tilde{\boldsymbol{\phi}}(m)) \\
&= \boldsymbol{\phi}^\top ((\boldsymbol{\phi}(1) \dots \boldsymbol{\phi}(m)) - \boldsymbol{\mu}_\phi) \\
&= ((\boldsymbol{\phi}^\top \boldsymbol{\phi}(1)) \dots (\boldsymbol{\phi}^\top \boldsymbol{\phi}(m))) - \frac{1}{m}\sum_{\nu=1}^m (\boldsymbol{\phi}^\top \boldsymbol{\phi}(\nu)),
\end{aligned}$$

or, in terms of the symmetric kernel function $k(\boldsymbol{g}, \boldsymbol{g}') = \boldsymbol{\phi}(\boldsymbol{g})^\top \boldsymbol{\phi}(\boldsymbol{g}')$,

$$\boldsymbol{\phi}^\top \tilde{\boldsymbol{\Phi}}^\top = (k(\boldsymbol{g}(1), \boldsymbol{g}) \dots k(\boldsymbol{g}(m), \boldsymbol{g})) - \frac{1}{m}\sum_{\nu=1}^m k(\boldsymbol{g}(\nu), \boldsymbol{g}) =: \mathcal{K}_g.$$

In a similar way (Exercise 6), one can show that

$$\begin{aligned}
\boldsymbol{\mu}_\phi^\top \tilde{\boldsymbol{\Phi}}^\top &= \frac{1}{m}\sum_{\nu=1}^m (k(\boldsymbol{g}(\nu), \boldsymbol{g}(1)) \dots k(\boldsymbol{g}(\nu), \boldsymbol{g}(m))) - \frac{1}{m^2}\sum_{\nu,\nu'=1}^m k(\boldsymbol{g}(\nu), \boldsymbol{g}(\nu')) \\
&=: \mathcal{K}_\mu.
\end{aligned} \tag{7.46}$$

The first term in $\mathcal{K}_\mu$ is the row vector of the column averages of the uncentered kernel matrix, and the second term is the overall average of the uncentered kernel matrix elements. Combining, we have finally

$$d_\phi(\boldsymbol{g}) = (\mathcal{K}_g - \mathcal{K}_\mu)\tilde{\mathcal{K}}^+(\mathcal{K}_g - \mathcal{K}_\mu)^\top \tag{7.47}$$

as our kernelized RX anomaly detector.

Excerpts from the ENVI/IDL extension `KRX_RUN.PRO` for kernel RX described in Appendix C are shown in Listings 7.3 and 7.4. After the usual preliminaries, sample observation vectors are placed in the array `G` and the kernel matrix and its centered version are determined, Listing 7.3, lines 1 and 2. In order to determine the number r of nonzero eigenvalues, the centered kernel matrix is diagonalized and its eigenvectors and eigenvalues sorted in decreasing order, lines 7–11. Then r is determined as the number of eigenvalues exceeding the machine tolerance, lines 12 and 13. The pseudoinverse of the centered kernel matrix, Equation (7.45), is calculated in line 16. In Listing 7.4, the row vector $\mathcal{K}_\mu$ (IDL array `Ku`), which depends only on the kernel matrix, is first determined, lines 2 and 3. Then the anomaly image (squared Mahalanobis distances, Equation (7.47)) is calculated row by row. If GPULib/CUDA is available (see Section 4.4.2), then the anomalies are calculated on the GPU in a separate code block (not shown). Figure 7.9 shows an example comparing the RX and kernel RX detectors. Qualitatively, the kernelized method performs better.

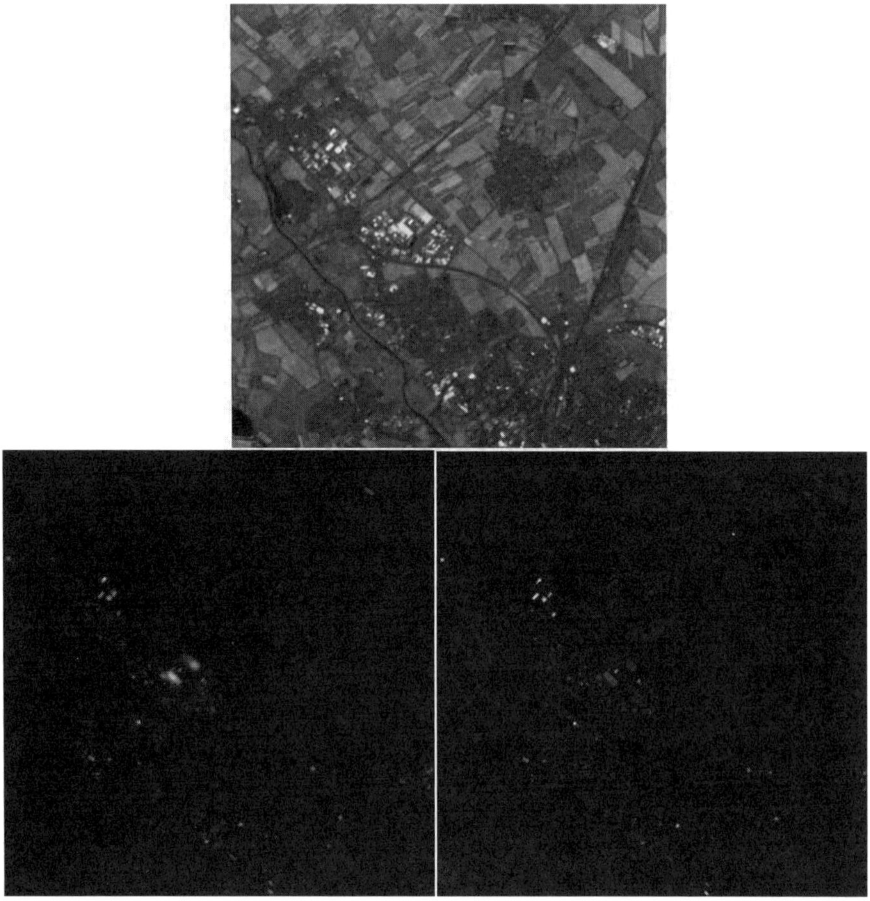

FIGURE 7.9
Anomaly detection: Top, RGB composite of bands 3, 2, and 1 of a spatial subset of the 9-band pan-sharpened ASTER image over Jülich. Bottom left, RX anomaly image calculated from all 9 bands using the script in Listing 7.2. Bottom right, kernel RX anomaly image calculated from all 9 bands using the ENVI/IDL extension `KRX_RUN.PRO`. **(See color insert.)**

7.6 Exercises

1. Derive the confidence limits on the misclassification rate θ given by Equation (7.14).

2. In *bootstrap aggregation* or *bagging* (Breiman, 1996; Polikar, 2006) an ensemble of classifiers is derived from a single classifier by training it re-

peatedly on *bootstrapped replicas* of the training dataset, i.e., on random samples drawn from the training set *with replacement*. In each replica, some training examples may appear more than once, and some may be missing altogether. A simple majority voting scheme is then used to classify new data with the ensemble. Breiman (1996) shows that improved classification accuracy is to be expected over that obtained by the original classifier/training dataset, especially for relatively unstable classifiers. Unstable classifiers are ones whose decision boundaries tend to be sensitive to small perturbations of the training data, and neural networks belong to this category.

(a) In the base object class `FFN` for a neural network, Listing 6.6, the function (method) `FFN::CLASSIFY(Gs,probs)` takes a row of pixels as input and returns their class labels in a byte array. For an ensemble of trained networks, one could use this function to build up a $c \times P$ array of class labels L, where c is the number of pixels in the row and P the number of classifiers. Write an IDL function `MAJORITY_VOTE(L)` which returns a $c \times 1$ array of those pixel labels which are chosen by the majority of the classifiers. You can use a loop over the class labels, but *not* over the pixels.

(b) Write an ENVI extension to implement bagging using the IDL object class `FFNKAL` described in Appendix B as the neural network classifier. Use the program `FFN_RUN.PRO` (Appendix C) as a reference. Your implementation should do the following:

- Get the image to be classified, the ROI training data, the number P of classifiers in the ensemble, and the size $n < m$ of the bootstrapped replicas, where m is the size of the training set.
- Create P instances of the neural network object class and train each of them on a different bootstrapped sample.
- Classify the entire image by constructing the array L (part (a) above) for each row of the image and then passing it to the function `MAJORITY_VOTE(L)`.
- Return the classified image to ENVI.

3. Derive the discriminant Equation (7.28) from the complex Wishart distribution, Equation (7.27).

4. Demonstrate that the mixing coefficients of Equation (7.34) minimize the standardized residual, Equation (7.31), under constraint Equation (7.32).

5. In searching for a single spectral signature of interest $\boldsymbol{d}$, we can write Equation (7.30) in the form

$$\boldsymbol{G} = \boldsymbol{d}\alpha_d + \boldsymbol{U\beta} + \boldsymbol{R}, \tag{7.48}$$

where $\boldsymbol{U}$ is the $N \times (K-1)$ matrix of unwanted (or perhaps unknown) spectra and $\boldsymbol{\beta}$ is the vector of their mixing coefficients. The terms *partial unmixing* or *matched filtering* are often used to characterize methods to eliminate or reduce the effect of $\boldsymbol{U}$.

(a) *Orthogonal subspace projection* (OSP) (Harsanyi and Chang, 1994). Show that the matrix

$$\boldsymbol{P} = \boldsymbol{I} - \boldsymbol{U}(\boldsymbol{U}^\top \boldsymbol{U})^{-1}\boldsymbol{U}^\top, \tag{7.49}$$

where $\boldsymbol{I}$ is the $N \times N$ identity matrix, "projects out" the unwanted components. (Hilger and Nielsen (2000) give an example of the use of this transformation for the suppression of cloud cover from a multispectral image.)

(b) Write an IDL or Python function `osp(G,U)` which takes a data matrix $\mathcal{G}$ and the matrix of undesired spectra $\boldsymbol{U}$ as input and returns the OSP projection $\boldsymbol{P}\mathcal{G}^\top$.

(c) *Constrained energy minimization* (CEM) (Harsanyi, 1993; Nielsen, 2001). OSP still requires knowledge of all of the end-member spectra. In fact, Settle (1996) shows that it is fully equivalent to linear unmixing. If the spectra $\boldsymbol{U}$ are unknown, CEM reduces the influence of the undesired spectra by finding a projection direction $\boldsymbol{w}$ for which

$$\boldsymbol{w}^\top \boldsymbol{d} = 1$$

while at the same time minimizing the "mean of the output energy" $\langle(\boldsymbol{w}^\top \boldsymbol{G})^2\rangle$. Assuming that the mean of the projection $\boldsymbol{w}^\top \boldsymbol{G}$ is approximately zero, show that the desired projection direction is

$$\boldsymbol{w} = \frac{\boldsymbol{\Sigma}_R^{-1}\boldsymbol{d}}{\boldsymbol{d}^\top \boldsymbol{\Sigma}_R^{-1}\boldsymbol{d}}. \tag{7.50}$$

(d) *Spectral angle mapping* (SAM) (Kruse et al., 1993). In order to establish a measure of closeness to a desired spectrum, one can compare the angle θ between the desired spectrum $\boldsymbol{d}$ and each pixel vector $\boldsymbol{g}$. Show that this is equivalent to CEM with a diagonal covariance matrix $\boldsymbol{\Sigma}_R = \sigma^2 \boldsymbol{I}$, i.e., to CEM without any allowance for covariance between the spectral bands.

6. Derive Equation (7.46).

8

Unsupervised Classification

Supervised classification of remote sensing imagery, the subject of the previous two chapters, involves the use of a training dataset consisting of labeled pixels representative of each land cover category of interest in an image. We saw how to use these data to generalize to a complete labeling, or thematic map, for an entire scene. The choice of training areas which adequately represent the spectral characteristics of each category is very important for supervised classification, as the quality of the training set has a profound effect on the validity of the result. Finding and verifying training areas can be laborious, since the analyst must select representative pixels for each of the classes by visual examination of the image and by information extraction from additional sources such as ground reference data (ground truth), aerial photos or existing maps.

Unlike supervised classification, unsupervised classification, or *clustering* as it is often called, requires no reference information at all. Instead, the attempt is made to find an underlying class structure automatically by organizing the data into groups sharing similar (e.g., spectrally homogeneous) characteristics. Often, one only needs to specify beforehand the number K of classes present. Unsupervised classification plays an especially important role when very little *a priori* information about the data is available. A primary objective of using clustering algorithms for multispectral remote sensing data is to obtain useful information for the selection of training regions in a subsequent supervised classification.

We can view the basic problem of unsupervised classification at the individual pixel level as the partitioning of a set of samples (the image pixel intensity vectors $\boldsymbol{g}(\nu)$, $\nu = 1 \dots m$) into K disjoint subsets, called classes or clusters. The members of each class are to be in some sense more similar to one another than to the members of the other classes. If one wishes to take the spatial context of the image into account, then the problem also becomes one of labeling a regular lattice. Here the random field concept introduced in Chapter 4 can be used to advantage.

Clearly a criterion is needed which will determine the quality of any given partitioning, along with a means to determine the partitioning which optimizes it. The *sum of squares* cost function will provide us with a sufficient basis to justify some of the most popular algorithms for clustering of multispectral imagery, so we begin with its derivation. For a broad overview of clustering techniques for multispectral images, see Tran et al. (2005). Segmentation and clustering methods for SAR imagery are reviewed in Oliver and Quegan (2004).

8.1 Simple cost functions

Let the (N-dimensional) observations that are to be partitioned by a clustering algorithm comprise the set

$$\{\boldsymbol{g}(\nu) \mid \nu = 1 \dots m\},$$

which we can conveniently represent as the $m \times N$ data matrix

$$\mathcal{G} = \begin{pmatrix} \boldsymbol{g}(1)^\top \\ \vdots \\ \boldsymbol{g}(m)^\top \end{pmatrix}.$$

A given partitioning may be written in the form

$$C = [C_1, \dots C_k, \dots C_K],$$

where C_k is the set of indices

$$C_k = \{\ \nu \mid \boldsymbol{g}(\nu) \text{ is in class } k\}.$$

Our strategy will be to maximize the posterior probability $\Pr(C \mid \mathcal{G})$ for observing the partitioning C given the data $\mathcal{G}$. From Bayes' Theorem we can write

$$\Pr(C \mid \mathcal{G}) = \frac{p(\mathcal{G} \mid C)\Pr(C)}{p(\mathcal{G})}. \tag{8.1}$$

$\Pr(C)$ is the prior probability for C. The quantity $p(\mathcal{G} \mid C)$ is the probability density function for the observations when the partitioning is C, also referred to as the likelihood of the partitioning C given the data $\mathcal{G}$, while $p(\mathcal{G})$ is a normalization independent of C.

Following Fraley (1996) we first of all make the strong assumption that the observations are chosen independently from K multivariate normally distributed populations corresponding, for instance, to the K land cover categories present in a satellite image. Under this assumption, the $\boldsymbol{g}(\nu)$ are realizations of random vectors

$$\boldsymbol{G}_k \sim N(\boldsymbol{\mu}_k, \boldsymbol{\Sigma}_k), \quad k = 1 \dots K,$$

with multivariate normal probability densities, which we denote $p(\boldsymbol{g} \mid k)$. The likelihood is the product of the individual probability densities given the partitioning, i.e.,

$$L(C) = p(\mathcal{G} \mid C) = \prod_{k=1}^{K} \prod_{\nu \in C_k} p(\boldsymbol{g}(\nu) \mid k)$$

$$= \prod_{k=1}^{K} \prod_{\nu \in C_k} (2\pi)^{-N/2} |\boldsymbol{\Sigma}_k|^{-1/2} \exp\left(-\frac{1}{2}(\boldsymbol{g}(\nu) - \boldsymbol{\mu}_k)^\top \boldsymbol{\Sigma}_k^{-1} (\boldsymbol{g}(\nu) - \boldsymbol{\mu}_k)\right).$$

Forming the product in this way is justified by the independence of the observations. Taking the logarithm gives the log-likelihood

$$\mathcal{L}(C) = \sum_{k=1}^{K} \sum_{\nu \in C_k} \left(-\frac{N}{2}\log(2\pi) - \frac{1}{2}\log|\boldsymbol{\Sigma}_k| - \frac{1}{2}(\boldsymbol{g}(\nu) - \boldsymbol{\mu}_k)^\top \boldsymbol{\Sigma}_k^{-1} (\boldsymbol{g}(\nu) - \boldsymbol{\mu}_k)\right). \tag{8.2}$$

Maximizing the log-likelihood is obviously equivalent to maximizing the likelihood. From Equation (8.1) we can then write

$$\log \Pr(C \mid \mathcal{G}) = \mathcal{L}(C) + \log \Pr(C) - \log p(\mathcal{G}). \tag{8.3}$$

Since the last term is independent of C, maximizing $\Pr(C \mid \mathcal{G})$ with respect to C is equivalent to maximizing $\mathcal{L}(C) + \log \Pr(C)$. If the even stronger assumption is now made that all K classes exhibit identical covariance matrices given by

$$\boldsymbol{\Sigma}_k = \sigma^2 \boldsymbol{I}, \quad k = 1 \dots K, \tag{8.4}$$

where $\boldsymbol{I}$ is the identity matrix, then $\mathcal{L}(C)$ is maximized when the last sum in Equation (8.2), namely the expression

$$\sum_{k=1}^{K} \sum_{\nu \in C_k} (\boldsymbol{g}(\nu) - \boldsymbol{\mu}_k)^\top (\frac{1}{2\sigma^2}\boldsymbol{I})(\boldsymbol{g}(\nu) - \boldsymbol{\mu}_k) = \sum_{k=1}^{K} \sum_{\nu \in C_k} \frac{\|\boldsymbol{g}(\nu) - \boldsymbol{\mu}_k\|^2}{2\sigma^2},$$

is minimized. Finally, with Equation (8.3), $\Pr(C \mid \mathcal{G})$ itself is maximized by minimizing the *cost function*

$$E(C) = \sum_{k=1}^{K} \sum_{\nu \in C_k} \frac{\|\boldsymbol{g}(\nu) - \boldsymbol{\mu}_k\|^2}{2\sigma^2} - \log \Pr(C). \tag{8.5}$$

Now let us introduce a "hard" class dependency in the form of a matrix $\boldsymbol{U}$ with elements

$$u_{k\nu} = \begin{cases} 1 & \text{if } \nu \in C_k \\ 0 & \text{otherwise.} \end{cases} \tag{8.6}$$

These matrix elements are required to satisfy the conditions

$$\sum_{k=1}^{K} u_{k\nu} = 1, \quad \nu = 1 \dots m, \tag{8.7}$$

meaning that each pixel $\boldsymbol{g}(\nu)$, $\nu = 1 \dots m$, belongs to precisely one class, and

$$\sum_{\nu=1}^{m} u_{k\nu} = m_k > 0, \quad k = 1 \dots K, \tag{8.8}$$

meaning that no class C_k is empty. The sum in Equation (8.8) is the number m_k of pixels in the kth class. Maximum likelihood estimates for the mean of the kth cluster can then be written in the form

$$\hat{\boldsymbol{\mu}}_k = \frac{1}{m_k} \sum_{\nu \in C_k} \boldsymbol{g}(\nu) = \frac{\sum_{\nu=1}^{m} u_{k\nu} \boldsymbol{g}(\nu)}{\sum_{\nu=1}^{m} u_{k\nu}}, \quad k = 1 \dots K, \tag{8.9}$$

and for the covariance matrix as

$$\hat{\boldsymbol{\Sigma}}_k = \frac{\sum_{\nu=1}^{m} u_{k\nu} (\boldsymbol{g}(\nu) - \hat{\boldsymbol{\mu}}_k)(\boldsymbol{g}(\nu) - \hat{\boldsymbol{\mu}}_k)^\top}{\sum_{\nu=1}^{m} u_{k\nu}}, \quad k = 1 \dots K; \tag{8.10}$$

see Section 2.4, Equations (2.69) and (2.70). The cost function, Equation (8.5), can also be expressed in terms of the class dependencies $u_{k\nu}$ as

$$E(C) = \sum_{k=1}^{K} \sum_{\nu=1}^{m} u_{k\nu} \frac{\|\boldsymbol{g}(\nu) - \hat{\boldsymbol{\mu}}_k\|^2}{2\sigma^2} - \log \Pr(C). \tag{8.11}$$

The parameter σ^2 can be thought of as the average within-cluster or image noise variance. If we have no prior information on the class structure, we can simply say that all partitionings C are *a priori* equally likely. Then the last term in Equation (8.11) is independent of C and, dropping it and the multiplicative constant $1/2\sigma^2$, we get the *sum of squares* cost function

$$E(C) = \sum_{k=1}^{K} \sum_{\nu=1}^{m} u_{k\nu} \|\boldsymbol{g}(\nu) - \hat{\boldsymbol{\mu}}_k\|^2. \tag{8.12}$$

8.2 Algorithms that minimize the simple cost functions

The problem to find the partitioning which minimizes the cost functions, Equation (8.11) or Equation (8.12), is unfortunately impossible to solve. The number of conceivable partitions, while obviously finite, is in any real situation astronomical. For example, for $m = 1000$ pixels and just $K = 2$ possible classes, there are $2^{1000-1} - 1 \approx 10^{300}$ possibilities (Duda and Hart, 1973). Direct enumeration is therefore not feasible. The line of attack most frequently taken is to start with some initial clustering and its associated cost function and then attempt to minimize the latter iteratively. This will always find a local minimum, but there is no guarantee that a global minimum for the cost function will be reached. Therefore, one can never know if the best solution has been found. Nevertheless, the approach is used because the computational burden is acceptable.

We shall follow the iterative approach in this section, beginning with the well-known K-means algorithm (including a kernelized version), followed by consideration of a variant due to Palubinskas (1998) which uses the cost function of Equation (8.11) and for which the number of clusters is determined automatically. Then we discuss a common example of bottom-up or agglomerative hierarchical clustering and conclude with a "fuzzy" version of the K-means algorithm.

8.2.1 K-means clustering

The K-means clustering algorithm (sometimes referred to as *basic ISODATA* (Duda and Hart, 1973) or *migrating means* (Richards and Jia, 2006)) is based on the sum of squares cost function, Equation (8.12). After some random initialization of the cluster centers $\hat{\boldsymbol{\mu}}_k$ and setting the class dependency matrix $\boldsymbol{U} = \boldsymbol{0}$, the distance measure corresponding to a minimization of Equation (8.12), namely

$$d(\boldsymbol{g}(\nu), k) = \|\boldsymbol{g}(\nu) - \hat{\boldsymbol{\mu}}_k\|^2, \tag{8.13}$$

is used to cluster the pixel vectors. Specifically, set $u_{k\nu} = 1$, where

$$k = \arg\min_k d(\boldsymbol{g}(\nu), k) \tag{8.14}$$

for $\nu = 1 \dots m$. Then Equation (8.9) is invoked to recalculate the cluster centers. This procedure is iterated until the class labels cease to change. A popular extension, referred to as ISODATA (Iterative Self-Organizing Data Analysis), involves splitting and merging of the clusters in repeated passes, albeit at the cost of setting additional parameters. K-means (and ISODATA) clustering may be performed within the ENVI Classic environment from the main menu:

```
Classification/Unsupervised/K-Means
Classification/Unsupervised/ISODATA
```

or from within the new GUI toolbox:

```
Classification/Unsupervised Classification/K-Means Classification
Classification/Unsupervised Classification/IsoData Classification
```

The `scipy` Python package includes a ready-made K-means function which can be used for clustering multispectral images, as illustrated in Listing 8.1. After reading in the chosen spatial/spectral image subsets, the K-means algorithm is called in line 30, returning the cluster centers and distortions (sums of the squared differences between the observations and the corresponding centroid), in this case in an anonymous variable since they are not used. The

Listing 8.1: K-means clustering of a multispectral image with Python.

```
#!/usr/bin/env python
#Name:  ex8_1.py
IMPORT auxil.auxil as auxil
FROM numpy IMPORT *
FROM osgeo IMPORT gdal
FROM osgeo.gdalconst IMPORT GA_ReadOnly, GDT_Byte
FROM scipy.cluster.vq IMPORT kmeans,vq

DEF main():
    gdal.AllRegister()
    infile = auxil.select_infile()
    IF infile:
        inDataset = gdal.Open(infile,GA_ReadOnly)
        cols = inDataset.RasterXSize
        rows = inDataset.RasterYSize
        bands = inDataset.RasterCount
    ELSE:
        RETURN
    pos =  auxil.select_pos(bands)
    bands = LEN(pos)
    x0,y0,rows,cols=auxil.select_dims([0,0,rows,cols])
    K = auxil.select_integer(6,msg='Number␣clusters')
    G = zeros((rows*cols,LEN(pos)))
    k = 0
    FOR b IN pos:
        band = inDataset.GetRasterBand(b)
        G[:,k] = band.ReadAsArray(x0,y0,cols,rows)\
                              .astype(FLOAT).ravel()
        k += 1
    centers, _ = kmeans(G,K)
    labels, _ = vq(G,centers)
    outfile,fmt = auxil.select_outfilefmt()
    IF outfile:
        driver = gdal.GetDriverByName(fmt)
        outDataset = driver.Create(outfile,
                        cols,rows,1,GDT_Byte)
        outBand = outDataset.GetRasterBand(1)
        outBand.WriteArray(reshape(labels,(rows,cols))\
                                                 ,0,0)
        outBand.FlushCache()
        outDataset = None
    inDataset = None

IF __name__ == '__main__':
    main()
```

image observations are then labeled with the `vq()` function (line 31, this function also returns the distortions) and written to disk.

8.2.2 Kernel K-means clustering

To kernelize the K-means algorithm (Shawe–Taylor and Cristianini, 2004), we require the dual formulation for Equation (8.13). With Equation (8.8), the matrix

$$\boldsymbol{M} = [\mathrm{Diag}(\boldsymbol{U}\boldsymbol{1}_m)]^{-1},$$

where $\boldsymbol{1}_m$ is an m-component vector of ones, can be seen to be a $K \times K$ diagonal matrix with inverse class populations along the diagonal,

$$\boldsymbol{M} = \begin{pmatrix} 1/m_1 & 0 & \dots & 0 \\ 0 & 1/m_2 & \dots & 0 \\ \vdots & \vdots & \ddots & \vdots \\ 0 & 0 & \dots & 1/m_K \end{pmatrix}.$$

Therefore, from the rule for matrix multiplication, Equation (1.14), we can express the class means given by Equation (8.9) as the columns of the $N \times K$ matrix

$$(\hat{\boldsymbol{\mu}}_1, \hat{\boldsymbol{\mu}}_2 \dots \hat{\boldsymbol{\mu}}_k) = \mathcal{G}^\top \boldsymbol{U}^\top \boldsymbol{M}.$$

We then obtain the dual formulation of Equation (8.13) as

$$\begin{aligned} d(\boldsymbol{g}(\nu), k) &= \|\boldsymbol{g}(\nu)\|^2 - 2\boldsymbol{g}(\nu)^\top \hat{\boldsymbol{\mu}}_k + \|\hat{\boldsymbol{\mu}}_k\|^2 \\ &= \|\boldsymbol{g}(\nu)\|^2 - 2\boldsymbol{g}(\nu)^\top [\mathcal{G}^\top \boldsymbol{U}^\top \boldsymbol{M}]_{\cdot k} + [\boldsymbol{M}\boldsymbol{U}\mathcal{G}\mathcal{G}^\top \boldsymbol{U}^\top \boldsymbol{M}]_{kk} \\ &= \|\boldsymbol{g}(\nu)\|^2 - 2[(\boldsymbol{g}(\nu)^\top \mathcal{G}^\top)\boldsymbol{U}^\top \boldsymbol{M}]_k + [\boldsymbol{M}\boldsymbol{U}\mathcal{G}\mathcal{G}^\top \boldsymbol{U}^\top \boldsymbol{M}]_{kk}\ . \end{aligned} \tag{8.15}$$

In the second line above, $[\ ]_{\cdot k}$ denotes the kth column. In the last line the observations appear only as inner products: $\boldsymbol{g}(\nu)^\top \boldsymbol{g}(\nu)$ (first term), in the Gram matrix $\mathcal{G}\mathcal{G}^\top$ (last term) and in $\boldsymbol{g}(\nu)^\top \mathcal{G}^\top$ (second term). The last expression is just the νth row of the Gram matrix, i.e.,

$$\boldsymbol{g}(\nu)^\top \mathcal{G}^\top = [\mathcal{G}\mathcal{G}^\top]_{\nu,\cdot}$$

For kernel K-means, where we work in an (implicit) nonlinear feature space $\phi(\boldsymbol{g})$, we substitute $\mathcal{G}\mathcal{G}^\top \to \mathcal{K}$ to get

$$d(\phi(\boldsymbol{g}(\nu)), k) = [\mathcal{K}]_{\nu\nu} - 2[\mathcal{K}_{\nu\cdot}\boldsymbol{U}^\top \boldsymbol{M}]_k + [\boldsymbol{M}\boldsymbol{U}\mathcal{K}\boldsymbol{U}^\top \boldsymbol{M}]_{kk}, \tag{8.16}$$

where $\mathcal{K}$ is the kernel matrix and $\mathcal{K}_{\nu\cdot}$ is its νth row. Since the first term in Equation (8.16) doesn't depend on the class index k, the clustering rule for kernel K-means is to assign observation $\boldsymbol{g}(\nu)$ to class k, where

$$k = \arg\min_k \left([\boldsymbol{M}\boldsymbol{U}\mathcal{K}\boldsymbol{U}^\top \boldsymbol{M}]_{kk} - 2[\mathcal{K}_{\nu\cdot}\boldsymbol{U}^\top \boldsymbol{M}]_k \right). \tag{8.17}$$

Listing 8.2: Excerpt from the program `KKMEANS_RUN.PRO`.

```
1  labels = ceil(randomu(seed,m)*K)-1
2  ; iteration
3  change = 1
4  iter = 0
5  ones = fltarr(1,m)+1
6  WHILE change AND (iter LT 100) DO BEGIN
7     progressbar->Update,iter
8     change = 0
9     U = fltarr(m,K)
10    FOR i=0,m-1 DO U[i,labels[i]] = 1
11    M1 = diag_matrix(1/(total(U,1)+1))
12    MU = M1##U
13 ; Z is a K by m array
14    Z = ones##diag_matrix(MU##KK##transpose(MU)) $
15                      - 2*KK##transpose(MU)
16    _ = min(Z,labels1,dimension=1)
17    labels1 = labels1 MOD K
18    IF total(labels1 NE labels) THEN change=1
19    labels=labels1
20    IF progressbar->CheckCancel() THEN BEGIN
21       PRINT,'interrupted...'
22       iter=99
23    ENDIF
24    iter++
25 ENDWHILE
```

An ENVI/IDL extension and a Python script for kernel K-means clustering are described in Appendices C and D. As in the case of kernel PCA (Chapter 4), memory restrictions require that one work with only a relatively small training sample of m pixel vectors. A portion of the IDL code is shown in Listing 8.2. The variable `Z` in line 14 is a $K \times m$ array of current values of the expression on the right-hand side of Equation (8.17), with one row for each of the training observations and one column for each class label. In line 16, the class labels corresponding to the minimum distance to the mean are extracted into the variable `labels1` (see the definition of the IDL function `MIN()`) and converted to k-values in line 17. This avoids an expensive `FOR`-loop over the training data. The iteration terminates when the class labels cease to change.

After convergence, unsupervised classification of the remaining pixels can again be achieved with Equation (8.17), merely replacing the row vector $\mathcal{K}_{\nu\cdot}$ by

$$\big(\kappa(\boldsymbol{g},\boldsymbol{g}(1)),\kappa(\boldsymbol{g},\boldsymbol{g}(2))\dots\kappa(\boldsymbol{g},\boldsymbol{g}(m))\big),$$

where $\boldsymbol{g}$ is a pixel to be classified. The processing of the entire image thus requires evaluation of the kernel for every image pixel with every training

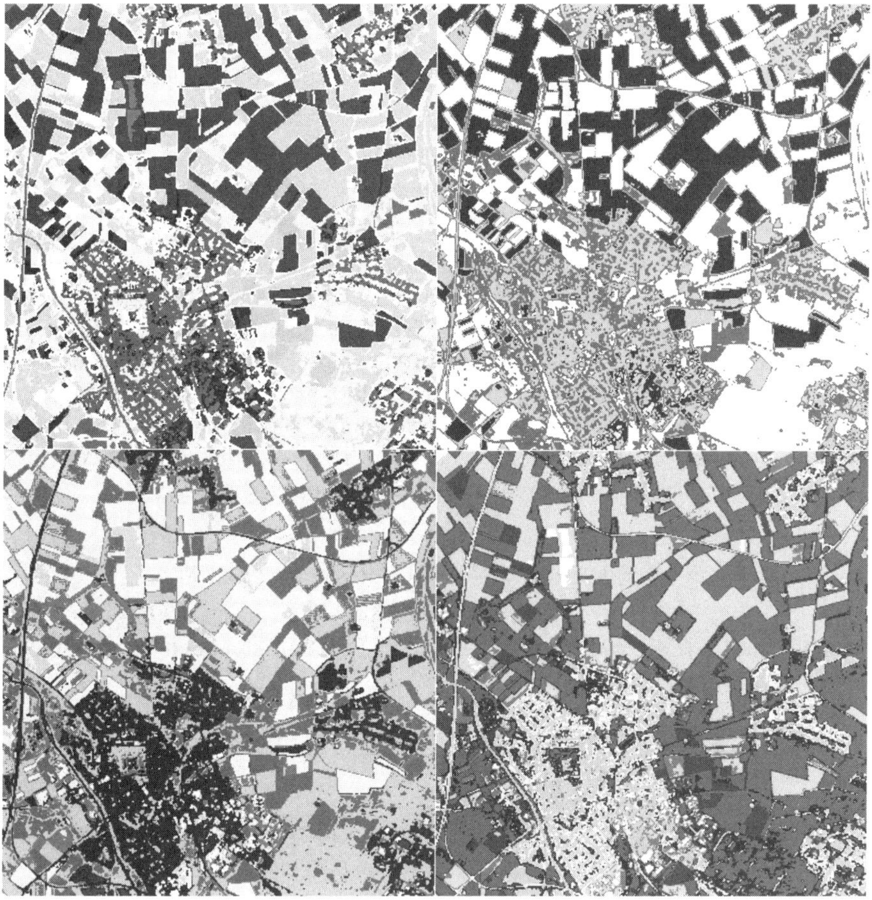

FIGURE 8.1
Unsupervised classification of a spatial subset of the principal components of the fused May 1, 2007 ASTER image over Jülich (Figure 6.1) using 8 classes. Upper left: kernel K-means clustering on the first four components; upper right: extended K-means clustering on the first component with $\tilde{K} = 8$, see Equation (8.22); lower left: agglomerative hierarchical clustering on the first four components; lower right: fuzzy K-means clustering on the first four components. **(See color insert.)**

pixel. Therefore, it is best to classify the image row by row, as was the case for kernel PCA and kernel RX. The ENVI/IDL extension makes use of CUDA/-GPULib if available (see Section 4.4.2), which accelerates the classification step substantially. Figure 8.1 (top left) shows an unsupervised classification of the Jülich image of Figure 6.1 with kernel K-means clustering.

8.2.3 Extended K-means clustering

Starting this time from the cost function Equation (8.11), denote by $p_k = \Pr(C_k)$ the prior probability for cluster C_k. The entropy H associated with this prior distribution is given by

$$H = -\sum_{k=1}^{K} p_k \log p_k, \tag{8.18}$$

see Equation (2.109). It was shown in Section 2.7 that distributions with high entropy are those for which the p_k are all similar. So in this case, high entropy corresponds to the pixels being distributed evenly over all available clusters, whereas low entropy means that most of the data are concentrated in very few clusters. Following Palubinskas (1998) we choose a prior distribution $\Pr(C)$ in the cost function of Equation (8.11) for which few clusters (low entropy) are more probable than many clusters (high entropy), namely

$$\Pr(C) = A\exp(-\alpha_E H) = A\exp\Big(\alpha_E \sum_{k=1}^{K} p_k \log p_k\Big),$$

where A and α_E are parameters. The cost function can then be written in the form

$$E(C) = \sum_{k=1}^{K}\sum_{\nu=1}^{m} u_{k\nu}\frac{\|\boldsymbol{g}(\nu) - \hat{\boldsymbol{\mu}}_k\|^2}{2\sigma^2} - \alpha_E \sum_{k=1}^{K} p_k \log p_k, \tag{8.19}$$

dropping the $\log A$ term, which is independent of the clustering. With

$$p_k = \frac{m_k}{m} = \frac{1}{m}\sum_{\nu=1}^{m} u_{k\nu}, \tag{8.20}$$

Equation (8.19) is equivalent to the cost function

$$E(C) = \sum_{k=1}^{K}\sum_{\nu=1}^{m} u_{k\nu}\left[\frac{\|\boldsymbol{g}(\nu) - \hat{\boldsymbol{\mu}}_k\|^2}{2\sigma^2} - \frac{\alpha_E}{m}\log p_k\right]. \tag{8.21}$$

An estimate for the parameter α_E in Equation (8.21) may be obtained as follows (Palubinskas, 1998). With the approximation

$$\sum_{\nu=1}^{m} u_{k\nu}\|\boldsymbol{g}(\nu) - \hat{\boldsymbol{\mu}}_k\|^2 \approx m_k\sigma^2 = m p_k \sigma^2,$$

we can write Equation (8.21) as

$$E(C) \approx \sum_{k=1}^{K}\left[\frac{m p_k}{2} - \alpha_E p_k \log p_k\right].$$

Equating the likelihood and prior terms in this expression to give $E(C) = 0$ and taking $p_k \approx 1/\tilde{K}$, where $\tilde{K}$ is some *a priori* expected number of clusters, then gives

$$\alpha_E \approx -\frac{m}{2\log(1/\tilde{K})}. \tag{8.22}$$

The parameter σ^2 in Equation (8.21) (the within-cluster variance) can be estimated from the data, see below.

The *extended K-means* (EKM) algorithm is then as follows. First an initial configuration $\boldsymbol{U}$ with a very large number of clusters K is chosen (for single-band data this might conveniently be the $K \leq 256$ gray values that an image with 8-bit quantization can possibly have) and initial values

$$\hat{\boldsymbol{\mu}}_k = \frac{1}{m_k}\sum_{\nu=1}^{m} u_{k\nu}\boldsymbol{g}(\nu), \ p_k = \frac{m_k}{m}, \quad k = 1 \ldots K, \tag{8.23}$$

are determined. Then the data are reclustered according to the distance measure which minimizes Equation (8.21):

$$k = \arg\min_k \left(\frac{\|\boldsymbol{g}(\nu) - \hat{\boldsymbol{\mu}}_k\|^2}{2\sigma^2} - \frac{\alpha_E}{m}\log p_k \right), \quad \nu = 1 \ldots m. \tag{8.24}$$

The prior term tends to put more observations into fewer clusters. Any cluster for which, in the course of the iteration, m_k equals zero (or is less than some threshold) is simply dropped from the calculation so that the final number of clusters is determined by the data. The condition that no class be empty, Equation (8.8), is thus relaxed. The explicit choice of the number of clusters K is replaced by the necessity of choosing a value for the "meta-parameter" α_E (or $\tilde{K}$ if Equation (8.22) is used). This has the advantage that one can use a single parameter for a wide variety of images and let the algorithm itself decide on the actual value of K in any given instance. Iteration of Equations (8.23) and (8.24) continues until the cluster labels cease to change.

An implementation of the extended K-means algorithm in IDL for gray-scale, or one-band, images is shown in Listings 8.3 and 8.4. The variance σ^2 is estimated (naively, see Exercise 5) by calculating the variance of the difference of the image with a copy of itself shifted by one pixel, program line 23. An *a priori* number of classes $\tilde{K} = 8$ determines the meta-parameter α_E, line 24. The initial number K of clusters is chosen as the number of nonempty bins in the 256-bin histogram of the linearly stretched image, while the initial cluster means m_k and prior probabilities p_k are the corresponding bin numbers, respectively the bin contents divided by the number of pixels, lines 27 to 30. The iteration of Equations (8.23) and (8.24) is carried out in the `WHILE` loop, lines 37 (Listing 8.3) to 24 (Listing 8.4). Clusters with prior probabilities less than 0.01 are discarded in line 44. The `REPLICATE` command in lines 8 and 9 in Listing 8.4 avoids the use of a `FOR` loop over the number of pixels. Instead, the array of distances $d(i, k)$ is manipulated directly in

Listing 8.3: Extended K-means clustering.

```
PRO ex8_1
; extended K-means

envi_select, title='Choose␣multispectral␣band', $
             fid=fid, dims=dims,pos=pos, /band_only
IF (fid EQ -1) THEN RETURN
cols = dims[2]-dims[1]+1
rows = dims[4]-dims[3]+1
m = float(cols*rows)

; map tie point
map_info = envi_get_map_info(fid=fid)
envi_convert_file_coordinates, fid, $
   dims[1], dims[3], e, n, /to_map
map_info.mc[2:3]= [e,n]

; image band
G = bytscl(envi_get_data(fid=fid,dims=dims,pos=0))
; classification image
labels = G*0

; parameters
sigma2 = stddev(float(G)-shift(G,[1,0]))^2
alphaE = -1/(2*alog(1.0/8))

; initial K, means and priors
hist = histogram(G,nbins=256)
indices = where(hist,K)
means = (findgen(256))[indices]
priors = hist[indices]/m

; iteration
progressbar = Obj_New('progressbar', Color='blue', $
  title='EKM␣clustering:␣delta...',xsize=250,ysize=20)
progressbar->start
delta = 100.0 & iter = 0
WHILE (delta GT 1.0) AND (iter LE 100) DO BEGIN
   IF progressbar->CheckCancel() THEN BEGIN
      PRINT,'clustering␣aborted'
      progressbar->Destroy & RETURN
   ENDIF
   progressbar->Update,iter,text=strtrim(delta,2)
; drop (almost) empty clusters
   indices = where(priors GT 0.01, K)
; distance array
   ds = fltarr(m,K)
```

Listing 8.4: Extended K-means clustering (continued).

```
1     means = means[indices]
2     priors = priors[indices]
3     means1 = means
4     priors1 = priors
5     means = means*0.0
6     priors = priors*0.0
7     FOR j=0,K-1 DO BEGIN
8        ms = replicate(means1[j],m)
9        logps = replicate(alog(priors1[j]),m)
10       ds[*,j] = (G-ms)^2/(2*sigma2) - alphaE*logps
11    ENDFOR
12    min_ds = min(ds,dimension=2)
13    FOR j=0,K-1 DO BEGIN
14       indices = where(ds[*,j] EQ min_ds,count)
15       IF count GT 0 THEN BEGIN
16          mj = n_elements(indices)
17          priors[j] = mj/m
18          means[j] = total(G[indices])/mj
19          labels[indices] = j+1
20       ENDIF
21    ENDFOR
22    delta = max(abs(means - means1))
23    iter++
24 ENDWHILE
25 progressbar->Destroy
26
27 envi_enter_data, labels, file_type=3, $
28   num_classes=K+1, $
29   class_names='class␣'+strtrim(indgen(K+1),2), $
30   lookup=class_lookup_table(indgen(K+1)), $
31   map_info=map_info
32
33 END
```

the variable DS with IDL's much more efficient array functions. The variable INDICES in line 14, Listing 8.4, locates the pixels, if any, which belong to cluster j for $j = 1 \ldots K$. For convenience, the termination condition is that there be no significant change in the cluster means, line 22 in Listing 8.4. A classification result is shown in Figure 8.1 (top right).

8.2.4 Agglomerative hierarchical clustering

The unsupervised classification algorithm that we consider next is, as for ordinary K-means clustering, based on the cost function of Equation (8.12).

Let us first write it in a more convenient form:

$$E(C) = \sum_{k=1}^{K} E_k, \tag{8.25}$$

where E_k is given by

$$E_k = \sum_{\nu \in C_k} \|\boldsymbol{g}(\nu) - \hat{\boldsymbol{\mu}}_k\|^2. \tag{8.26}$$

The algorithm is initialized by assigning each observation to its own class. At this stage, the cost function is zero, since $C_\nu = \{\nu\}$ and $\boldsymbol{g}(\nu) = \hat{\boldsymbol{\mu}}_\nu$. Every agglomeration of clusters to form a smaller number will increase $E(C)$. We therefore seek a prescription for choosing two clusters for combination that increases $E(C)$ *by the smallest amount possible* (Duda and Hart, 1973).

Suppose that, at some stage of the algorithm, clusters k with m_k members and j with m_j members are merged, where $k < j$, and the new cluster is labeled k. Then the mean of the merged cluster is the weighted average of the original cluster means,

$$\hat{\boldsymbol{\mu}}_k \rightarrow \frac{m_k \hat{\boldsymbol{\mu}}_k + m_j \hat{\boldsymbol{\mu}}_j}{m_k + m_j} = \bar{\boldsymbol{\mu}}.$$

Thus, after the agglomeration, E_k increases to

$$E_k = \sum_{\nu \in C_k \cup C_j} \|\boldsymbol{g}(\nu) - \bar{\boldsymbol{\mu}}\|^2$$

and E_j disappears. The net change in $E(C)$ is therefore, after some algebra (Exercise 7),

$$\begin{aligned}\Delta(k,j) &= \sum_{\nu \in C_k \cup C_j} \|\boldsymbol{g}(\nu) - \bar{\boldsymbol{\mu}}\|^2 - \sum_{\nu \in C_k} \|\boldsymbol{g}(\nu) - \hat{\boldsymbol{\mu}}_k\|^2 - \sum_{\nu \in C_j} \|\boldsymbol{g}(\nu) - \hat{\boldsymbol{\mu}}_j\|^2 \\ &= \frac{m_k m_j}{m_k + m_j} \cdot \|\hat{\boldsymbol{\mu}}_k - \bar{\boldsymbol{\mu}}_j\|^2.\end{aligned} \tag{8.27}$$

The minimum increase in $E(C)$ is achieved by combining those two clusters k and j which minimize the above expression. Given two alternative candidate cluster pairs with similar combined memberships $m_k + m_j$ and whose means have similar separations $\|\hat{\boldsymbol{\mu}}_k - \hat{\boldsymbol{\mu}}_j\|$, this prescription obviously favors combining that pair having the larger discrepancy between m_k and m_j. Thus similar-sized clusters are preserved and smaller clusters are absorbed by larger ones.

Let $\langle k, j\rangle$ represent the cluster formed by combination of the clusters k and j. Then the increase in cost incurred by combining this cluster with some other cluster r can be determined from Equation (8.27) as (Fraley, 1996)

$$\Delta(\langle k,j\rangle, r) = \frac{(m_k + m_r)\Delta(k,r) + (m_j + m_r)\Delta(j,r) - m_r\Delta(k,j)}{m_k + m_j + m_r}. \tag{8.28}$$

To see this, note that with Equation (8.27) the right-hand side of Equation (8.28) is, apart from the factor $1/(m_k + m_j + m_r)$, equivalent to

$$m_k m_r \|\hat{\boldsymbol{\mu}}_k - \hat{\boldsymbol{\mu}}_r\|^2 + m_j m_r \|\hat{\boldsymbol{\mu}}_j - \hat{\boldsymbol{\mu}}_r\|^2 - \frac{m_r m_k m_j}{m_k + m_j} \|\hat{\boldsymbol{\mu}}_k - \hat{\boldsymbol{\mu}}_j\|^2$$

and the left-hand side similarly is equivalent to

$$m_r(m_k + m_j) \left\| \frac{m_k \hat{\boldsymbol{\mu}}_k + m_j \hat{\boldsymbol{\mu}}_j}{m_k + m_j} - \hat{\boldsymbol{\mu}}_r \right\|^2 .$$

The identity can be established by comparing coefficients of the vector products. For example, the coefficient of $\|\hat{\boldsymbol{\mu}}_k\|^2$ is $m_r m_k^2/(m_k + m_j)$ on both sides.

Once the quantities $\Delta(k, j)$ have been initialized from Equation (8.27), that is,

$$\Delta(k, j) = \frac{1}{2} \|\boldsymbol{g}(k) - \boldsymbol{g}(j)\|^2$$

for all possible combinations of observations, $k = 2 \dots m,\ j = 1 \dots k - 1$, the recursive Equation (8.28) can be used to calculate the cost function efficiently for any further merging without reference to the original data. The algorithm terminates when the desired number of clusters has been reached or continues until a single cluster has been formed. Assuming that the data consist of $\tilde{K}$ compact and well-separated clusters, the slope of $E(C)$ vs. the number of clusters K should decrease (become more negative) for $K \le \tilde{K}$. This provides, at least in principle, a means to decide on the optimal number of clusters.* An ENVI/IDL extension for agglomerative hierarchic clustering is given in Appendix C. Figure 8.1 (bottom left) shows an example of a classified image.

8.2.5 Fuzzy K-means clustering

Introducing the parameter $q > 1$, we can write the cluster means and the sum of squares cost function equivalently as (Dunn, 1973)

$$\hat{\boldsymbol{\mu}}_k = \frac{\sum_{i=\nu}^m u_{k\nu}^q \boldsymbol{g}(\nu)}{\sum_{\nu=1}^m u_{k\nu}^q}, \quad k = 1 \dots K, \tag{8.29}$$

$$E(C) = \sum_{\nu=1}^m \sum_{k=1}^K u_{k\nu}^q \|\boldsymbol{g}(\nu) - \hat{\boldsymbol{\mu}}_k\|^2. \tag{8.30}$$

Since $u_{k\nu} \in \{0, 1\}$, these equations are identical to Equations (8.9) and (8.12). The transition from "hard" to "fuzzy" clustering is effected by label relaxation, that is, by replacing the class memberships by continuous variables

$$0 \le u_{k\nu} \le 1, \quad k = 1 \dots K,\ \nu = 1 \dots m, \tag{8.31}$$

*For multispectral imagery, however, well-separated clusters seldom occur.

but retaining the requirements in Equations (8.7) and (8.8). The parameter q now has an effect and determines the "degree of fuzziness". It is often chosen as $q = 2$. The matrix $\boldsymbol{U}$ is referred to as a *fuzzy class membership* matrix. The classification problem is to find that value of $\boldsymbol{U}$ which minimizes Equation (8.30). Minimization can be achieved by finding values for the $u_{k\nu}$ which solve the minimization problems

$$E_\nu = \sum_{k=1}^K u_{k\nu}^q \|\boldsymbol{g}(\nu) - \hat{\boldsymbol{\mu}}_k\|^2 \;\to\; \min, \quad \nu = 1\ldots m,$$

under the constraints imposed by Equation (8.7). Accordingly, we define a Lagrange function which takes the constraints into account, i.e.,

$$L_\nu = E_\nu - \lambda\left(\sum_{k=1}^K u_{k\nu} - 1\right),$$

and solve the unconstrained problem $L_\nu \;\to\; \min$ by setting the derivative with respect to $u_{k\nu}$ equal to zero,

$$\frac{\partial L_\nu}{\partial u_{k\nu}} = q(u_{k\nu})^{q-1}\|\boldsymbol{g}_\nu - \hat{\boldsymbol{\mu}}_k\|^2 - \lambda = 0, \quad k = 1\ldots K.$$

This gives an expression for fuzzy memberships in terms of λ,

$$u_{k\nu} = \left(\frac{\lambda}{q}\right)^{1/(q-1)}\left(\frac{1}{\|\boldsymbol{g}_\nu - \hat{\boldsymbol{\mu}}_k\|^2}\right)^{1/(q-1)}. \tag{8.32}$$

The Lagrange multiplier λ, in turn, is determined by the constraint

$$1 = \sum_{k=1}^K u_{k\nu} = \left(\frac{\lambda}{q}\right)^{1/(q-1)}\sum_{k=1}^K\left(\frac{1}{\|\boldsymbol{g}(\nu) - \hat{\boldsymbol{\mu}}_k\|^2}\right)^{1/(q-1)}.$$

If we solve for λ and substitute it into Equation (8.32), we obtain

$$u_{k\nu} = \frac{\left(\frac{1}{\|\boldsymbol{g}(\nu)-\hat{\boldsymbol{\mu}}_k\|^2}\right)^{1/(q-1)}}{\sum_{k'=1}^K\left(\frac{1}{\|\boldsymbol{g}(\nu)-\hat{\boldsymbol{\mu}}_{k'}\|^2}\right)^{1/(q-1)}}, \quad k = 1\ldots K,\ \nu = 1\ldots m. \tag{8.33}$$

The *fuzzy K-means* (FKM) algorithm consists of a simple iteration of Equations (8.29) and (8.33). The iteration terminates when the cluster centers $\hat{\boldsymbol{\mu}}_k$ — or alternatively the matrix elements $u_{k\nu}$ — cease to change significantly. After convergence, each pixel vector is labeled with the index of the class with the highest membership probability,

$$\ell_\nu = \arg\max_k u_{k\nu}, \quad \nu = 1\ldots m.$$

This algorithm should give similar results to the K-means algorithm, Section 8.2.1. However, because every pixel "belongs" to some extent to all of the classes, one expects the algorithm to be less likely to become trapped in a local minimum of the cost function. An ENVI/IDL extension for fuzzy K-means clustering is given in Appendix C and an example is shown in Figure 8.1 (bottom right).

We might interpret the fuzzy memberships as "reflecting the relative proportions of each category within the spatially and spectrally integrated multispectral vector of the pixel" (Schowengerdt, 1997). This interpretation will be more plausible when, in the next section, we replace the fuzzy memberships $u_{k\nu}$ by posterior class membership probabilities.

8.3 Gaussian mixture clustering

The unsupervised classification algorithms of the preceding section are based on the cost functions of Equation (8.11) or Equation (8.12). These favor the formation of hyperspherical clusters having similar radii, due to the assumption made in Equation (8.4). The use of multivariate Gaussian probability densities to model the classes allows for ellipsoidal clusters of arbitrary extent and is considerably more flexible. Rather than deriving an algorithm from first principles (see Exercise 9), we can obtain one directly from the FKM algorithm. We merely replace Equation (8.33) for the class memberships $u_{k\nu}$ by the posterior probability $\Pr(k \mid \boldsymbol{g}(\nu))$ for class k given the observation $\boldsymbol{g}(\nu)$ (Gath and Geva, 1989):

$$u_{k\nu} \to \Pr(k \mid \boldsymbol{g}(\nu)),$$

Then, using Bayes' Theorem,

$$u_{k\nu} \propto p(\boldsymbol{g}(\nu) \mid k)\Pr(k),$$

where $p(\boldsymbol{g}(\nu) \mid k)$ will be chosen to be a multivariate normal density function. Its estimated mean $\hat{\boldsymbol{\mu}}_k$ and covariance matrix $\hat{\boldsymbol{\Sigma}}_k$ are given by Equations (8.9) and (8.10), respectively. Thus, apart from a normalization factor, we have

$$\begin{aligned} u_{k\nu} &= p(\boldsymbol{g}(\nu) \mid k)\Pr(k) \\ &= \frac{1}{\sqrt{|\hat{\boldsymbol{\Sigma}}_k|}} \exp\left[-\frac{1}{2}(\boldsymbol{g}(\nu) - \hat{\boldsymbol{\mu}}_k)^\top \hat{\boldsymbol{\Sigma}}_k^{-1}(\boldsymbol{g}(\nu) - \hat{\boldsymbol{\mu}}_k)\right] p_k. \end{aligned} \tag{8.34}$$

In the above expression the prior probability $\Pr(k)$ has been replaced by

$$p_k = \frac{m_k}{m} = \frac{1}{m}\sum_{\nu=1}^{m} u_{k\nu}. \tag{8.35}$$

The algorithm consists of an iteration of Equations (8.9), (8.10), (8.34), and (8.35) with the same termination condition as for the fuzzy K-means algorithm. After each iteration, the columns of $\boldsymbol{U}$ are normalized according to Equation (8.7).

Unlike the FKM classifier, the memberships $u_{k\nu}$ are now functions of the directionally sensitive Mahalanobis distance

$$d = \sqrt{(\boldsymbol{g}(\nu) - \hat{\boldsymbol{\mu}}_k)^\top \hat{\boldsymbol{\Sigma}}_k^{-1} (\boldsymbol{g}(\nu) - \hat{\boldsymbol{\mu}}_k)} \ .$$

Because of the exponential dependence of the memberships on d^2 in Equation (8.34), the computation is very sensitive to initialization conditions and can even become unstable. To avoid this problem, one can first obtain initial values for $\boldsymbol{U}$ by preceding the calculation with the fuzzy K-means algorithm (see Gath and Geva (1989), who referred to their algorithm as *fuzzy maximum likelihood expectation* (FMLE) clustering). Explicitly, then, the algorithm is as follows:

Algorithm (FMLE clustering)

1. Determine starting values for the initial memberships $u_{k\nu}$, e.g., from the FKM algorithm.
2. Determine the cluster centers $\hat{\boldsymbol{\mu}}_k$ with Equation (8.9), the covariance matrices $\hat{\boldsymbol{\Sigma}}_k$ with Equation (8.10), and the priors p_k with Equation (8.35).
3. Calculate with Equation (8.34) the new class membership probabilities $u_{k\nu}$. Normalize the columns of $\boldsymbol{U}$.
4. If $\boldsymbol{U}$ has not changed significantly, stop, else go to 2.

8.3.1 Expectation maximization

Since the above algorithm has nothing to do with the simple cost functions of Section 8.1, we might ask why it should converge at all. The FMLE algorithm is in fact exactly equivalent to the application of the *expectation maximization* (EM) algorithm (Redner and Walker, 1984) to a Gaussian mixture model of the clustering problem. In that formulation, the probability density $p(\boldsymbol{g})$ for observing a value $\boldsymbol{g}$ in a given clustering configuration is modeled as a superposition of class-specific, multivariate normal probability density functions $p(\boldsymbol{g} \mid k)$ with mixing coefficients p_k:

$$p(\boldsymbol{g}) = \sum_{k=1}^{K} p(\boldsymbol{g} \mid k) p_k. \tag{8.36}$$

This expression has the same form as Equation (2.65), with the mixing coefficients p_k playing the role of the prior probabilities. To obtain the parameters of the model, the likelihood

$$\prod_{\nu=1}^{m} p(\boldsymbol{g}(\nu)) = \prod_{\nu=1}^{m} \left[\sum_{k=1}^{K} p(\boldsymbol{g}(\nu) \mid k) p_k\right] \tag{8.37}$$

is maximized. The maximization takes place in alternating *expectation* and *maximization* steps. In step 3 of the FMLE algorithm, the class membership probabilities $u_{k\nu} = p(k \mid \boldsymbol{g}(\nu))$ are computed. This corresponds to the expectation step of the EM algorithm in which the terms $p(\boldsymbol{g} \mid k)p_k$ of the mixture model are recalculated. In step 2 of the FMLE algorithm, the parameters for $p(\boldsymbol{g} \mid k)$, namely $\hat{\boldsymbol{\mu}}_k$ and $\hat{\boldsymbol{\Sigma}}_k$, and the mixture coefficients p_k are estimated for $k = 1 \ldots K$. This corresponds to the EM maximization step; see also Exercise 9.

The likelihood must in fact increase on each iteration and therefore the algorithm will indeed converge to a (local) maximum in the likelihood. To demonstrate this, we follow Bishop (2006) and write Equation (8.37) in the more general form

$$p(\mathcal{G} \mid \theta) = \prod_{\nu=1}^{m} \left[\sum_{k=1}^{K} p(\boldsymbol{g}(\nu) \mid \theta_k) p_k\right], \tag{8.38}$$

where $\mathcal{G}$ is the observed dataset, θ_k represents the parameters $\mu_k, \boldsymbol{\Sigma}_k$ of the kth Gaussian distribution and θ is the set of all parameters of the model, $\theta = \{\mu_k, \boldsymbol{\Sigma}_k, p_k \mid k = 1 \ldots K\}$.

Now suppose that the class labels were known. That is, in the notation of Chapter 6, suppose we were given

$$\ell = \{\boldsymbol{\ell}_1 \ldots \boldsymbol{\ell}_m\}, \quad \boldsymbol{\ell}_\nu = (0, \ldots 1, \ldots 0)^\top,$$

where, if observation ν is in class k, $(\boldsymbol{\ell}_\nu)_k = \ell_{\nu k} = 1$ and the other components are 0. Then it would only be necessary to maximize the likelihood function

$$p(\mathcal{G}, \ell \mid \theta) = \prod_{\nu=1}^{m} \left[\sum_{k=1}^{K} \ell_{\nu k} p(\boldsymbol{g}(\nu) \mid \theta_k) p_k\right] = \prod_{\nu=1}^{m} \prod_{k=1}^{K} p(\boldsymbol{g}(\nu) \mid \theta_k)^{\ell_{\nu k}} p_k^{\ell_{\nu k}}, \tag{8.39}$$

which is straightforward, since it can be done separately for each class. Thus, taking the logarithm of Equation (8.39), we get the log-likelihood

$$\log p(\mathcal{G}, \ell \mid \theta) = \sum_{\nu} \sum_{k} \log\left[p(\boldsymbol{g}(\nu) \mid \theta_k) p_k\right] \ell_{\nu k} = \sum_{k} \sum_{\nu \in C_k} \log\left[p(\boldsymbol{g}(\nu) \mid \theta_k) p_k\right].$$

However the variables ℓ are unknown; they are referred to as *latent* variables (Bishop, 2006). Let us postulate an unknown mass function $q(\ell)$ for the latent variables. Then the likelihood function $p(\mathcal{G} \mid \theta)$, Equation (8.38), can be expressed as an average of $p(\mathcal{G}, \ell \mid \theta)$ over $q(\ell)$,

$$p(\mathcal{G} \mid \theta) = \sum_{\ell} p(\mathcal{G} \mid \ell, \theta) q(\ell) = \sum_{\ell} p(\mathcal{G}, \ell \mid \theta). \tag{8.40}$$

The second equality above follows from the definition of conditional probability, Equation (2.59). The log-likelihood may now be separated into two terms,

$$\log p(\mathcal{G} \mid \theta) = \mathcal{L}(q, \theta) + \mathrm{KL}(q, p), \tag{8.41}$$

where

$$\mathcal{L}(q, \theta) = \sum_{\ell} q(\ell) \log \left(\frac{p(\mathcal{G}, \ell \mid \theta)}{q(\ell)} \right) \tag{8.42}$$

and

$$\mathrm{KL}(q, p) = -\sum_{\ell} q(\ell) \log \left(\frac{p(\ell \mid \mathcal{G}, \theta)}{q(\ell)} \right). \tag{8.43}$$

This decomposition can be seen immediately by writing

$$p(\mathcal{G}, \ell \mid \theta) = p(\ell \mid \mathcal{G}, \theta) p(\mathcal{G} \mid \theta),$$

which again follows from the definition of conditional probability, and expanding Equation (8.42):

$$\begin{aligned} \mathcal{L}(q, \theta) &= \sum_{\ell} q(\ell)[\log p(\ell \mid \mathcal{G}, \theta) + \log p(\mathcal{G} \mid \theta) - \log q(\ell)] \\ &= -\mathrm{KL}(q, p) + \sum_{\ell} q(\ell) \log p(\mathcal{G} \mid \theta) = -\mathrm{KL}(q, p) + \log p(\mathcal{G} \mid \theta). \end{aligned}$$

This is just Equation (8.41).

From Section 2.7.1 we recognize $\mathrm{KL}(q, p)$ as the Kullback-Leibler divergence between $q(\ell)$ and the posterior density $p(\ell \mid \mathcal{G}, \theta)$. Since $\mathrm{KL}(q, p) \geq 0$, with equality only when the two densities are equal, it follows that

$$\log p(\mathcal{G} \mid \theta) \geq \mathcal{L}(q, \theta),$$

that is, $\mathcal{L}(q, \theta)$ is a lower bound for $\log p(\mathcal{G} \mid \theta)$. The EM procedure is then as follows:

- **E-step:** Let the current best set of parameters be θ'. The lower bound on the log-likelihood is maximized by estimating $q(\ell)$ as $p(\ell \mid \mathcal{G}, \theta')$, causing the $\mathrm{KL}(q, p)$ term to vanish and not affecting the log-likelihood since the likelihood does not depend on $q(\ell)$; see Equation (8.40). At this stage $\log p(\mathcal{G} \mid \theta') = \mathcal{L}(q, \theta')$.

- **M-step:** Now $q(\ell)$ is held fixed and $p(\mathcal{G}, \ell \mid \theta)$ is maximized wrt θ (which, as we saw above, is straightforward), giving a new set of parameters θ'', and causing the lower bound $\mathcal{L}(q, \theta'')$ again to increase (unless it is already at a maximum). This will necessarily cause the log-likelihood to increase, since $q(\ell)$ is not the same as $p(\ell \mid \mathcal{G}, \theta'')$ and therefore the Kullback-Leibler divergence will be positive. Now the E-step can be repeated. Iteration of the two steps will therefore always cause the log-likelihood to increase until a maximum is reached.

At any stage of the iteration of the FMLE algorithm, the current estimate for the log-likelihood for the dataset can be shown to be given by (Bishop, 1995)

$$\sum_{\nu=1}^{m} \log p(\boldsymbol{g}(\nu)) = \sum_{k=1}^{K}\sum_{\nu=1}^{m} [u_{k\nu}]^o \log u_{k\nu}. \tag{8.44}$$

Here $[u_{k\nu}]^o$ is the "old" value of the posterior class membership probability, i.e., the value determined on the previous step.

8.3.2 Simulated annealing

Notwithstanding the initialization procedure, the FMLE (or EM) algorithm, like all iterative methods, may be trapped in a local optimum. A remedial scheme is to apply a technique called *simulated annealing.* The membership probabilities in the early iterations are given a strong random component and only gradually are the calculated class values allowed to influence the estimation of the class means and covariance matrices. The rate of reduction of randomness may be determined by a temperature parameter T, e.g.,

$$u_{k\nu} \to u_{k\nu}(1 - r^{1/T}) \tag{8.45}$$

on each iteration, where $r \in [0, 1]$ is a uniformly distributed random number and where T is initialized to some maximum value T_0 (Hilger, 2001). The temperature T is reduced at each iteration by a factor $c < 1$ according to $T \to cT$. As T approaches zero, $u_{ki\nu}$ will be determined more and more by the probability distribution parameters in Equation (8.34) alone.

8.3.3 Partition density

Since the sum of squares cost function $E(C)$ in Equation (8.12) is no longer relevant, we choose with Gath and Geva (1989) the *partition density* as a possible criterion for selecting the best number of clusters. To obtain it, first of all define the *fuzzy hypervolume* as

$$FHV = \sum_{k=1}^{K} \sqrt{|\hat{\boldsymbol{\Sigma}}_k|}. \tag{8.46}$$

This expression is proportional to the volume in feature space occupied by the ellipsoidal clusters generated by the algorithm. For instance, for a two-dimensional cluster with a normal (elliptical) probability density we have, in its principal axis coordinate system,

$$\sqrt{|\hat{\boldsymbol{\Sigma}}|} = \sqrt{\begin{vmatrix} \sigma_1^2 & 0 \\ 0 & \sigma_2^2 \end{vmatrix}} = \sigma_1\sigma_2 \approx \text{ area (volume) of the ellipse.}$$

Listing 8.5: Excerpt from the IDL program module EM.PRO.

```
; loop over the cluster index
  IF NN GT 1 THEN FOR j=0,K-1 DO BEGIN
    Ms[j,*] = temporary(Ms[j,*])/ns[j]
    W = (fltarr(n)+1)##transpose(Ms[j,*])
    Ds = G - W
;   covariance matrix
    FOR i=0,NN-1 DO W[i,*] = $
         sqrt(U[*,j])*Ds[i,*]
    C = matrix_multiply(W,W,/btranspose)/ns[j]
    Cs[*,*,j] = C
    sqrtdetC = sqrt(determ(C,/double))
    Cinv = invert(C,/double)
    qf = total(Ds*(Ds##Cinv),1)
;   class hypervolume and partition density
    fhv[j] = sqrtdetC
    indices = where(qf LT 1.0, count)
    IF (count GT 0) THEN BEGIN
       pdens[j] = total(U[indices,j])/fhv[j]
    ENDIF
;   new memberships
    U[unfrozen,j] = $
          exp(-qf[unfrozen]/2.0)*(Ps[j]/sqrtdetC)
;   random membership for annealing
    IF T GT 0.0 THEN BEGIN
      Ur=1.0-randomu(seed,n_elements(unfrozen))^(1.0/T)
      U[unfrozen,j] = temporary(U[unfrozen,j])*Ur
    ENDIF
  ENDFOR $
```

Next, let s be the sum of all membership probabilities for observations which lie within unit Mahalanobis distance of a cluster center:

$$s = \sum_{\nu\in\mathcal{N}}\sum_{k=1}^{K} u_{\nu k}, \quad \mathcal{N} = \{\nu \mid (\boldsymbol{g}(\nu) - \hat{\boldsymbol{\mu}}_k)^\top \hat{\boldsymbol{\Sigma}}_k^{-1}(\boldsymbol{g}(\nu) - \hat{\boldsymbol{\mu}}_k) < 1\}.$$

Then the partition density is defined as

$$P_D = s/FHV. \tag{8.47}$$

Assuming that the data consist of $\tilde{K}$ well-separated clusters of approximately multivariate normally distributed pixels, the partition density should exhibit a maximum at $K = \tilde{K}$.

8.3.4 Implementation notes

An ENVI/IDL extension EM_RUN.PRO and a Python script em.py for FMLE (or EM) clustering, are described in Appendices C and D, respectively. The need

Listing 8.6: Excerpt from the Python script `em.py`.

```
 1 #        loop over the cluster index
 2          FOR k IN RANGE(K):
 3              Ms[:,k] = Ms[:,k]/ns[k]
 4              W = np.tile(Ms[:,k].ravel(),(n,1))
 5              Ds = G - W
 6 #           covariance matrix
 7              FOR i IN RANGE(N):
 8                  W[:,i] = np.sqrt(U[k,:]) \
 9                      .ravel()*Ds[:,i].ravel()
10              C = np.mat(W).T*np.mat(W)/ns[k]
11              Cs[k,:,:] = C
12              sqrtdetC = np.sqrt(np.linalg.det(C))
13              Cinv = np.linalg.inv(C)
14              qf = np.asarray(np.SUM(np.multiply(Ds \
15                           ,np.mat(Ds)*Cinv),1)).ravel()
16 #           class hypervolume and partition density
17              fhv[k] = sqrtdetC
18              idx = np.where(qf < 1.0)
19              pdens[k] = np.SUM(U[k,idx])/fhv[k]
20 #           new memberships
21              U[k,unfrozen] = np.exp(-qf[unfrozen]/2.0)\
22                      *(Ps[k]/sqrtdetC)
23 #           random membership for annealing
24              IF T > 0.0:
25                  Ur = 1.0 - np.random\
26                       .random(LEN(unfrozen))**(1.0/T)
27                  U[k,unfrozen] = U[k,unfrozen]*Ur
```

for continual re-estimation of $\mathbf{\Sigma}_k$ is computationally expensive, so the algorithm is not suited for clustering large and/or high-dimensional datasets. The necessity to invert the covariance matrices can also occasionally lead to instability. Listing 8.5 shows an excerpt from the program module `EM.PRO` which implements the EM algorithm in IDL. In the listing, `NN` is the dimension of the observations and `K` is the number of classes. By means of the keyword `UNFROZEN`, the observations partaking in the clustering process can be additionally specified. In program lines 21–22, only the membership probabilities $u_{j\nu}$ for "unfrozen" observations are recalculated, while the remaining observations are "frozen" to their current values; see, e.g., Bruzzone and Prieto (2000). This will be made use of in Section 8.4.2 below. The code snippet also reveals methods used to avoid `FOR` loops over the pixels. For instance, after line 5 the variable `Ds` is an array of distance vectors $(\boldsymbol{g}(\nu) - \hat{\boldsymbol{\mu}}_j)^\top$, $\nu = 1 \dots m$, from the mean of cluster j. In lines 7 to 10 the covariance matrix $\hat{\mathbf{\Sigma}}_j$ (the IDL array `Cs[*,*,j]`) is then calculated using array operations involving `Ds` without a loop over the pixel index ν. A similar technique is used in line 13 to calculate the quadratic forms $(\boldsymbol{g}(\nu) - \hat{\boldsymbol{\mu}}_j)^\top \hat{\boldsymbol{\Sigma}}_j^{-1} (\boldsymbol{g}(\nu) - \hat{\boldsymbol{\mu}}_j)$, $\nu = 1 \dots m$. The Python script excerpt in Listing 8.6 uses similar techniques.

Since the cluster memberships are multivariate probability densities, the IDL and Python scripts optionally generate a rule image; see Chapters 6 and 7. The rule image can be postprocessed, for example, with the probabilistic label relaxation filter discussed in Section 7.1.2. The IDL script will also take advantage of CUDA/GPULib (Section 4.4.2) whenever available.

8.4 Including spatial information

All of the clustering algorithms described so far make use exclusively of the spectral properties of the individual observations (pixel vectors). Spatial relationships within an image, such as scale, coherent regions or textures, are not taken into account. In the following we look at two refinements which deal with spatial context. They will be illustrated in connection with Gaussian mixture classification, but could equally well be applied to other clustering approaches.

8.4.1 Multiresolution clustering

Prior to performing Gaussian mixture clustering, an image pyramid for the chosen scene may be created using, for example, the discrete wavelet transform (DWT) filter bank described in Chapter 4. Clustering will then begin at the coarsest resolution and proceed to finer resolutions, with the membership probabilities passed on to each successive scale (Hilger, 2001). In our implementation, the membership probabilities from preceding scales which, after upsampling to the next scale, exceed some threshold (e.g., 0.9) are frozen in the manner discussed in Section 8.3.4. The depth of the pyramid can be chosen by the user.

The relevant program code is shown in Listing 8.7. The initial clustering carried out at the lowest resolution (the call to the procedure `gpuEM` or `EM` in lines 4 to 9, depending on CUDA/GPULib availability) is gradually refined in the `FOR`-loop beginning at line 16, whereby pixel vectors determined to have sufficiently high membership probabilities (> 0.9, line 33) at the lower resolutions are not reclassified. In line 13, the clusters are sorted according to decreasing partition density.* Apart from improving the rate of convergence of the calculation, scaling has the intended effect that the "spatial awareness" extracted at coarse resolution is passed up to the finer scales.

*This will be particularly useful in Chapter 9 when the algorithm is used to cluster multispectral change images, since the no-change pixels usually form the most dense cluster.

Listing 8.7: Excerpt from the program module EM_RUN.PRO.

```
; initial clustering of maximally compressed image
window,11,xsize=600,ysize=400,title='Log␣Likelihood'
IF gpu_detect() AND (num_bands GT 1) THEN $
  gpuEM, s_image, U, Ms, Ps, Cs, T0=T0, wnd=11, $
      num_cols=s_num_cols, num_rows=s_num_rows, $
      beta=beta, pdens=pdens,status=status,/verbose $
ELSE EM,  s_image, U, Ms, Ps, Cs, T0=T0, wnd=11, $
      num_cols=s_num_cols, num_rows=s_num_rows, $
      beta=beta, pdens=pdens,status=status,/verbose
IF status THEN message, 'EM␣algoritm␣failed'

; sort clusters wrt partition density
U = U[*,reverse(sort(pdens))]

; clustering at increasing scales
FOR scale=1,max_scale-min_scale DO BEGIN
; upsample class memberships (bilinear interpolation)
   U = reform(U,s_num_cols,s_num_rows,K,/overwrite)
   s_num_cols = s_num_cols*2
   s_num_rows = s_num_rows*2
   s_num_pixels = s_num_cols*s_num_rows
   U = shift(rebin(U,s_num_cols,s_num_rows,K),1,1,0)
   U = reform(U,s_num_pixels,K,/overwrite)

; expand the image
   s_image = fltarr(num_bands,s_num_pixels)
   FOR i=0,num_bands-1 DO BEGIN
      DWTBands[i]->invert
      s_image[i,*] = DWTbands[i]->get_quadrant(0)
   ENDFOR

; cluster
   unfrozen = where(max(U,dimension=2) LE 0.90, count)
   IF count GT 0 THEN $
     IF gpu_detect() AND (num_bands GT 1) THEN $
       gpuEM,s_image,U,Ms,Ps,Cs,miniter=5,wnd=11, $
          unfrozen=unfrozen,T0=0.0, $
          num_cols=s_num_cols, num_rows=s_num_rows, $
          beta=beta, status=status,/verbose $
     ELSE EM,s_image,U,M,Ps,Cs,miniter=5,wnd=11, $
            unfrozen=unfrozen, T0=0.0, $
            num_cols=s_num_cols,num_rows=s_num_rows, $
            beta=beta, status=status,/verbose
   IF status THEN RETURN
ENDFOR
```

8.4.2 Spatial clustering

As described in Chapter 4, class labels for multispectral images can be represented by realizations of a Markov random field, for which the label of a given pixel may be influenced only by the class labels of other pixels in its immediate neighborhood. According to Gibbs-Markov equivalence, Theorem 4.3, the probability density for any complete labeling ℓ of the image is given by

$$p(\ell) = \frac{1}{Z}\exp(-\beta U(\ell)), \tag{8.48}$$

where Z is a normalization and the energy function $U(\ell)$ is given by a sum over clique potentials,

$$U(\ell) = \sum_{c \in \mathcal{C}} V_c(\ell), \tag{8.49}$$

relative to a neighborhood system $\mathcal{N}$. If we restrict discussion to 4-neighborhoods, the only possible cliques are singletons and pairs of vertically or horizontally adjacent pixels, see Figure 4.15(a). If, furthermore, the potential for singleton cliques is set to zero and the random field is isotropic (independent of clique orientation), then we can write Equation (8.49) in the form

$$U(\ell) = \sum_{\nu \in \mathcal{I}} \sum_{\nu' \in \mathcal{N}_\nu} V_2(\ell_\nu, \ell_{\nu'}), \tag{8.50}$$

where $\mathcal{I}$ is the complete image lattice, $\mathcal{N}_\nu$ is the neighborhood of pixel ν and $V_2(\ell_\nu, \ell_{\nu'})$ is the clique potential for two neighboring sites. Let us now choose

$$V_2(\ell_\nu, \ell_{\nu'}) = \frac{1}{4}(1 - u_{\ell_\nu \nu'}), \tag{8.51}$$

where $u_{\ell_\nu \nu'}$ is an element of the cluster membership probability matrix $\boldsymbol{U}$. This says that, when the probability $u_{\ell_\nu \nu'}$ is large that neighboring site ν' has the same label ℓ_ν as site ν, then the clique potential is small and the configuration is favored. Combining Equations (8.50) and (8.51),

$$U(\ell) = \sum_{\nu \in \mathcal{I}} \frac{1}{4}\left(4 - \sum_{\nu' \in \mathcal{N}_\nu} u_{\ell_\nu \nu'}\right) = \sum_{\nu \in \mathcal{I}} (1 - u_{\ell_\nu \mathcal{N}_\nu}), \tag{8.52}$$

where $u_{\ell_\nu \mathcal{N}_\nu}$ is the averaged membership probability for ℓ_ν within the neighborhood,

$$u_{\ell_\nu \mathcal{N}_\nu} = \frac{1}{4} \sum_{\nu' \in \mathcal{N}_\nu} u_{\ell_\nu \nu'} \ . \tag{8.53}$$

Substituting Equation (8.52) into Equation (8.48), we obtain

$$p(\ell) = \frac{1}{Z} \prod_{\nu \in \mathcal{I}} \exp(-\beta(1 - u_{\ell_\nu \mathcal{N}_\nu})). \tag{8.54}$$

Listing 8.8: Excerpt from the program module EM.PRO.

```
; determine spatial membership matrix
  IF beta GT 0 THEN BEGIN
    FOR j=0,K-1 DO BEGIN
      U_N=1.0-convol(reform(U[*,j],num_cols, $
                             num_rows),Nb,/center)
      V[*,j] = exp(-beta*U_N)
    ENDFOR
;   combine spectral/spatial for unfrozen
    U[unfrozen,*]=temporary(U[unfrozen,*])* $
                             V[unfrozen,*]
  ENDIF
; normalize all
  a = 1/total(U,2)
  void=where(finite(a),complement=complement, $
    ncomplement=ncomplement)
  IF ncomplement GT 0 THEN a[complement]=0.0
  FOR j=0,K-1 DO U[*,j]=temporary(U[*,j])*a
```

Equation (8.54) is reminiscent of the likelihood functions that we have been using for pixel-based clustering and suggests the following heuristic ansatz (Hilger, 2001): Along with the *spectral* class membership probabilities $u_{k\nu}$ that characterize the FMLE algorithm and which are given by Equation (8.34), introduce a *spatial* class membership probability $v_{k\nu}$,

$$v_{k\nu} \propto \exp(-\beta(1 - u_{k\mathcal{N}_\nu})). \tag{8.55}$$

A combined *spectral-spatial* class membership probability for the νth observation is then determined by replacing $u_{k\nu}$ with

$$\frac{u_{k\nu} v_{k\nu}}{\sum_{k'=1}^{K} u_{k'\nu} v_{k'\nu}}, \tag{8.56}$$

apart from which the algorithm proceeds as before. Note that, from Equations (8.34) and (8.55), the combined membership probability above is now proportional to

$$\frac{1}{\sqrt{|\hat{\boldsymbol{\Sigma}}_k|}} \cdot \frac{m_k}{m} \cdot \exp\Big(-\frac{1}{2}(\boldsymbol{g}(\nu) - \hat{\boldsymbol{\mu}}_k)^\top \boldsymbol{C}_k^{-1}(\boldsymbol{g}(\nu) - \hat{\boldsymbol{\mu}}_k) - \beta(1 - u_{k\mathcal{N}_\nu})\Big).$$

This way of folding spatial information with the spectral similarity measure (here the Mahalanobis distance) is referred to in Tran et al. (2005) as the *addition form.*

The quantity $u_{k\mathcal{N}_\nu}$ appearing in Equation (8.55) can be determined for an entire image simply by convolving the two-dimensional kernel

$$\begin{array}{ccc} 0 & \frac{1}{4} & 0 \\ \frac{1}{4} & 0 & \frac{1}{4} \\ 0 & \frac{1}{4} & 0 \end{array}$$

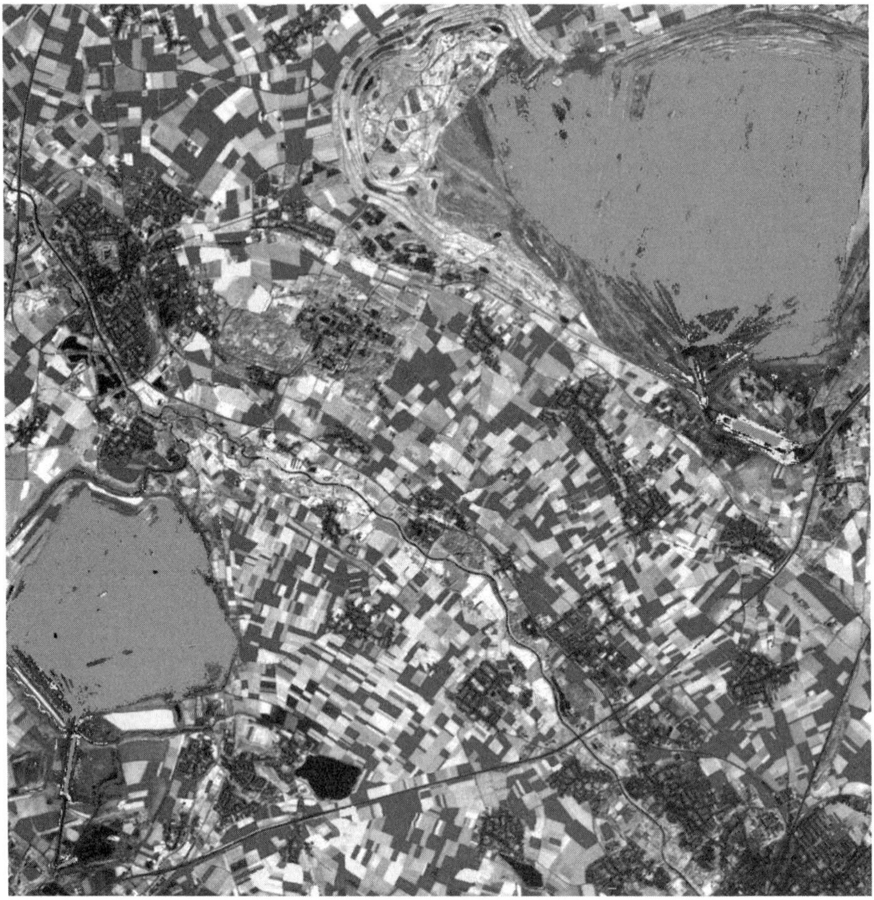

FIGURE 8.2
Gaussian mixture clustering of the first four principal components of the Jülich ASTER scene, eight clusters. Three clusters associated with new sugar beet plantation (maroon), rapeseed (yellow), and open cast coal mining (coral) are shown superimposed on VNIR spectral band 3N (gray). **(See color insert.)**

with the array $\boldsymbol{U}$, after reforming the latter to the correct image dimensions; see Listing 8.8, lines 2 to 11. (The kernel is stored in the variable `Nb`.)

Figure 8.2 shows an example, again using the Jülich ASTER image. The classification was determined with an image pyramid of depth 1 (i.e., two levels) and with $\beta = 1.0$.

8.5 A benchmark

Together with the built-in algorithms K-means and ISODATA, we now have no less than seven clustering methods at our disposal in the ENVI environment, three with Python. So which one should we use? The degree of success of unsupervised image classification is notoriously difficult to quantify; see, e.g., Duda and Canty (2002). This is because there is, by definition, no prior information with regard to what one might expect to be a "reasonable" result. Ultimately, judgment is qualitative, almost a question of esthetics.

Listing 8.9: Generating a toy image.

```
PRO toy_image
 image = fltarr (800 ,800 ,3)
 b = 2.0
 image [99:699 ,299:499 ,*] = b
 image [299:499 ,99:699 ,*] = b
 image [299:499 ,299:499 ,*] = 2*b
 image = smooth (image ,[13 ,13 ,1] ,/ edge_truncate )
 n1 = randomu (seed ,800 ,800 ,/ normal )
 n2 = randomu (seed ,800 ,800 ,/ normal )
 n3 = randomu (seed ,800 ,800 ,/ normal )
 image [* ,* ,0] = image [* ,* ,0] + n1
 image [* ,* ,1] = image [* ,* ,1] + n2 + n1
 image [* ,* ,2] = image [* ,* ,2] + n3 + n1 /2 + n2 /2
 envi_enter_data , image
END
```

Nevertheless, in order to compare the algorithms we have looked at so far more objectively, we will generate a "toy" benchmark image with the code in Listing 8.9. The image is shown in the upper left-hand corner of Figure 8.3. It consists of three "spectral bands" and three classes, namely the dark background, the four points of the cross and the cross center. The cluster sizes differ considerably (approximately in the ratio 12:4:1) with a large overlap, and the bands are strongly correlated. The program in Listing 3.8 or 3.9 estimates the image noise covariance matrix as

```
ENVI> EX3_4
Noise covariance matrix, file [Memory1] (800x800x3)
         1.0041422        1.0038484        0.5020555
         1.0038484        2.0039095        1.0036373
         0.5020555        1.0036373        1.5035802
```

Some results are shown in Figure 8.3. Neither the K-means variants nor the Gaussian mixture algorithm without scaling identify all three classes. Agglom-

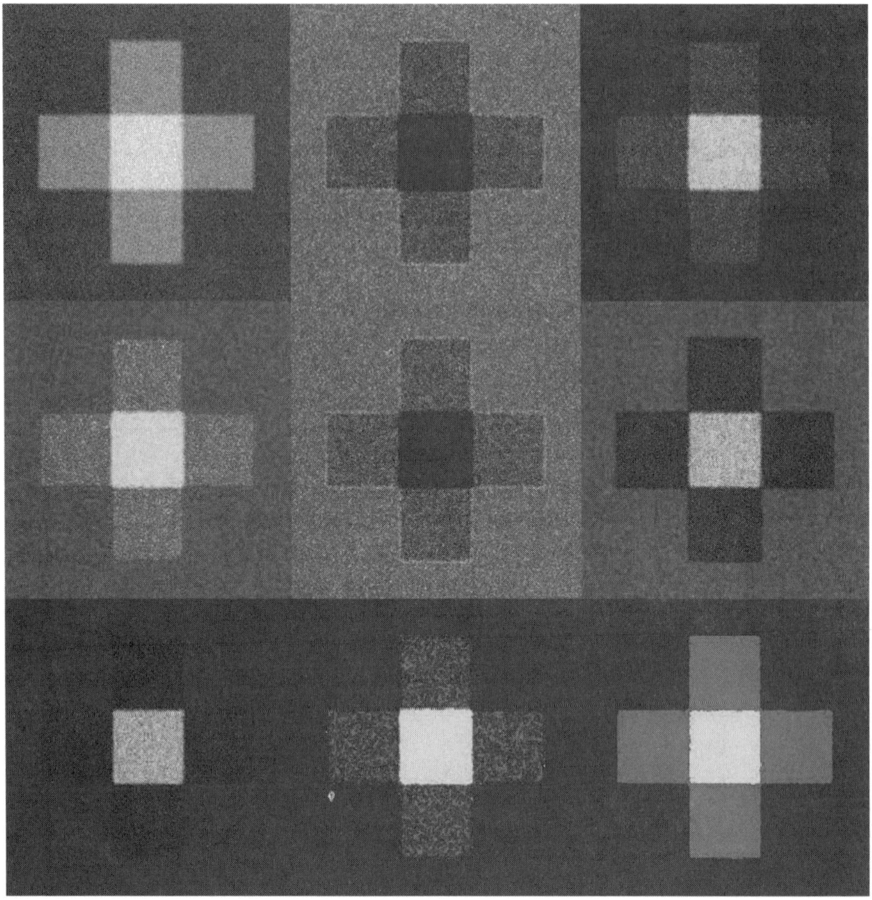

FIGURE 8.3
Unsupervised classification of a toy image. Row-wise, left to right, top to bottom: The toy image, K-means, kernel K-means, extended K-means (on first principal component), fuzzy K-means, agglomerative hierarchical, Gaussian mixture with depth 0 and $\beta = 1.0$, Gaussian mixture with depth 2 and $\beta = 0.0$, Gaussian mixture with depth 2 and $\beta = 1.0$. **(See color insert.)**

erative hierarchical clustering succeeds reasonably well. The "best" classification is obtained with the Gaussian mixture model when both multi-resolution and spatial clustering are employed. The success of Gaussian mixture classification is not particularly surprising since the toy image classes are multivariate normally distributed. However, if the other methods exhibit inferior performance on normal distributions, then they might be expected to be similarly inferior when presented with real, non-Gaussian data.

8.6 The Kohonen self-organizing map

The *Kohonen self-organizing map* (SOM), a simple example of which is sketched in Figure 8.4 , belongs to a class of neural networks which are trained by *competitive learning* (Hertz et al., 1991; Kohonen, 1989). It is very useful as a visualization tool for exploring the class structure of multispectral imagery. The layer of neurons shown in the figure can have any geometry, but usually a one-, two-, or three-dimensional array is chosen. The input signal is the observation vector $\boldsymbol{g} = (g_1, g_2 \dots g_N)^\top$, where, in the figure, $N = 2$. Each input to a neuron is associated with a *synaptic weight* so that, for K neurons, the synaptic weights can be represented as an $(N \times K)$ matrix

$$\boldsymbol{w} = \begin{pmatrix} w_{11} & w_{12} & \cdots & w_{1K} \\ w_{21} & w_{22} & \cdots & w_{2K} \\ \vdots & \vdots & \vdots & \vdots \\ w_{N1} & w_{N2} & \cdots & w_{NK} \end{pmatrix}. \tag{8.57}$$

The components of the vector $\boldsymbol{w}_k = (w_{1k}, w_{2k} \dots w_{Nk})^\top$ are the synaptic weights of the kth neuron. The set of observations $\{\boldsymbol{g}(\nu) \mid \nu = 1 \dots m\}$, where

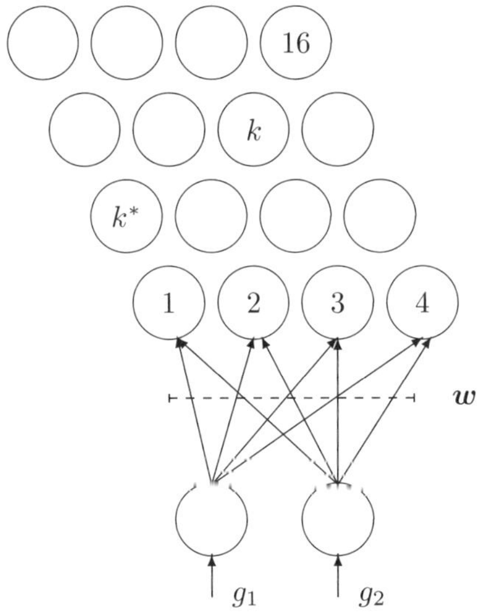

FIGURE 8.4
The Kohonen self-organizing map in two dimensions with a two-dimensional input. The inputs are connected to all 16 neurons (only the first four connections are shown).

$m > K$, comprise the training data for the network. The synaptic weight vectors are to be adjusted so as to reflect in some way the class structure of the training data in an N-dimensional feature space.

The training procedure is as follows. First of all, the neurons' weights are initialized from a random subset of K training observations, setting

$$\boldsymbol{w}_k = \boldsymbol{g}(k), \quad k = 1 \dots K.$$

Then, when a new training vector $\boldsymbol{g}(\nu)$ is presented to the input of the network, the neuron whose weight vector $\boldsymbol{w}_k$ lies nearest to $\boldsymbol{g}(\nu)$ is designated to be the "winner." Distances are given by $\|\boldsymbol{g}(\nu) - \boldsymbol{w}_k\|$. Suppose that the winner's index is k^*. Its weight vector is then shifted a small amount in the direction of the training vector:

$$\boldsymbol{w}_{k^*}(\nu + 1) = \boldsymbol{w}_{k^*}(\nu) + \eta(\boldsymbol{g}(\nu) - \boldsymbol{w}_{k^*}(\nu)), \tag{8.58}$$

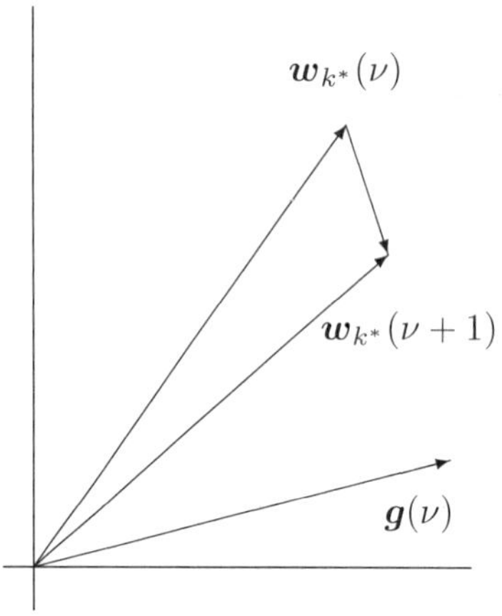

FIGURE 8.5
Movement of a synaptic weight vector in the direction of a training vector.

where $\boldsymbol{w}_{k^*}(\nu+1)$ is the weight vector after presentation of the νth training vector; see Figure 8.5. The parameter η is called the *learning rate* of the network. The intention is to repeat this learning procedure until the synaptic weight vectors reflect the class structure of the training data, thereby achieving what is often referred to as a *vector quantization* of the feature space (Hertz et al., 1991). In order for this method to converge, it is necessary to allow the learning rate to decrease gradually during the training process. A convenient function for this is

$$\eta(\nu) = \eta_{max}\left(\frac{\eta_{min}}{\eta_{max}}\right)^{\nu/m}.$$

However, the SOM algorithm goes a step further and attempts to map the *topology* of the feature space onto the network as well. This is achieved by defining a neighborhood function for the winner neuron on the network of neurons. Usually a Gaussian of the form

$$\mathcal{N}(k^*, k) = \exp(-d^2(k^*, k)/2\sigma^2)$$

is chosen, where $d^2(k^*, k)$ is the square of the distance between neurons k^* and k in the network array. For example, for a two-dimensional array of $n \times n$

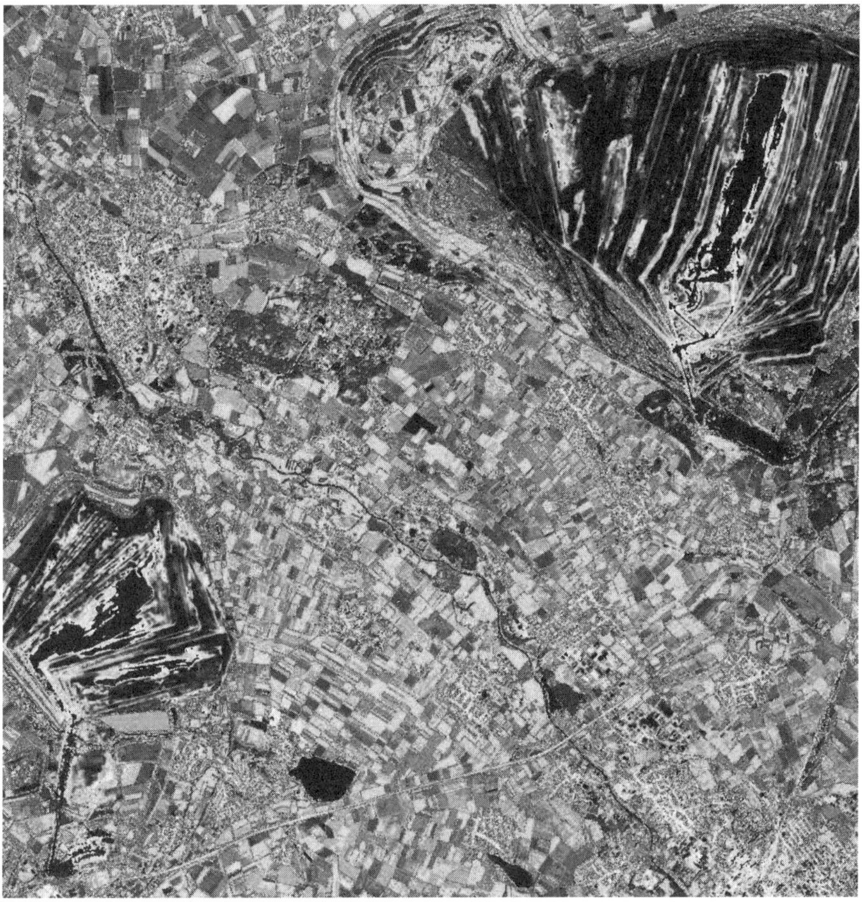

FIGURE 8.6
Kohonen self-organizing map of the nine VNIR and sharpened SWIR spectral bands of the Jülich ASTER scene. The network is a cube having dimensions $6 \times 6 \times 6$. **(See color insert.)**

neurons $d^2(k^*, k)$ can by calculated in integer arithmetic as

$$d^2(k^*, k) = [(k^* - 1) \bmod n - (k - 1) \bmod n]^2 + [(k^* - 1)/n - (k - 1)/n]^2. \tag{8.59}$$

During the learning phase, not only the winner neuron, but also the neurons in its neighborhood, are moved in the direction of the training vectors by an amount proportional to the value of the neighborhood function:

$$\boldsymbol{w}_k(\nu + 1) = \boldsymbol{w}_k(\nu) + \eta(\nu)\mathcal{N}(k^*, k)(\boldsymbol{g}(\nu) - \boldsymbol{w}_k(\nu)), \quad k = 1 \ldots K. \tag{8.60}$$

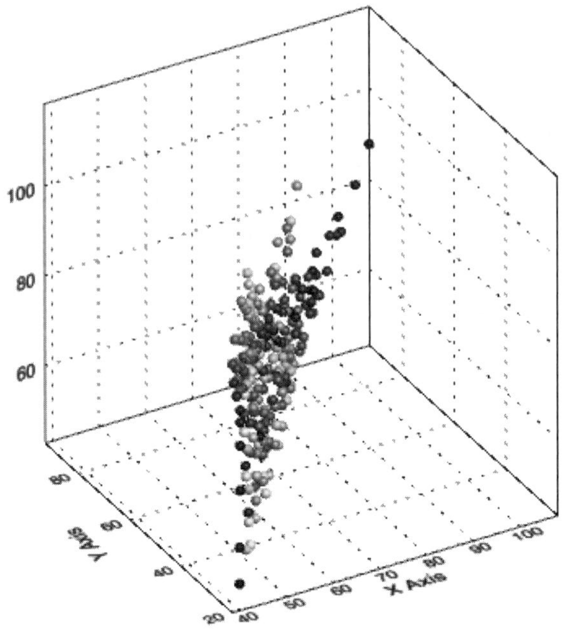

FIGURE 8.7
The $6^3 = 216$ SOM neurons in the feature space of the three VNIR spectral bands of the Jülich ASTER scene; see Figure 8.6. **(See color insert.)**

Finally, the extent of the neighborhood is allowed to shrink steadily as well:

$$\sigma(\nu) = \sigma_{max} \left(\frac{\sigma_{min}}{\sigma_{max}} \right)^{\nu/m}.$$

Typically, $\sigma_{max} \approx n/2$ and $\sigma_{min} \approx 1/2$. The neighborhood is initially the entire network, but toward the end of training it becomes very localized.

For clustering of multispectral satellite imagery a cubic network geometry is useful (Groß and Seibert, 1993). After training on some representative sample of pixel vectors, the entire image is classified by associating each pixel vector with the neuron having the closest synaptic weight vector. Then the pixel is colored by mapping the position of that neuron in the cube to coordinates in RGB color space. Thus, the N-dimensional feature space is "projected" onto the three-dimensional RGB color cube and pixels that are close together in feature space are given similar colors. An ENVI/IDL extension for the Kohonen self-organizing map is given in Appendix C and an example is shown in Figure 8.6. The positions and colors of the neurons in feature space after completion of training are shown in Figure 8.7.

8.7 Image segmentation

The term *image segmentation* refers, in its broadest sense, to the process of partitioning a digital image into multiple regions. Certainly all of the clustering algorithms that we have considered till now fall within this category. However, in the present section, we will use the term in a more restrictive way, namely as referring to the partitioning of pixels not only by spectral similarity (essentially what has been discussed so far) but also by spatial proximity. Segmentation plays a major role in low-level computer vision and in autonomous image processing in general. Popular methods include the use of edge detectors, watershed segmentation, region growing algorithms and morphology; see Gonzalez and Woods (2002), Chapter 10, for a good overview.

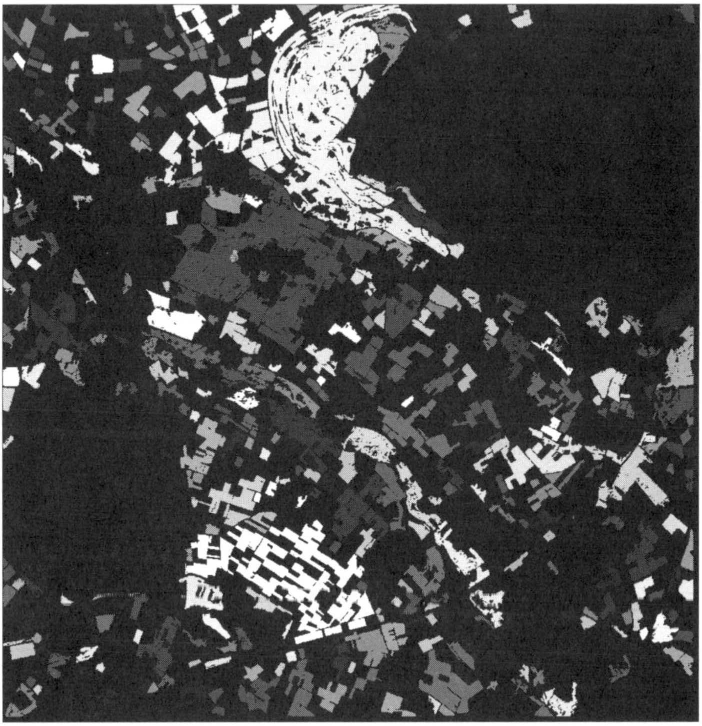

FIGURE 8.8
Postclassification segmentation of the Gaussian mixture cluster associated with green vegetation in the Jülich ASTER scene. Each connected region has been assigned a random label for display purposes.

8.7.1 Segmenting a classified image

The ENVI Classic menu command

`Classification/Post Classification/Segmentation Image`

provides a simple mechanism to convert a classification image into a segmented image by means of *blobbing*, i.e., finding connected regions. Unfortunately, this command does not respect class labels. The IDL program `SEGMENT_CLASS_RUN`, which is described in Appendix C and an excerpt of which is shown in Listing 8.10, may be used as a replacement. In lines 2 to 12 of the listing, the IDL function `LABEL_REGION()` is used to segment each selected class successively. Connected regions smaller than a user-defined value `MIN_SIZE` are then eliminated in lines 14 to 21. Here use is made of the powerful, if somewhat intimidating,[*] `HISTOGRAM()` function.

Listing 8.10: Excerpt from the program `SEGMENT_CLASS_RUN.PRO`.

```
1  ; segment first class
2  limg = label_region(climg EQ cl[0],/ULONG, $
3           all_neighbors=all_neighbors)
4  ; segment others, if any
5  FOR i = 1, K - 1 DO BEGIN
6    mxlb   = max(limg)
7    climgi = climg EQ cl[i]
8    limg   = temporary(limg) + $
9           label_region(climgi, /ULONG, $
10           all_neighbors=all_beighbors) $
11          + climgi*mxlb
12 ENDFOR
13 ; re-label segments larger than min_size
14 indices = where(histogram(limg,reverse_indices=r) $
15                            GE min_size,count)
16 limg=limg*0
17 IF count GT 0 THEN FOR i=1L,count-1 DO BEGIN
18    j = indices[i]        ;label of segment
19    p = r[r[j]:r[j+1]-1] ;pixels in segment
20    limg[p]=i             ;new label
21 ENDFOR
```

Figure 8.8 shows an example. A single cluster associated with a mixture of green vegetation (mixed forest, winter wheat, grassland) has been segmented into about 350 connected regions. Note that the final segmentation has been achieved in two distinct steps: spectral clustering (in this case with the FMLE

*See the highly recommended and, for insiders, notorious "HISTOGRAM: The Breathless Horror and Disgust" by J. D. Smith, `http://www.astro.virginia.edu/-class/oconnell/astr511/IDLresources/histogram_tutorial-JDSmith.html`

Listing 8.11: Clustering segments on the basis of their invariant moments.

```
PRO Ex8_2
; number of clusters
   K = 3
; select segment file
   envi_select,fid=fid,pos=pos,dims=dims, $
     /no_dims,/no_spec,/band_only, $
     title='Select␣segment␣file'
   IF fid EQ -1 THEN RETURN
   num_cols = dims[2]-dims[1]+1
   seg_im = envi_get_data(fid=fid,dims=dims,pos=pos)
   num_segs = max(seg_im)
; calculate table of invariant moments
   moments = fltarr(7,num_segs)
   FOR i=0L, num_segs-1 DO BEGIN
      indices = where(seg_im EQ i+1,count)
      IF count GT 0 THEN BEGIN
         X = indices MOD num_cols
         Y = indices/num_cols
         A = seg_im[min(X):max(X)+1,min(Y):max(Y)+1]
         indices1=where(A EQ i+1, complement=comp,$
                                  ncomplement=ncomp)
         A[indices1]=1.0
         IF ncomp GT 0 THEN A[comp]=0.0
         moments[*,i] = hu_moments(A,/log)
      ENDIF
   ENDFOR
; agglomerative hierarchic clustering
   HCL,standardize(moments),K,Ls
; re-label segments with cluster label
   _ = histogram(seg_im,reverse_indices=r)
   FOR i=1L,num_segs DO BEGIN
       p = r[r[i]:r[i+1]-1]  ;pixels in segment
       seg_im[p]=Ls[i-1]+1   ;new label
   ENDFOR
; output to memory
   envi_enter_data, seg_im, $
      file_type=3,num_classes=K+1, $
      class_names=string(indgen(K+1)), $
      lookup=class_lookup_table(indgen(K+1))
END
```

or EM algorithm) followed by identification of connected regions. In Section 8.7.3 below these steps are merged into a single procedure.

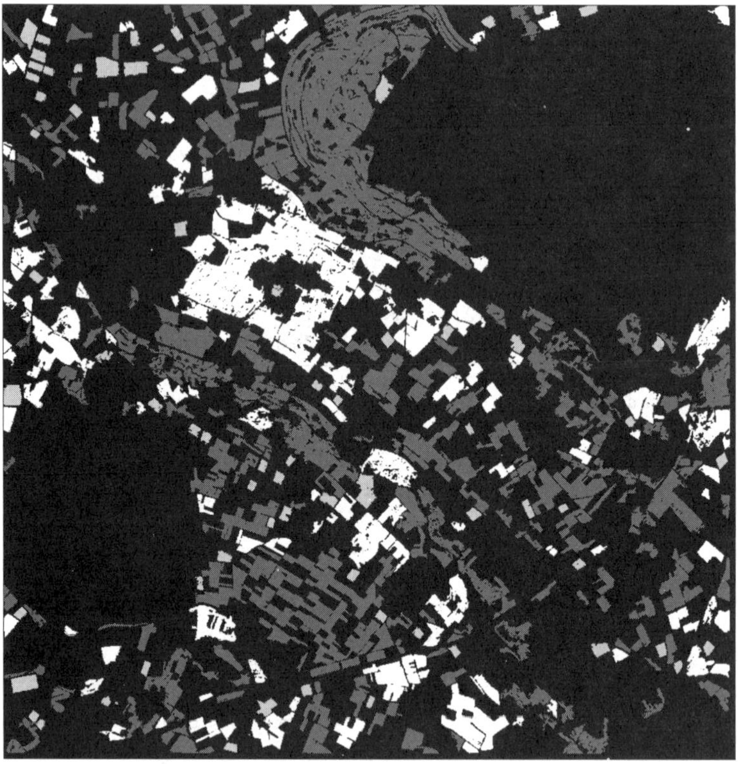

FIGURE 8.9
Classification of the segments in Figure 8.8 on the basis of their invariant moments. **(See color insert.)**

8.7.2 Object-based classification

Segmentation of a classification image, or segmentation by any other means, offers the possibility of a refinement of classification on the basis of characteristics of the individual segments not necessarily related to pixel intensities, such as shape, compactness, proximity to other segments, etc. One speaks generally of *object-based* classification. Probably the best-known commercial implementation of object-based classification is the Definiens AG software suite; see, e.g., Benz et al. (2004). In the simple example shown in Listing 8.11, the segments in Figure 8.8 are clustered into three classes on the basis of their Hu invariant moments (see Section 5.2.4). After reading in a segmented image (line 10), a table of the logarithms of the Hu moments is constructed (lines 14 to 26). Then the agglomerative hierarchical clustering algorithm of Section 8.2.4 is called on the standardized (zero mean, unit variance) logarithms (line 28). The result is shown in Figure 8.9. The three clusters are seen to vary from compact and nearly square-shaped (green) to extended and multiply connected (red) with the white segment cluster intermediate between the two. The segments are plotted in the feature space of the standardized logarithms of the first three invariant moments in Figure 8.10.

8.7.3 The mean shift

We conclude the present chapter with the description of an especially popular, non-parametric segmentation algorithm called the *mean shift* (Fukunaga and Hostetler, 1975; Comaniciu and Meer, 2002). The algorithm is non-parametric in the sense that no assumptions are made regarding the probability density of the observations being clustered. The mean shift partitions pixels in an N-dimensional feature space by associating each pixel with a local maximum in the estimated probability density, called a *mode.* For each pixel, the associated mode is determined by defining a (hyper-)sphere of radius $r_{spectral}$ centered at the pixel and calculating the mean of the pixels that lie within the sphere. Then the sphere's center is shifted to that mean. This continues until convergence, i.e., until the mean shift is less than some threshold. At each iteration, the sphere moves to a region of higher probability density until a mode is reached. The pixel is assigned that mode.

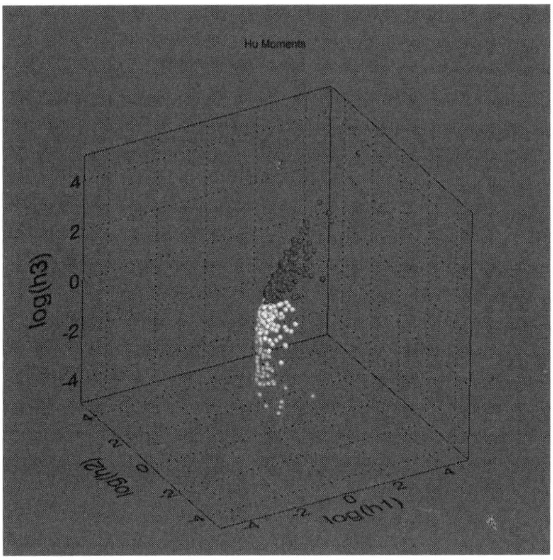

FIGURE 8.10
Standardized logarithms of the first three Hu moments of the classified segments of Figure 8.9. **(See color insert.)**

It will be apparent from the above that mean shift clustering *per se* will not lead to image segmentation in the restricted sense that we are discussing here. However, simply by extending the feature space to include the spatial position of the pixels, an elegant segmentation algorithm emerges. We only need distinguish, additionally to $r_{spectral}$, a spatial radius $r_{spatial}$ for determining the mean shift. After an appropriate normalization of the spectral and spatial distances, for example by re-scaling the pixel intensities according to

$$\boldsymbol{g}(\nu) \rightarrow \boldsymbol{g}(\nu)\frac{r_{spatial}}{r_{spectral}}, \quad \nu = 1 \ldots m,$$

FIGURE 8.11
Left, a spatial subset of the Jülich principal component image of Figure 6.1. Right, the mean shift filtered image and the segment boundaries. **(See color insert.)**

the mean shift procedure is carried out in the concatenated, $N+2$-dimensional, spectral-spatial feature space using a hyper-sphere of radius $r = r_{spatial}$.

Specifically, let $\boldsymbol{y}_\nu = \begin{pmatrix} \boldsymbol{g}(\nu) \\ \boldsymbol{x}_\nu \end{pmatrix}$, $\nu = 1 \dots m$, represent the image pixels in the combined feature space, let $\boldsymbol{z}_\nu$, $\nu = 1 \dots m$, denote the mean shift filtered pixels after the segmentation has concluded, and define $S(\boldsymbol{w})$ as the set of pixels within a radius r of a point $\boldsymbol{w}$ (cardinality $|S(\boldsymbol{w})|$). Then the algorithm is as follows:

Algorithm (Mean shift segmentation)

For each $\nu = 1 \dots m$, do the following:

1. Set $k = 1$ and $\boldsymbol{w}_k = \boldsymbol{y}_\nu$.
2. Repeat:
   $\boldsymbol{w}_{k+1} = \frac{1}{|S(\boldsymbol{w}_k)|} \sum_{\boldsymbol{y}_\nu \in S(\boldsymbol{w}_k)} \boldsymbol{y}_\nu$
   $k = k + 1$
   until convergence to $\boldsymbol{w} = \begin{pmatrix} \boldsymbol{f} \\ \boldsymbol{x} \end{pmatrix}$.
3. Assign $\boldsymbol{z}_\nu = \begin{pmatrix} \boldsymbol{f} \\ \boldsymbol{x}_\nu \end{pmatrix}$.

The last step assigns the filtered spectral component $\boldsymbol{f}$ (the spectral mode) to the original pixel location $\boldsymbol{x}_\nu$. Segmentation simply involves identifying all pixels with the same mode as a segment.

A direct implementation of this algorithm would be prohibitively slow, since mean shifts are to be computed for all of the pixels. For a partial solution in Python using the `OpenCV` package, see Exercise 12(c).

In the ENVI/IDL extension `MEAN_SHIFT_RUN` described in Appendix C, two approximations are employed to speed things up. First, all pixels within radius r of convergence point $\boldsymbol{w}$ are assumed also to converge to that mode and are excluded from further processing. Second, all pixels lying sufficiently close (within a distance $r/2$) to the path traversed from the initial vector $\boldsymbol{y}_\nu$ to the mode $\boldsymbol{w}$ are similarly assigned to that mode and excluded from the calculation. Additionally, a user-defined minimum segment size can be specified. Segments with smaller extent than this minimum are assigned to the nearest segment in the combined spatial/spectral feature space. An example of the mean shift applied to the first three principal components of the Jülich ASTER scene is shown in Figure 8.11.

8.8 Exercises

1. (a) Show that the sum of squares cost function, Equation (8.12), can be expressed equivalently as

$$E(C) = \frac{1}{2}\sum_{k=1}^{K} \bar{s}_k, \tag{8.61}$$

where $\bar{s}_k$ is the average squared distance between points in the kth cluster,

$$\bar{s}_k = \frac{1}{m_k}\sum_{i=1}^{m}\sum_{i'=1}^{m} u_{ki}u_{ki'}\|\boldsymbol{g}_i - \boldsymbol{g}_{i'}\|^2. \tag{8.62}$$

(b) Recall that, in the notation of Section 8.2.2,

$$\mathcal{G}^\top \boldsymbol{U}^\top \boldsymbol{M} = (\hat{\boldsymbol{\mu}}_1 \ldots \hat{\boldsymbol{\mu}}_k).$$

It follows that $(\mathcal{G}^\top \boldsymbol{U}^\top \boldsymbol{M})\boldsymbol{U}$ is an $N \times m$ matrix whose νth column is the mean vector associated with the νth observation. Use this to demonstrate that the sum of squares cost function can also be written in the form

$$E(C) = \mathrm{tr}(\mathcal{G}\mathcal{G}^\top) - \mathrm{tr}(\boldsymbol{U}^\top \boldsymbol{M}\boldsymbol{U}\mathcal{G}\mathcal{G}^\top). \tag{8.63}$$

2. (a) The ENVI/IDL extension `KKMEANS_RUN.PRO` and the Python script `kkmeans.py` make use of the Gaussian kernel only. Modify one or the other to include the option of using polynomial kernels; see Chapter 4, Exercise 8.

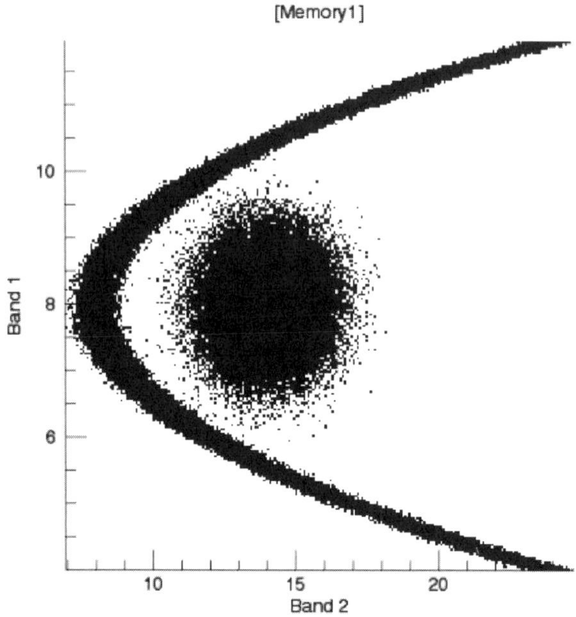

FIGURE 8.12
Two clusters which are not linearly separable.

(b) The program in Listing 8.12 generates a two-band toy image with the two nonlinearly separable clusters shown in Figure 8.12. Experiment with the IDL or Python kernel K-means scripts to achieve a good clustering of the image. Compare your result with ENVI's K-means tool or the Python K-means implementation in Listing 8.1.

Listing 8.12: Generating another toy image.

```
PRO toy_image1
 image = fltarr (400 ,400 ,2)
 n = randomu(seed ,400 ,400,/normal )
 n1 = 8*randomu(seed, 400, 400)-4
 image[*,*,0] = n1+8
 image[*,*,1]=n1^2+0.3*randomu(seed,400,400,/normal )+8
 image[0:199,*,0]=randomu(seed, 200,400,/normal )/2+8
 image[0:199,*,1]=randomu(seed, 200,400,/normal )+14
 envi_enter_data,image
END
```

3. Kernel K-means clustering is closely related to *spectral clustering* (see, e.g., Dhillon et al. (2005)). Let $\mathcal{K}$ be an $m \times m$ Gaussian kernel matrix. Define the diagonal *degree* matrix

$$\boldsymbol{D} = \mathrm{Diag}(d_1 \ldots d_m),$$

where $d_i = \sum_{j=1}^{m} (\mathcal{K})_{ij}$, and the symmetric *Laplacian* matrix

$$\boldsymbol{L} = \boldsymbol{D} - \mathcal{K}. \tag{8.64}$$

Minimization of the kernelized version of the sum of squares cost function, namely

$$E(C) = \mathrm{tr}(\mathcal{K}) - \mathrm{tr}(\boldsymbol{U}^\top \boldsymbol{M} \boldsymbol{U} \mathcal{K}),$$

see Equation (8.63), can be shown to be equivalent to ordinary (linear) clustering of the m training observations in the space of the first K eigenvectors of $\boldsymbol{L}$ corresponding to the K smallest eigenvalues (von Luxburg, 2006; Shawe–Taylor and Cristianini, 2004).

(a) Show that $\boldsymbol{L}$ is positive semi-definite and that its smallest eigenvalue is zero with corresponding eigenvector $\mathbf{1}_m$.

(b) Write an IDL or Python function `spectral_cluster(G,K)` to implement the following algorithm: (The input observations $\boldsymbol{g}(\nu)$, $\nu = 1 \ldots m$, are passed to the function in the form of a data matrix $\mathcal{G}$ along with K, the number of clusters.)

Algorithm (Unnormalized spectral clustering)

1. Determine the Laplacian matrix $\boldsymbol{L}$ as given by Equation (8.64).
2. Compute the $m \times K$ matrix $\boldsymbol{V} = (\boldsymbol{v}_1 \ldots \boldsymbol{v}_K)$, the columns of which are the eigenvectors of $\boldsymbol{L}$ corresponding to the K smallest eigenvalues. Let $\boldsymbol{y}(\nu)$ be the K-component vector consisting of the νth row of $\boldsymbol{V}$, $\nu = 1 \ldots m$.
3. Partition the vectors $\boldsymbol{y}(\nu)$ with the K-means algorithm into clusters $C_1 \ldots C_K$. (*Hint for IDL:* see the IDL library functions `CLT_WTS` and `CLUSTER()`.)
4. Output the cluster labels for $\boldsymbol{g}(\nu)$ as those for $\boldsymbol{y}(\nu)$.

(c) Thinking of $\mathcal{G}$ in the algorithm as a training dataset sampled from a multispectral image, suggest how the clustering result might be generalized to the whole image.

4. Write an IDL or Python routine to give a first-order estimate of the entropy of an image spectral band.* Use the definition in Equation (8.18)

* *First order* means that one assumes that all pixel intensities are statistically independent.

and interpret p_i as the probability that a pixel in the image has intensity i, $i = 0 \ldots 255$. (*Hint:* Some intensities may have zero probabilities. The logarithm of zero is undefined, but the product $p \cdot \log p \to 0$ as $p \to 0$.)

5. Modify the IDL program in Listings 8.3 and 8.4 to give a more robust estimation of the variance σ^2 by masking out edge pixels with a thresholded Sobel filter; see Section 5.2.1.

6. Using the modified program from Exercise 5 as a starting point, write an IDL extension to the ENVI environment to perform extended K-means clustering on image data with more than one spectral band.

7. Derive the expression in Equation (8.27) for the net change in the sum of squares cost function when clusters k and j are merged.

8. Recall the definition of fuzzy hypervolume in Section 8.3.3, namely

$$FHV = \sum_{k=1}^{K} \sqrt{|\hat{\boldsymbol{\Sigma}}_k|},$$

where $\hat{\boldsymbol{\Sigma}}_k$ is the estimated covariance matrix for the kth cluster, Equation (8.10). This quantity might itself be used as a cost function, since it measures the compactness of the clusters corresponding to some partitioning C. Suppose that the observations $\boldsymbol{g}(\nu)$ are re-scaled by some arbitrary nonsingular transformation matrix $\boldsymbol{T}$, i.e., $\boldsymbol{g}'(\nu) = \boldsymbol{T}\boldsymbol{g}(\nu)$. Show that any algorithm which achieves partitioning C on the basis of the original observations $\boldsymbol{g}(\nu)$ by minimizing FHV will produce the same partitioning with the transformed observations $\boldsymbol{g}'(\nu)$.

9. (EM algorithm) Consider a set of one-dimensional observations $\{g(\nu) \mid \nu = 1 \ldots m\}$ which are to be clustered into K classes according to a Gaussian mixture model having the form

$$p(g) = \sum_{k=1}^{K} p(g \mid k)\Pr(k),$$

where the $\Pr(k)$ are weighting factors, $\sum_{k=1}^{K} \Pr(k) = 1$, and

$$p(g \mid k) = \frac{1}{\sqrt{2\pi}\sigma_k} \exp\left(-\frac{(g-\mu_k)^2}{2\sigma_k^2}\right).$$

The log-likelihood for the observations is

$$\mathcal{L} = \log \prod_{\nu=1}^{m} p(g(\nu)) = \sum_{\nu=1}^{m} \log\left(\sum_{k=1}^{K} p(g(\nu) \mid k)\Pr(k)\right).$$

(a) Show, with the help of Bayes' Theorem, that this expression is maximized by the following values for the model parameters μ_k, σ_k and $\Pr(k)$:

$$\begin{aligned}
\mu_k &= \frac{\sum_\nu \Pr(k \mid g(\nu))g(\nu)}{\sum_\nu \Pr(k \mid g(\nu))} \\
\sigma_k^2 &= \frac{\sum_\nu \Pr(k \mid g(\nu))(g(\nu) - \mu_k)^2}{\sum_\nu \Pr(k \mid g(\nu))} \\
\Pr(k) &= \frac{1}{m}\sum_\nu \Pr(k \mid g(\nu)).
\end{aligned}$$

(b) The EM algorithm consists of iterating these three equations (in the order given), together with

$$\Pr(k \mid g(\nu)) \propto p(g(\nu) \mid k)\Pr(k), \quad \sum_k \Pr(k \mid g(\nu)) = 1,$$

until convergence. Explain why this is identical to the FMLE algorithm for one-dimensional data.

10. Give an integer arithmetic expression for $d^2(k^*, k)$ for a cubic geometry self-organizing map.

11. The traveling salesperson problem is to find the shortest closed route between n points on a map (cities) which visits each point exactly once. Program an approximate solution to the TSP using a one-dimensional self-organizing map in the form of a closed loop.

12. (a) The mean shift segmentation algorithm of Section 8.7.3 is said to be "edge-preserving". Explain why this is so.

(b) A uniform kernel was used for simplicity to determine the mean shift, the hypersphere S in the algorithm, which we can represent as

$$S(\boldsymbol{w} - \boldsymbol{y}_\nu) = \begin{cases} 1 & \text{for } \|\boldsymbol{w} - \boldsymbol{y}_\nu\| \le r \\ 0 & \text{otherwise} \end{cases}.$$

We could alternatively have used the radial basis (Gaussian) kernel

$$k(\boldsymbol{w} - \boldsymbol{y}_\nu) = \exp\left(-\frac{\|\boldsymbol{w} - \boldsymbol{y}_\nu\|^2}{2r^2}\right).$$

The density estimate at the point $\boldsymbol{w}$ is (see Section 6.4)

$$p(\boldsymbol{w}) = \frac{1}{m(2\pi r^2)^{(N+2)/2}}\sum_{\nu=1}^{m}\exp\left(-\frac{\|\boldsymbol{w} - \boldsymbol{y}_\nu\|^2}{2r^2}\right)$$

and the estimated gradient at $\boldsymbol{w}$, i.e., the direction of maximum change in $p(\boldsymbol{w})$, is

$$\frac{\partial p(\boldsymbol{w})}{\partial \boldsymbol{w}}.$$

Show that the mean shift is always in the direction of the gradient. (This demonstrates that the mean shift algorithm will find the modes of the distribution without actually estimating its density, and can be shown to be a general result (Comaniciu and Meer, 2002).)

(c) The `OpenCV` Python package exports the function `PyrMeanShiftFiltering()`, which performs mean shift filtering (but not segmentation) of RGB color images. In the code snippet:

```
import cv2.cv as cv
    src = cv.fromarray(src)
    dst = cv.CreateMat(rows, cols, cv.CV_8UC3)
    cv.PyrMeanShiftFiltering(src, dst, sp, sr)
```

the variables are:

`src`: the source 8-bit, 3-channel image in a `numpy` array
`dst`: the destination image, same format and same size as the source
`sp`: the spatial window radius
`sr`: the color window radius.

Using this, and the `OpenCV` documentation, as a starting point, write a Python script to perform mean shift segmentation of a three-band multispectral image.

9

Change Detection

To quote a well-known, if now rather dated, review article on change detection (Singh, 1989),

> The basic premise in using remote sensing data for change detection is that changes in land cover must result in changes in radiance values ... [which] must be large with respect to radiance changes from other factors.

When comparing multispectral images of a given scene taken at different times, it is therefore desirable to correct the pixel intensities as much as possible for uninteresting differences such as those due to solar illumination, atmospheric conditions or sensor calibration. In the case of SAR imagery, solar illumination of course plays no role, but the other considerations are similarly important. If comparison is on a pixel-by-pixel basis, then the images must also be co-registered to high accuracy in order to avoid spurious signals resulting from misaligned pixels. Some of the required preprocessing steps were discussed in Chapter 5. Two co-registered satellite images are shown in Figure 9.1.

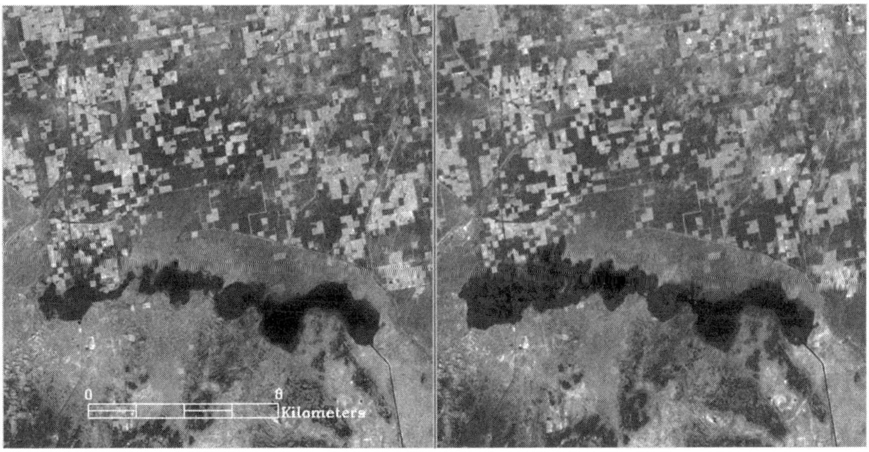

FIGURE 9.1
LANDSAT 5 TM images (band 4) over a water reservoir in India; left image acquired on March 29, 1998, right image on May 16, 1998.

After having performed the necessary preprocessing, it is common to examine various functions of the spectral bands involved (differences, ratios or linear combinations) which in some way bring the change information contained within them to the fore. The shallow flooding at the western edge of the reservoir in Figure 9.1 is evident at a glance. However, other changes have occurred between the two acquisition times and require more image processing to be made visible. In the present chapter we will mention some commonly used techniques for enhancing "change signals" in bitemporal satellite images. Then we will focus our particular attention on the *multivariate alteration detection* (MAD) algorithm of Nielsen et al. (1998) for visible/infrared imagery and on a change statistic for polarimetric SAR data based on the complex Wishart distribution (Conradsen et al., 2003). The chapter concludes with an "inverse" application of change detection, in which *unchanged* pixels are used for automatic relative radiometric normalization of multi-temporal imagery (Canty and Nielsen, 2008). For more recent reviews of change detection in a general context; see Radke et al. (2005) or Coppin et al. (2004).

9.1 Algebraic methods

A simple way to detect changes in two corrected and coregistered multispectral images, represented by N-dimensional random vectors $\boldsymbol{F}$ and $\boldsymbol{G}$, is to subtract them from each other component by component and then examine the N difference images

$$D_i = F_i - G_i, \quad i = 1 \dots N. \tag{9.1}$$

Small intensity differences indicate no change, large positive or negative values indicate change, and decision thresholds can be set to define significance. The thresholds are usually expressed in terms of standard deviations from the mean difference value, which is taken to correspond to no change. If the images are uncorrelated, the variances of the difference images are simply

$$\mathrm{var}(D_i) = \mathrm{var}(G_i) + \mathrm{var}(F_i), \quad i = 1 \dots N,$$

or about twice as noisy as the individual image bands. When the significant difference signatures in the spectral channels are then combined so as to try to characterize the kinds of changes that have taken place, one speaks of *spectral change vector analysis* (Jensen, 2005).

Alternatively, ratios of intensities

$$\frac{F_k}{G_k}, \quad k = 1 \dots N \tag{9.2}$$

are sometimes formed between successive images. Ratios near unity correspond to no change, while small and large values indicate change. A disadvantage of this method when applied to optical/infrared data is that ratios of random variables are not normally distributed even if the random variables themselves are, so that symmetric threshold values defined in terms of standard deviations are not valid. More formally, for data with additive errors like multispectral image pixels, the best statistic for rejecting a no-change hypothesis is a linear combination of observations, not a ratio.

For SAR imagery, the situation is fundamentally different. From Equation (5.27), the variance of the difference of two uncorrelated m-look intensity images is

$$\text{var}(G - F) = \frac{\langle G\rangle^2 + \langle F\rangle^2}{m}.$$

Simple thresholding of the difference image will yield larger errors for a given change in a bright area (large mean intensity) than in a darker area (small mean intensity). Indeed, it turns out that image ratios are a much better choice for detection of changes in multi-look SAR intensity images. This will be illustrated in Section 9.7.1 in the present chapter. Oliver and Quegan (2004) give a thorough discussion in Chapter 12 of their book.

More complicated algebraic combinations, such as differences in vegetation indices or tasseled cap transforms, are also in use. All of these band manipulations can be performed conveniently within the ENVI environment using the `BAND MATH` facility, as well as with the main menu command (ENVI Classic)

```
Basic Tools/Change Detection/Compute Difference Map
```

9.2 Postclassification comparison

If two co-registered satellite images have been classified to yield thematic maps, then the class labels can be compared to determine land cover changes. If classification is carried out at the pixel level (as opposed to using segments or objects), then classification errors (typically $> 5\%$) may corrupt or even dominate the true change signal, depending on the strength of the latter. The ENVI Classic interface, for example, offers the function

```
Basic Tools/Change Detection/Change Detection Statistics
```

for the statistical analysis of postclassification change detection.

9.3 Principal components analysis (PCA)

In a scatterplot of pixel intensities in band i of two multispectral images $\boldsymbol{F}$ and $\boldsymbol{G}$ acquired at different times, each point is a realization of the random

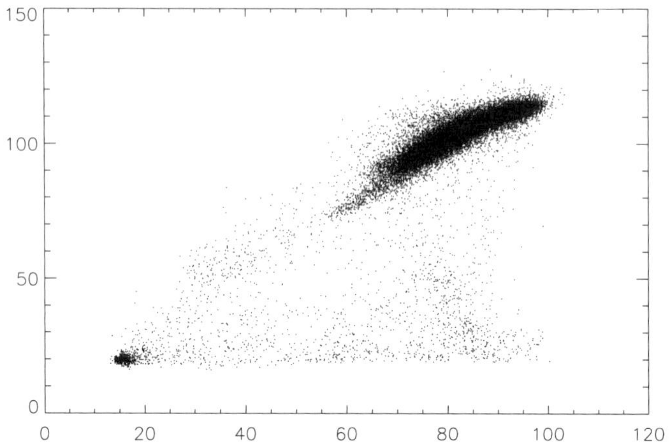

FIGURE 9.2
Scatterplot of the two spectral bands of Figure 9.1.

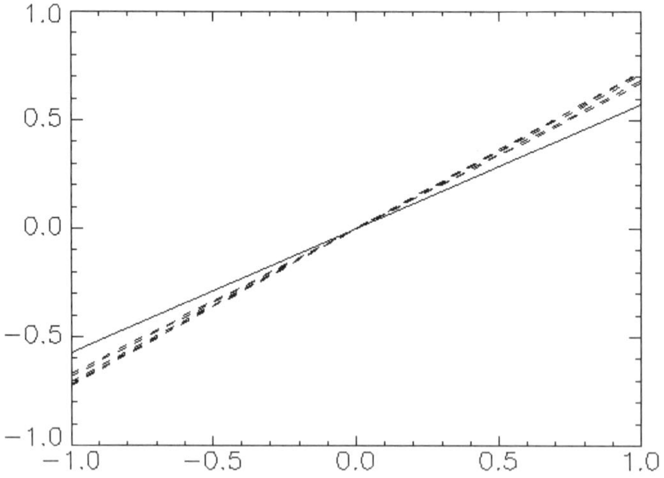

FIGURE 9.3
Iterated PCA. The solid line is the initial principal axis, the dashed lines correspond to five subsequent iterations.

Listing 9.1: Iterated principal components.

```
1  PRO ex9_1
2
3  ; get an image band for T1
4  envi_select, title='Choose⎵multispectral⎵band⎵for⎵T1',$
5              fid=fid1, dims=dims1,pos=pos1, /band_only
6  IF (fid1 EQ -1) THEN RETURN
7
8  ; get an image band for T2
9  envi_select, title='Choose⎵multispectral⎵band⎵for⎵T2',$
10             fid=fid2, dims=dims2,pos=pos2, /band_only
11 IF (fid2 EQ -1) THEN RETURN
12
13 IF (dims1[2]-dims1[1] NE dims2[2]-dims2[1]) OR $
14    (dims1[4]-dims1[3] NE dims2[4]-dims2[3]) THEN RETURN
15 num_cols = dims1[2]-dims1[1]+1
16 num_rows = dims1[4]-dims1[3]+1
17 num_pixels = num_cols*num_rows
18 bitemp = fltarr(2,num_pixels)
19 bitemp[0,*] = envi_get_data(fid=fid1,dims=dims1,pos=0)
20 bitemp[1,*] = envi_get_data(fid=fid2,dims=dims2,pos=0)
21 bitemp[0,*] = bitemp[0,*] - mean(bitemp[0,*])
22 bitemp[1,*] = bitemp[1,*] - mean(bitemp[1,*])
23
24 ; initial principal components transformation
25 C = correlate(bitemp,/covariance)
26 eivs = eigenql(C, eigenvectors=V, /double)
27 PCs = bitemp##transpose(V)
28 window,11,xsize=600,ysize=400, $
29    title='Iterated⎵principal⎵components'
30 PLOT, [-1,1],[-abs(V[0,1]/V[0,0]),abs(V[0,1]/V[0,0])],$
31   xrange=[-1,1],yrange=[-1,1],color=0, $
32         background='FFFFFF'XL
33
34 ; iterated principal components
35 covpm = Obj_New('CPM',2)
36 iter = 0L
37 niter = 5
38 ; begin iteration
39 REPEAT BEGIN
40 ; determine weights via clustering
41    sigma1 = sqrt(eivs[1])
42    U = randomu(seed,num_pixels,2)
43    unfrozen = where( abs(PCs[1,*]) GE sigma1, $
44        complement = frozen)
```

Listing 9.2: Iterated principal components (continued).

```
1      U[frozen,0] = 1.0
2      U[frozen,1] = 0.0
3      FOR j=0,1 DO U[*,j]=U[*,j]/total(U,2)
4      EM, PCs[1,*], U, Ms, Ps, Fs, T0=0, $
5         unfrozen=unfrozen
6   ; weighted covariance matrix
7      covpm->update, bitemp, weights=U[*,0]
8      C = covpm->covariance()
9      eivs = eigenql(C, eigenvectors=V, /double)
10     OPLOT, [-1,1],[-abs(V[0,1]/V[0,0]), $
11         abs(V[0,1]/V[0,0])],linestyle=2, color=0
12     PCs = bitemp##transpose(V)
13     iter++
14  END UNTIL iter GE niter
15
16  ; return result to ENVI
17  out_array = fltarr(num_cols,num_rows,2)
18  FOR i=0,1 DO out_array[*,*,i] = $
19       reform(U[*,i],num_cols,num_rows)
20  envi_enter_data, out_array
21
22  END
```

vector $(F_i, G_i)^\top$. Since unchanged pixels will be highly correlated over time, they will lie in a narrow, elongated cluster along the principal axis, whereas changed pixels will be scattered some distance away from it; see Figure 9.2. The second principal component, which measures intensities at right angles to the first principal axis, will therefore quantify the degree of change associated with a given pixel and may serve as a change image. A change detector based on PCA is implemented in the ENVI SPEAR Tools menu

```
Spectral/SPEAR Tools/Change Detection
```

9.3.1 Iterated PCA

Since the principal axes are determined by diagonalization of the covariance matrix for all of the pixels, the no-change axis may be poorly defined. To overcome this problem, the principal components can be calculated iteratively using weights for each pixel determined by the probability of no change obtained from the preceding iteration.

Listings 9.1 and 9.2 give an IDL routine for performing change detection with iterated PCA. After an initial principal components transformation, lines 25 to 27 in Listing 9.1, the EM routine of Section 8.3 is used to cluster the second principal component into two classes, lines 4 and 5 in Listing 9.2. The probability array `U[*,0]` of membership of the pixels to the central cluster is

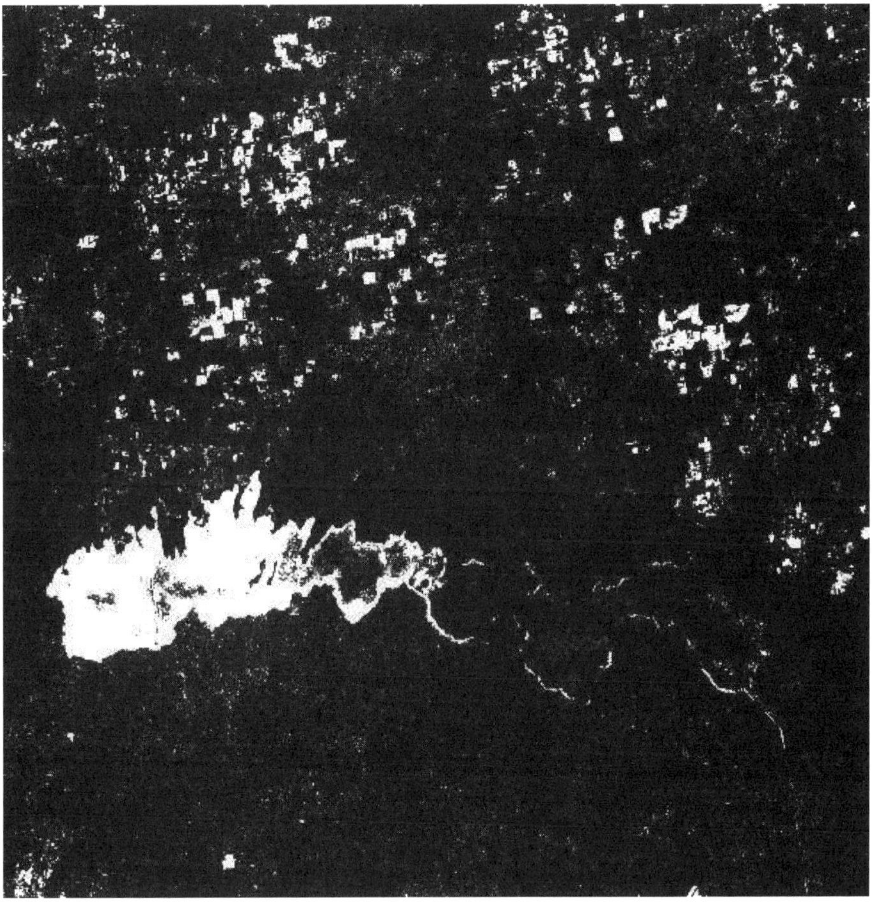

FIGURE 9.4
The change probability for the bitemporal image in Figure 9.1 after five iterations of principal components analysis.

then roughly their probability of no change. The complement array `U[*,1]` contains the change probabilities. The probabilities of no change for observations having second principal component within one standard deviation of the first principal axis are "frozen" to the value 1 (lines 1 and 2 in Listing 9.2) in order to accelerate convergence. Using the no-change probabilities as weights, the covariance matrix is recalculated (lines 7 and 8, Listing 9.2) and the PCA repeated. This procedure is iterated five times in the example. Results are shown in Figures 9.3 and 9.4. In Figure 9.4, the quantity `U[*,1]` is displayed rather than the second principal component, as it better highlights the changes. The method can be generalized to treat all multispectral bands together (Wiemker, 1997).

9.3.2 Kernel PCA

For highly nonlinear data, change detection based on linear transformations such as PCA will we expected to give inferior results. As an illustration (Nielsen and Canty, 2008), consider the bitemporal scene in Figure 9.5. These images were recorded with the airborne DLR 3K-camera system from the German Aerospace Center (DLR) (Kurz et al., 2007), a system consisting of three off-the-shelf cameras arranged on a mount with one camera looking in the nadir direction and two cameras tilted approximately 35° across track. The 600×600 pixel sub-images shown in the figure were acquired 0.7 seconds apart over a busy motorway near Munich, Germany. The images were registered to one another with subpixel accuracy. The only physical changes on the ground are due to the motion of the vehicles.

FIGURE 9.5
Two traffic scenes taken 0.7 seconds apart.

The upper row in Figure 9.6 shows an ordinary PCA (left) and kernel PCA (right). Kernel PCA was discussed in Chapter 4. In the upper left-hand image, the second principal component is displayed, signaling changes as bright and dark pixels, with no-change indicated as middle gray. In the upper right-hand image, the fifth principal component in the kernel-induced nonlinear feature space is displayed and is seen to return the same change information. In the lower row of the figure, the data were artificially "nonlinearized" by raising the intensities in the second image to the third power, while leaving those of the first image unchanged. On the lower left (linear PCA) there are now bright and dark pixels falsely signaling change, whereas on the lower right (kernel PCA), the fifth kernel principal component still indicates changes correctly. Figure 9.7 is a contour plot of the fifth kernel principal component with the sampled training pixels superimposed and marking the no-change axis.

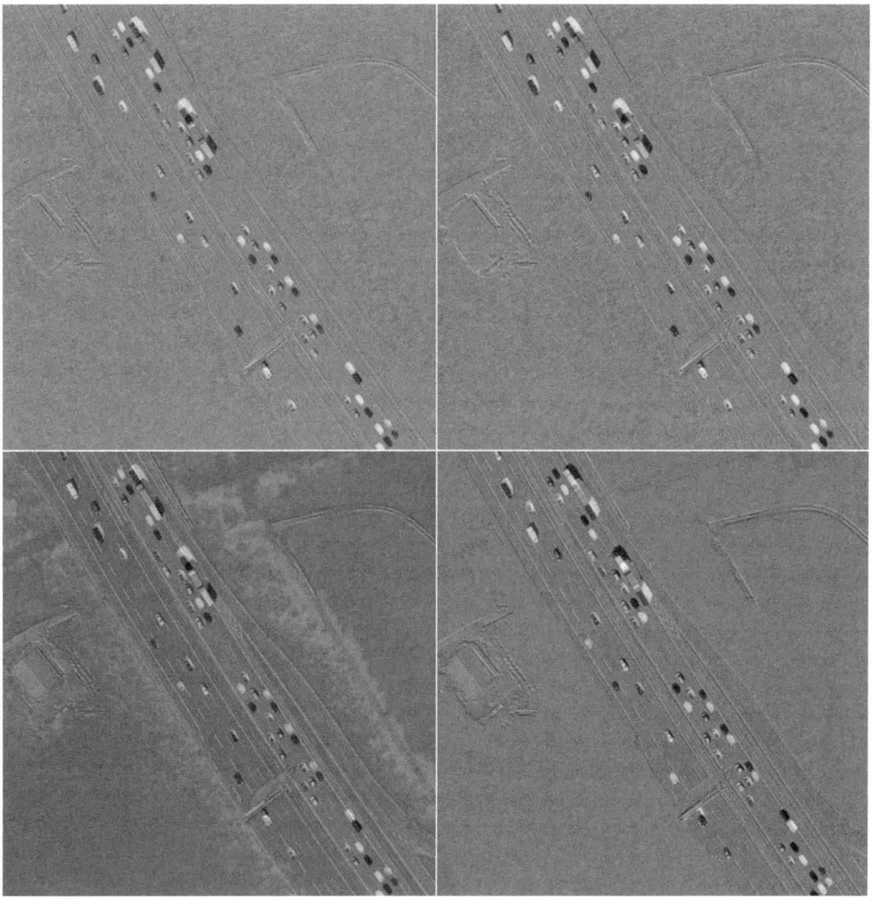

FIGURE 9.6
Change detection with nonlinear PCA. Top row: linear data, bottom row: artificial nonlinear data, left column: second principal component for linear PCA, right column: fifth principal component for kernel PCA.

9.4 Multivariate alteration detection (MAD)

Let us consider two N-band multi- or hyperspectral images of the same scene acquired at different times, between which ground reflectance changes have occurred at some locations but not everywhere. We make a linear combination of the intensities for all N bands in the first image, represented by the random vector $\boldsymbol{G}_1$, thus creating a scalar image characterized by the random variable

$$U = \boldsymbol{a}^\top \boldsymbol{G}_1.$$

FIGURE 9.7
The fifth kernel principal component for the nonlinearized bitemporal traffic scene. The brightness (darkness) indicates the value of the projection onto the fifth principal axis. The dots are the sampled pixels.

The vector of coefficients $\boldsymbol{a}$ is as yet unspecified. We do the same for the second image, represented by $\boldsymbol{G}_2$, forming the linear combination

$$V = \boldsymbol{b}^\top \boldsymbol{G}_2,$$

and then look at the scalar difference image $U - V$. Change information is now contained in a single image. One has, of course, to choose the coefficients $\boldsymbol{a}$ and $\boldsymbol{b}$ in some suitable way. In Nielsen et al. (1998) it is suggested that they be determined by applying standard canonical correlation analysis (CCA), first described by Hotelling (1936), to the two sets of random variables represented by vectors $\boldsymbol{G}_1$ and $\boldsymbol{G}_2$. The resulting linear combinations are then, as we shall see, ordered by similarity (correlation) rather than, as in the original images, by wavelength. This provides a more natural framework in which to look for change. Forming linear combinations has an additional advantage that, due to the Central Limit Theorem (Theorem 2.3), the quantities involved are increasingly well described by the normal distribution.

To anticipate somewhat, in performing CCA on a bitemporal image, one maximizes the correlation ρ between the random variables U and V. The correlation is given by (see Chapter 2)

$$\rho = \frac{\mathrm{cov}(U,V)}{\sqrt{\mathrm{var}(U)}\sqrt{\mathrm{var}(V)}}. \tag{9.3}$$

Arbitrary multiples of U and V would clearly have the same correlation, so a constraint must be chosen. A convenient one is

$$\mathrm{var}(U) = \mathrm{var}(V) = 1. \tag{9.4}$$

Note that, under this constraint, the variance of the difference image is

$$\mathrm{var}(U-V) = \mathrm{var}(U) + \mathrm{var}(V) - 2\mathrm{cov}(U,V) = 2(1-\rho). \tag{9.5}$$

Therefore, the vectors $\boldsymbol{a}$ and $\boldsymbol{b}$ which maximize the correlation, Equation (9.3), under the constraints of Equation (9.4) will in fact minimize the variance of the difference image.

The philosophy of making the images as similar (correlated) as possible before taking their difference is followed in a number of so-called *anomalous change detection* approaches, including the *chronochrome* method of Schaum and Stocker (1997). See Theiler and Matsekh (2009) for a unifying overview.

9.4.1 Canonical correlation analysis (CCA)

Canonical correlation analysis entails a linear transformation of each set of image bands $(G_{1_1} \ldots G_{1_N})$ and $(G_{2_1} \ldots G_{2_N})$ such that, rather than being ordered according to wavelength, the transformed components are ordered according to their mutual correlation; see especially Anderson (2003).

The bitemporal, multispectral image may be represented by the combined random vector $\begin{pmatrix} \boldsymbol{G}_1 \\ \boldsymbol{G}_2 \end{pmatrix}$. This random vector has a $2N \times 2N$ covariance matrix which can be written in block form:

$$\boldsymbol{\Sigma} = \begin{pmatrix} \boldsymbol{\Sigma}_{11} & \boldsymbol{\Sigma}_{12} \\ \boldsymbol{\Sigma}_{12}^\top & \boldsymbol{\Sigma}_{22} \end{pmatrix}.$$

Assuming that the means have been subtracted from the image data, $\boldsymbol{\Sigma}_{11} = \langle \boldsymbol{G}_1\boldsymbol{G}_1^\top \rangle$ is the covariance matrix of the first image, $\boldsymbol{\Sigma}_{22} = \langle \boldsymbol{G}_2\boldsymbol{G}_2^\top \rangle$ that of the second image, and $\boldsymbol{\Sigma}_{12} = \langle \boldsymbol{G}_1\boldsymbol{G}_2^\top \rangle$ is the matrix of covariances between the two. We then have, for the transformed variables U and V,

$$\mathrm{var}(U) = \boldsymbol{a}^\top\boldsymbol{\Sigma}_{11}\boldsymbol{a}, \quad \mathrm{var}(V) = \boldsymbol{b}^\top\boldsymbol{\Sigma}_{22}\boldsymbol{b}, \quad \mathrm{cov}(U,V) = \boldsymbol{a}^\top\boldsymbol{\Sigma}_{12}\boldsymbol{b}.$$

CCA now consists of maximizing the covariance $\boldsymbol{a}^\top \boldsymbol{\Sigma}_{12}\boldsymbol{b}$ under constraints $\boldsymbol{a}^\top \boldsymbol{\Sigma}_{11}\boldsymbol{a} = 1$ and $\boldsymbol{b}^\top \boldsymbol{\Sigma}_{22}\boldsymbol{b} = 1$. If, following the usual procedure, we introduce the Lagrange multipliers $\nu/2$ and $\mu/2$ for each of the two constraints, then the problem becomes one of maximizing the unconstrained Lagrange function

$$L = \boldsymbol{a}^\top \boldsymbol{\Sigma}_{12}\boldsymbol{b} - \frac{\nu}{2}(\boldsymbol{a}^\top \boldsymbol{\Sigma}_{11}\boldsymbol{a} - 1) - \frac{\mu}{2}(\boldsymbol{b}^\top \boldsymbol{\Sigma}_{22}\boldsymbol{b} - 1).$$

Setting derivatives with respect to $\boldsymbol{a}$ and $\boldsymbol{b}$ equal to zero gives

$$\begin{aligned}\frac{\partial L}{\partial \boldsymbol{a}} &= \boldsymbol{\Sigma}_{12}\boldsymbol{b} - \nu\boldsymbol{\Sigma}_{11}\boldsymbol{a} = \boldsymbol{0}\\ \frac{\partial L}{\partial \boldsymbol{b}} &= \boldsymbol{\Sigma}_{12}^\top\boldsymbol{a} - \mu\boldsymbol{\Sigma}_{22}\boldsymbol{b} = \boldsymbol{0}.\end{aligned} \tag{9.6}$$

Multiplying the first of the above equations from the left with the vector $\boldsymbol{a}^\top$, the second with $\boldsymbol{b}^\top$, and using the constraints leads immediately to

$$\nu = \mu = \boldsymbol{a}^\top \boldsymbol{\Sigma}_{12}\boldsymbol{b} = \rho.$$

Therefore we can write Equations (9.6) in the form

$$\begin{aligned}\boldsymbol{\Sigma}_{12}\boldsymbol{b} - \rho\boldsymbol{\Sigma}_{11}\boldsymbol{a} &= \boldsymbol{0}\\ \boldsymbol{\Sigma}_{12}^\top\boldsymbol{a} - \rho\boldsymbol{\Sigma}_{22}\boldsymbol{b} &= \boldsymbol{0}.\end{aligned} \tag{9.7}$$

Next, multiply the first of Equations (9.7) by ρ and the second from the left by $\boldsymbol{\Sigma}_{22}^{-1}$. This gives

$$\rho\boldsymbol{\Sigma}_{12}\boldsymbol{b} = \rho^2\boldsymbol{\Sigma}_{11}\boldsymbol{a} \tag{9.8}$$

and

$$\boldsymbol{\Sigma}_{22}^{-1}\boldsymbol{\Sigma}_{12}^\top\boldsymbol{a} = \rho\boldsymbol{b}. \tag{9.9}$$

Finally, combining Equations (9.8) and (9.9), we obtain the following equation for the transformation coefficient $\boldsymbol{a}$,

$$\boldsymbol{\Sigma}_{12}\boldsymbol{\Sigma}_{22}^{-1}\boldsymbol{\Sigma}_{12}^\top\boldsymbol{a} = \rho^2\boldsymbol{\Sigma}_{11}\boldsymbol{a}. \tag{9.10}$$

A similar argument (Exercise 3) leads to the corresponding equation for $\boldsymbol{b}$, namely

$$\boldsymbol{\Sigma}_{12}^\top\boldsymbol{\Sigma}_{11}^{-1}\boldsymbol{\Sigma}_{12}\boldsymbol{b} = \rho^2\boldsymbol{\Sigma}_{22}\boldsymbol{b}. \tag{9.11}$$

Solution of these equations will indeed lead to a maximum in the correlation ρ since, from Equations (9.6), we have

$$\frac{\partial^2 L}{\partial \boldsymbol{a}^\top \partial \boldsymbol{a}} = -\rho\boldsymbol{\Sigma}_{11} \qquad \frac{\partial^2 L}{\partial \boldsymbol{b}^\top \partial \boldsymbol{b}} = -\rho\boldsymbol{\Sigma}_{22}$$

so that the Hessian of L is negative definite.

Equations (9.10) and (9.11) are generalized eigenvalue problems similar to those that we already met for the minimum noise fraction (MNF) and maximum autocorrelation factor (MAF) transformations of Chapter 3; see Equations (3.55) and (3.72). Note, however, that they are coupled via the eigenvalue ρ^2. The desired projections $U = \boldsymbol{a}^\top \boldsymbol{G}_1$ are given by the eigenvectors $\boldsymbol{a}_1 \dots \boldsymbol{a}_N$ of Equation (9.10) corresponding to eigenvalues $\rho_1^2 \geq \rho_2^2 \geq \dots \geq \rho_N^2$. Similarly, the desired projections $V = \boldsymbol{b}^\top \boldsymbol{G}_2$ are given by the eigenvectors $\boldsymbol{b}_1 \dots \boldsymbol{b}_N$ of Equation (9.11) corresponding to the *same* eigenvalues.

Solution of the eigenvalue problems generates new multispectral images $\boldsymbol{U} = (U_1 \dots U_N)^\top$ and $\boldsymbol{V} = (V_1 \dots V_N)^\top$, the components of which are called the *canonical variates* (CVs). The CVs are ordered by similarity (correlation) rather than, as in the original images, by wavelength. The canonical correlations $\rho_i = \text{corr}(U_i, V_i)$, $i = 1 \dots N$, are the square roots of the eigenvalues of the coupled eigenvalue problem. The pair (U_1, V_1) is maximally correlated, the pair (U_2, V_2) is maximally correlated subject to being orthogonal to (uncorrelated with) both U_1 and V_1, and so on. Taking paired differences (in reverse order) then generates a sequence of transformed difference images

$$M_i = U_{N-i+1} - V_{N-i+1}, \quad i = 1 \dots N, \tag{9.12}$$

referred to as the *multivariate alteration detection* (MAD) *variates* (Nielsen et al., 1998). Since we are dealing with change detection, we want the pairs of canonical variates U_i and V_i to be positively correlated, just like the original image bands. This is easily achieved by appropriate choice of the relative signs of the eigenvector pairs $\boldsymbol{a}_i, \boldsymbol{b}_i$ so as to ensure that $\boldsymbol{a}_i^\top \boldsymbol{\Sigma}_{12} \boldsymbol{b}_i > 0$, $i = 1 \dots N$.

An ENVI/IDL extension `MAD_RUN.PRO` and Python script `imad.py` for multivariate alteration detection are described in Appendices C and D. Figure 9.8 gives an example of an application to the bitemporal data of Figure 9.1. Only the last (least correlated) and first (most correlated) CVs are shown. Corresponding scatterplots are displayed in Figure 9.9.

9.4.2 Orthogonality properties

Equations (9.10) and (9.11) are of the form

$$\boldsymbol{\Sigma}_1 \boldsymbol{a} = \rho^2 \boldsymbol{\Sigma} \boldsymbol{a}, \tag{9.13}$$

where both $\boldsymbol{\Sigma}_1$ and $\boldsymbol{\Sigma}$ are symmetric and $\boldsymbol{\Sigma}$ is positive definite. Equation (9.13) can be solved in the same way as for the MNF transformation in Chapter 3. We repeat the procedure here for convenience. First, write Equation (9.13) in the form

$$\boldsymbol{\Sigma}_1 \boldsymbol{a} = \rho^2 \boldsymbol{L} \boldsymbol{L}^\top \boldsymbol{a},$$

where $\boldsymbol{\Sigma}$ has been replaced by its Cholesky decomposition $\boldsymbol{L}\boldsymbol{L}^\top$. The matrix $\boldsymbol{L}$ is positive definite, lower triangular. Equivalently,

$$\boldsymbol{L}^{-1} \boldsymbol{\Sigma}_1 (\boldsymbol{L}^\top)^{-1} \boldsymbol{L}^\top \boldsymbol{a} = \rho^2 \boldsymbol{L}^\top \boldsymbol{a}$$

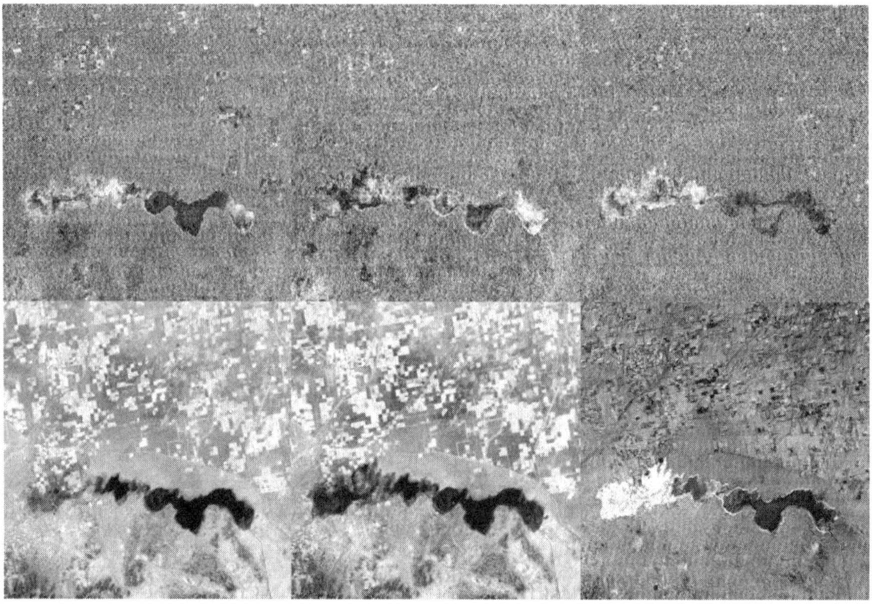

FIGURE 9.8
MAD change detection using the six nonthermal bands of the bitemporal scene of Figure 9.1. The top row shows, left to right, the canonical variates U_6 and V_6 and their difference M_1. The bottom row shows U_1, V_1 and M_6. All images are displayed with a 2% saturated linear histogram stretch.

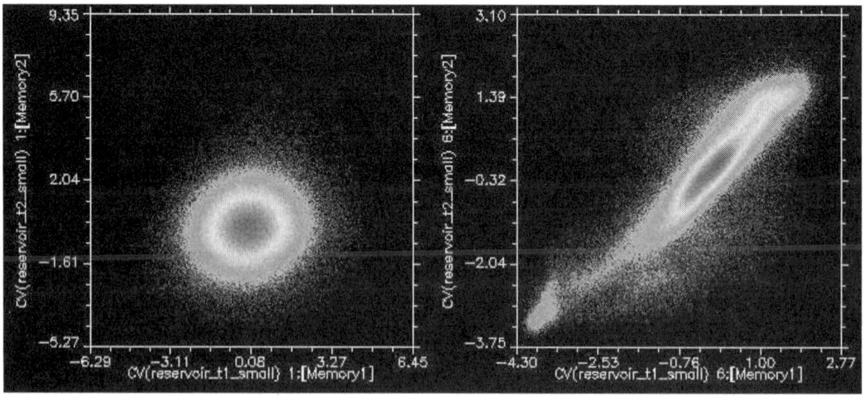

FIGURE 9.9
Scatterplots of V_6 vs. U_6 (left) and of V_1 vs. U_1 (right) from Figure 9.8. The last two canonical variates are virtually uncorrelated, the first two are maximally correlated.

or, with $\boldsymbol{d} = \boldsymbol{L}^\top \boldsymbol{a}$ and the commutativity of inverse and transpose,

$$[\boldsymbol{L}^{-1}\boldsymbol{\Sigma}_1(\boldsymbol{L}^{-1})^\top]\boldsymbol{d} = \rho^2 \boldsymbol{d},$$

a standard eigenvalue problem for the symmetric matrix $\boldsymbol{L}^{-1}\boldsymbol{\Sigma}_1(\boldsymbol{L}^{-1})^\top$. Let its eigenvectors be $\boldsymbol{d}_i$. Since they are orthogonal and normalized, we have

$$\delta_{ij} = \boldsymbol{d}_i^\top \boldsymbol{d}_j = \boldsymbol{a}_i^\top \boldsymbol{L}\boldsymbol{L}^\top \boldsymbol{a}_j = \boldsymbol{a}_i^\top \boldsymbol{\Sigma} \boldsymbol{a}_j. \tag{9.14}$$

From Equation (9.14), taking $\boldsymbol{\Sigma} = \boldsymbol{\Sigma}_{11}$ and then $\boldsymbol{\Sigma} = \boldsymbol{\Sigma}_{22}$, it follows that

$$\begin{aligned} \text{cov}(U_i, U_j) &= \boldsymbol{a}_i^\top \boldsymbol{\Sigma}_{11} \boldsymbol{a}_j = \delta_{ij} \\ \text{cov}(V_i, V_j) &= \boldsymbol{b}_i^\top \boldsymbol{\Sigma}_{22} \boldsymbol{b}_j = \delta_{ij}. \end{aligned} \tag{9.15}$$

Furthermore, according to Equation (9.9),

$$\boldsymbol{b}_i = \frac{1}{\rho_i}\boldsymbol{\Sigma}_{22}^{-1}\boldsymbol{\Sigma}_{21}\boldsymbol{a}_i.$$

Therefore we have, with Equation (9.10),

$$\text{cov}(U_i, V_j) = \boldsymbol{a}_i^\top \boldsymbol{\Sigma}_{12}\boldsymbol{b}_j = \boldsymbol{a}_i^\top \frac{1}{\rho_j}\boldsymbol{\Sigma}_{12}\boldsymbol{\Sigma}_{22}^{-1}\boldsymbol{\Sigma}_{21}\boldsymbol{a}_j = \rho_j\ \boldsymbol{a}_i^\top \boldsymbol{\Sigma}_{11}\boldsymbol{a}_j = \rho_j\ \delta_{ij}. \tag{9.16}$$

Thus we see that the canonical variates are *all mutually uncorrelated* except for the pairs (U_i, V_i), and these are ordered by decreasing correlation. The MAD variates themselves are consequently also mutually uncorrelated, their covariances being given by (recalling the reversed order in Equation (9.12))

$$\text{cov}(M_i, M_j) = \text{cov}(U_{N-i+1} - V_{N-i+1}, U_{N-j+1} - V_{N-j+1}) = 0,\ i \neq j = 1 \ldots N, \tag{9.17}$$

and their variances by

$$\sigma_{M_i}^2 = \text{var}(U_{N-i+1} - V_{N-i+1}) = 2(1 - \rho_{N-i+1}), \quad i = 1 \ldots N. \tag{9.18}$$

The first MAD variate has maximum variance in its pixel intensities. The second MAD variate has maximum spread subject to the condition that its pixel intensities are statistically uncorrelated with those in the first variate; the third has maximum spread subject to being uncorrelated with the first two, and so on. Depending on the type of change present, any of the components may exhibit significant change information. Interesting small-scale anthropogenic changes, for instance, will generally be unrelated to dominating seasonal vegetation changes or stochastic image noise, so it is quite common that such changes will be concentrated in higher-order MAD variates. This is apparently the situation in Figure 9.8, where the flooding shows up more strongly in M_6 than in M_1. In fact, one of the nicest aspects of the method is that it sorts different categories of change into different, uncorrelated image components.

9.4.3 Scale invariance

An additional advantage of the MAD procedure stems from the fact that the calculations involved are invariant under linear and affine transformations of the original image intensities. This implies insensitivity to linear differences in atmospheric conditions or sensor calibrations at the two acquisition times. Scale invariance also offers the possibility of using the MAD transformation to perform relative radiometric normalization of multi-temporal imagery in a fully automatic way, as will be explained in Section 9.8 below.

We can illustrate the invariance as follows. Suppose the second image $\boldsymbol{G}_2$ is transformed according to some linear transformation

$$\boldsymbol{H} = \boldsymbol{T}\boldsymbol{G}_2,$$

where $\boldsymbol{T}$ is a constant non-singular matrix.* The relevant covariance matrices for the coupled generalized eigenvalue problems, Equations (9.10) and (9.11), are then

$$\begin{aligned} \boldsymbol{\Sigma}_{1h} &= \langle \boldsymbol{G}_1 \boldsymbol{H}^\top \rangle = \boldsymbol{\Sigma}_{12}\boldsymbol{T}^\top \\ \boldsymbol{\Sigma}_{11} &\quad \text{unchanged} \\ \boldsymbol{\Sigma}_{hh} &= \langle \boldsymbol{H}\boldsymbol{H}^\top \rangle = \boldsymbol{T}\boldsymbol{\Sigma}_{22}\boldsymbol{T}^\top. \end{aligned}$$

The coupled eigenvalue problems now read

$$\begin{aligned} \boldsymbol{\Sigma}_{12}\boldsymbol{T}^\top(\boldsymbol{T}\boldsymbol{\Sigma}_{22}\boldsymbol{T}^\top)^{-1}\boldsymbol{T}\boldsymbol{\Sigma}_{12}^\top \boldsymbol{a} &= \rho^2 \boldsymbol{\Sigma}_{11}\boldsymbol{a} \\ \boldsymbol{T}\boldsymbol{\Sigma}_{12}^\top \boldsymbol{\Sigma}_{11}^{-1}\boldsymbol{\Sigma}_{12}\boldsymbol{T}^\top \boldsymbol{c} &= \rho^2 \boldsymbol{T}\boldsymbol{\Sigma}_{22}\boldsymbol{T}^\top \boldsymbol{c}, \end{aligned}$$

where $\boldsymbol{c}$ is the desired projection for the transformed image $\boldsymbol{H}$. These equations are easily seen to be equivalent to

$$\begin{aligned} \boldsymbol{\Sigma}_{12}\boldsymbol{\Sigma}_{22}^{-1}\boldsymbol{\Sigma}_{12}^\top \boldsymbol{a} &= \rho^2 \boldsymbol{\Sigma}_{11}\boldsymbol{a} \\ \boldsymbol{\Sigma}_{12}^\top \boldsymbol{\Sigma}_{11}^{-1}\boldsymbol{\Sigma}_{12}(\boldsymbol{T}^\top \boldsymbol{c}) &= \rho^2 \boldsymbol{\Sigma}_{22}(\boldsymbol{T}^\top \boldsymbol{c}), \end{aligned}$$

which are identical to Equations (9.10) and (9.11) with $\boldsymbol{b}$ replaced by $\boldsymbol{T}^\top \boldsymbol{c}$. Therefore, the MAD components in the transformed situation are, as before,

$$\boldsymbol{a}_i^\top \boldsymbol{G}_1 - \boldsymbol{c}_i^\top \boldsymbol{H} = \boldsymbol{a}_i^\top \boldsymbol{G}_1 - \boldsymbol{c}_i^\top \boldsymbol{T}\boldsymbol{G}_2 = \boldsymbol{a}_i^\top \boldsymbol{G}_1 - (\boldsymbol{T}^\top \boldsymbol{c}_i)^\top \boldsymbol{G}_2 = \boldsymbol{a}_i^\top \boldsymbol{G}_1 - \boldsymbol{b}_i^\top \boldsymbol{G}_2.$$

9.4.4 Iteratively re-weighted MAD

Let us now imagine two images of the same scene, acquired at different times under similar conditions, but for which no ground reflectance changes have occurred whatsoever. Then the only differences between them will be due to random effects like instrument noise and atmospheric fluctuation. In such a

*A translation $\boldsymbol{H} = \boldsymbol{G}_2 + \boldsymbol{C}$, where $\boldsymbol{C}$ is a constant vector, will clearly make no difference, since the data matrices are always centered prior to the transformation.

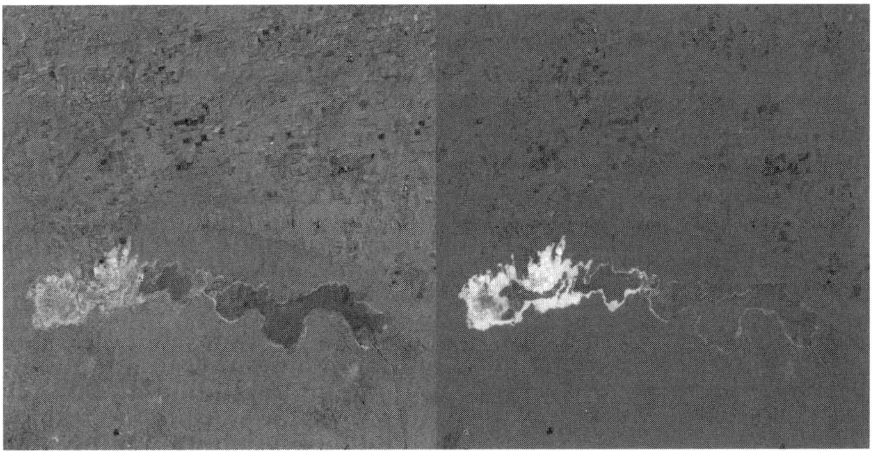

FIGURE 9.10
The sixth MAD variate for the bitemporal image in Figure 9.1, shown in an unsaturated linear histogram stretch, left without iteration, right after iteration to convergence. Bright and dark pixels indicate change.

case we would expect that the histogram of any difference component that we generate will be very nearly Gaussian. In particular, the MAD variates, being uncorrelated, should follow a multivariate, zero mean normal distribution with diagonal covariance matrix. Change observations would deviate more or less strongly from such a distribution. Just as for the iterated principal components method of Section 9.3.1, we might therefore expect an improvement in the sensitivity of the MAD transformation if we can establish an increasingly better background of no change against which to detect change (Nielsen, 2007). This can again be done in an iteration scheme in which, when calculating the means and covariance matrices for the next iteration of the MAD transformation, observations are weighted by the probability of no change determined on the preceding iteration.*

One way to determine the no-change probabilities is to perform an unsupervised classification of the MAD variates using, for example, the Gaussian mixture algorithm of Chapter 8 with K *a priori* clusters, one for no change and $K-1$ clusters for different categories of change. The pixel class membership probabilities for no change then provide the necessary weights for iteration. This method, being unsupervised, will not distinguish which clusters correspond to change and which to no change, but the no-change cluster can be identified by its compactness, e.g., by its partition density; see Section 8.3.3 and also Figure 9.17 later in the present chapter. This strategy has, however,

*This was in fact the main motivation for allowing for weights in the provisional means algorithm described in Chapter 2.

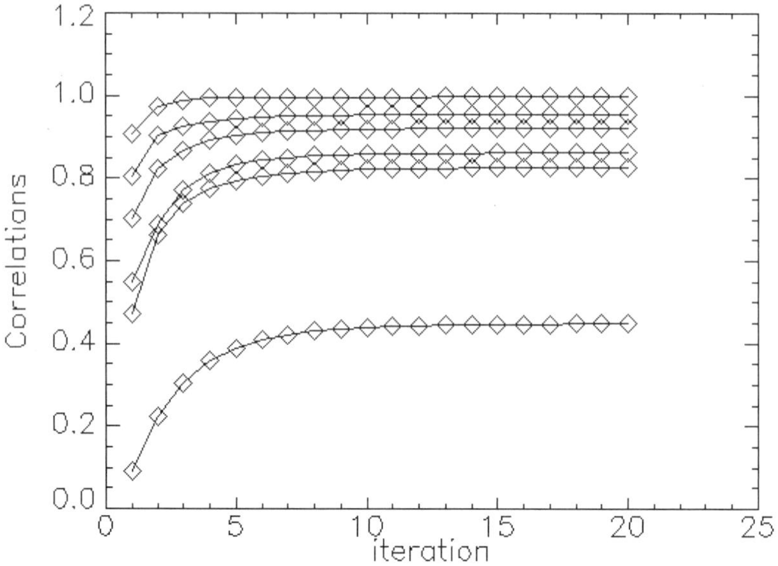

FIGURE 9.11
The canonical correlations ρ_i, $i = 1 \ldots 6$, under 20 iterations of the MAD transformation of the bitemporal image in Figure 9.1.

the disadvantage that a computationally expensive clustering procedure must be repeated for every iteration. Moreover, one has to make a more or less arbitrary decision as to how many no-change clusters are present and tolerate the unpredictability of clustering results generally.

An alternative scheme is to examine the MAD variates directly. Let the random variable Z represent the sum of the squares of the standardized MAD variates,

$$Z = \sum_{i=1}^{N} \left(\frac{M_i}{\sigma_{M_i}} \right)^2, \tag{9.19}$$

where σ_{M_i} is given by Equation (9.18). Then, since the no-change observations are expected to be normally distributed and uncorrelated, the random variable Z corresponding to such observations should be chi-square distributed with N degrees of freedom (Theorem 2.5). For each iteration, the observations can thus be given weights determined by the chi-square distribution, namely

$$\Pr(\text{no change}) = 1 - P_{\chi^2;N}(z); \tag{9.20}$$

see Equation (2.37). Here Pr(no change) is the probability that a sample z drawn from the chi-square distribution could be that large or larger. A small z implies a correspondingly large probability. Iteration of the MAD transformation continues until some stopping criterion is met, such as lack of significant change in the canonical correlations ρ_i, $i = 1 \dots N$. Nielsen (2007) refers to this procedure as the *iteratively reweighted* MAD (IR-MAD or iMAD) algorithm.

The ENVI/IDL extension `MAD_RUN` and the Python script `imad.py` described in Appendices C and D include the possibility of iteration of the MAD transformation on the basis of Equation (9.20). Figure 9.10 indicates the kind of improvement that can be achieved through iteration. The iterated canonical correlations are shown in Figure 9.11.

9.4.5 Correlation with the original observations

As in the case of principal components analysis, the linear transformations which comprise multivariate alteration detection mix the spectral bands of the images, making physical interpretation of observed changes more difficult. Interpretation can be facilitated by examining the correlation of the MAD variates with the original spectral bands (Nielsen et al., 1998). In a more compact matrix notation for the transformation coefficients, define

$$\boldsymbol{A} = (\boldsymbol{a}_i \dots \boldsymbol{a}_N), \quad \boldsymbol{B} = (\boldsymbol{b}_i \dots \boldsymbol{b}_N),$$

so that $\boldsymbol{A}$ is an $N \times N$ matrix whose columns are the eigenvectors $\boldsymbol{a}_i$ of the MAD transformation, and similarly for $\boldsymbol{B}$. Furthermore, let

$$\boldsymbol{M} = (M_N \dots M_1)^\top$$

be the random vector of MAD variates (in the same ordering as the CVs). Then the covariances of the MAD variates with the original bands are given by

$$\begin{aligned} \langle \boldsymbol{G}_1 \boldsymbol{M}^\top \rangle &= \langle \boldsymbol{G}_1 (\boldsymbol{A}^\top \boldsymbol{G}_1 - \boldsymbol{B}^\top \boldsymbol{G}_2)^\top \rangle = \boldsymbol{\Sigma}_{11} \boldsymbol{A} - \boldsymbol{\Sigma}_{12} \boldsymbol{B} \\ \langle \boldsymbol{G}_2 \boldsymbol{M}^\top \rangle &= \langle \boldsymbol{G}_2 (\boldsymbol{A}^\top \boldsymbol{G}_1 - \boldsymbol{B}^\top \boldsymbol{G}_2)^\top \rangle = \boldsymbol{\Sigma}_{12}^\top \boldsymbol{A} - \boldsymbol{\Sigma}_{22} \boldsymbol{B}, \end{aligned} \tag{9.21}$$

from which the correlations can be calculated by dividing by the square roots of the variances, Equation (2.21). Figure 9.12 shows the correlation of the fourth MAD variate with the spectral bands of the first acquisition of Figure 9.1. The strong positive correlation with spectral band 4 indicates that the change signal is due primarily to changes in vegetation. This is confirmed in Figure 9.13, showing the fourth IR-MAD component in which significant changes in the agricultural fields to the north of the reservoir are evident.

9.4.6 Regularization

To avoid possible near singularity problems in the solution of Equations (9.10) and (9.11), it is useful to introduce some form of regularization into the MAD

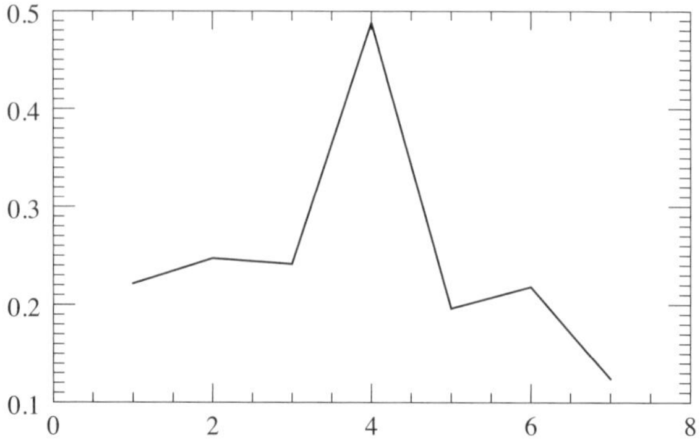

FIGURE 9.12
Correlation of the fourth iteratively reweighted MAD (IR-MAD) component with the seven spectral bands for the scene in Figure 9.1 acquired on March 29, 1998.

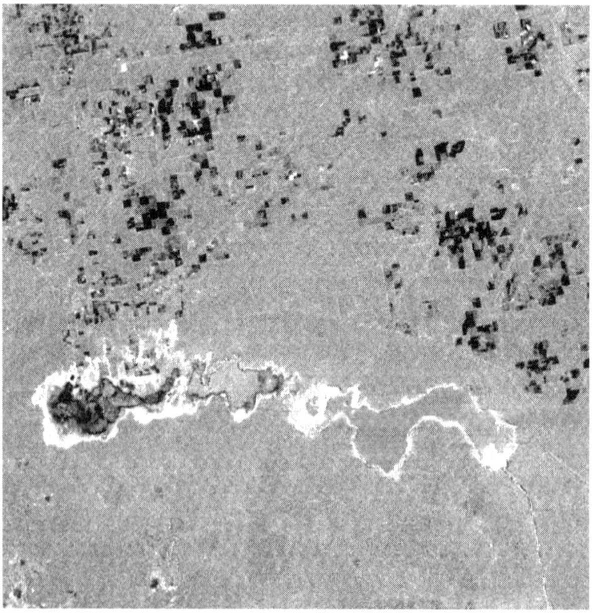

FIGURE 9.13
The fourth IR-MAD component, 2% saturated linear histogram stretch.

Listing 9.3: Excerpt from the Python script `imad.py`.

```
#      weighted covariance matrices and means
         S = cpm.covariance()
         means = cpm.means()
#      reset prov means object
         cpm.__init__(2*bands)
         s11 = S[0:bands,0:bands]
         s11 = (1-lam)*s11 + lam*np.eye(bands)
         s22 = S[bands:,bands:]
         s22 = (1-lam)*s22 + lam*np.eye(bands)
         s12 = S[0:bands,bands:]
         s21 = S[bands:,0:bands]
         c1 = s12*linalg.inv(s22)*s21
         b1 = s11
         c2 = s21*linalg.inv(s11)*s12
         b2 = s22
#      solution of generalized eigenproblems
         IF bands>1:
             mu2a,A = auxil.geneiv(c1,b1)
             mu2b,B = auxil.geneiv(c2,b2)
#           sort a
             idx = np.argsort(mu2a)
             A = A[:,idx]
#           sort b
             idx = np.argsort(mu2b)
             B = B[:,idx]
             mu2 = mu2b[idx]
         ELSE:
             mu2 = c1/b1
             A = 1/np.sqrt(b1)
             B = 1/np.sqrt(b2)
#       canonical correlations
         mu = np.sqrt(mu2)
         a2 = np.diag(A.T*A)
         b2 = np.diag(B.T*B)
         sigma = np.sqrt( (2-lam*(a2+b2))/(1-lam)-2*mu )
         rho=mu*(1-lam)/np.sqrt( (1-lam*a2)*(1-lam*b2) )
```

algorithm (Nielsen, 2007). This may in particular be necessary when the number of spectral bands N is large, as is the case for change detection involving hyperspectral imagery. The concept of regularization by means of length penalization was met in Section 2.6.4. Here we allow for length penalization of the eigenvectors $\boldsymbol{a}$ and $\boldsymbol{b}$ in CCA by replacing the constraint (9.4) by

$$(1-\lambda)\mathrm{var}(U) + \lambda\|\boldsymbol{a}\|^2 = (1-\lambda)\mathrm{var}(V) + \lambda\|\boldsymbol{b}\|^2 = 1, \tag{9.22}$$

where $\mathrm{var}(U) = \boldsymbol{a}^\top\boldsymbol{\Sigma}_{11}\boldsymbol{a}$, $\mathrm{var}(V) = \boldsymbol{b}^\top\boldsymbol{\Sigma}_{22}\boldsymbol{b}$ and λ is a regularization parameter. To maximize the covariance $\mathrm{cov}(U, V) = \boldsymbol{a}^\top\boldsymbol{\Sigma}_{12}\boldsymbol{b}$ under this constraint, we now maximize the unconstrained Lagrange function

$$L = \boldsymbol{a}^\top\boldsymbol{\Sigma}_{12}\boldsymbol{b} - \frac{\nu}{2}((1-\lambda)\boldsymbol{a}^\top\boldsymbol{\Sigma}_{11}\boldsymbol{a} + \lambda\|\boldsymbol{a}\|^2 - 1) - \frac{\mu}{2}((1-\lambda)\boldsymbol{b}^\top\boldsymbol{\Sigma}_{22}\boldsymbol{b} + \lambda\|\boldsymbol{b}\|^2 - 1).$$

Setting the derivatives equal to zero, we obtain

$$\begin{aligned} \frac{\partial L}{\partial \boldsymbol{a}} &= \boldsymbol{\Sigma}_{12}\boldsymbol{b} - \nu(1-\lambda)\boldsymbol{\Sigma}_{11}\boldsymbol{a} - \nu\lambda\boldsymbol{a} = 0 \\ \frac{\partial L}{\partial \boldsymbol{b}} &= \boldsymbol{\Sigma}_{12}^\top\boldsymbol{a} - \mu(1-\lambda)\boldsymbol{\Sigma}_{22}\boldsymbol{b} - \mu\lambda\boldsymbol{b} = 0. \end{aligned} \tag{9.23}$$

Multiplying the first equation above from the left with $\boldsymbol{a}^\top$ gives

$$\boldsymbol{a}^\top\boldsymbol{\Sigma}_{12}\boldsymbol{b} - \nu((1-\lambda)\boldsymbol{a}^\top\boldsymbol{\Sigma}_{11}\boldsymbol{a} + \lambda\|\boldsymbol{a}\|^2) = 0$$

and, from (9.22), $\nu = \boldsymbol{a}^\top\boldsymbol{\Sigma}_{12}\boldsymbol{b}$. Similarly, multiplying the second equation from the left with $\boldsymbol{b}^\top$, we have $\mu = \boldsymbol{b}^\top\boldsymbol{\Sigma}_{12}^\top\boldsymbol{a} = \nu$. Equation (9.23) can now be written as the single generalized eigenvalue problem ($\boldsymbol{I}$ is the $N \times N$ identity matrix)

$$\begin{pmatrix} \boldsymbol{0} & \boldsymbol{\Sigma}_{12} \\ \boldsymbol{\Sigma}_{21} & \boldsymbol{0} \end{pmatrix} \begin{pmatrix} \boldsymbol{a} \\ \boldsymbol{b} \end{pmatrix} = \mu \begin{pmatrix} (1-\lambda)\boldsymbol{\Sigma}_{11} + \lambda\boldsymbol{I} & \boldsymbol{0} \\ \boldsymbol{0} & (1-\lambda)\boldsymbol{\Sigma}_{22} + \lambda\boldsymbol{I} \end{pmatrix} \begin{pmatrix} \boldsymbol{a} \\ \boldsymbol{b} \end{pmatrix}. \tag{9.24}$$

Note that this equation is equivalent to Equations (9.8) and (9.9) for $\lambda = 0$. Note also that the eigenvalue μ is the covariance of U and V and not the correlation. Thus, with Equation (9.22),

$$\sigma^2_{MAD} = \text{var}(U) + \text{var}(V) - 2\text{cov}(U, V) = \frac{2 - \lambda(\|\boldsymbol{a}\|^2 + \|b\|^2)}{1-\lambda} - 2\mu \tag{9.25}$$

and the correlation is

$$\rho = \frac{\mu}{\sqrt{\text{var}(U)\text{var}(V)}} = \frac{\mu(1-\lambda)}{\sqrt{(1-\lambda\|\boldsymbol{a}\|^2)(1-\lambda\|\boldsymbol{b}\|^2)}}, \tag{9.26}$$

which reduces to $\rho = \mu$ only when $\lambda = 0$. Regularization is included as an option in the ENVI/IDL and Python IR-MAD scripts.

Canonical correlation with regularization is illustrated in the Python script excerpt of Listing 9.3. After collecting image statistics with the weighted provisional means method (Section 2.3.2), the covariance matrix and mean vector of the bitemporal image are extracted from the `cpm` class instance in lines 2 and 3. The (regularized) covariance matrices required for CCA are then constructed (lines 6 to 11) and the generalized eigenvalue problems, Equations (9.10) and (9.11), are solved in lines 18 and 19, or for single band images, in lines 28 to 30. Then, in lines 32 to 36, the MAD standard deviations and canonical correlations are determined as given, respectively, by Equations (9.25) and (9.26).

9.4.7 Postprocessing

The MAD transformation can be augmented by subsequent application of the MAF transformation in order to improve the spatial coherence of the MAD variates (Nielsen et al., 1998). When image noise is estimated as the difference between intensities of neighboring pixels, the MAF transformation is equivalent to the MNF transformation, as was discussed in Chapter 3. MAD components postprocessed in this way will be referred to as MAD/MNF variates in the remainder of this chapter.

9.5 Decision thresholds

Since the MAD (or MAD/MNF) variates are approximately normally distributed about zero and uncorrelated, thresholds for deciding between change and no change can be set in terms of standard deviations about the mean for each variate separately, as is done, for example, in spectral change vector analysis (Section 9.1). Thresholds may be chosen in an *ad hoc* fashion, for example by saying that all pixels in a MAD component M_i whose intensities are within $\pm 2\sigma_{M_i}$ of zero, where σ_{M_i} is given by Equation (9.18), are no-change pixels.

We can do better than this, however. Let us consider a simple, three-Gaussian mixture model for a random variable M representing one of the MAD components. The probability density of M is modeled as

$$p(m) = p(m \mid NC)\Pr(NC) + p(m \mid C-)\Pr(C-) + p(m \mid C+)\Pr(C+), \quad (9.27)$$

where $C+$, $C-$ and NC denote positive change, negative change and no change, respectively, and where $p(m \mid \cdot\ )$ is a normal density function. The set $S = \{m(\nu)\}$ of realizations of M may be partitioned into four disjoint subsets:

$$S_{NC}, \quad S_{C-}, \quad S_{C+}, \quad S_U = S \backslash (S_{NC} \cup S_{C-} \cup S_{C+}),$$

with S_U denoting a set of ambiguous pixels.* From the sample mean and sample variance, we can estimate initial values for the mean of the no-change distribution:

$$\mu_{NC} = \frac{1}{|S_{NC}|} \cdot \sum_{\nu \in S_{NC}} m(\nu),$$

*The symbols $\cup$ and $\backslash$ denote set union and set difference, respectively. These sets can be determined in practice by setting generous, scene-independent thresholds for change and no-change pixel intensities; see Bruzzone and Prieto (2000).

and for the variance:

$$(\sigma_{NC})^2 = \frac{1}{|S_{NC}|} \cdot \sum_{\nu \in S_{NC}} (m(\nu) - \mu_{NC})^2$$

($|S_{NC}|$ denotes set cardinality) and similarly for $C-$ and $C+$. Following a suggestion by Bruzzone and Prieto (2000), these estimates can be improved by applying the expectation maximization (EM) algorithm. The EM algorithm recalculates the parameters of the mixture model according to (see Exercise 9, Chapter 8)

$$\begin{aligned}
\mu'_{NC} &= \sum_{\nu \in S} \Pr(NC \mid m(\nu)) \cdot m(\nu) \Big/ \sum_{\nu \in S} \Pr(NC \mid m(\nu)) \\
(\sigma'_{NC})^2 &= \sum_{\nu \in S} \Pr(NC \mid m(\nu)) \cdot (m(\nu) - \mu'_{NC})^2 \Big/ \sum_{\nu \in S} \Pr(NC \mid m(\nu)) \\
\Pr(NC)' &= \frac{1}{|S|} \cdot \sum_{\nu \in S} \Pr(NC \mid m(\nu)) \ ,
\end{aligned} \tag{9.28}$$

where the primes denote improved estimates and the three equations are iterated to convergence. In Equation (9.28), $\Pr(NC \mid m(\nu))$ is the posterior probability for a no-change pixel conditional on measurement $m(\nu)$. We have

$$\begin{aligned}
i \in S_{NC} : &\qquad \Pr(NC \mid m(\nu)) = 1 \\
i \in S_{C\pm} : &\qquad \Pr(NC \mid m(\nu)) = 0.
\end{aligned}$$

Equations (9.28) can therefore be written in the form

$$\begin{aligned}
\mu'_{NC} &= \frac{\sum_{\nu \in S_U} \Pr(NC \mid m(\nu)) m(\nu) + \sum_{\nu \in S_{NC}} m(\nu)}{\sum_{\nu \in S_U} \Pr(NC \mid m(\nu)) + |S_{NC}|} \\
(\sigma'_{NC})^2 &= \frac{\sum_{\nu \in S_U} \Pr(NC \mid m(\nu))(m(\nu) - \mu'_{NC})^2 + \sum_{\nu \in S_{NC}} (m(\nu) - \mu'_{NC})^2}{\sum_{\nu \in S_U} \Pr(NC \mid m(\nu)) + |S_{NC}|} \\
\Pr(NC)' &= \frac{1}{|S|} \left(\sum_{\nu \in S_U} \Pr(NC \mid m(\nu)) + |S_{NC}| \right),
\end{aligned}$$

where with Bayes' Theorem,

$$\Pr(NC \mid m(\nu)) = p(m(\nu) \mid NC) \Pr(NC) / p(m(\nu))$$

and

$$p(m(\nu) \mid NC) = \frac{1}{\sqrt{2\pi} \cdot \sigma_{NC}} \cdot \exp\left(\frac{-(m(\nu) - \mu_{NC})^2}{2\sigma_{NC}^2} \right).$$

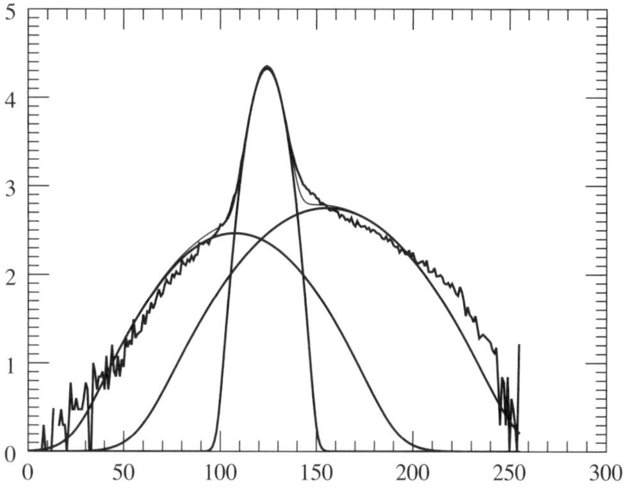

FIGURE 9.14
Gaussian mixture fit to a MAD/MNF component. Optimal decision thresholds correspond to the upper and lower intersections of the central no-change Gaussian with the positive and negative change Gaussians, respectively.

A similar set of equations hold for $C+$ and $C-$. They can be iterated to obtain final estimates of the distributions.* A typical fit to a MAD/MNF variate is shown in Figure 9.14. Note the logarithmic scale.

Once a fit to the Gaussian mixture model has been obtained, one can determine the upper change threshold as the appropriate solution of

$$p(m \mid NC)\Pr(NC) = p(m \mid C+)\Pr(C+).$$

Taking logarithms,

$$\frac{1}{2\sigma_{C+}^2}(m-\mu_{C+})^2 - \frac{1}{2\sigma_{NC}^2}(m-\mu_{NC})^2 = \log\left[\frac{\sigma_{NC}}{\sigma_{C+}} \cdot \frac{\Pr(C+)}{\Pr(NC)}\right] =: A$$

*This is in fact just the Gaussian mixture algorithm of Chapter 8 for one-dimensional observations, including "frozen" memberships (Section 8.3.4).

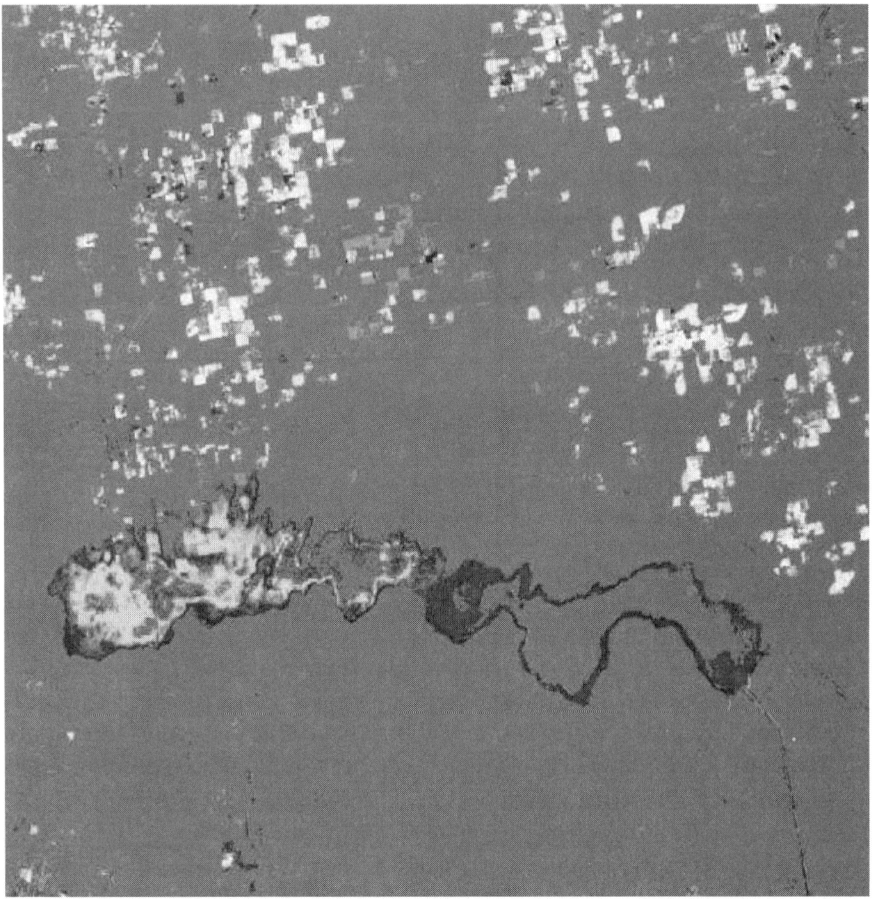

FIGURE 9.15
Color composite of MAD/MNF components 1 (red), 2 (green), and 3 (blue) for the bitemporal image of Figure 9.1. Image generated with `MAD_VIEW_RUN` using decision thresholds obtained from the Gaussian mixture model fit, linear stretch over ±16 standard deviations of the no-change observations. **(See color insert.)**

with solutions

$$m = \frac{\mu_{C+}\sigma_{NC}^2 - \mu_{NC}\sigma_{C+}^2 \pm \sigma_{NC}\sigma_{C+}\sqrt{(\mu_{NC}-\mu_{C+})^2 + 2A(\sigma_{NC}^2 - \sigma_{C+}^2)}}{\sigma_{NC}^2 - \sigma_{C+}^2}. \tag{9.29}$$

A corresponding expression obtains for the lower threshold. These thresholds minimize the overall error probabilities in deciding between change and no change; see Exercise 6. A simple IDL widget, called `MAD_VIEW_RUN`, for determining change thresholds and displaying the thresholded MAD or MAD/MNF variates in various color combinations and histogram stretches, is described in Appendix C. An example is shown in Figure 9.15. The image is stretched over ±16 standard deviations of the no-change observations. This very large dynamic range typifies the sensitivity of the IR-MAD method.

9.6 Unsupervised change classification

Figure 9.15 provides a rather nice visualization of the significant changes that have taken place. However, it constitutes a marginal analysis in the sense that the decision thresholds are determined band-wise. We might go a step

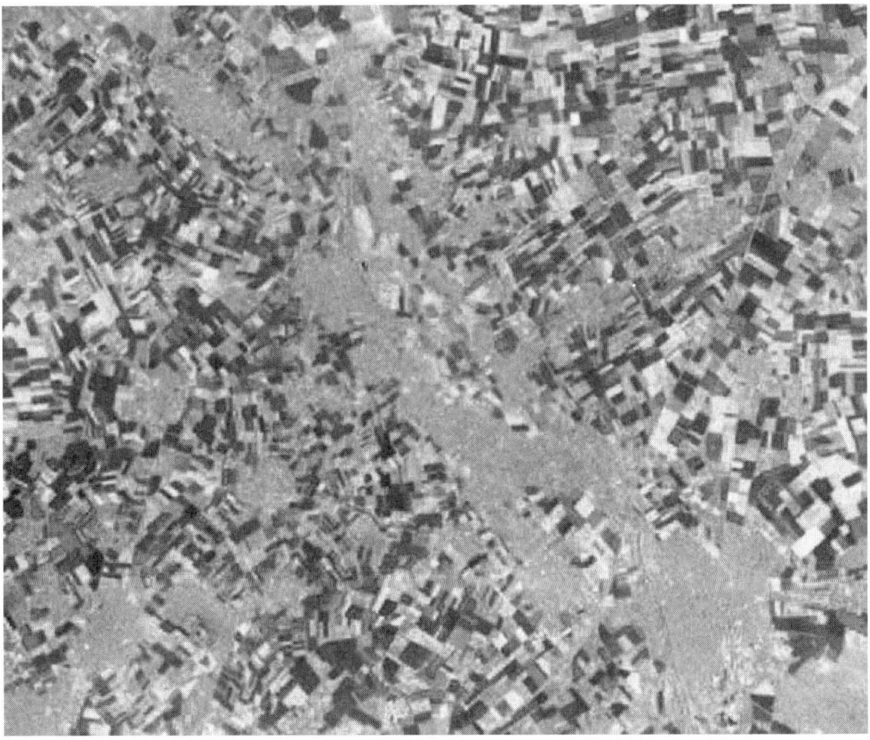

FIGURE 9.16
RGB composite of an iteratively re-weighted MAD image (MAD components 4,5,6 in a linear 2% stretch) obtained from LANDSAT ETM+ scenes acquired in June and August 2001. **(See color insert.)**

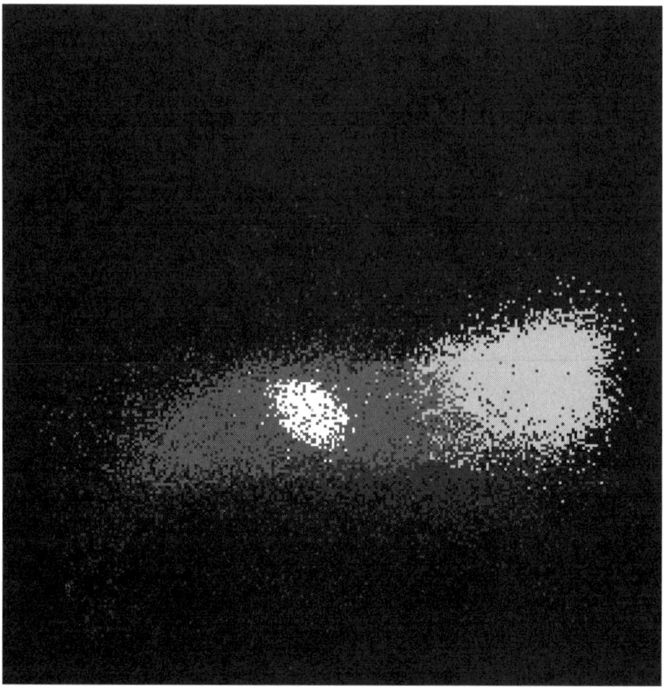

FIGURE 9.17
Four clusters in MAD feature space projected onto the plane of variates 5 and 6. The partially obscured white cluster is no change. **(See color insert.)**

further (Canty and Nielsen, 2006) and cluster the change and no-change pixels in multidimensional MAD or MAD/MNF feature space, again using the Gaussian mixture or FMLE algorithm discussed in Chapter 8. Since the no-change cluster will tend to be very dominant, it is to be expected that this will only be sensible for situations in which a good deal of change has taken place and some *a priori* information about the number of change categories exists.

The image shown in Figure 9.16 is an RGB composite of three MAD components showing changes in a region northwest of Jülich, Germany. The MAD components were derived from two LANDSAT 7 ETM+ scenes acquired in the months of June and August 2001. The light blue-green (featureless) regions correspond to built-up areas and to forest canopy, neither of which have changed significantly between acquisitions. The significant changes, indicated by the various brightly colored areas, are quite heterogeneous and not easy to interpret, but from their form they are clearly associated with cultivated fields. The main crops in this area are sugar beets, corn and cereal grain. The former two are still maturing in August, whereas the grain fields have

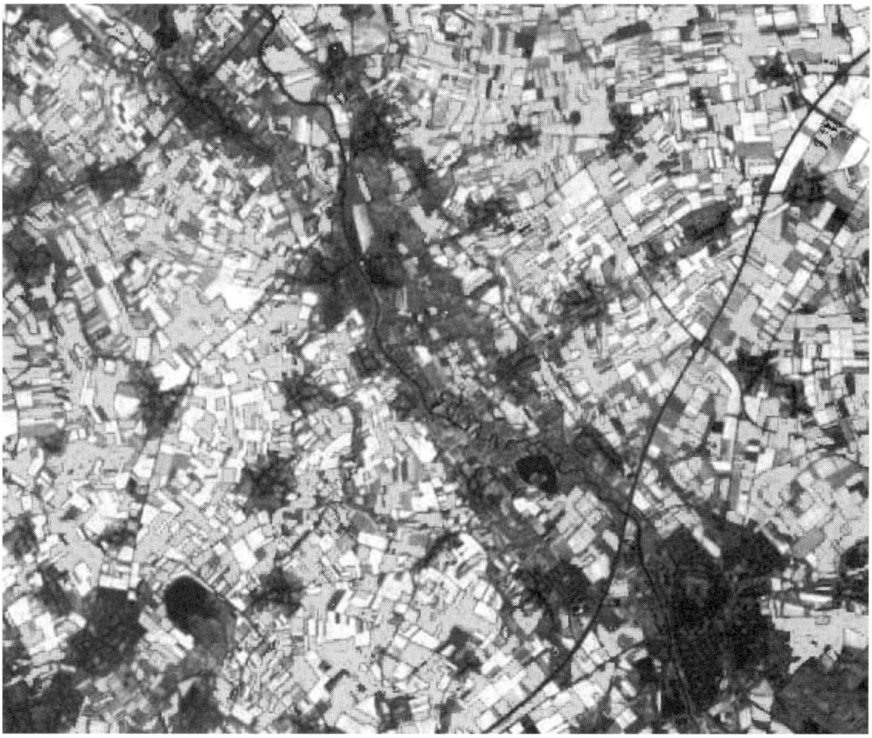

FIGURE 9.18
Overlay of the harvested grain cluster onto spectral band 4 of the August 2001 LANDSAT ETM+ image. **(See color insert.)**

been harvested. So we might postulate two main classes of change: maturing crops and harvested crops. Allowing for a no-change class and a "catch-all" class for other kinds of change, we can try to classify the MAD image with the Gaussian mixture clustering algorithm by assuming four classes in all. The resulting clusters, in the feature space of the MAD variates, are shown in Figure 9.17.

There are three fairly well-defined clusters (the central (white) one is no change) and a diffuse cluster (blue). Identifying the red and green classes as the two postulated ones and overlaying the green cluster onto band 4 of the August image, we obtain the result shown in Figure 9.18. The green pixels correspond to low reflectance in band 4 (near infrared) and therefore may be associated with harvested grain.

9.7 Change detection with polarimetric SAR imagery

Polarimetric SAR image pixels were represented in Chapter 5 in the form of look-averaged complex covariance matrices (Equation (5.30)). Assuming that the measured scattering amplitudes,

$$\boldsymbol{s} = \begin{pmatrix} s_{hh} \\ \sqrt{2}s_{hv} \\ s_{vv} \end{pmatrix}, \tag{9.30}$$

are indeed zero-mean, complex multivariate normally distributed, we expect that this representation will completely characterize the image statistics. A change detection scheme which compares the covariance matrices in bitemporal images will then best exploit the polarimetric information available. Conradsen et al. (2003) developed a per-pixel likelihood ratio test for changes which is based on the complex Wishart distribution. In the following discussion, their change detection algorithm will be explained and, as usual, programmed. We begin with the easier case of single polarimetry (Conradsen et al. (2003), Appendix C) and then proceed to the multivariate situation.

9.7.1 Single polarimetry: The gamma distribution

Let us represent the pixels in an m look-averaged, single polarimetric SAR intensity image acquired at some initial time by the random variable G_1, with mean $\langle G_1 \rangle = x_1$. Here, x_1 is understood to be the underlying signal; see Section 5.4.3. The density function for G_1, conditional on the value of x_1, is, with Equation (5.34), the gamma density

$$p(g_1 \mid x_1) = \frac{1}{(x_1/m)^m \Gamma(m)} g_1^{m-1} e^{-g_1 m/x_1}. \tag{9.31}$$

For a second image G_2, acquired at a later time, with m looks and $\langle G_2 \rangle = x_2$, we have

$$p(g_2 \mid x_2) = \frac{1}{(x_2/m)^m \Gamma(m)} g_2^{m-1} e^{-g_2 m/x_2}. \tag{9.32}$$

We now set up a likelihood ratio test as described in Definition 2.7. Under the null hypothesis H_0 that the pixels are unchanged, $x_1 = x_2 = x$, the likelihood for x is

$$L(x) = p(g_1 \mid x)p(g_2 \mid x) = \frac{1}{(x/m)^{2m} \Gamma(m)^2}\, g_1^{m-1} g_2^{m-1} e^{-(g_1+g_2)m/x}, \tag{9.33}$$

and under the alternative hypothesis H_1, $x_1 \neq x_2$, the likelihood for x_1 and x_2 is

$$\begin{aligned} L(x_1, x_2) &= p(g_1 \mid x_1) p(g_2 \mid x_2) \\ &= \frac{1}{(x_1/m)^m (x_2/m)^m \Gamma(m)^2}\, g_1^{m-1} g_2^{m-1} e^{-(g_1 m/x_1 + g_2 m/x_2)}. \end{aligned} \tag{9.34}$$

By taking derivatives of the log-likelihoods, it is easy to show (Exercise 9) that the likelihood $L(x)$ is maximized by

$$\hat{x} = \frac{g_1 + g_2}{2}$$

and that $L(x_1, x_2)$ is maximized by

$$\hat{x}_1 = g_1, \quad \hat{x}_2 = g_2.$$

Then, according to Equation (2.73), the likelihood ratio test has the critical region

$$Q = \frac{L(\hat{x})}{L(\hat{x}_1, \hat{x}_2)} = \frac{(g_1/m)^m (g_2/m)^n}{\left(\frac{g_1+g_2}{2m}\right)^{2m}} = 2^{2m} \frac{g_1^m g_2^m}{(g_1+g_2)^{2m}} \leq k.$$

Equivalently,

$$\left(\frac{g_1 g_2}{(g_1+g_2)^2}\right)^m \leq \frac{k}{2^{2m}}$$

or

$$\frac{g_1 g_2}{(g_1+g_2)^2} \leq \tilde{k},$$

where $\tilde{k}$ depends on k. Inverting both sides of the above inequality and simplifying, we obtain

$$\frac{g_1}{g_2} + \frac{g_2}{g_1} \geq \frac{1}{\tilde{k}} - 2.$$

It follows that the critical region has the form

$$\frac{g_1}{g_2} \leq c_1 \text{ or } \frac{g_1}{g_2} \geq c_2, \tag{9.35}$$

where the decision thresholds c_1 and c_2 depend on $\tilde{k}$.

The test statistic is thus the random variable G_1/G_2 corresponding to the ratio of SAR pixel intensities. We still require its distribution. Both G_1 and G_2 have gamma distributions of form:

$$p(g \mid x) = \frac{1}{(x/m)^m \Gamma(m)} g^{m-1} e^{-gm/x}.$$

Making the change of variable $z = 2gm/x$, we have for the random variable Z, from Theorem 2.1,

$$p(z \mid x) = p(g(z) \mid x) \left|\frac{dg}{dz}\right| = \frac{1}{(x/m)^m \Gamma(m)} \left(\frac{zx}{2m}\right)^{m-1} e^{-z/2} \left|\frac{dg}{dz}\right|.$$

With $|dg/dz| = x/2m$ we obtain

$$p(z \mid x) = \frac{1}{2^m \Gamma(m)} z^{m-1} e^{-z/2}.$$

Comparison with Equation (2.37) shows that Z, and hence the random variables $2G_1 m/x_1$ and $2G_2 m/x_2$, are chi-square distributed with $2m$ degrees of freedom. Under the null hypothesis, we have $x_1 = x_2$, so that the test statistic G_1/G_2 is a ratio of two chi-square distributed random variables. Our test statistic is therefore F-distributed with $2m$ and $2m$ degrees of freedom; see Equation (2.79). The percentiles of the F-distribution can therefore be used to set change/no-change decision thresholds to any desired degree of significance.

9.7.2 Quad polarimetry: The complex Wishart distribution

In the multivariate situation, we represent a pixel vector in an m look-averaged polarimetric image by the random matrix $\boldsymbol{X}$ with a complex Wishart distribution and with realization (Equation (5.31))

$$\boldsymbol{x} = \sum_{\nu=1}^{m} \boldsymbol{s}(\nu)\boldsymbol{s}(\nu)^\dagger = m\bar{\boldsymbol{c}}.$$

The components of the image pixel vectors are the components of the matrix $\bar{\boldsymbol{c}}$. We need the following result (Goodman, 1963): If

$$\boldsymbol{X} = \sum_{\nu=1}^{m} \boldsymbol{Z}(\nu)\boldsymbol{Z}(\nu)^\dagger, \quad \boldsymbol{Z}(\nu) \sim \mathcal{N}_C(\boldsymbol{0}, \boldsymbol{\Sigma}), \ \nu = 1 \ldots m,$$

is complex Wishart distributed with covariance matrix $\boldsymbol{\Sigma}$ and m degrees of freedom, then the maximum likelihood estimate for $\boldsymbol{\Sigma}$ is

$$\hat{\boldsymbol{\Sigma}} = \frac{1}{m}\sum_{\nu=1}^{m} \boldsymbol{z}(\nu)\boldsymbol{z}(\nu)^\dagger = \frac{1}{m}\boldsymbol{x}. \tag{9.36}$$

This closely parallels the real random variable case, Equation (2.71). We conclude that $\bar{\boldsymbol{c}}$ is the maximum likelihood estimate of $\boldsymbol{\Sigma}$.

The density function for the random matrix $\boldsymbol{X}$ is given by Theorem 2.9. To simplify the notation a little, let us define

$$\Gamma_N(m) = \pi^{N(N-1)/2} \prod_{i=1}^{N} \Gamma(m+1-i).$$

Then, for two m-look quad polarimetric covariance images $\boldsymbol{X}_1$ and $\boldsymbol{X}_2$, the multivariate densities are*

$$p(\boldsymbol{x}_1 \mid m, \boldsymbol{\Sigma}_1) = \frac{|\boldsymbol{x}_1|^{m-3}\exp(-\mathrm{tr}(\boldsymbol{\Sigma}_1^{-1}\boldsymbol{x}_1))}{|\boldsymbol{\Sigma}_1|^m \Gamma_3(m)}$$
$$p(\boldsymbol{x}_2 \mid m, \boldsymbol{\Sigma}_2) = \frac{|\boldsymbol{x}_2|^{m-3}\exp(-\mathrm{tr}(\boldsymbol{\Sigma}_2^{-1}\boldsymbol{x}_2))}{|\boldsymbol{\Sigma}_2|^m \Gamma_3(m)}.$$

We now define the null (or no-change) simple hypothesis

$$H_0: \quad \boldsymbol{\Sigma}_1 = \boldsymbol{\Sigma}_2 = \boldsymbol{\Sigma},$$

against the alternative composite hypothesis

$$H_1: \quad \boldsymbol{\Sigma}_1 \neq \boldsymbol{\Sigma}_2.$$

Under H_0, the likelihood for $\boldsymbol{\Sigma}$ is given by

$$L(\boldsymbol{\Sigma}) = p(\boldsymbol{x}_1 \mid m, \boldsymbol{\Sigma})p(\boldsymbol{x}_2 \mid m, \boldsymbol{\Sigma}) = \frac{|\boldsymbol{x}_1|^{m-3}|\boldsymbol{x}_2|^{m-3}\exp(-\mathrm{tr}(\boldsymbol{\Sigma}^{-1}(\boldsymbol{x}_1+\boldsymbol{x}_2)))}{|\boldsymbol{\Sigma}|^{2m}\Gamma_3(m)^2}.$$

According to Theorem 2.10, $\boldsymbol{X}_1 + \boldsymbol{X}_2$ is complex Wishart distributed with $2m$ degrees of freedom and therefore, with Equation (9.36), the maximum likelihood estimate of $\boldsymbol{\Sigma}$ is

$$\hat{\boldsymbol{\Sigma}} = \frac{1}{2m}(\boldsymbol{x}_1 + \boldsymbol{x}_2).$$

Hence the maximum likelihood under the null hypothesis is

$$L(\hat{\boldsymbol{\Sigma}}) = \frac{|\boldsymbol{x}_1|^{m-3}|\boldsymbol{x}_2|^{m-3}\exp(-2m\cdot\mathrm{tr}(\boldsymbol{I}))}{\left(\frac{1}{2m}\right)^{3\cdot 2m}|\boldsymbol{x}_1+\boldsymbol{x}_2|^{2m}\Gamma_3(m)^2},$$

where $\boldsymbol{I}$ is the 3×3 identity matrix and $\mathrm{tr}(\boldsymbol{I}) = 3$. (Note that $|a\boldsymbol{x}| = a^3|\boldsymbol{x}|$ for constant a and a 3×3 matrix $\boldsymbol{x}$.) Under H_1 we obtain, similarly, the maximum likelihood for $\boldsymbol{\Sigma}_1$ and $\boldsymbol{\Sigma}_2$ as

$$L(\hat{\boldsymbol{\Sigma}}_1, \hat{\boldsymbol{\Sigma}}_2) = \frac{|\boldsymbol{x}_1|^{m-3}|\boldsymbol{x}_2|^{m-3}\exp(-2m\cdot\mathrm{tr}(\boldsymbol{I}))}{\left(\frac{1}{m}\right)^{3m}\left(\frac{1}{m}\right)^{3m}|\boldsymbol{x}_1|^m|\boldsymbol{x}_2|^m\Gamma_3(m)^2}. \tag{9.37}$$

Then, again according to Equation (2.72), the likelihood ratio test has the critical region

$$Q = \frac{L(\hat{\boldsymbol{\Sigma}})}{L(\hat{\boldsymbol{\Sigma}}_1, \hat{\boldsymbol{\Sigma}}_2)} = 2^{6m}\frac{|\boldsymbol{x}_1|^m|\boldsymbol{x}_2|^m}{|\boldsymbol{x}_1+\boldsymbol{x}_2|^{2m}} \le k. \tag{9.38}$$

*We set $N = 3$ to correspond to the quad polarimetric case with reciprocity, Equation (9.30).

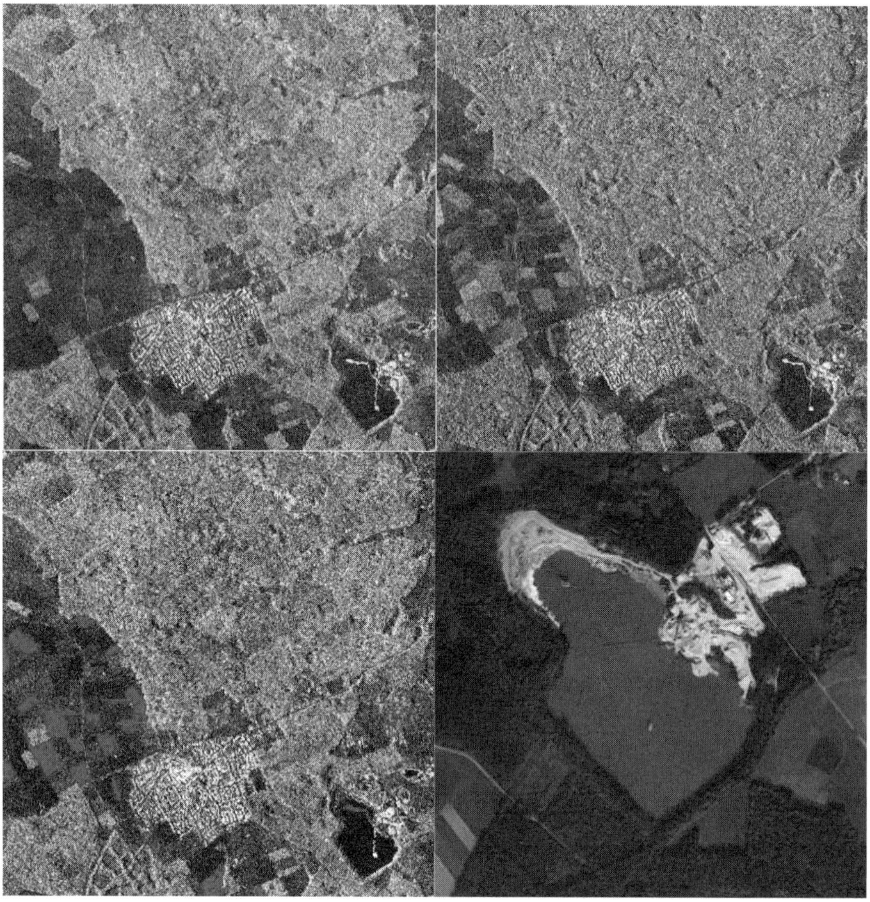

FIGURE 9.19
Wishart change detection. Upper left: RGB composite $(\langle|s_{hh}|^2\rangle, \langle|s_{hv}|^2\rangle, \langle|s_{vv}|^2\rangle)$ of a TerraSAR-X strip map, quad polarimetric image over a region southwest of Bonn, Germany, acquired on April 21, 2010 (logarithmic intensity scale). Upper right: the same scene acquired on May 13, 2010. Lower left: rejection of the no-change hypothesis at the 1% significance level (red) overlaid onto the HH intensity band of the April image. Lower right: an aerial view of the sand quarry in the bottom right-hand corner of the images, taken at an unknown date. **(See color insert.)**

For a large number of looks m, we can apply Theorem 2.13 to obtain the following asymptotic distribution for the test statistic: As $m \to \infty$, the quantity

$$-2\log Q = -2m(6\log 2 + \log|\boldsymbol{x}_1| + \log|\boldsymbol{x}_2| - 2\log|\boldsymbol{x}_1 + \boldsymbol{x}_2|)$$

is a realization of a chi-square random variable with 9 degrees of freedom:

$$\Pr(-2\log Q \leq z) \simeq P_{\chi^2;9}(z). \tag{9.39}$$

This follows from the fact that the number of parameters required to specify a 3×3 complex covariance matrix is 9: three for the real diagonal elements and 2×3 for the three complex elements above the diagonal.* Thus, in the notation of Theorem 2.13, the parameter space ω for $(\mathbf{\Sigma}_1, \mathbf{\Sigma}_2)$ has dimension $q = 2 \times 9$, and the subspace ω_0 for the simple null hypothesis has dimension $r = 9$, so $q - r = 9$.

In practice, m may be rather small for typical look-averaged polarimetric SAR images. Conradsen et al. (2003) give a better (and more complicated) approximation to the distribution of the test statistic based on large sample distribution theory. For finite m and an $N \times N$ covariance matrix:

$$\Pr(-2\rho\log Q \leq z) \simeq P_{\chi^2;N^2}(z) + \omega_2\left[P_{\chi^2;N^2+4}(z) - P_{\chi^2;N^2}(z)\right], \tag{9.40}$$

where

$$\rho = 1 - \frac{2N^2-1}{6N} \cdot \frac{3}{2m}$$

and

$$\omega_2 = -\frac{N^2}{4} \cdot \left(1 - \frac{1}{\rho}\right)^2 + \frac{N^2(N^2-1)}{24\rho^2} \cdot \frac{7}{4m^2}.$$

The quantities $\rho \to 1$ and $\omega_2 \to 0$ as $m \to \infty$, so that Equation (9.40) converges to Equation (9.39) for $N = 3$ and for large m. The ENVI/IDL extension `WISHART_CHANGE_RUN.PRO` and the Python script `wishart_change.py` given in Appendices C and D, respectively, use this more exact approximation. They also allow for single, dual and quad polarimetric image pairs with differing numbers of looks. An example of change detection with TerraSAR-X imagery is displayed in Figure 9.19. The images shown have been look-averaged to a ground resolution of 10 m, and the equivalent number of looks was determined to be ENL $\approx$ 14. Most of the scene comprises mixed forest, where little change in the canopy has taken place. Significant change due to crop growth in the agricultural fields on the left and lower edge are evident. One of the dredging arms in the sand quarry in the lower right corner has evidently moved slightly between the acquisitions.

9.8 Radiometric normalization of multispectral imagery

Ground reflectance determination from satellite imagery requires, among other things, an atmospheric correction algorithm and the associated atmospheric

*The three elements below the diagonal are their complex conjugates.

properties at the time of image acquisition. For most historical satellite scenes such data are not available and even for planned acquisitions they may be difficult to obtain. A relative normalization based on the radiometric information intrinsic to the images themselves is an alternative whenever absolute surface reflectances are not required.

In performing relative radiometric normalization, one usually makes the assumption that the relationship between the at-sensor radiances recorded at two different times from regions of constant reflectance can be approximated by linear functions. The critical aspect is the determination of suitable time-invariant features upon which to base the normalization (Schott et al., 1988; Yang and Lo, 2000; Du et al., 2002). We begin by illustrating this with the simple technique of scatterplot matching for red/near-infrared spectral bands. Then we go on to demonstrate how to take advantage of the linear invariance of the MAD transformation to perform fully automatic radiometric normalization across the visual/infrared spectrum.

9.8.1 Scatterplot matching

Maas and Rajan (2010) describe a method for radiometric normalization of the RED and NIR bands of multi-temporal LANDSAT TM and ETM+ images which exploits the characteristic shape of NIR vs. RED scatterplots (bands 4 and 3, respectively). The so-called *bare soil line* (BSL) and *full canopy point* (FCP) derived from the scatterplots of target and reference images are used as invariant features to re-scale the RED and NIR bands of the target to match those of the reference. The procedure works for other platforms too, for instance, with bands 2 and 3 of the ASTER VNIR sensor as illustrated in Figure 9.20. The reference and target scenes in the figure were acquired in different years (July 2001 and September 2005) and with different sensor gains.

With reference to Figure 9.21, one can derive the following linear transformation, which relates a point (x, y) in the target scatterplot to its transform $(\tilde{x}, \tilde{y})$ in the reference scatterplot (Exercise 13):

$$\begin{aligned} \tilde{x} &= X_R + (x - X_T)\frac{L_R}{L_T}\frac{a_T}{a_R} \\ \tilde{y} &= Y_R - (Y_T - y)\frac{L_R}{L_T}. \end{aligned} \tag{9.41}$$

The reference and target features can be extracted from the scatterplots by constructing a histogram of the intensity ratios NIR/RED and using the first percentile for the bare soil line and the 99.9th percentile for the full canopy point. This is illustrated for the reference image in listing 9.4, showing an excerpt from the ENVI/IDL extension `BSLFCPNORM_RUN.PRO` (Appendix C). In line 2, the NIR/RED ratio image is calculated and its histogram is formed in line 5. Points with the smallest ratio (first percentile) are extracted in

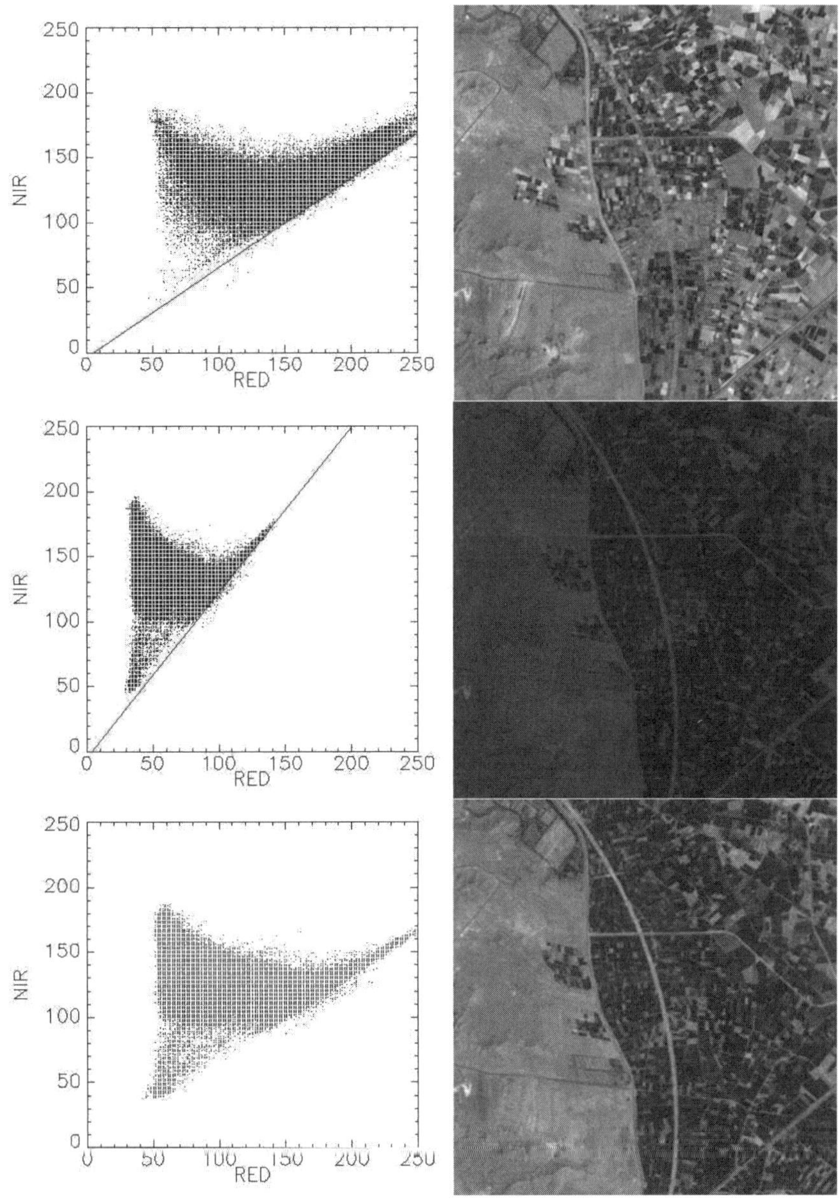

FIGURE 9.20
Scatterplot matching of two ASTER images. Top row: RGB composite of the July 2001 reference image (bands 2,3,2 in a 0-255 byte linear stretch) along with the NIR vs. RED scatterplot showing the full canopy point (cross) and the bare soil line. Middle row: the September 2005 target image. Bottom row: the normalized target. **(See color insert.)**

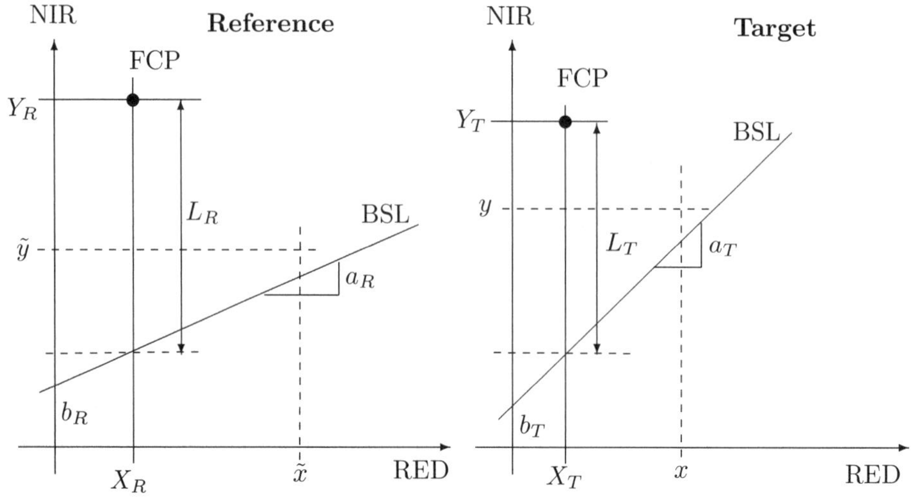

FIGURE 9.21
Principle of scatterplot matching. The coordinates of the full canopy point for reference and target are (X_R, Y_R) and (X_T, Y_T), respectively, and the respective slopes and intercepts of the bare soil line are a_R, b_R and a_T, b_T.

lines 9 through 13, taking advantage of the `reverse_indices` keyword of IDL's `histogram` function, and the slope and intercept are extracted using orthogonal regression (see the next section) in lines 14 and 15. Similarly, in lines 19 to 21, the points with the largest ratio (99.9th percentile) are determined. Their mean coordinates determine the position of the full canopy point (lines 22 to 24). The script was used to generate the results shown in Figure 9.20.

9.8.2 IR-MAD normalization

As we have seen in Section 9.4.3, the MAD transformation is invariant under arbitrary linear transformations of the pixel intensities for the images involved. Thus, if one uses MAD for change detection applications, preprocessing by linear radiometric normalization is superfluous. However, radiometric normalization of imagery is important for many other applications, such as mosaicking, tracking vegetation indices over time, comparison of supervised and unsupervised land cover classifications, etc. Furthermore, if some other, non-invariant change detection procedure is preferred, it must generally be preceded by radiometric normalization. Taking advantage of invariance, one can apply the MAD transformation to select the no-change pixels in un-normalized bitemporal images, and then use them for relative radiometric normalization. The procedure is simple, fast, and completely automatic and compares very favorably with normalization using hand-selected,

Listing 9.4: Scatterplot matching. Excerpt from `BSLFCPNORM_RUN.PRO`.

```
    idx=where(r_image[*,0] GT 0 AND r_image[*,1] LE 255)
    ratio=r_image[idx,1]/r_image[idx,0]
    nbins=1000
    binsize=(max(ratio) - min(ratio))/nbins
    hist=histogram(ratio,binsize=binsize, $
                                reverse_indices=ri)
    i = 1
; 0.01 quantile
    thresh = num_pixels1/100L
    WHILE ri[i]-ri[0] LT thresh DO i++
    idx1 = idx[ ri[ri[0]:ri[i-1]-1] ]
    X = r_image[idx1,0]
    Y = r_image[idx1,1]
    ortho_regress,transpose(X),transpose(Y),A_R,xm,ym
    B_R = ym-A_R*xm
    oplot, [0,255],[B_R,255*A_R+B_R],color='0000FF'XL
; 0.999 quantile
    thresh = num_pixels1/1000L
    i= 999
    WHILE ri[1000]-ri[i] LT thresh DO i--
    idx2 = idx[ ri[ri[i]:*] ]
    FCP = mean( r_image[idx2,*],dimension=1 )
    X_R = FCP[0]
    Y_R = FCP[1]
    oplot, [X_R],[Y_R],psym=1,color='0000FF'XL
```

time-invariant features (Canty et al., 2004; Schroeder et al., 2006; Canty and Nielsen, 2008). See also Philpot and Ansty (2013) for a physical interpretation of the invariant features detected by the IR-MAD algorithm.

An ENVI/IDL extension `RADCAL_RUN.PRO` for radiometric normalization with the MAD transformation is given in Appendix C. A Python script with the same functionality is documented in Appendix D. The programs read the output from a previous IR-MAD transformation which has been performed on overlapping portions of the images to be normalized. Then Equations (9.19) and (9.20) are used to select pixels with a high no-change probability, typically ≥ 0.95. By regressing the reference image onto the target image at the no-change locations, slope and intercept parameters are obtained with which to perform the linear normalization. The preferred regression method is in this case *orthogonal linear regression*, see Appendix A, as both variables involved have similar uncertainties. Figure 9.22 shows a mosaic of two LANDSAT 7 ETM+ images taken over the same area. The October 2000 image (target) is to be normalized to the December 1999 scene (reference). The orthogonal regression coefficients obtained in a fit to 15,696 no-change pixels identified by the IR-MAD procedure are given in Table 9.1. Figure 9.23 shows the

FIGURE 9.22
Mosaic of two LANDSAT 7 ETM+ images over Morocco, spectral band 5. The left side was acquired December 19, 1999, the right side October 18, 2000.

TABLE 9.1
Orthogonal regression coefficients and statistics for radiometric normalization of the images of Figure 9.22.

Band	Intercept a	σ_a	Slope b	σ_b	Correlation	RMSE
1	−5.06	0.18	1.285	0.003	0.962	0.708
2	−6.84	0.13	1.226	0.002	0.976	0.917
3	−11.20	0.16	1.222	0.002	0.981	1.418
4	−9.69	0.11	1.285	0.002	0.985	1.012
5	−4.74	0.17	1.176	0.002	0.978	1.535
6	−4.30	0.17	1.182	0.002	0.973	1.357

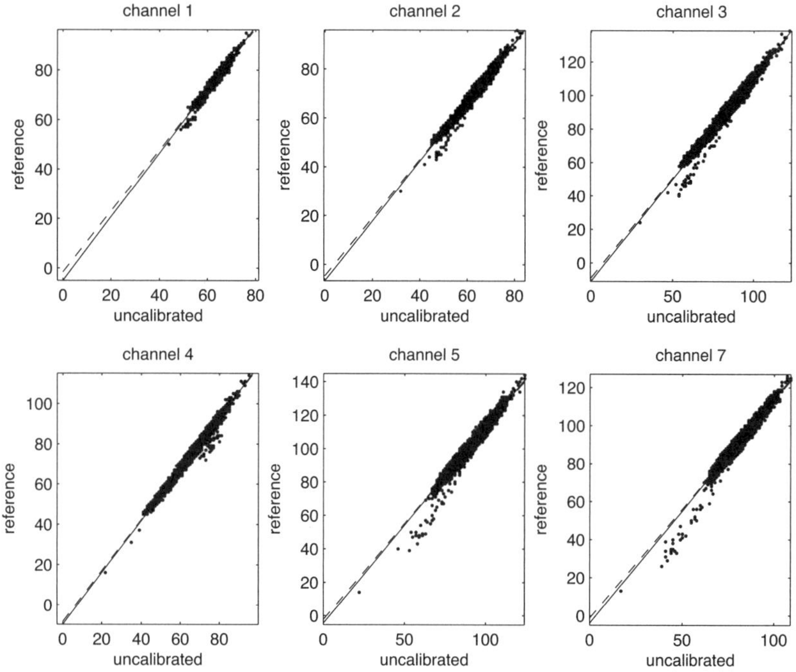

FIGURE 9.23
Regressions of the December 1999 reference scene on the October 2000 target (uncalibrated) scene. Solid line: orthogonal linear regression, dashed line: ordinary linear regression (Canty et al., 2004).

TABLE 9.2
Comparison of means and variances for hold-out test pixels, with t- and F-tests for unequal means and variances.

	Band 1	Band 2	Band 3	Band 4	Band 5	Band 7
Target mean	62.684	61.460	83.761	64.462	88.000	80.009
Reference mean	75.509	68.501	91.131	73.193	98.775	90.335
Normalized mean	75.518	68.508	91.143	73.198	98.776	90.327
t-statistic	−0.144	−0.0688	−0.0672	−0.036	−0.007	0.054
P-value	0.885	0.945	0.945	0.975	1.000	0.957
Target variance	10.61	28.83	87.78	55.46	95.24	59.63
Reference variance	17.10	42.72	129.64	90.83	129.82	81.98
Normalized variance	17.54	43.34	131.16	91.70	131.80	83.42
F-statistic	1.025	1.014	1.011	1.009	1.015	1.017
P-value	0.255	0.524	0.606	0.672	0.504	0.440

FIGURE 9.24
Figure 9.22 after automatic radiometric normalization.

regression lines obtained for the six nonthermal bands. The mosaic after radiometric normalization is shown in Figure 9.24.

In order to evaluate the normalization procedure the program holds back one third of the no-change pixels for testing purposes. These are used to calculate means and variances before and after normalization and also to perform statistical hypothesis tests for equal means and variances of the invariant pixels in the reference and normalized target images. These tests were discussed in Section 2.5. Results are given in Table 9.2 for 7849 no-change test pixels. The significance values (P-values) for the t-test for equal means and for the F-test for equal variances indicate that the hypotheses of equality cannot be rejected for any of the spectral bands.

9.9 Exercises

1. Use the ENVI `BAND MATH` facility to calculate NDVI (normalized difference vegetation index) images for two LANDSAT images and to perform change detection by subtracting one from the other.

2. Show that, for one-dimensional, zero-mean images represented by random variables F and G, the MAD transformation, Equation (9.12), is just
$$M = F/\sigma_F - G/\sigma_G.$$

3. Demonstrate the validity of Equation (9.11) for the MAD transformation vector $\boldsymbol{b}$.

4. The requirement that the correlations of the canonical variates be positive, namely
$$\boldsymbol{a}_i^\top \boldsymbol{\Sigma}_{12}\boldsymbol{b}_i > 0, \quad i = 1\ldots N,$$
does not completely remove the ambiguity in the signs of the transformation vectors $\boldsymbol{a}_i$ and $\boldsymbol{b}_i$, since if we invert both their signs simultaneously, the condition is still met. The ambiguity can be resolved by requiring that the sum of the correlations of the first image (represented here by $\boldsymbol{G}$) with each of the canonical variates $U_j = \boldsymbol{a}_i^\top \boldsymbol{G}$, $j = 1\ldots N$, be positive:
$$\sum_{\nu=1}^{N} \mathrm{corr}(G_i, U_j) > 0, \quad j = 1\ldots N. \tag{9.42}$$
This condition is implemented in the ENVI/IDL extension `MAD_RUN` and in the Python script `imad.py`.

   (a) Show that the matrix of correlations
$$\boldsymbol{C} = \begin{pmatrix} \mathrm{corr}(G_1,U_1) & \mathrm{corr}(G_1,U_2) & \cdots & \mathrm{corr}(G_1,U_N) \\ \mathrm{corr}(G_2,U_1) & \mathrm{corr}(G_2,U_2) & \cdots & \mathrm{corr}(G_2,U_N) \\ \vdots & \vdots & \ddots & \vdots \\ \mathrm{corr}(G_N,U_1) & \mathrm{corr}(G_N,U_2) & \cdots & \mathrm{corr}(G_N,U_N) \end{pmatrix}$$
is given by
$$\boldsymbol{C} = \boldsymbol{D}\boldsymbol{\Sigma}_{11}\boldsymbol{A},$$
where $\boldsymbol{A} = (\boldsymbol{a}_1, \boldsymbol{a}_2 \ldots \boldsymbol{a}_N)$, $\boldsymbol{\Sigma}_{11}$ is the covariance matrix for $\boldsymbol{G}$ and
$$\boldsymbol{D} = \begin{pmatrix} \frac{1}{\sqrt{\mathrm{var}(G_1)}} & 0 & \cdots & 0 \\ 0 & \frac{1}{\sqrt{\mathrm{var}(G_2)}} & \cdots & 0 \\ \vdots & \vdots & \ddots & \vdots \\ 0 & 0 & \cdots & \frac{1}{\sqrt{\mathrm{var}(G_N)}} \end{pmatrix}.$$

Listing 9.5: Copying a spatial subset between images.

```
PRO subset_copy
   COMPILE_OPT IDL2

   dim = 256L

; get an image band for T1
   envi_select, title='Choose␣T1␣image', $
                fid=fid1, dims=dims1,pos=pos1
   IF (fid1 EQ -1) THEN RETURN

; get an image band for T2
   envi_select, title='Choose␣T2␣image', $
             fid=fid2, dims=dims2,pos=pos2
   IF (fid2 EQ -1) THEN RETURN

; read images
   num_cols = dims1[2]-dims1[1]+1
   num_rows = dims1[4]-dims1[3]+1
   num_bands = n_elements(pos1)
   num_pixels = num_cols*num_rows
   im1 = fltarr(num_cols,num_rows,num_bands)
   im2 = im1*0.0
   FOR i=0,num_bands-1 DO BEGIN
     im1[*,*,i] = envi_get_data(fid=fid1,dims=dims1,$
                    pos=pos1[i])
     im2[*,*,i] = envi_get_data(fid=fid2,dims=dims2, $
                    pos=pos2[i])
   ENDFOR

; copy subset with Gaussian noise
   im2[0:dim-1,0:dim-1,*] = im1[0:dim-1,0:dim-1,*] $
      + 0.1*randomu(seed,dim,dim,num_bands,/normal)
   envi_enter_data, im2

END
```

(b) Let $s_j = \sum_i C_{ij}$, $j = 1 \dots N$, be the column sums of the correlation matrix. Show that Equation (9.42) is fulfilled by replacing $\boldsymbol{A}$ by $\boldsymbol{AS}$, where

$$\boldsymbol{S} = \begin{pmatrix} \frac{s_1}{|s_1|} & 0 & \cdots & 0 \\ 0 & \frac{s_2}{|s_2|} & \cdots & 0 \\ \vdots & \vdots & \ddots & \vdots \\ 0 & 0 & \cdots & \frac{s_N}{|s_N|} \end{pmatrix}.$$

5. Consider the following experiment: The ENVI/IDL extension `MAD_RUN`

is used to generate MAD variates from two co-registered multispectral images. Then a principal components transformation of one of the images is performed and the MAD transformation is repeated. Will the MAD variates have changed? Why or why not?

6. Given the validity of the three Gaussian mixture models of Section 9.5, show that the choice of threshold given in Equation (9.29) minimizes the overall error (incorrectly interpreting a no-change observation as change and vice versa).

7. The program of Listing 9.5 simulates no-change pixels by copying a spatial subset of one image to another, and adding some Gaussian noise. Ideally, the iteratively reweighted MAD scheme should identify these pixels unambiguously. Experiment with a bitemporal image (in byte format) and different subset sizes to see the extent to which this is the case. Use both the `MAD_RUN` and `RADCAL_RUN` ENVI extensions or their Python equivalents.

8. The more recent versions of ENVI include, among the so-called SPEAR tools, a change detection "wizard" which may be called from the ENVI main menu with

   `Spectral/SPEAR Tools/Change Detection`

   Experiment with the different algorithms offered by the wizard (image transform, subtractive, spectral angle), and compare them with the iteratively re-weighted MAD method.

9. Show that the likelihood functions, Equations (9.33) and (9.34), are maximized by $\hat{x} = (g_1 + g_2)/2$ and by $x_1 = g1,\ x_2 = g_2$, respectively.

10. Write down the critical region Equation (9.38) for the dual polarimetric case. What is the asymptotic distribution of the test statistic?

11. The Wishart change detection algorithm can be used as a simple SAR image edge detector simply by comparing an image with a copy of itself shifted by one row and one column. Modify the ENVI/IDL extension `WISHART_CHANGE_RUN.PRO` or the Python script `wishart_run.py` documented in the appendices to implement edge detection.

12. Derive Equations (9.41) from the geometry of Figure 9.21.

A

Mathematical Tools

A.1 Cholesky decomposition

Cholesky decomposition is used in some of the IDL and Python routines in this book to solve generalized eigenvalue problems associated with the maximum autocorrelation factor (MAF) and maximum noise fraction (MNF) transformations as well as with canonical correlation analysis. We sketch its justification in the following.

THEOREM A.1
If the $p \times p$ matrix $\boldsymbol{A}$ is symmetric positive definite and if the $p \times q$ matrix $\boldsymbol{B}$, where $q \le p$, has rank q, then $\boldsymbol{B}^\top \boldsymbol{A}\boldsymbol{B}$ is positive definite and symmetric.

Proof. Choose any q-dimensional vector $\boldsymbol{y} \ne \boldsymbol{0}$ and let $\boldsymbol{x} = \boldsymbol{B}\boldsymbol{y}$. We can write this as

$$\boldsymbol{x} = y_1 \boldsymbol{b}_1 + \ldots + y_q \boldsymbol{b}_q,$$

where $\boldsymbol{b}_i$ is the ith column of $\boldsymbol{B}$. Since $\boldsymbol{B}$ has rank q, we conclude that $\boldsymbol{x} \ne \boldsymbol{0}$ as well, for otherwise the column vectors would be linearly dependent. But

$$\boldsymbol{y}^\top (\boldsymbol{B}^\top \boldsymbol{A}\boldsymbol{B})\boldsymbol{y} = (\boldsymbol{B}\boldsymbol{y})^\top \boldsymbol{A}(\boldsymbol{B}\boldsymbol{y}) = \boldsymbol{x}^\top \boldsymbol{A}\boldsymbol{x} > 0,$$

since $\boldsymbol{A}$ is positive definite. So $\boldsymbol{B}^\top \boldsymbol{A}\boldsymbol{B}$ is positive definite (and clearly symmetric). □

A square matrix $\boldsymbol{A}$ is *diagonal* if $a_{ij} = 0$ for $i \ne j$. It is *lower triangular* if $a_{ij} = 0$ for $i < j$ and *upper triangular* if $a_{ij} = 0$ for $i > j$. The product of two lower(upper) triangular matrices is lower(upper) triangular. The inverse of a lower(upper) triangular matrix is lower(upper) triangular. If $\boldsymbol{A}$ is diagonal and positive definite, then all of its diagonal elements are positive. Otherwise if, say, $a_{ii} \le 0$ then, for $\boldsymbol{x} = (0\ldots, 1, \ldots 0)^\top$ with the 1 at the ith position,

$$\boldsymbol{x}^\top \boldsymbol{A}\boldsymbol{x} \le 0$$

contradicting the fact that $\boldsymbol{A}$ is positive definite. Now we state without proof Theorem A.2.

THEOREM A.2
If $\boldsymbol{A}$ *is nonsingular, then there exists a nonsingular lower triangular matrix* $\boldsymbol{F}$ *such that* $\boldsymbol{FA}$ *is nonsingular upper triangular.*

The proof is straightforward, but somewhat lengthy; see e.g., Anderson (2003), Appendix A.

It follows directly that, if $\boldsymbol{A}$ is symmetric and positive definite, there exists a lower triangular matrix $\boldsymbol{F}$ such that $\boldsymbol{FAF}^\top$ is diagonal and positive definite. That is, from Theorem A.2, $\boldsymbol{FA}$ is upper triangular and nonsingular. But then, since $\boldsymbol{F}^\top$ is upper triangular, $\boldsymbol{FAF}^\top$ is also upper triangular. But it is clearly also symmetric, so it must be diagonal. Finally, since $\boldsymbol{F}$ is nonsingular it has full rank, so that $\boldsymbol{FAF}^\top$ is also positive definite by Theorem A.1.

One can now go one step further and claim that, if $\boldsymbol{A}$ is positive definite, then there exists a lower triangular matrix $\boldsymbol{G}$ such that $\boldsymbol{GAG}^\top = \boldsymbol{I}$. To show this, choose an $\boldsymbol{F}$ such that $\boldsymbol{FAF}^\top = \boldsymbol{D}$ is diagonal and positive definite. Let $\boldsymbol{D}'$ be the diagonal matrix whose diagonal elements are the positive square roots of the diagonal elements of $\boldsymbol{D}$. Choose $\boldsymbol{G} = \boldsymbol{D}'^{-1}\boldsymbol{F}$. Then $\boldsymbol{GAG}^\top = \boldsymbol{I}$.

THEOREM A.3
(Cholesky Decomposition) *If* $\boldsymbol{A}$ *is symmetric and positive definite, there exists a lower triangular matrix* $\boldsymbol{L}$ *such that* $\boldsymbol{A} = \boldsymbol{LL}^\top$.

Proof. Since there exists a lower triangular matrix $\boldsymbol{G}$ such that $\boldsymbol{GAG}^\top = \boldsymbol{I}$, we must have

$$\boldsymbol{A} = \boldsymbol{G}^{-1}(\boldsymbol{G}^\top)^{-1} = \boldsymbol{G}^{-1}(\boldsymbol{G}^{-1})^\top = \boldsymbol{LL}^\top,$$

where $\boldsymbol{L} = \boldsymbol{G}^{-1}$ is lower triangular. □

Cholesky decomposition on a positive definite symmetric matrix is analogous to finding the square root of a positive real number. The IDL procedure `CHOLDC, A, D` performs Cholesky decomposition of a matrix `A` returning `L` in the lower diagonal of `A` except for its diagonal elements, which are returned in `D`. For example:

```
PRO chol
    A = [[2,1],[1,3]]
    choldc, A, D, /double
    L = diag_matrix(D)
    L[0,1]=A[0,1]
    PRINT, L##transpose(L)
END
ENVI> chol
         2.0000000        1.0000000
         1.0000000        3.0000000
```

The Python function `scipy.linalg.cholesky()` provides the similar functionality.

A.2 Vector and inner product spaces

The real column vectors of dimension N, which were introduced in Chapter 1 to represent multispectral pixel intensities, provide the standard example of the more general concept of a *vector space.*

DEFINITION A.1 *A set S is a* (real) vector space *if the operations addition and scalar multiplication are defined on S so that, for $\boldsymbol{x}, \boldsymbol{y} \in S$ and $\alpha, \beta \in \mathbb{R}$,*

$$
\begin{aligned}
\boldsymbol{x} + \boldsymbol{y} &\in S \\
\alpha \boldsymbol{x} &\in S \\
1\boldsymbol{x} &= \boldsymbol{x} \\
0\boldsymbol{x} &= \boldsymbol{0} \\
\boldsymbol{x} + \boldsymbol{0} &= \boldsymbol{x} \\
\alpha(\boldsymbol{x} + \boldsymbol{y} &= \alpha\boldsymbol{x} + \alpha\boldsymbol{y} \\
(\alpha + \beta)\boldsymbol{x} &= \alpha\boldsymbol{x} + \beta\boldsymbol{y}.
\end{aligned}
$$

The elements of S are called vectors.

Another example of a vector space satisfying the above definition is the set of continuous functions $f(x)$ on the real interval $[a, b]$, which is encountered in Chapter 3 in connection with the discrete wavelet transform. Unlike the column vectors of Chapter 1, the elements of this vector space have infinite dimension. Both vector spaces are *inner product spaces* according to Definition A.2.

DEFINITION A.2 *A vector space S is an* inner product space *if there is a function mapping two elements $\boldsymbol{x}, \boldsymbol{y} \in S$ to a real number $\langle \boldsymbol{x}, \boldsymbol{y} \rangle$ such that*

$$
\begin{aligned}
&\langle \boldsymbol{x}, \boldsymbol{y} \rangle = \langle \boldsymbol{y}, \boldsymbol{x} \rangle \\
&\langle \boldsymbol{x}, \boldsymbol{x} \rangle \geq 0 \\
&\langle \boldsymbol{x}, \boldsymbol{x} \rangle = 0 \quad \textit{if and only if} \quad \boldsymbol{x} = 0.
\end{aligned}
$$

In the case of the real column vectors, the inner product is defined by Equation (1.8), i.e., $\langle \boldsymbol{x}, \boldsymbol{y} \rangle = \boldsymbol{x}^\top \boldsymbol{y}$. For the vector space of continuous functions, we define

$$
\langle f, g \rangle = \int_a^b f(x)g(x)dx.
$$

DEFINITION A.3 *Two elements $\boldsymbol{x}$ and $\boldsymbol{y}$ of an inner product space S*

are said to be orthogonal *if* $\langle \boldsymbol{x}, \boldsymbol{y} \rangle = 0$. *The set of elements* $\boldsymbol{x}_i$, $i = 1 \ldots n$ *is* orthonormal *if* $\langle \boldsymbol{x}_i, \boldsymbol{x}_j \rangle = \delta_{ij}$.

A finite set S of linearly independent vectors (see Definition 1.2) constitutes a *basis* for the vector space V comprising all vectors that can be expressed as a linear combination of the vectors in S. The number of vectors in the basis is called the *dimension* of V. An orthogonal basis for a finite-dimensional inner product space can always be constructed by the *Gram–Schmidt orthogonalization procedure* (Press et al., 2002; Shawe–Taylor and Cristianini, 2004). If $\boldsymbol{v}_i$, $i = 1 \ldots N$, is an orthogonal basis for V, then for any $\boldsymbol{x} \in V$,

$$\boldsymbol{x} = \sum_{i=1}^{N} \frac{\langle \boldsymbol{x}, \boldsymbol{v}_i \rangle}{\langle \boldsymbol{v}_i, \boldsymbol{v}_i \rangle} \boldsymbol{v}_i.$$

Let W be a subset of vector space V. Then it will have an orthogonal basis $\{\boldsymbol{w}_1, \boldsymbol{w}_2 \ldots \boldsymbol{w}_K\}$. For any $\boldsymbol{x} \in V$, the *projection* $\boldsymbol{y}$ of $\boldsymbol{x}$ onto the subspace W is given by

$$\boldsymbol{y} = \sum_{i=1}^{K} \frac{\langle \boldsymbol{x}, \boldsymbol{w}_i \rangle}{\langle \boldsymbol{w}_i, \boldsymbol{w}_i \rangle} \boldsymbol{w}_i,$$

so that $\boldsymbol{y} \in W$. We define the *orthogonal complement* $W^\perp$ of W as the set

$$W^\perp = \{\boldsymbol{x} \in V \mid \langle \boldsymbol{x}, \boldsymbol{y} \rangle = 0 \text{ for all } \boldsymbol{y} \in W\}.$$

It is then easy to show that the *residual vector* $\boldsymbol{y}_\perp = \boldsymbol{x} - \boldsymbol{y}$ is in $W^\perp$, i.e., that $\langle \boldsymbol{y}_\perp, \boldsymbol{y} \rangle = 0$ for all $\boldsymbol{y} \in W$. Thus we can always write $\boldsymbol{x}$ as

$$\boldsymbol{x} = \boldsymbol{y} + \boldsymbol{y}_\perp,$$

where $\boldsymbol{y} \in W$ and $\boldsymbol{y}_\perp \in W^\perp$.

THEOREM A.4

(Orthogonal Decomposition Theorem) *If* W *is a finite-dimensional subspace of an inner product space* V, *then any* $\boldsymbol{x} \in V$ *can be written uniquely as* $\boldsymbol{x} = \boldsymbol{y} + \boldsymbol{y}_\perp$, *where* $\boldsymbol{y} \in W$ *and* $\boldsymbol{y}_\perp \in W^\perp$.

A.3 Complex numbers, vectors and matrices

A complex number is an expression of the form $z = a + \mathbf{i}b$, where $\mathbf{i}^2 = -1$. The *real part* of z is a and the *imaginary part* is b. The number z can be represented as a point or vector in the *complex plane* with the real part along the x-axis and the imaginary part along the y-axis. Complex number addition then corresponds to addition of two-dimensional vectors:

$$(a_1 + \mathbf{i}b_1) + (a_2 + \mathbf{i}b_2) = (a_1 + a_2) + \mathbf{i}(b_1 + b_2).$$

Multiplication is also straightforward, e.g.,

$$(a_1 + \mathbf{i}b_1)(a_2 + \mathbf{i}b_2) = a_1a_2 - b_1b_2 + \mathbf{i}(a_1b_2 - a_2b_1).$$

The *complex conjugate* of z is $z^* = a - \mathbf{i}b$. Thus

$$z^*z = (a - \mathbf{i}b)(a + \mathbf{i}b) = a^2 + b^2 = |z|^2,$$

where $|z| = \sqrt{a^2 + b^2}$ is the *magnitude* of the complex number z. If θ is the angle that z makes with the real axis, then

$$z = a + \mathbf{i}b = |z|\cos(\theta) + \mathbf{i}|z|\sin(\theta).$$

From the well-known *Euler's Theorem* we can write this as

$$z = |z|e^{\mathbf{i}\theta}.$$

A *complex vector* is a vector of complex numbers,

$$\boldsymbol{z} = \begin{pmatrix} a_1 + \mathbf{i}b_1 \\ \vdots \\ a_N + \mathbf{i}b_N \end{pmatrix}.$$

Complex matrices similarly are matrices with complex elements. The operations of vector and matrix addition, multiplication and scalar multiplication carry over straightforwardly from real vector spaces. The operation of transposition is replaced by the *conjugate transpose*

$$\boldsymbol{A}^\dagger = (\boldsymbol{A}^*)^\top.$$

For example, the inner product of two complex vectors $\boldsymbol{x}$ and $\boldsymbol{y}$ is

$$\boldsymbol{x}^\dagger\boldsymbol{y} = (x_1^* \dots x_N^*)\begin{pmatrix} y_1 \\ \vdots \\ y_N \end{pmatrix} = x_1^*y_1 + \dots x_N^*y_N$$

and the length or Euclidean norm of $\boldsymbol{x}$ is

$$\|\boldsymbol{x}\| = \sqrt{\boldsymbol{x}^\dagger\boldsymbol{x}} = \sqrt{|x_1|^2 + \dots + |x_N|^2}.$$

The complex analog of a symmetric real matrix is the *Hermitian matrix*, with the property

$$\boldsymbol{A}^\dagger = \boldsymbol{A}.$$

A Hermitian matrix $\boldsymbol{A}$ is positive definite if $\boldsymbol{x}^\dagger\boldsymbol{A}\boldsymbol{x} > 0$ for all nonzero complex vectors $\boldsymbol{x}$. Like positive definite symmetric matrices, positive definite Hermitian matrices have real, positive eigenvalues. The matrix $\boldsymbol{A}$ is said to be *unitary* if $\boldsymbol{A}^\dagger = \boldsymbol{A}^{-1}$.

A.4 Least squares procedures

In this section, ordinary linear regression is extended to a recursive procedure for sequential data. This forms the basis of one of the neural network training algorithms derived in Appendix B. In addition, the orthogonal linear regression procedure used in Chapter 9 for radiometric normalization is explained.

A.4.1 Recursive linear regression

Consider the statistical model given by Equation (2.92), now in a slightly different notation:

$$Y(j) = \sum_{i=0}^{N} w_j x_i(j) + R(j), \quad j = 1 \dots \nu. \tag{A.1}$$

This model relates the independent variables $\boldsymbol{x}(j) = (1, x_1(j) \dots x_N(j))^\top$ to a measured quantity $Y(j)$ via the parameters $\boldsymbol{w} = (w_0, w_1 \dots w_N)^\top$. The index ν is now intended to represent the number of measurements that have been made *so far*. The random variables $R(j)$ represent the measurement uncertainty in the realizations $y(j)$ of $Y(j)$. We assume that they are uncorrelated and normally distributed with zero mean and unit variance ($\sigma^2 = 1$), whereas the values $\boldsymbol{x}(j)$ are exact. We wish to determine the best values for parameters $\boldsymbol{w}$. Equation (A.1) can be written in the terms of a data matrix $\boldsymbol{\mathcal{X}}_\nu$ as

$$\boldsymbol{Y}_\nu = \boldsymbol{\mathcal{X}}_\nu \boldsymbol{w} + \boldsymbol{R}_\nu, \tag{A.2}$$

where

$$\boldsymbol{\mathcal{X}}_\nu = \begin{pmatrix} \boldsymbol{x}(1)^\top \\ \vdots \\ \boldsymbol{x}(\nu)^\top \end{pmatrix},$$

$\boldsymbol{Y}_\nu = (Y(1) \dots Y(\nu))^\top$ and $\boldsymbol{R}_\nu = (R(1) \dots R(\nu))^\top$. As was shown in Chapter 2, the best solution in the least squares sense for the parameter vector $\boldsymbol{w}$ is given by

$$\boldsymbol{w}(\nu) = [(\boldsymbol{\mathcal{X}}_\nu^\top \boldsymbol{\mathcal{X}}_\nu)^{-1} \boldsymbol{\mathcal{X}}_\nu^\top] \boldsymbol{y}_\nu = \boldsymbol{\Sigma}(\nu) \boldsymbol{\mathcal{X}}_\nu^\top \boldsymbol{y}_\nu, \tag{A.3}$$

where the expression in square brackets is the pseudoinverse of $\boldsymbol{\mathcal{X}}_\nu$ and where $\boldsymbol{\Sigma}(\nu)$ is an estimate of the covariance matrix of $\boldsymbol{w}$,

$$\boldsymbol{\Sigma}(\nu) = (\boldsymbol{\mathcal{X}}_\nu^\top \boldsymbol{\mathcal{X}}_\nu)^{-1}. \tag{A.4}$$

Suppose a new observation $(\boldsymbol{x}(\nu+1), y(\nu+1))$ becomes available. Now we must solve the least squares problem

$$\begin{pmatrix} \boldsymbol{Y}_\nu \\ Y(\nu+1) \end{pmatrix} = \begin{pmatrix} \boldsymbol{\mathcal{X}}_\nu \\ \boldsymbol{x}(\nu+1)^\top \end{pmatrix} \boldsymbol{w} + \boldsymbol{R}_{\nu+1}. \tag{A.5}$$

With Equation (A.3), the solution is

$$\boldsymbol{w}(\nu+1) = \boldsymbol{\Sigma}(\nu+1)\begin{pmatrix}\boldsymbol{\mathcal{X}}_\nu \\ \boldsymbol{x}(\nu+1)^\top\end{pmatrix}^\top \begin{pmatrix}\boldsymbol{y}_\nu \\ y(\nu+1)\end{pmatrix}. \tag{A.6}$$

Inverting Equation (A.4) with $\nu \to \nu+1$, we obtain a recursive formula for the new covariance matrix $\boldsymbol{\Sigma}(\nu+1)$:

$$\boldsymbol{\Sigma}(\nu+1)^{-1} = \begin{pmatrix}\boldsymbol{\mathcal{X}}_\nu \\ \boldsymbol{x}(\nu+1)^\top\end{pmatrix}^\top \begin{pmatrix}\boldsymbol{\mathcal{X}}_\nu \\ \boldsymbol{x}(\nu+1)^\top\end{pmatrix} = \boldsymbol{\mathcal{X}}_\nu^\top \boldsymbol{\mathcal{X}}_\nu + \boldsymbol{x}(\nu+1)\boldsymbol{x}(\nu+1^\top)$$

or

$$\boldsymbol{\Sigma}(\nu+1)^{-1} = \boldsymbol{\Sigma}(\nu)^{-1} + \boldsymbol{x}(\nu+1)\boldsymbol{x}(\nu+1)^\top. \tag{A.7}$$

To obtain a similar recursive formula for $\boldsymbol{w}(\nu+1)$ we multiply Equation (A.6) out, giving

$$\boldsymbol{w}(\nu+1) = \boldsymbol{\Sigma}(\nu+1)(\boldsymbol{\mathcal{X}}_\nu^\top \boldsymbol{y}_\nu + \boldsymbol{x}(\nu+1)y(\nu+1)),$$

and replace $\boldsymbol{y}_\nu$ with $\boldsymbol{\mathcal{X}}_\nu \boldsymbol{w}(\nu)$ to obtain

$$\boldsymbol{w}(\nu+1) = \boldsymbol{\Sigma}(\nu+1)\Big(\boldsymbol{\mathcal{X}}_\nu^\top \boldsymbol{\mathcal{X}}_\nu \boldsymbol{w}(\nu) + \boldsymbol{x}(\nu+1)y(\nu+1)\Big).$$

Using Equations (A.4) and (A.7),

$$\begin{aligned}\boldsymbol{w}(\nu+1) &= \boldsymbol{\Sigma}(\nu+1)\Big(\boldsymbol{\Sigma}(\nu)^{-1}\boldsymbol{w}(\nu) + \boldsymbol{x}(\nu+1)y(\nu+1)\Big)\\ &= \boldsymbol{\Sigma}(\nu+1)\Big[\boldsymbol{\Sigma}(\nu+1)^{-1}\boldsymbol{w}(\nu) - \boldsymbol{x}(\nu+1)\boldsymbol{x}(\nu+1)^\top \boldsymbol{w}(\nu) + \boldsymbol{x}(\nu+1)y(\nu+1)\Big].\end{aligned}$$

This simplifies to

$$\boldsymbol{w}(\nu+1) = \boldsymbol{w}(\nu) + \boldsymbol{K}(\nu+1)\Big[y(\nu+1) - \boldsymbol{x}(\nu+1)^\top \boldsymbol{w}(\nu)\Big], \tag{A.8}$$

where the *Kalman gain* $\boldsymbol{K}(\nu+1)$ is given by

$$\boldsymbol{K}(\nu+1) = \boldsymbol{\Sigma}(\nu+1)\boldsymbol{x}(\nu+1). \tag{A.9}$$

Equations (A.7–A.9) define a so-called *Kalman filter* for the least squares problem of Equation (A.1). For observations

$$\boldsymbol{x}(\nu+1) \quad \text{and} \quad y(\nu+1)$$

the *system response* $\boldsymbol{x}(\nu+1)^\top \boldsymbol{w}(\nu)$ is calculated and compared in Equation (A.8) with the measurement $y(\nu+1)$. Then the *innovation*, that is to say the difference between the measurement and system response, is multiplied by the Kalman gain $\boldsymbol{K}(\nu+1)$ determined by Equations (A.9) and (A.7) and the old estimate $\boldsymbol{w}(\nu)$ for the parameter vector $\boldsymbol{w}$ is corrected to the new value $\boldsymbol{w}(\nu+1)$.

Relation (A.7) is inconvenient as it calculates the inverse of the covariance matrix $\mathbf{\Sigma}(\nu+1)$, whereas we require the noninverted form in order to determine the Kalman gain in Equation (A.9). However, equations (A.7) and (A.9) can be reformed as follows:

$$\begin{aligned}\mathbf{\Sigma}(\nu+1) &= \Big[\boldsymbol{I} - \boldsymbol{K}(\nu+1)\boldsymbol{x}(\nu+1)^\top\Big]\mathbf{\Sigma}(\nu)\\ \boldsymbol{K}(\nu+1) &= \mathbf{\Sigma}(\nu)\boldsymbol{x}(\nu+1)\Big[\boldsymbol{x}(\nu+1)^\top\mathbf{\Sigma}(\nu)\boldsymbol{x}(\nu+1)+1\Big]^{-1}.\end{aligned} \tag{A.10}$$

To see this, first of all note that the second equation above is a consequence of the first equation and Equation (A.9). Therefore it suffices to show that the first equation is indeed the inverse of Equation (A.7):

$$\begin{aligned}&\mathbf{\Sigma}(\nu+1)\mathbf{\Sigma}(\nu+1)^{-1} = \Big[\boldsymbol{I} - \boldsymbol{K}(\nu+1)\boldsymbol{x}(\nu+1)^\top\Big]\mathbf{\Sigma}(\nu)\mathbf{\Sigma}(\nu+1)^{-1}\\ &= \boldsymbol{I} - \boldsymbol{K}(\nu+1)\boldsymbol{x}(\nu+1)^\top + \Big[\boldsymbol{I} - \boldsymbol{K}(\nu+1)\boldsymbol{x}(\nu+1)^\top\Big]\mathbf{\Sigma}(\nu)\boldsymbol{x}(\nu+1)\boldsymbol{x}(\nu+1)^\top\\ &= \boldsymbol{I} - \boldsymbol{K}(\nu+1)\boldsymbol{x}(\nu+1)^\top + \mathbf{\Sigma}(\nu)\boldsymbol{x}(\nu+1)\boldsymbol{x}(\nu+1)^\top\\ &\quad - \boldsymbol{K}(\nu+1)\boldsymbol{x}(\nu+1)^\top\mathbf{\Sigma}(\nu)\boldsymbol{x}(\nu+1)\boldsymbol{x}(\nu+1)^\top.\end{aligned}$$

The second equality above follows from Equation (A.7). But from the second of Equations (A.10) we have

$$\boldsymbol{K}(\nu+1)\boldsymbol{x}(\nu+1)^\top\mathbf{\Sigma}(\nu)\boldsymbol{x}(\nu+1) = \mathbf{\Sigma}(\nu)\boldsymbol{x}(\nu+1) - \boldsymbol{K}(\nu+1)$$

and therefore

$$\begin{aligned}\mathbf{\Sigma}(\nu+1)\mathbf{\Sigma}(\nu+1)^{-1} &= \boldsymbol{I} - \boldsymbol{K}(\nu+1)\boldsymbol{x}(\nu+1)^\top + \mathbf{\Sigma}(\nu)\boldsymbol{x}(\nu+1)\boldsymbol{x}(\nu+1)^\top\\ &\quad - (\mathbf{\Sigma}(\nu)\boldsymbol{x}(\nu+1) - \boldsymbol{K}(\nu+1))\boldsymbol{x}(\nu+1)^\top = \boldsymbol{I}\end{aligned}$$

as required.

A.4.2 Orthogonal linear regression

In the model for ordinary linear regression described in Chapter 2, the independent variable x is assumed to be error-free. If we are regressing one spectral band against another, for example, then this is manifestly not the case. If we impose the model

$$Y(\nu) - R(\nu) = a + b(X(\nu) - S(\nu)), \quad i = 1\ldots m, \tag{A.11}$$

with $R(\nu)$ and $S(\nu)$ being uncorrelated, normally distributed random variables with mean zero and equal variances σ^2, then we might consider the analog of Equation (2.45) as a starting point:

$$z(a,b) = \sum_{\nu=1}^{m}\frac{(y(\nu) - a - bx(\nu))^2}{\sigma^2 + b^2\sigma^2}. \tag{A.12}$$

Finding the minimum of Equation (A.12) with respect to a and b is now more difficult because of the nonlinear dependence on b. Let us begin with the derivative with respect to a:

$$\frac{\partial z(a,b)}{\partial a} = 0 = -\frac{2}{\sigma^2(1+b^2)}\sum_\nu (y(\nu) - a - bx(\nu))$$

which leads to the estimate

$$\hat{a} = \bar{y} - b\bar{x}. \tag{A.13}$$

Differentiating with respect to b, we obtain

$$0 = \frac{2b}{1+b^2}\sum_\nu (y(\nu) - a - bx(\nu))^2 + 2\sum_\nu (y(\nu) - a - bx(\nu))x(\nu),$$

which simplifies to

$$\sum_\nu (y(\nu) - a - bx(\nu))[b(y(\nu) - a) + x(\nu)] = 0.$$

Now substitute $\hat{a}$ for a using Equation (A.13). This gives

$$\sum_\nu [y(\nu) - \bar{y} - b(x(\nu) - \bar{x})][b(y(\nu) - \bar{y} + b\bar{x}) + x(\nu)] = 0.$$

This equation is in fact only quadratic in b, since the cubic term is

$$-b^3\bar{x}\sum_\nu (x(\nu) - \bar{x}) = 0.$$

The quadratic term is, with the definition of s_{xy} given in Equation (2.86),

$$b^2\sum_\nu \big((y(\nu) - \bar{y})\bar{x} - (x(\nu) - \bar{x})(y(\nu) - \bar{y}) - (x(\nu) - \bar{x})\bar{x}\big) = -mb^2 s_{xy}.$$

The linear term is, defining

$$s_{yy} = \frac{1}{m}\sum_{\nu=1}^{m}(y(\nu) - \bar{y})^2,$$

given by

$$b\sum_\nu (y(\nu) - \bar{y})^2 - (x(\nu) - \bar{x})x(\nu) = mb(s_{yy} - s_{xx}),$$

because of the inequality $\sum_\nu (x(\nu) - \bar{x})x(\nu) = \sum_\nu (x(\nu) - \bar{x})(x(\nu) - \bar{x})$. Similarly, the constant term is

$$\sum_\nu (y(\nu) - \bar{y})x(\nu) = \sum_\nu (y(\nu) - \bar{y})(x(\nu) - \bar{x}) = ms_{xy}.$$

Listing A.1: An IDL procedure for orthogonal linear regression.

```
PRO ortho_regress, X, Y, b, Xm, Ym, $
                   sigma_a, sigma_b, sigma, rank=rank
   m = n_elements(X)
   Xm = mean(X)
   Ym = mean(Y)
   S = correlate([X,Y],/covariance,/double)
   Sxx = S[0,0]
   Syy = S[1,1]
   Sxy = S[1,0]
   void = eigenql(S,eigenvectors=eigenvectors,/double)
; slope
   b = eigenvectors[1,0]/eigenvectors[0,0]
; standard errors
   sigma2 = m*(Syy-2*b*Sxy+b*b*Sxx)/((m-2)*(1+b^2))
   tau = sigma2*b/((1+b^2)*Sxy)
   sigma_b=sqrt( sigma2*b*(1+b^2)*(1+tau)/(m*Sxy))
   sigma_a=sqrt( sigma2*b*(1+b^2)*(Xm^2*(1+tau)+Sxy/b)$
                                              /(m*Sxy) )
   sigma = sqrt(sigma2)
   rank = r_correlate(X,Y)
END
```

Thus b is a solution of the quadratic equation

$$b^2 s_{xy} + b(s_{xx} - s_{yy}) - s_{xy} = 0.$$

The solution (for positive slope) is

$$\hat{b} = \frac{(s_{yy} - s_{xx}) + \sqrt{(s_{yy} - s_{xx})^2 + 4s_{xy}^2}}{2s_{xy}}. \tag{A.14}$$

According to Patefield (1977) and Bilbo (1989), the variances in the regression parameters are given by

$$\begin{aligned} \sigma_a^2 &= \frac{\sigma^2 \hat{b}(1+\hat{b}^2)}{m s_{xy}} \left(\bar{x}^2(1+\hat{\tau}) + \frac{s_{xy}}{\hat{b}} \right) \\ \sigma_b^2 &= \frac{\sigma^2 \hat{b}(1+\hat{b}^2)}{m s_{xy}} \, (1+\hat{\tau}) \end{aligned} \tag{A.15}$$

with

$$\hat{\tau} = \frac{\sigma^2 \hat{b}}{(1+\hat{b}^2) s_{xy}}. \tag{A.16}$$

If σ^2 is not known *a priori*, it can be estimated by (Kendall and Stuart, 1979)

$$\hat{\sigma}^2 = \frac{m}{(m-2)(1+\hat{b}^2)} (s_{yy} - 2\hat{b} s_{xy} + \hat{b}^2 s_{xx}). \tag{A.17}$$

It is easy to see that the estimate $\hat{b}$, Equation (A.14), can be calculated as the slope of the first principal component vector $\boldsymbol{u} = (U-1, u_2)^\top$ of the covariance matrix

$$\boldsymbol{s} = \begin{pmatrix} s_{xx} & s_{xy} \\ s_{yx} & s_{yy} \end{pmatrix},$$

that is,

$$\hat{b} = u_2/u_1.$$

Thus orthogonal linear regression on one independent variable is equivalent to principal components analysis. This is the basis for the IDL procedure shown in Listing A.1, which performs orthogonal regression on the input arrays `X` and `Y`. The routine is used in some of the ENVI/IDL extensions described in Appendix C. The Python counterpart `orthoregress()` can be found in the module `auxil.py`.

A.5 Proof of Theorem 7.1

We need the following inequality:

$$\alpha^r \le 1 - (1-\alpha)r, \tag{A.18}$$

which holds for any $\alpha \ge 0$ and $r \in [0,1]$. To see this, note that α^r is convex, that is, its second derivative is

$$\frac{d^2}{dr^2}(\alpha^r) = \alpha^r \log(\alpha)^2 \ge 0,$$

so we have the situation shown in Figure A.1.

Proof:(Freund and Shapire, 1997) Applying inequality (A.18) to step (e) in the AdaBoost algorithm (Section 7.3), we obtain

$$\begin{aligned}\sum_{\nu=1}^m w_{i+1}(\nu) &= \sum_{\nu=1}^m w_i(\nu)\beta_i^{1-[[h_i(\nu)\ne k(\nu)]]} \\ &\le \sum_{\nu=1}^m w_i(\nu)[1-(1-\beta_i)(1-[[h_i(\nu)\ne k(\nu)]])] \\ &= \sum_{\nu=1}^m w_i(\nu) - (1-\beta_i)\left[\sum_{\nu=1}^m w_i(\nu)(1-[[h_i(\nu)\ne k(\nu)]])\right].\end{aligned}$$

From steps (a) and (c) in the algorithm, we can write

$$\sum_{\nu=1}^m w_i(\nu)[[h_i(\nu)\ne k(\nu)]] = \sum_{\nu=1}^m\sum_{\nu'=1}^m w_i(\nu')p_i(\nu)[[h_i(\nu)\ne k(\nu)]] = \sum_{\nu'=1}^m w_i(\nu')\epsilon_i.$$

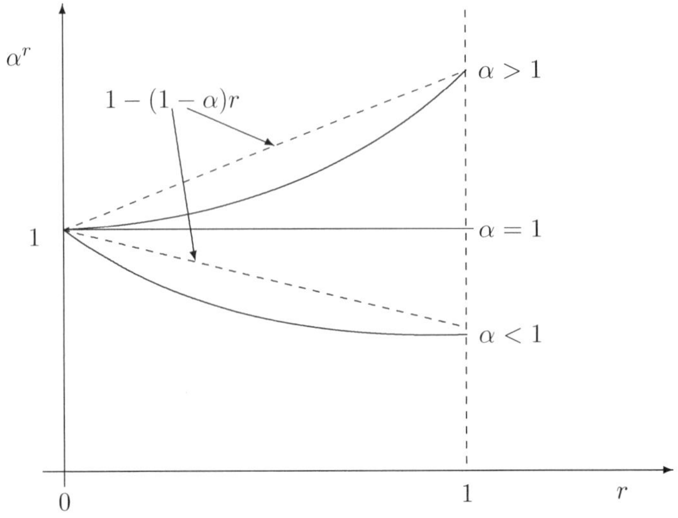

FIGURE A.1
Illustrating Inequality (A.18).

Combining the last two equations then gives

$$\sum_{\nu=1}^{m} w_{i+1}(\nu) \leq \sum_{\nu=1}^{m} w_i(\nu)[1-(1-\beta_i)(1-\epsilon_i)]. \tag{A.19}$$

If we apply this inequality successively for $i = 1 \ldots N_c$, it follows that

$$\begin{aligned} \sum_{\nu=1}^{m} w_{N_c+1}(\nu) &\leq \sum_{\nu=1}^{m} w_1(\nu) \prod_{i=1}^{m} [1-(1-\beta_i)(1-\epsilon_i)] \\ &= \prod_{i=1}^{m} [1-(1-\beta_i)(1-\epsilon_i)], \end{aligned} \tag{A.20}$$

since, in the algorithm, the initial weights sum to unity.

The final voting procedure in step (3) will make an error on training example ν if

$$\sum_{\{i|h_i(\nu)\neq k(\nu)\}} \log(1/\beta_i) \;\geq \sum_{\{i|h_i(\nu)=k(\nu)\}} \log(1/\beta_i).$$

Adding $\sum_{\{i|h_i(\nu)\neq k(\nu)\}} \log(1/\beta_i)$ to both sides,

$$2\left(\sum_{\{i|h_i(\nu)\neq k(\nu)\}} \log(1/\beta_i) \right) \geq \sum_{i=1}^{m} \log(1/\beta_i),$$

or equivalently

$$-2\left(\sum_{\{i|h_i(\nu)\neq k(\nu)\}} \log(\beta_i)\right) \geq -\sum_{i=1}^{m} \log(\beta_i),$$

which we can write in the form

$$-\log\left(\prod_{\{i|h_i(\nu)\neq k(\nu)\}} \beta_i\right) \geq -\frac{1}{2}\log\left(\prod_{i=1}^{N_c} \beta_i\right)$$

or, since the logarithm is a monotonic increasing function of its argument, equivalently as

$$\prod_{i=1}^{N_c} \beta_i^{[[h_i(\nu)\neq k(\nu)]]} \geq \left(\prod_{i=1}^{N_c} \beta_i\right)^{-1/2}. \tag{A.21}$$

From step (e) we have further

$$w_{N_c+1} = w_1(\nu)\prod_{i=1}^{N_c} \beta_i^{1-[[h_i(\nu)\neq k(\nu)]]}. \tag{A.22}$$

Now let U be the set of incorrectly classified training examples for the sequence of N_c classifiers. Then

$$\begin{aligned}
\sum_{\nu=1}^{m} w_{N_c+1}(\nu) &\geq \sum_{\nu\in U} w_{N_c+1}(\nu) \\
&= \sum_{\nu\in U} w_1(\nu)\prod_{i=1}^{N_c} \beta_i^{1-[[h_i(\nu)\neq k(\nu)]]} \quad \text{from } (A.22) \\
&= \sum_{\nu\in U} w_1(\nu)\prod_{i=1}^{N_c} \beta_i \prod_{i=1}^{N_c} \beta_i^{-[[h_i(\nu)\neq k(\nu)]]} \\
&\geq \sum_{\nu\in U} w_1(\nu)\left(\prod_{i=1}^{N_c} \beta_i\right)^{1/2} \quad \text{from } (A.21).
\end{aligned}$$

But $\sum_{\nu\in U} w_1(\nu) = \epsilon$ and, combining this last inequality with Inequality (A.20), we have

$$\epsilon\left(\prod_{i=1}^{N_c} \beta_i\right)^{1/2} \leq \sum_{\nu=1}^{m} w_{N_c+1}(\nu) \leq \prod_{i=1}^{m}[1-(1-\beta_i)(1-\epsilon_i)]$$

or, solving for ϵ,

$$\epsilon \leq \prod_{i=1}^{N_c}\left(\frac{1-(1-\beta_i)(1-\epsilon_i)}{\sqrt{\beta_i}}\right). \tag{A.23}$$

Since each of the N_c factors is positive, the upper bound will be minimized when

$$\frac{d}{d\beta_i}\left(\frac{1-(1-\beta_i)(1-\epsilon_i)}{\sqrt{\beta_i}}\right) = 0, \quad i = 1 \dots N_c,$$

with solution

$$\beta_i = \frac{\epsilon_i}{1-\epsilon_i}.$$

Substituting this back into Equation (A.23) gives the upper bound Equation (7.26) and the proof is complete. □

B

Efficient Neural Network Training Algorithms

The standard backpropagation algorithm introduced in Chapter 6 is notoriously slow to converge. In this appendix we will develop two additional training algorithms for the two-layer, feed-forward neural network of Figure 6.10. The first of these, *scaled conjugate gradient*, makes use of the second derivatives of the cost function with respect to the synaptic weights, i.e., of the Hessian matrix. The second, the *extended Kalman filter* method, takes advantage of the statistical properties of the weight parameters themselves. Both techniques are considerably more efficient than backpropagation.

B.1 The Hessian matrix

We begin with a detailed discussion of the Hessian matrix and how to calculate it efficiently. The Hessian matrix $\boldsymbol{H}$ for a neural network training cost function $E(\boldsymbol{w})$ is given by

$$(\boldsymbol{H})_{ij} = \frac{\partial^2 E(\boldsymbol{w})}{\partial w_i \partial w_j}; \tag{B.1}$$

see Equation (1.56). It is the (symmetric) matrix of second-order partial derivatives of the cost function with respect to the synaptic weights, the latter being thought of as a single column vector

$$\boldsymbol{w} = \begin{pmatrix} \boldsymbol{w}_1^h \\ \vdots \\ \boldsymbol{w}_L^h \\ \boldsymbol{w}_1^o \\ \vdots \\ \boldsymbol{w}_K^o \end{pmatrix}$$

of length $n_w = L(N+1) + K(L+1)$ for the network architecture of Figure 6.10. Since $\boldsymbol{H}$ is symmetric, it is positive definite if and only if its eigenvalues are positive; see Section 1.4. Thus a good way to check if one is at or near a local minimum in the cost function is to examine the eigenvalues of the Hessian matrix.

The scaled conjugate gradient algorithm makes explicit use of the Hessian matrix for more efficient convergence to a minimum in the cost function. The disadvantage of using $\boldsymbol{H}$ is that it is difficult to compute. For example, for a typical classification problem with $N = 3$-dimensional input data, $L = 8$ hidden neurons and $K = 12$ land use categories, there are

$$\big(L(N+1)+K(L+1)\big)\big(L(N+1)+K(L+1)+1\big)/2 = 9870$$

matrix elements to determine at each iteration (allowing for symmetry). We develop, in the following, an efficient method (Bishop, 1995) not to calculate $\boldsymbol{H}$ directly, but rather the product $\boldsymbol{v}^\top \boldsymbol{H}$ for any vector $\boldsymbol{v}$ having n_w components.

B.1.1 The R-operator

Let us begin by summarizing some results from Chapter 6 for the two-layer, feed-forward network, changing the notation slightly to simplify what follows:

$\boldsymbol{g}' = (g'_1 \dots g'_N)^\top$	input observation vector
$\boldsymbol{g} = \begin{pmatrix} 1 \\ \boldsymbol{g}' \end{pmatrix}$	biased input observation
$\boldsymbol{\ell} = (0 \dots 1 \dots 0)^\top$	class label
$\boldsymbol{I}^h = \boldsymbol{W}^{h\top} \boldsymbol{g}$	activation vector for the hidden layer
$n'_j = f(I^h_j),\ j = 1 \dots L$	output signal vector from the hidden layer
$\boldsymbol{n} = \begin{pmatrix} 1 \\ \boldsymbol{n}' \end{pmatrix}$	biased output signal vector
$\boldsymbol{I}^o = \boldsymbol{W}^{o\top} \boldsymbol{n}$	activation vector for the output layer
$m_k(\boldsymbol{I}^o),\ k = 1 \dots K$	softmax output signal from kth output neuron.

(B.2)

The corresponding activation functions are, for the hidden neurons,

$$f(I^h_j) = \frac{1}{1 + e^{-I^h_j}}, \quad j = 1 \dots L, \tag{B.3}$$

and for the output neurons,

$$m_k(\boldsymbol{I}^o) = \frac{e^{I^o_k}}{\sum_{k'=1}^K e^{I^o_{k'}}}, \quad k = 1 \dots K. \tag{B.4}$$

The first derivatives of the local cross-entropy cost function, Equation (6.34), with respect to the output and hidden weights, Equations (6.36) and (6.40), can be written concisely as

$$\begin{aligned}\frac{\partial E}{\partial \boldsymbol{W}^o} &= -\boldsymbol{n}\boldsymbol{\delta}^{o\top} \\ \frac{\partial E}{\partial \boldsymbol{W}^h} &= -\boldsymbol{g}\boldsymbol{\delta}^{h\top},\end{aligned} \tag{B.5}$$

where, see Equations (6.38) and (6.42),

$$\boldsymbol{\delta}^o = \boldsymbol{\ell} - \boldsymbol{m} \tag{B.6}$$

and

$$\begin{pmatrix} 0 \\ \boldsymbol{\delta}^h \end{pmatrix} = \boldsymbol{n} \cdot (1 - \boldsymbol{n}) \cdot \boldsymbol{W}^o \boldsymbol{\delta}^o. \tag{B.7}$$

(The dot denotes component-by-component multiplication.) Following Bishop (1995), we introduce the *R-operator* according to the definition

$$R_v\{\boldsymbol{x}\} := \boldsymbol{v}^\top \frac{\partial}{\partial \boldsymbol{w}} \boldsymbol{x}, \quad \boldsymbol{v}^\top = (v_1 \dots v_{n_w}).$$

We have

$$R_v\{\boldsymbol{w}\} = \boldsymbol{v}^\top \frac{\partial}{\partial \boldsymbol{w}} \boldsymbol{w} = \sum_j v_j \frac{\partial \boldsymbol{w}}{\partial w_j} = \boldsymbol{v}.$$

Note that we are now taking derivatives of vectors. This shouldn't confuse us. For example, the above result in two dimensions is

$$R_v\{\boldsymbol{w}\} = v_1 \frac{\partial}{\partial w_1}(w_1 \boldsymbol{i} + w_2 \boldsymbol{j}) + v_2 \frac{\partial}{\partial w_2}(w_1 \boldsymbol{i} + w_2 \boldsymbol{j}) = v_1 \boldsymbol{i} + v_2 \boldsymbol{j} = \boldsymbol{v}.$$

We adopt the convention that the result of applying the R-operator has the same structure as the argument to which it is applied. Thus, for example,

$$R_v\{\boldsymbol{W}^h\} = \boldsymbol{V}^h,$$

where $\boldsymbol{V}^h$, like $\boldsymbol{W}^h$, is an $(N+1) \times L$ matrix consisting of the first $(N+1)L$ components of the n_w-dimensional vector $\boldsymbol{v}$. Implicitly we set the last $n_w - (N+1)L$ components of $\boldsymbol{v}$ equal to zero.

Next we derive an expression for $\boldsymbol{v}^\top \boldsymbol{H}$ in terms of the R-operator.

$$(\boldsymbol{v}^\top \boldsymbol{H})_j = \sum_{i=1}^{n_w} v_i H_{ij} = \sum_{i=1}^{n_w} v_i \frac{\partial^2 E}{\partial w_i \partial w_j} = \sum_{i=1}^{n_w} v_i \frac{\partial}{\partial w_i}\left(\frac{\partial E}{\partial w_j}\right)$$

or

$$(\boldsymbol{v}^\top \boldsymbol{H})_j = \boldsymbol{v}^\top \frac{\partial}{\partial \boldsymbol{w}}\left(\frac{\partial E}{\partial w_j}\right) = R_v\left\{\frac{\partial E}{\partial w_j}\right\}, \quad j = 1 \dots n_w.$$

Since $\boldsymbol{v}^\top \boldsymbol{H}$ is a row vector, this can be written

$$\boldsymbol{v}^\top \boldsymbol{H} = R_v\left\{\frac{\partial E}{\partial \boldsymbol{w}^\top}\right\} \cong \left(R_v\left\{\frac{\partial E}{\partial \boldsymbol{W}^h}\right\}, R_v\left\{\frac{\partial E}{\partial \boldsymbol{W}^o}\right\}\right). \tag{B.8}$$

Note the reorganization of the structure in the argument of R_v, namely $\boldsymbol{w}^\top \to (\boldsymbol{W}^h, \boldsymbol{W}^o)$. This is merely for convenience of evaluation. Once the expressions on the right have been evaluated, the result must be "flattened" back to a row vector. Note also that Equation (B.8) is understood to involve the local cost function. In order to complete the calculation, we must sum over all training pairs; see Equation (6.33).

Applying R_v to Equations (B.5),

$$\begin{aligned} R_v\left\{\frac{\partial E}{\partial \boldsymbol{W}^o}\right\} &= -\boldsymbol{n} R_v\{\boldsymbol{\delta}^{o\top}\} - R_v\{\boldsymbol{n}\}\boldsymbol{\delta}^{o\top} \\ R_v\left\{\frac{\partial E}{\partial \boldsymbol{W}^h}\right\} &= -\boldsymbol{g} R_v\{\boldsymbol{\delta}^{h\top}\}, \end{aligned} \tag{B.9}$$

so that, in order to evaluate Equation (B.8), we need expressions for

$$R_v\{\boldsymbol{n}\},\ R_v\{\boldsymbol{\delta}^{o\top}\} \text{ and } R_v\{\boldsymbol{\delta}^{h\top}\}.$$

This is somewhat tedious, but well worth the effort.

B.1.1.1 Determination of $R_v\{\boldsymbol{n}\}$

From Equation (B.2) we can write

$$R_v\{\boldsymbol{n}\} = \begin{pmatrix} 0 \\ R_v\{\boldsymbol{n}'\} \end{pmatrix}, \tag{B.10}$$

and from the chain rule,

$$R_v\{\boldsymbol{n}'\} = \boldsymbol{n}' \cdot (1 - \boldsymbol{n}') \cdot R_v\{\boldsymbol{I}^h\}, \tag{B.11}$$

where, by differentiation of $\boldsymbol{I}^h$, we evaluate

$$R_v\{\boldsymbol{I}^h\} = \boldsymbol{V}^{h\top}\boldsymbol{g}. \tag{B.12}$$

Note that, according to our convention, $\boldsymbol{V}^{h\top}$ must be interpreted as an $L \times (N+1)$-dimensional matrix, since the argument $\boldsymbol{I}^h$ is a vector of length L and the result must have the same structure.

B.1.1.2 Determination of $R_v\{\boldsymbol{\delta}^o\}$

With Equations (B.2) and Equation (B.6) we get

$$R_v\{\boldsymbol{\delta}^o\} = -R_v\{\boldsymbol{m}\} = -\boldsymbol{v}^\top \frac{\partial \boldsymbol{m}}{\partial \boldsymbol{w}} = -\boldsymbol{v}^\top \frac{\partial \boldsymbol{m}}{\partial \boldsymbol{I}^o} \cdot \frac{\partial \boldsymbol{I}^o}{\partial \boldsymbol{w}} = -\frac{\partial \boldsymbol{m}}{\partial \boldsymbol{I}^o} \cdot R_v\{\boldsymbol{I}^o\},$$

But from Equation (B.4) it is easy to see that

$$\frac{\partial \boldsymbol{m}}{\partial \boldsymbol{I}^o} = \boldsymbol{m} \cdot (\boldsymbol{1} - \boldsymbol{m})$$

and therefore

$$R_v\{\boldsymbol{\delta}^o\} = -\boldsymbol{m}\cdot(\mathbf{1}-\boldsymbol{m})\cdot R_v\{\boldsymbol{I}^o\}. \tag{B.13}$$

Similarly, with the expression for $\boldsymbol{I}^o$ in Equations (B.2) we get

$$R_v\{\boldsymbol{I}^o\} = \boldsymbol{W}^{o\top} R_v\{\boldsymbol{n}\} + \boldsymbol{V}^{o\top}\boldsymbol{n}, \tag{B.14}$$

where $R_v\{\boldsymbol{n}\}$ is given by Equations (B.10) to (B.12).

B.1.1.3 Determination of $R_v\{\boldsymbol{\delta}^h\}$

We begin by writing Equation (B.7) in the form

$$\begin{pmatrix} 0 \\ \boldsymbol{\delta}^h \end{pmatrix} = \begin{pmatrix} 0 \\ f'(\boldsymbol{I}^h) \end{pmatrix} \cdot \boldsymbol{W}^o\boldsymbol{\delta}^o,$$

where

$$f'(\boldsymbol{I}^h) = (f'(I_1^h)\ldots f'(I_L^h))^\top$$

and where the prime on f denotes differentiation with respect to its argument, $f'(x) = df(x)/dx$. Operating with $R_v\{\cdot\}$ and applying the chain rule, we obtain

$$\begin{pmatrix} 0 \\ R_v\{\boldsymbol{\delta}^h\} \end{pmatrix} = \begin{pmatrix} 0 \\ f''(\boldsymbol{I}^h) \end{pmatrix} \cdot \begin{pmatrix} 0 \\ R_v\{\boldsymbol{I}^h\} \end{pmatrix} \cdot \boldsymbol{W}^o\boldsymbol{\delta}^o + \begin{pmatrix} 0 \\ f'(\boldsymbol{I}^h) \end{pmatrix} \cdot \boldsymbol{V}^o\boldsymbol{\delta}^o + \begin{pmatrix} 0 \\ f'(\boldsymbol{I}^h) \end{pmatrix} \cdot \boldsymbol{W}^o R_v\{\boldsymbol{\delta}^o\}. \tag{B.15}$$

Finally, substitute the derivatives of the logistic function

$$f'(\boldsymbol{I}^h) = \boldsymbol{n}'(\mathbf{1}-\boldsymbol{n}')$$
$$f''(\boldsymbol{I}^h) = \boldsymbol{n}'(\mathbf{1}-\boldsymbol{n}')(\mathbf{1}-2\boldsymbol{n}')$$

into Equation (B.15) to obtain

$$\begin{pmatrix} 0 \\ R_v\{\boldsymbol{\delta}^h\} \end{pmatrix} = \boldsymbol{n}\cdot(\mathbf{1}-\boldsymbol{n})\cdot\left[(\mathbf{1}-2\boldsymbol{n})\cdot\begin{pmatrix} 0 \\ R_v\{\boldsymbol{I}^h\} \end{pmatrix}\cdot\boldsymbol{W}^o\boldsymbol{\delta}^o + \boldsymbol{V}^o\boldsymbol{\delta}^o + \boldsymbol{W}^o R_v\{\boldsymbol{\delta}^o\}\right], \tag{B.16}$$

in which all of the terms on the right have now been determined. As already mentioned, we have done everything so far in terms of the local cost function. The final step in the calculation involves summing over all of the training examples. This concludes the evaluation of Equation (B.8).

B.1.2 Calculating the Hessian

To calculate the Hessian matrix for the neural network, we evaluate Equation (B.8) successively for the vectors

$$\boldsymbol{v}_1^\top = (1, 0, 0\ldots 0) \quad \ldots \quad \boldsymbol{v}_{n_w}^\top = (0, 0, 0\ldots 1)$$

and build up $\boldsymbol{H}$ row for row:

$$\boldsymbol{H} = \begin{pmatrix} \boldsymbol{v}_1^\top \boldsymbol{H} \\ \vdots \\ \boldsymbol{v}_{n_w}^\top \boldsymbol{H} \end{pmatrix}.$$

The excerpt from the IDL program `FFNCG__DEFINE.PRO` shown in Listing B.1 extends the object class `FFN` introduced in Chapter 6 and implements a vectorized version of the above determination of of $\boldsymbol{v}^\top \boldsymbol{H}$ (method `FFNCG::ROP`) and $\boldsymbol{H}$ (method `FFNCG::HESSIAN`). The eigenvalues of $\boldsymbol{H}$ are calculated in method `FFNCG::EIGENVALUES`. The corresponding Python class `Ffncg` is coded in the Python module `supervisedclass.py`.

B.2 Scaled conjugate gradient training

The backpropagation algorithm of Chapter 6 attempts to minimize the cost function *locally*, that is, weight updates are made immediately after presentation of a single training pair to the network. We will now consider a *global* approach aimed at minimization of the full cost function, Equation (6.33), which we denote in the following by $E(\boldsymbol{w})$. The symbol $\boldsymbol{w}$ is, as before, the n_w-component vector of synaptic weights.

Let the gradient of the cost function at the point $\boldsymbol{w}$ be $\mathbf{g}(\boldsymbol{w})$,* i.e.,

$$\mathbf{g}(\boldsymbol{w}) = \frac{\partial}{\partial \boldsymbol{w}} E(\boldsymbol{w}).$$

The Hessian matrix

$$(\boldsymbol{H})_{ij} = \frac{\partial^2 E(\boldsymbol{w})}{\partial w_i \partial w_j} \quad i, j = 1 \dots n_w$$

can then be expressed conveniently as the outer product

$$\boldsymbol{H} = \frac{\partial}{\partial \boldsymbol{w}} \mathbf{g}(\boldsymbol{w})^\top. \tag{B.17}$$

B.2.1 Conjugate directions

The search for a minimum in the cost function can be visualized as tracing out a series of points in the space of synaptic weight parameters,

$$\boldsymbol{w}^1, \boldsymbol{w}^2 \dots \boldsymbol{w}^{k-1}, \boldsymbol{w}^k, \boldsymbol{w}^{k+1} \dots ,$$

*The symbol $\mathbf{g}$ for gradient should not to be confused with the observation vector $\boldsymbol{g}$.

Listing B.1: Excerpt from the object class FFNCG.

```
FUNCTION FFNCG::Rop, V
COMPILE_OPT STRICTARR
; reform V to dimensions of Wh and Wo and transpose
  VhT=transpose(reform(V[0:self.LL*(self.NN+1)-1], $
                  self.LL,self.NN+1))
  Vo=reform(V[self.LL*(self.NN+1):*],self.KK,self.LL+1)
  VoT = transpose(Vo)
; transpose the weights
  Wo  = *self.Wo
  WoT = transpose(Wo)
; vectorized forward pass
  M = self->vForwardPass(*self.GTs,N)
; evaluation of v^T.H
  Zeroes = fltarr(self.ntp)
  D_o=*self.LTs-M                               ;d^o
  RIh=VhT##(*self.GTs)                          ;Rv{I^h}
  RN=N*(1-N)*[[Zeroes],[RIh]]                   ;Rv{n}
  RIo=WoT##RN + VoT##N                          ;Rv{I^o}
  Rd_o=-M*(1-M)*RIo                             ;Rv{d^o}
  Rd_h=N*(1-N)*((1-2*N)*[[Zeroes],[RIh]]*(Wo##D_o) $
         + Vo##D_o + Wo##Rd_o)
  Rd_h=Rd_h[*,1:*]                              ;Rv{d^h}
  REo=-N##transpose(Rd_o)-RN##transpose(D_o);Rv{dE/dWo}
  REh=-*self.GTs##transpose(Rd_h)               ;Rv{dE/dWh}
  RETURN, [REh[*],REo[*]]                       ;v^T.H
END

FUNCTION FFNCG::Hessian
COMPILE_OPT STRICTARR
   nw = self.LL*(self.NN+1)+self.KK*(self.LL+1)
   v = diag_matrix(fltarr(nw)+1.0)
   H = fltarr(nw,nw)
   FOR i=0,nw-1 DO H[*,i] = self -> Rop(v[*,i])
   RETURN, H
END

FUNCTION FFNCG::Eigenvalues
COMPILE_OPT STRICTARR
   H = self->Hessian()
   H = (H+transpose(H))/2
   RETURN, eigenql(H,/double)
END
```

where the point $\boldsymbol{w}^k$ is determined by minimizing $E(\boldsymbol{w})$ along some *search direction* $\boldsymbol{d}^{k-1}$ which originated at the preceding point $\boldsymbol{w}^{k-1}$. This is illustrated

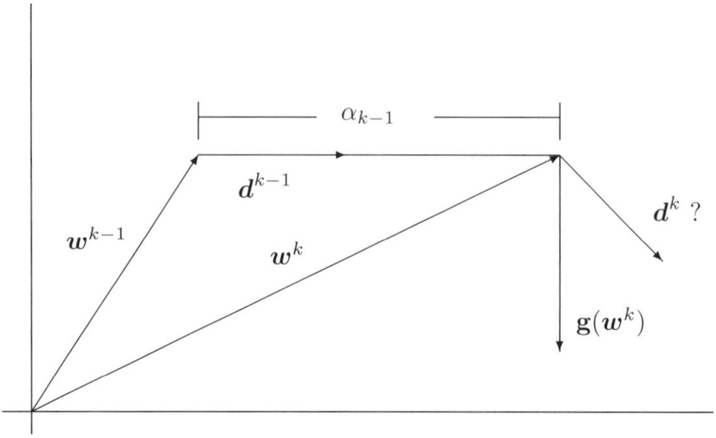

FIGURE B.1
Search directions in weight space.

in Figure B.1 and corresponds to the vector equation

$$\boldsymbol{w}^k = \boldsymbol{w}^{k-1} + \alpha_{k-1}\boldsymbol{d}^{k-1}. \tag{B.18}$$

Here $\boldsymbol{d}^{k-1}$ is a unit vector along the chosen search direction and the scalar α_{k-1} minimizes the cost function along that direction:

$$\alpha_{k-1} = \arg\min_{\alpha} E\big(\boldsymbol{w}^{k-1} + \alpha\boldsymbol{d}^{k-1}\big).$$

If, starting from $\boldsymbol{w}^k$, we now wish to take the next minimizing step in the weight space, it is not efficient simply to choose, as in backpropagation, the direction of the local gradient $\mathbf{g}(\boldsymbol{w}_k)$ at the new starting point $\boldsymbol{w}^k$. Since the cost function has been minimized along the direction $\boldsymbol{d}^{k-1}$ at the point $\boldsymbol{w}^{k-1} + \alpha_{k-1}\boldsymbol{d}^{k-1}$, its gradient along that direction is zero,

$$\mathbf{g}(\boldsymbol{w}^k)^\top \boldsymbol{d}^{k-1} = 0, \tag{B.19}$$

as indicated in Figure B.1. Since the algorithm has just succeeded in reducing the gradient of the cost function along $\boldsymbol{d}^{k-1}$ to zero, we prefer to choose the new search direction $\boldsymbol{d}^k$ so that the component of the gradient along the old search direction remains as small as possible. Otherwise, we are undoing what we have just accomplished. Therefore we choose $\boldsymbol{d}^k$ according to the condition

$$\mathbf{g}\big(\boldsymbol{w}^k + \alpha\boldsymbol{d}^k\big)^\top \boldsymbol{d}^{k-1} = 0. \tag{B.20}$$

But to first order in α we have, with Equation (B.17),

$$\mathbf{g}\big(\boldsymbol{w}^k + \alpha\boldsymbol{d}^k\big)^\top = \mathbf{g}(\boldsymbol{w}^k)^\top + \alpha\boldsymbol{d}^{k\top}\frac{\partial}{\partial\boldsymbol{w}}\mathbf{g}(\boldsymbol{w}^k)^\top = \mathbf{g}(\boldsymbol{w}^k)^\top + \alpha\boldsymbol{d}^{k\top}\boldsymbol{H}$$

and Equation (B.20) is, together with Equation (B.19), equivalent to

$$\boldsymbol{d}^{k\top} \boldsymbol{H} \boldsymbol{d}^{k-1} = 0. \tag{B.21}$$

Directions which satisfy Equation (B.21) are referred to as *conjugate directions.*

B.2.2 Minimizing a quadratic function

Although the neural network cost function is not quadratic in the synaptic weights, within a sufficiently small region of weight space it can be approximated as a quadratic function. We describe in the following an efficient procedure to find the global minimum of a quadratic function of $\boldsymbol{w}$ having the general form (see Equation (1.55))

$$E(\boldsymbol{w}) = E_0 + \boldsymbol{b}^\top \boldsymbol{w} + \frac{1}{2} \boldsymbol{w}^\top \boldsymbol{H} \boldsymbol{w}, \tag{B.22}$$

where $\boldsymbol{b}$ and $\boldsymbol{H}$ are constant and the matrix $\boldsymbol{H}$ is positive definite. This will form the basis of the neural network training algorithm presented in the next subsection.

The local gradient of $E(\boldsymbol{w})$ at the point $\boldsymbol{w}$ is given by

$$\mathrm{g}(\boldsymbol{w}) = \frac{\partial}{\partial \boldsymbol{w}} E(\boldsymbol{w}) = \boldsymbol{b} + \boldsymbol{H}\boldsymbol{w},$$

and it vanishes at the global minimum $\boldsymbol{w}^*$,

$$\boldsymbol{b} + \boldsymbol{H}\boldsymbol{w}^* = \boldsymbol{0}. \tag{B.23}$$

Now let $\{\boldsymbol{d}^k \mid k = 1 \dots n_w\}$ be a set of n_w conjugate directions satisfying Equation (B.21),*

$$\boldsymbol{d}^{k\top} \boldsymbol{H} \boldsymbol{d}^{\ell} = 0 \quad \text{for } k \neq \ell, \quad k, \ell = 1 \dots n_w. \tag{B.24}$$

The search directions $\boldsymbol{d}^k$ are in fact linearly independent. In order to demonstrate this, let us assume the contrary, that is, that there exists an index k and constants $\alpha_{k'}$, $k' \neq k$, not all of which are zero, such that

$$\boldsymbol{d}^k = \sum_{\substack{k'=1 \\ k' \neq k}}^{n_w} \alpha_{k'} \boldsymbol{d}^{k'}.$$

Substituting this into Equation (B.24), we have at once

$$\alpha_{k'} \boldsymbol{d}^{k'\top} \boldsymbol{H} \boldsymbol{d}^{k'} = 0 \quad \text{for } k' \neq k$$

*It can be shown that such a set always exists; see, e.g., Bishop (1995).

and, since $\boldsymbol{H}$ is positive definite,

$$\alpha_{k'} = 0 \quad \text{for } k' \neq k.$$

The assumption leads to a contradiction, hence the $\boldsymbol{d}^k$ are indeed linearly independent. The conjugate directions thus constitute a (nonorthogonal) vector basis for the entire weight space.

In the search for the global minimum, suppose we begin at an arbitrary point $\boldsymbol{w}^1$ and express the vector $\boldsymbol{w}^* - \boldsymbol{w}^1$ spanning the distance to the global minimum as a linear combination of the basis vectors $\boldsymbol{d}^k$:

$$\boldsymbol{w}^* - \boldsymbol{w}^1 = \sum_{k=1}^{n_w} \alpha_k \boldsymbol{d}^k. \tag{B.25}$$

Further, define

$$\boldsymbol{w}^k = \boldsymbol{w}^1 + \sum_{\ell=1}^{k-1} \alpha_\ell \boldsymbol{d}^\ell \tag{B.26}$$

and split Equation (B.25) up into n_w steps

$$\boldsymbol{w}^{k+1} = \boldsymbol{w}^k + \alpha_k \boldsymbol{d}^k, \quad k = 1 \dots n_w. \tag{B.27}$$

At the kth step, the search starts at the point $\boldsymbol{w}^k$ and proceeds a distance α_k along the conjugate direction $\boldsymbol{d}^k$. After n_w such steps, the global minimum $\boldsymbol{w}^*$ is reached, since from Equations (B.25) to (B.27) it follows that

$$\boldsymbol{w}^* = \boldsymbol{w}^1 + \sum_{k=1}^{n_w} \alpha_k \boldsymbol{d}^k = \boldsymbol{w}^2 + \sum_{k=2}^{n_w} \alpha_k \boldsymbol{d}^k = \dots = \boldsymbol{w}^{n_w} + \alpha_{n_w} \boldsymbol{d}^{n_w} = \boldsymbol{w}^{n_w+1}.$$

We get the necessary step sizes α_k from Equation (B.25) by multiplying from the left with ${\boldsymbol{d}^\ell}^\top \boldsymbol{H}$,

$${\boldsymbol{d}^\ell}^\top \boldsymbol{H}\boldsymbol{w}^* - {\boldsymbol{d}^\ell}^\top \boldsymbol{H}\boldsymbol{w}^1 = \sum_{k=1}^{n_w} \alpha_k {\boldsymbol{d}^\ell}^\top \boldsymbol{H}\boldsymbol{d}^k.$$

From Equations (B.23) and (B.24) we can write this as

$$-{\boldsymbol{d}^\ell}^\top (\boldsymbol{b} + \boldsymbol{H}\boldsymbol{w}^1) = \alpha_\ell {\boldsymbol{d}^\ell}^\top \boldsymbol{H}\boldsymbol{d}^\ell,$$

so that an explicit formula for the step sizes is given by

$$\alpha_\ell = -\frac{{\boldsymbol{d}^\ell}^\top (\boldsymbol{b} + \boldsymbol{H}\boldsymbol{w}^1)}{{\boldsymbol{d}^\ell}^\top \boldsymbol{H}\boldsymbol{d}^\ell}, \quad \ell = 1 \dots n_w.$$

But with Equations (B.24) and (B.26),

$${\boldsymbol{d}^k}^\top \boldsymbol{H}\boldsymbol{w}^k = {\boldsymbol{d}^k}^\top \boldsymbol{H}\boldsymbol{w}^1 + 0,$$

and therefore, replacing index k by ℓ,

$$\boldsymbol{d}^{\ell^\top} \boldsymbol{H} \boldsymbol{w}^\ell = \boldsymbol{d}^{\ell^\top} \boldsymbol{H} \boldsymbol{w}^1.$$

The step lengths are thus

$$\alpha_\ell = -\frac{\boldsymbol{d}^{\ell^\top} (\boldsymbol{b} + \boldsymbol{H}\boldsymbol{w}^\ell)}{\boldsymbol{d}^{\ell^\top} \boldsymbol{H} \boldsymbol{d}^\ell}, \quad \ell = 1 \ldots n_w.$$

Finally, using the notation $\mathbf{g}^\ell = \mathbf{g}(\boldsymbol{w}^\ell) = \boldsymbol{b} + \boldsymbol{H}\boldsymbol{w}^\ell$ and substituting $\ell \to k$,

$$\alpha_k = -\frac{\boldsymbol{d}^{k^\top} \mathbf{g}^k}{\boldsymbol{d}^{k^\top} \boldsymbol{H} \boldsymbol{d}^k}, \quad k = 1 \ldots n_w. \tag{B.28}$$

For want of a better alternative, we can choose the first search direction along the negative local gradient

$$\boldsymbol{d}^1 = -\mathbf{g}^1 = -\frac{\partial}{\partial \boldsymbol{w}} E(\boldsymbol{w}^1).$$

(Note that $\boldsymbol{d}^1$ is not a unit vector.) We move, according to Equation (B.28), a distance

$$\alpha_1 = \frac{\boldsymbol{d}^{1^\top} \boldsymbol{d}^1}{\boldsymbol{d}^{1^\top} \boldsymbol{H} \boldsymbol{d}^1}$$

along this direction to the point $\boldsymbol{w}^2$, at which the local gradient $\mathbf{g}^2$ is orthogonal to $\boldsymbol{d}^1$. We then choose the new conjugate search direction $\boldsymbol{d}^2$ as a linear combination of the two:

$$\boldsymbol{d}^2 = -\mathbf{g}^2 + \beta_1 \boldsymbol{d}^1$$

or, at the kth step,

$$\boldsymbol{d}^{k+1} = -\mathbf{g}^{k+1} + \beta_k \boldsymbol{d}^k. \tag{B.29}$$

We get the coefficient β_k from Equations (B.29) and (B.21) by multiplication on the left with $\boldsymbol{d}^{k^\top} \boldsymbol{H}$:

$$0 = -\boldsymbol{d}^{k^\top} \boldsymbol{H} \mathbf{g}^{k+1} + \beta_k \boldsymbol{d}^{k^\top} \boldsymbol{H} \boldsymbol{d}^k,$$

from which follows

$$\beta_k = \frac{\mathbf{g}^{k+1^\top} \boldsymbol{H} \boldsymbol{d}^k}{\boldsymbol{d}^{k^\top} \boldsymbol{H} \boldsymbol{d}^k}. \tag{B.30}$$

Equations (B.27 to B.30) constitute a recipe with which, starting at an arbitrary point $\boldsymbol{w}^1$ in weight space, the global minimum of the quadratic function, Equation (B.22), is found in precisely n_w steps.

B.2.3 The algorithm

Returning now to the nonquadratic neural net cost function $E(\boldsymbol{w})$, we will apply the above method to minimize it. We must take two things into consideration.

First of all, the Hessian matrix $\boldsymbol{H}$ is neither constant nor everywhere positive definite. When $\boldsymbol{H}$ is not positive definite, it can happen that Equation (B.28) leads to a step along the wrong direction — the denominator might turn out to be negative. Therefore we replace Equation (B.28) with*

$$\alpha_k = -\frac{\boldsymbol{d}^{k\top}\mathbf{g}^k}{\boldsymbol{d}^{k\top}\boldsymbol{H}\boldsymbol{d}^k + \lambda_k\|\boldsymbol{d}^k\|^2}, \quad k = 1 \dots n_w. \tag{B.31}$$

The constant λ_k is supposed to ensure that the denominator in Equation (B.31) is always positive. It is initialized for $k = 1$ with a small numerical value. If, at the kth iteration, it is determined that

$$\delta_k := \boldsymbol{d}^{k\top}\boldsymbol{H}\boldsymbol{d}^k + \lambda_k\|\boldsymbol{d}^k\|^2 < 0,$$

then λ_k is replaced by the larger value $\bar{\lambda}_k$ given by

$$\bar{\lambda}_k = 2\left(\lambda_k - \frac{\delta_k}{\|\boldsymbol{d}^k\|^2}\right). \tag{B.32}$$

This ensures that the denominator in Equation (B.31) becomes positive again. Note that this increase in λ_k has the effect of *decreasing* the step size α_k, as is apparent from Equation (B.31).

Second, we must take into account any deviation of the cost function from its local quadratic approximation. Such deviations are to be expected for large step sizes α_k. As a measure of the *quadricity* of $E(\boldsymbol{w})$ along the chosen step length, we can use the ratio

$$\Delta_k = -\frac{2\big(E(\boldsymbol{w}^k) - E(\boldsymbol{w}^k + \alpha_k\boldsymbol{d}^k)\big)}{\alpha_k\boldsymbol{d}^{k\top}\mathbf{g}^k}. \tag{B.33}$$

This quantity is precisely 1 for a strictly quadratic function like Equation (B.22). Therefore we can use the following heuristic: For the $k+1$st iteration

$$\begin{aligned} &\text{if } \Delta_k > 3/4, && \lambda_{k+1} := \lambda_k/2 \\ &\text{if } \Delta_k < 1/4, && \lambda_{k+1} := 4\lambda_k \\ &\quad\text{else,} && \lambda_{k+1} := \lambda_k. \end{aligned}$$

In other words, if the local quadratic approximation looks good according to criterion of Equation (B.33), then the step size can be increased (λ_{k+1} is

*This corresponds to the substitution $\boldsymbol{H} \to \boldsymbol{H} + \lambda_k\boldsymbol{I}$, where $\boldsymbol{I}$ is the identity matrix.

reduced relative to λ_k). If this is not the case, then the step size is decreased (λ_{k+1} is made larger).

All of which leads us, at last, to the following algorithm (Moeller, 1993):

Algorithm (Scaled Conjugate Gradient)

1. Initialize the synaptic weights $\boldsymbol{w}$ with random numbers, set $k = 0$, $\lambda = 0.001$ and $\boldsymbol{d} = -\mathbf{g} = -\partial E(\boldsymbol{w})/\partial \boldsymbol{w}$.

2. Set $\delta = \boldsymbol{d}^\top \boldsymbol{H}\boldsymbol{d} + \lambda\|\boldsymbol{d}\|^2$. If $\delta < 0$, set $\lambda = 2(\lambda - \delta/\|\boldsymbol{d}\|^2)$ and $\delta = -\boldsymbol{d}^\top \boldsymbol{H}\boldsymbol{d}$. Save the current cost function $E1 = E(\boldsymbol{w})$.

3. Determine the step size $\alpha = -\boldsymbol{d}^\top \mathbf{g}/\delta$ and new synaptic weights $\boldsymbol{w} = \boldsymbol{w} + \alpha\boldsymbol{d}$.

4. Calculate the quadricity $\Delta = -(E1 - E(\boldsymbol{w}))/(\alpha\boldsymbol{d}^\top\mathbf{g})$. If $\Delta < 1/4$, restore the old weights: $\boldsymbol{w} = \boldsymbol{w} - \alpha\boldsymbol{d}$, set $\lambda = 4\lambda$, $\boldsymbol{d} = -\mathbf{g}$ and go to 2.

5. Set $k = k + 1$. If $\Delta > 3/4$, set $\lambda = \lambda/2$.

6. Determine the new local gradient $\mathbf{g} = \partial E(\boldsymbol{w})/\partial \boldsymbol{w}$ and the new search direction $\boldsymbol{d} = -\mathbf{g} + \beta\boldsymbol{d}$, whereby, if $k \bmod n_w \neq 0$ then $\beta = \mathbf{g}^\top \boldsymbol{H}\boldsymbol{d}/(\boldsymbol{d}^\top \boldsymbol{H}\boldsymbol{d})$ else $\beta = 0$.

7. If $E(\boldsymbol{w})$ is small enough, stop, else go to 2.

A few remarks on this algorithm:

1. The integer k counts the total number of iterations. Whenever $k \bmod n_w = 0$, then exactly n_w weight updates have been carried out and the minimum of a truly quadratic function would have been reached. This is taken as a good stage at which to restart the search along the negative local gradient $-\boldsymbol{g}$ rather than continuing along the current conjugate direction $\boldsymbol{d}$. One expects that approximation errors will gradually corrupt the determination of the conjugate directions and the "fresh start" is intended to counter this.

2. Whenever the quadricity condition is not filled, i.e., whenever $\Delta < 1/4$, the last weight update is cancelled and the search again restarted along $-\mathbf{g}$.

3. Since the Hessian only occurs in the forms $\boldsymbol{d}^\top \boldsymbol{H}$, and $\mathbf{g}^\top \boldsymbol{H}$, these quantities can be determined efficiently with the R-operator method.

Listings B.2 and B.3 show an excerpt from `FFNGC__DEFINE.PRO` extending the object class `FFN`. They list the training method which implements the scaled conjugate gradient algorithm. Again, see the module `supervisedclass.py` for the corresponding Python implementation.

Listing B.2: Excerpt from the object class FFNCG.

```
PRO FFNCG::Train, key=key
COMPILE_OPT STRICTARR
   IF n_elements(key) EQ 0 THEN key=0
; if validation option then train only on odd examples
   IF key THEN void = $
   where((lindgen(self.np) MOD 2) EQ 0, $
                        complement=indices) $
   ELSE indices = lindgen(self.np)
   self.ntp = n_elements(indices)
   self.GTs = ptr_new((*self.Gs)[indices,*])
   self.LTs = ptr_new((*self.Ls)[indices,*])
   w = [(*self.Wh)[*],(*self.Wo)[*]]
   nw = n_elements(w)
   g = self->gradient()
   d = -g     ; search direction, row vector
   k = 0L
   lambda = 0.001
   window,12,xsize=600,ysize=400, $
      title='FFN(scaled␣conjugate␣gradient)'
   wset,12
   progressbar = Obj_New('cgprogressbar', $
     Color='blue', Text='0', $
     title='Training:␣epoch␣No...',xsize=250,ysize=20)
   progressbar->start
   eivminmax = '?'
   REPEAT BEGIN
      IF progressbar->CheckCancel() THEN BEGIN
         PRINT,'Training␣interrupted'
         progressbar->Destroy
         RETURN
      ENDIF
      d2 = total(d*d)                   ; d^2
      dTHd = total(self->Rop(d)*d)   ; d^T.H.d
      delta = dTHd+lambda*d2
      IF delta LT 0 THEN BEGIN
         lambda = 2*(lambda-delta/d2)
         delta = -dTHd
      ENDIF
      E1 = self->cost(key)              ; E(w)
      (*self.cost_array)[k] = E1
      IF key THEN $
          (*self.valid_cost_array)[k]=self->cost(2)
      dTg = total(d*g)                  ; d^T.g
      alpha = -dTg/delta
```

Listing B.3: Excerpt from the object class `FFNCG` (continued).

```
      dw = alpha*d
      w = w+dw
      *self.Wh = reform(w[0:self.LL*(self.NN+1)-1], $
         self.LL,self.NN+1)
      *self.Wo = reform(w[self.LL*(self.NN+1):*], $
         self.KK,self.LL+1)
      E2 = self->cost(key)           ; E(w+dw)
      Ddelta = -2*(E1-E2)/(alpha*dTg); quadricity
      IF Ddelta LT 0.25 THEN BEGIN
         w = w - dw   ; undo weight change
         *self.Wh = reform(w[0:self.LL*(self.NN+1)-1],$
            self.LL,self.NN+1)
         *self.Wo = reform(w[self.LL*(self.NN+1):*], $
            self.KK,self.LL+1)
         lambda = 4*lambda        ; decrease step size
         IF lambda GT 1e20 THEN $; if step too small
           k=self.epochs   $      ;  then give up
         ELSE d = -g              ;  else restart
      END ELSE BEGIN
         k++
         IF Ddelta GT 0.75 THEN lambda = lambda/2
         g = self->gradient()
         IF k MOD nw EQ 0 THEN BEGIN
             beta = 0
             eivs = self->eigenvalues()
             eivminmax = string(min(eivs)/max(eivs),$
                format='(F10.6)')
         END ELSE beta = total(self->Rop(g)*d)/dTHd
         d = beta*d-g
         IF k MOD 10 EQ 0 THEN PLOT,$
            *self.cost_array,xrange=[0,k>100L], $
            color=0, background='FFFFFF'XL, $
            ytitle='cross␣entropy', $
            xtitle='Epoch␣['+ $
             'min(lambda)/max(lambda)='+ $
             eivminmax+']'
         IF key THEN oplot, $
            *self.valid_cost_array,color=0,linestyle=2
      ENDELSE
      progressbar->Update,k*100/self.epochs
   ENDREP UNTIL k GE self.epochs
   progressbar->Destroy
END
```

B.3 Kalman filter training

In this section the recursive linear regression method described in Appendix A will be applied to train the feed-forward neural network of Figure 6.10. The appropriate cost function in this case is the quadratic function, Equation (6.30), or more specifically, its local version

$$E(\nu) = \frac{1}{2}\|\boldsymbol{\ell}(\nu) - \boldsymbol{m}(\nu)\|^2, \tag{B.34}$$

however our algorithm will also minimize the cross-entropy cost function, as will be mentioned later.

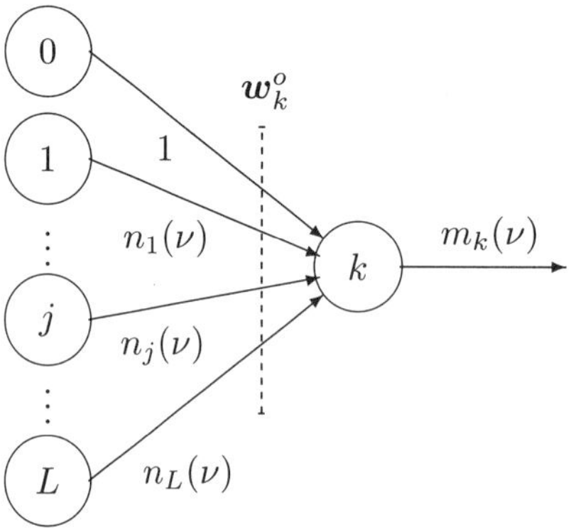

FIGURE B.2
An isolated output neuron.

We begin with consideration of the training process of an isolated neuron. Figure B.2 depicts an output neuron in the network during presentation of the νth training pair $(\boldsymbol{g}(\nu), \boldsymbol{\ell}(\nu))$. The neuron receives its input from the hidden layer (input vector $\boldsymbol{n}(\nu)$) and generates the softmax output signal

$$m_k(\nu) = \frac{e^{\boldsymbol{w}_k^{o\top}(\nu)\boldsymbol{n}(\nu)}}{\sum_{k'=1}^{K} e^{\boldsymbol{w}_{k'}^{o\top}(\nu)\boldsymbol{n}(\nu)}},$$

for $k = 1 \ldots K$, which is compared to the desired output $\ell_k(\nu)$. It is easy to show that differentiation of m_k with respect to $\boldsymbol{w}_k^o$ yields

$$\frac{\partial}{\partial \boldsymbol{w}_k^o} m_k(\nu) = m_k(\nu)(1 - m_k(\nu))\boldsymbol{n}(\nu) \tag{B.35}$$

and with respect to $\boldsymbol{n}$,

$$\frac{\partial}{\partial \boldsymbol{n}} m_k(\nu) = m_k(\nu)(1 - m_k(\nu))\boldsymbol{w}_k^o(\nu). \tag{B.36}$$

B.3.1 Linearization

We shall drop, for the time being, the indices on $\boldsymbol{w}_k^o$, m_k, and ℓ_k, writing them simply as $\boldsymbol{w}$, m and ℓ. The weight vectors for the other $K-1$ output neurons are considered to be frozen, so that m can be thought of as being a function of $\boldsymbol{w}$ only:

$$m(\nu) = m(\boldsymbol{w}(\nu)^\top \boldsymbol{n}(\nu)).$$

The weight vector $\boldsymbol{w}(\nu)$ is an approximation to the desired synaptic weight vector for our isolated output neuron, one which has been achieved so far in the training process, after presentation of the first ν labeled training observations. A linear approximation to $m(\nu+1)$ can be obtained by expanding in a first-order Taylor series about the point $\boldsymbol{w}(\nu)$,

$$m(\nu+1) \approx m(\boldsymbol{w}(\nu)^\top \boldsymbol{n}(\nu+1)) + \left(\frac{\partial}{\partial \boldsymbol{w}} m(\boldsymbol{w}(\nu)^\top \boldsymbol{n}(\nu+1))\right)^\top (\boldsymbol{w} - \boldsymbol{w}(\nu)).$$

From Equation (B.35) we then have

$$m(\nu+1) \approx \hat{m}(\nu+1) + \hat{m}(\nu+1)(1-\hat{m}(\nu+1))\boldsymbol{n}(\nu+1)^\top(\boldsymbol{w} - \boldsymbol{w}(\nu)), \quad \text{(B.37)}$$

where $\hat{m}(\nu+1)$ is given by

$$\hat{m}(\nu+1) = m(\boldsymbol{w}(\nu)^\top \boldsymbol{n}(\nu+1)).$$

The caret indicates that the signal is calculated from the next (i.e., the $\nu+$ 1st) training input, but using the current (i.e., the νth) weights. With the definition of the *linearized input*

$$\boldsymbol{a}(\nu) = \hat{m}(\nu)(1-\hat{m}(\nu))\boldsymbol{n}(\nu)^\top, \quad \text{(B.38)}$$

we can write Equation (B.37) in the form

$$m(\nu+1) \approx \boldsymbol{a}(\nu+1)\boldsymbol{w} + [\hat{m}(\nu+1) - \boldsymbol{a}\nu+1)\boldsymbol{w}(\nu)].$$

The term in square brackets is — to first order — the error that arises from the fact that the neuron's output signal is *not* simply linear in $\boldsymbol{w}$. If we neglect it altogether, then we get the *linearized* neuron output signal

$$m(\nu+1) = \boldsymbol{a}(\nu+1)\boldsymbol{w}.$$

Note that $\boldsymbol{a}$ has been defined in Equation (B.38) as a row vector. In order to calculate the synaptic weight vector $\boldsymbol{w}$, we can now apply the theory of recursive linear regression developed in Appendix A. We simply identify the parameter vector $\boldsymbol{w}$ with the synaptic weight vector, y with the desired output ℓ and observation $\boldsymbol{x}(\nu+1)^T$ with $\boldsymbol{a}(\nu+1)$. We then have the least squares problem

$$\begin{pmatrix} \boldsymbol{\ell}_\nu \\ \ell(\nu+1) \end{pmatrix} = \begin{pmatrix} \boldsymbol{\mathcal{A}}_\nu \\ \boldsymbol{a}(\nu+1) \end{pmatrix} \boldsymbol{w} + \boldsymbol{R}_{\nu+1};$$

see Equation (A.5). The Kalman filter equations for the recursive solution of this problem are unchanged:

$$\begin{aligned} \boldsymbol{\Sigma}(\nu+1) &= \left[\boldsymbol{I} - \boldsymbol{K}(\nu+1)\boldsymbol{a}(\nu+1)\right]\boldsymbol{\Sigma}(\nu) \\ \boldsymbol{K}(\nu+1) &= \boldsymbol{\Sigma}(\nu)\boldsymbol{a}(\nu+1)^\top\left[\boldsymbol{a}(\nu+1)\boldsymbol{\Sigma}(\nu)\boldsymbol{a}(\nu+1)^\top + 1\right]^{-1}, \end{aligned} \quad \text{(B.39)}$$

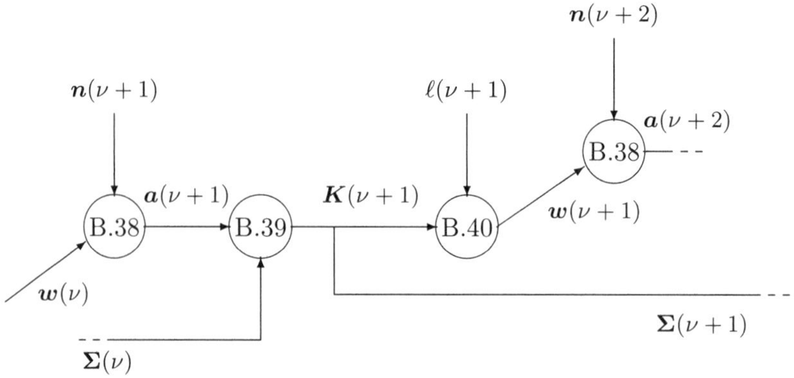

FIGURE B.3
Determination of the synaptic weights for an isolated neuron with the Kalman filter.

while the recursive expression for the parameter vector, Equation (A.8), can be improved somewhat by replacing the linear approximation to the neuron output $\boldsymbol{a}(\nu+1)\boldsymbol{w}(\nu)$ by the actual output for the $\nu+1$st training observation, namely $\hat{m}(\nu+1)$, so we have

$$\boldsymbol{w}(\nu+1) = \boldsymbol{w}(\nu) + \boldsymbol{K}(\nu+1)\big[\ell(\nu+1) - \hat{m}(\nu+1)\big]. \qquad \text{(B.40)}$$

B.3.2 The algorithm

The recursive calculation of $\boldsymbol{w}$ is depicted in Figure B.3. The input is the current weight vector $\boldsymbol{w}(\nu)$, its covariance matrix $\boldsymbol{\Sigma}_\nu$, and the output vector of the hidden layer $\boldsymbol{n}(\nu+1)$ obtained by propagating the next input observation $\boldsymbol{g}(\nu+1)$ through the network. After determining the linearized input $\boldsymbol{a}(\nu+1)$ from Equation (B.38), the Kalman gain $\boldsymbol{K}_{\nu+1}$ and the new covariance matrix $\boldsymbol{\Sigma}_{\nu+1}$ are calculated with Equation (B.39). Finally, the weights are updated according to Equation (B.40) to give $\boldsymbol{w}(\nu+1)$ and the procedure is repeated.

To make our notation explicit for the output neurons, we substitute

$$\begin{aligned}
\ell(\nu) &\to \ell_k(\nu) \\
\boldsymbol{w}(\nu) &\to \boldsymbol{w}^o_k(\nu) \\
\hat{m}(\nu+1) &\to m_k\left(\boldsymbol{w}^{o\top}_k(\nu)\boldsymbol{n}(\nu+1)\right) \\
\boldsymbol{a}(\nu+1) &\to \boldsymbol{a}^o_k(\nu+1) = \hat{m}_k(\nu+1)(1-\hat{m}_k(\nu+1))\boldsymbol{n}(\nu+1)^\top \\
\boldsymbol{K}(\nu) &\to \boldsymbol{K}^o_k(\nu) \\
\boldsymbol{\Sigma}(\nu) &\to \boldsymbol{\Sigma}^o_k(\nu),
\end{aligned}$$

for $k = 1 \dots K$. Then Equation (B.40) becomes

$$\boldsymbol{w}^o_k(\nu+1) = \boldsymbol{w}^o_k(\nu) + \boldsymbol{K}^o_k(\nu+1)\big[\ell_k(\nu+1) - \hat{m}_k(\nu+1)\big], \quad k = 1 \dots K. \qquad \text{(B.41)}$$

Recalling that we wish to minimize the local quadratic cost function, Equation (B.34), note that the expression in square brackets in Equation (B.41) is in fact the negative derivative of $E(\nu)$ with respect to the output signal of the neuron, i.e.,

$$\ell_k(\nu+1) - m_k(\nu+1) = -\frac{\partial E(\nu+1)}{\partial m_k(\nu+1)},$$

so that Equation (B.41) can be expressed in the form

$$\boldsymbol{w}_k^o(\nu+1) = \boldsymbol{w}_k^o(\nu) - \boldsymbol{K}_k^o(\nu+1)\left[\frac{\partial E(\nu+1)}{\partial m_k(\nu+1)}\right]_{\hat{m}_k(\nu+1)}. \tag{B.42}$$

With this observation, we can turn consideration to the hidden neurons, making the substitutions

$$\begin{aligned}
\boldsymbol{w}(\nu) &\rightarrow \boldsymbol{w}_j^h(\nu)\\
\hat{m}(\nu+1) &\rightarrow \hat{n}_j(\nu+1) = f\left(\boldsymbol{w}_j^{h\top}(\nu)\boldsymbol{g}(\nu+1)\right)\\
\boldsymbol{a}(\nu+1) &\rightarrow \boldsymbol{a}_j^h(\nu+1) = \hat{n}_j(\nu+1)(1-\hat{n}_j(\nu+1))\boldsymbol{g}(\nu+1)^\top\\
\boldsymbol{K}(\nu) &\rightarrow \boldsymbol{K}_j^h(\nu)\\
\boldsymbol{\Sigma}(\nu) &\rightarrow \boldsymbol{\Sigma}_j^h(\nu),
\end{aligned}$$

for $j = 1 \ldots L$. Then, analogously to Equation (B.42), the update equation for the weight vector of the jth hidden neuron is

$$\boldsymbol{w}_j^h(\nu+1) = \boldsymbol{w}_j^h(\nu) - \boldsymbol{K}_j^h(\nu+1)\left[\frac{\partial E(\nu+1)}{\partial n_j(\nu+1)}\right]_{\hat{n}_j(\nu+1)}. \tag{B.43}$$

To obtain the partial derivative in Equation (B.43), we differentiate the cost function, Equation (B.34), applying the chain rule:

$$\frac{\partial E(\nu+1)}{\partial n_j(\nu+1)} = -\sum_{k=1}^{K}(\ell_k(\nu+1) - m_k(\nu+1))\frac{\partial m_k(\nu+1)}{\partial n_j(\nu+1)}.$$

From Equation (B.36), noting that $(\boldsymbol{w}_k^o)_j = W_{jk}^o$, we have

$$\frac{\partial m_k(\nu+1)}{\partial n_j(\nu+1)} = m_k(\nu+1)(1-m_k(\nu+1))W_{jk}^o(\nu+1).$$

Combining the last two equations,

$$\frac{\partial E(\nu+1)}{\partial n_j(\nu+1)} = -\sum_{k=1}^{K}(\ell_k(\nu+1)-m_k(\nu+1))m_k(\nu+1)(1-m_k(\nu+1))W_{jk}^o(\nu+1),$$

which we can write more compactly as

$$\frac{\partial E(\nu+1)}{\partial n_j(\nu+1)} = -\boldsymbol{W}_{j\cdot}^o(\nu+1)\boldsymbol{\beta}^o(\nu+1), \tag{B.44}$$

where $\boldsymbol{W}^o_{j\cdot}$ is the jth *row* (!) of the output-layer weight matrix, and where

$$\boldsymbol{\beta}^o(\nu+1) = (\boldsymbol{\ell}(\nu+1) - \boldsymbol{m}(\nu+1)) \cdot \boldsymbol{m}(\nu+1) \cdot (\boldsymbol{1} - \boldsymbol{m}(\nu+1)).$$

The correct update relation for the weights of the jth hidden neuron is therefore

$$\boldsymbol{w}^h_j(\nu+1) = \boldsymbol{w}^h_j(\nu) + \boldsymbol{K}^h_j(\nu+1)\big[\boldsymbol{W}^o_{j\cdot}(\nu+1)\boldsymbol{\beta}^o(\nu+1)\big]. \qquad \text{(B.45)}$$

Apart from initialization of the covariance matrices $\boldsymbol{\Sigma}^h_j(0)$, $\boldsymbol{\Sigma}^o(0)$, the Kalman training procedure has no adjustable parameters whatsoever. The initial covariance matrices are simply taken to be proportional to the corresponding identity matrices:

$$\boldsymbol{\Sigma}^h_j(0) = Z\boldsymbol{I}^h, \quad \boldsymbol{\Sigma}^o_k(0) = Z\boldsymbol{I}^o, \quad Z \gg 1,\ j = 1\ldots L,\ k = 1\ldots K,$$

where I^h is the $(N+1)\times(N+1)$ and I^o the $(L+1)\times(L+1)$ identity matrix. We choose $Z = 100$ and obtain the following algorithm:

Algorithm (Kalman Filter Training)

1. Set $\nu = 0$, $\boldsymbol{\Sigma}^h_j(0) = 100\boldsymbol{I}^h, j = 1\ldots L$, $\boldsymbol{\Sigma}^o_k(0) = 100\boldsymbol{I}^o$, $k = 1\ldots K$ and initialize the synaptic weight matrices $\boldsymbol{W}^h(0)$ and $\boldsymbol{W}^o(0)$ with random numbers.

2. Choose a training pair $(\boldsymbol{g}(\nu+1), \boldsymbol{\ell}(\nu+1))$ and determine the hidden layer output vector

$$\hat{\boldsymbol{n}}(\nu+1) = \begin{pmatrix} 1 \\ f\left(\boldsymbol{W}^h(\nu)^\top \boldsymbol{g}(\nu+1)\right) \end{pmatrix},$$

and with it the quantities

$$\boldsymbol{a}^h_j(\nu+1) = \hat{n}_j(\nu+1)(1-\hat{n}_j(\nu+1))\boldsymbol{g}(\nu+1)^\top, \quad j = 1\ldots L,$$

$$\hat{m}_k(\nu+1) = m_k\left(\boldsymbol{w}^{o\top}_k(\nu)\hat{\boldsymbol{n}}(\nu+1)\right)$$

$$\boldsymbol{a}^o_k(\nu+1) = \hat{m}_k(\nu+1)(1-\hat{m}_k(\nu+1))\hat{\boldsymbol{n}}(\nu+1)^\top, \quad k = 1\ldots K$$

and

$$\boldsymbol{\beta}^o(\nu+1) = (\boldsymbol{\ell}(\nu+1) - \hat{\boldsymbol{m}}(\nu+1)) \cdot \hat{\boldsymbol{m}}(\nu+1) \cdot (\boldsymbol{1} - \hat{\boldsymbol{m}}(\nu+1)).$$

3. Determine the Kalman gains for all of the neurons according to

$$\boldsymbol{K}^o_k(\nu+1) = \boldsymbol{\Sigma}^o_k(\nu)\boldsymbol{a}^o_k(\nu+1)^\top\left[\boldsymbol{a}^o_k(\nu+1)\boldsymbol{\Sigma}^o_k(\nu)\boldsymbol{a}^o_k(\nu+1)^\top + 1\right]^{-1}, \quad k = 1\ldots K$$

$$\boldsymbol{K}^h_j(\nu+1) = \boldsymbol{\Sigma}^h_j(\nu)\boldsymbol{a}^h_j(\nu+1)^\top\left[\boldsymbol{a}^h_j(\nu+1)\boldsymbol{\Sigma}^h_j(\nu)\boldsymbol{a}^h_j(\nu+1)^\top + 1\right]^{-1}, \quad j = 1\ldots L$$

Listing B.4: Excerpt from the object class FFNKAL.

```
PRO FFNKAL:: train, key=key
   IF n_elements(key) EQ 0 THEN key=0
; define update matrices for Wh and Wo
   dWh = dblarr(self.LL,self.NN+1)
   dWo = dblarr(self.KK,self.LL+1)
   iter = 0L
   iter100 = 0L
   progressbar = Obj_New('cgprogressbar', $
     Color='blue', Text='0',$
     title='Training:␣example␣number...', $
     xsize=250,ysize=20)
   progressbar->start
   window,12,xsize=600,ysize=400, $
     title='FFN(Kalman␣filter)'
   wset,12
   REPEAT BEGIN
      IF progressbar->CheckCancel() THEN BEGIN
         PRINT,'Training␣interrupted'
         progressbar->Destroy
         RETURN
      ENDIF
; select training pair at random
      ell = long(self.np*randomu(seed))
; use only odd ones if validation option is set
      IF key EQ 1 THEN ell = (ell*2 MOD (self.np-2))+1
      x=(*self.Gs)[ell,*]
      y=(*self.Ls)[ell,*]
; send it through the network
      m=self->forwardPass(x)
; error at output
      e=y-m
; loop over the output neurons
      FOR k=0,self.KK-1 DO BEGIN
;     linearized input (column vector)
         Ao = m[k]*(1-m[k])*(*self.N)
;     Kalman gain
         So = (*self.So)[*,*,k]
         SA = So##Ao
         Ko = SA/((transpose(Ao)##SA)[0]+1)
;     determine delta for this neuron
         dWo[k,*] = Ko*e[k]
;     update its covariance matrix
         So = So - Ko##transpose(Ao)##So
         (*self.So)[*,*,k] = So
      ENDFOR
```

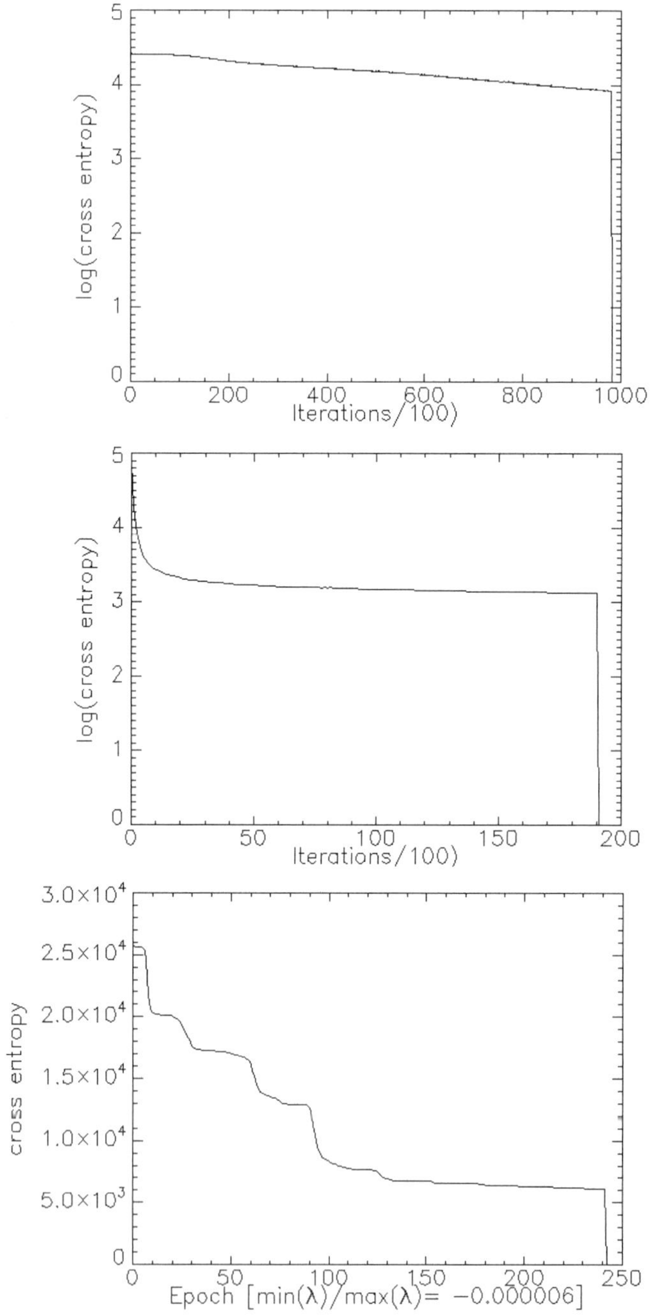

FIGURE B.4
Top: Training with backpropagation, center: with Kalman filter, bottom: with scaled conjugate gradient. Note that the upper two plots are logarithmic.

Listing B.5: Excerpt from the object class FFNKAL (continued).

```
 1 ; update the output weights
 2       *self.Wo = *self.Wo + dWo
 3 ; backpropagated error
 4       beta_o =e*m*(1-m)
 5 ; loop over the hidden neurons
 6       FOR j=0,self.LL-1 DO BEGIN
 7 ;    linearized input (column vector)
 8          Ah = X*(*self.N)[j+1]*(1-(*self.N)[j+1])
 9 ;    Kalman gain
10          Sh = (*self.Sh)[*,*,j]
11          SA = Sh##Ah
12          Kh = SA/((transpose(Ah)##SA)[0]+1)
13 ;    determine delta for this neuron
14          dWh[j,*] = Kh*((*self.Wo)[*,j+1]##beta_o)[0]
15 ;    update its covariance matrix
16          Sh = Sh - Kh##transpose(Ah)##Sh
17          (*self.Sh)[*,*,j] = Sh
18      ENDFOR
19 ; update the hidden weights
20       *self.Wh = *self.Wh + dWh
21 ; record cost history
22       IF iter MOD 100 EQ 0 THEN BEGIN
23          IF key THEN BEGIN
24             (*self.cost_array)[iter100]= $
25              alog10(self->cost(1)) ; training cost
26             (*self.valid_cost_array)[iter100]= $
27              alog10(self->cost(2)) ; validation cost
28          END ELSE (*self.cost_array)[iter100]= $
29              alog10(self->cost(0)) ; full training cost
30          iter100 = iter100+1
31          progressbar->Update,iter*100/self.iterations
32          PLOT,*self.cost_array,xrange=[0,iter100], $
33             color=0,background='FFFFFF'XL,$
34             xtitle='Iterations/100)', $
35             ytitle='log(cross entropy)'
36          IF key THEN oplot, *self.valid_cost_array, $
37             color=0,linestyle=2
38       END
39       iter=iter+1
40    ENDREP UNTIL iter EQ self.iterations
41    progressbar->Destroy
42 END
```

4. Update the synaptic weight matrices:

$$\boldsymbol{w}_k^o(\nu+1) = \boldsymbol{w}_k^o(\nu) + \boldsymbol{K}_k^o(\nu+1)[\ell_k(\nu+1) - \hat{m}_k(\nu+1)], \quad k=1\ldots K$$

$$\boldsymbol{w}_j^h(\nu+1) = \boldsymbol{w}_j^h(\nu) + \boldsymbol{K}_j^h(\nu+1)\big[\boldsymbol{W}_{j\cdot}^o(\nu+1)\boldsymbol{\beta}^o(\nu+1)\big], \quad j=1\ldots L.$$

5. Determine the new covariance matrices:

$$\boldsymbol{\Sigma}_k^o(\nu+1) = \left[\boldsymbol{I}^o - \boldsymbol{K}_k^o(\nu+1)\boldsymbol{a}_k^o(\nu+1)\right]\boldsymbol{\Sigma}_k^o(\nu), \quad k = 1\dots K$$
$$\boldsymbol{\Sigma}_j^h(\nu+1) = \left[\boldsymbol{I}^h - \boldsymbol{K}_j^h(\nu+1)\boldsymbol{a}_j^h(\nu+1)\right]\boldsymbol{\Sigma}_j^h(\nu), \quad j = 1\dots L.$$

6. If the overall cost function, Equation (6.30), is sufficiently small, stop, else set $\nu = \nu + 1$ and go to 2.

This method was originally suggested by Shah and Palmieri (1990), who called it the *multiple extended Kalman algorithm* (MEKA),* and explained in detail by Hayken (1994). Its IDL implementation is given in Listings B.4 and B.5 showing an excerpt from `FFNKAL__DEFINE.PRO`, which extends `FFN`.

B.4 A neural network classifier with hybrid training

The Kalman filter algorithm is designed for minimization of the quadratic cost function of Equation (6.30) rather than of the cross entropy, Equation (6.33), which is more appropriate to the softmax output signal generated by the neural network. However, when one is minimized, so is the other. The ENVI/IDL extension `FFN_RUN` described in Appendix C implements both the Kalman filter and scaled conjugate gradient training algorithms, beginning with the former in order to approach a minimum quickly, and then using the latter to refine the synaptic weights. (The scaled conjugate gradient algorithm is guaranteed to reduce the overall cost function at each iteration.) The switch is achieved with the `GET_WEIGHTS()` and `PUT_WEIGHTS()` functions defined in the parent object class `FFN`. Convergence is extremely fast when compared to backpropagation; see Figure B.4.

* *Multiple*, because the algorithm is applied to multiple neurons, the adjective *extended* characterizes Kalman filter methods that linearize nonlinear models by using a first-order Taylor expansion.

C

ENVI Extensions in IDL

This appendix provides installation instructions and documentation for all of the ENVI/IDL extensions and their associated routines described in the text.

C.1 Installation

To install the complete extension package:

1. Unpack the ENVI/IDL package (containing the software directory CRC) available at

   `http://ms-image-analysis.appspot.com/static/homepage/software.html`

   or at

   `http://mcanty.homepage.t-online.de/software.html`

   to any convenient location and point the IDL path to it. Be sure to include its subdirectories.

2. Many of the routines make use of David Fanning's object class

   `CGPROGRESSBAR__DEFINE`

   and associated dependencies, so his `COYOTE` library must also be in the IDL path. The library may be downloaded from

   `http://http://www.idlcoyote.com/`.

3. The IR-MAD scripts require the C-library `PROVMEANS.DLL`. If you are running on a Windows OS, simply point the IDL DLM path to the subdirectory CRC/AUXIL. In addition, if you are using 64-bit IDL and don't have Microsoft Visual C++ installed on your computer, you will need to download and install the freely available Microsoft Visual C++ 2010 Redistributable Package (x64). Unix flavors and Mac OS users will need to create the DLL from the C source code `PROVMEANS.C` provided. See the IDL documentation for `MAKE_DLL`.

4. Under the ENVI Classic GUI, the programs with filenames of the form

   `<program name>_RUN`

   will create their own menu items automatically, so if you wish to call an extension directly from the ENVI Classic menu system, copy (or move) it to the ENVI `SAVE_ADD` directory before invoking ENVI.

5. Under the new ENVI 5 GUI, if you wish the extensions to appear in the Toolbox, copy the programs with filenames of the form

   `<program name>_RUN5`

   to the ENVI `Extensions File Directory`. The programs will appear in the Toolbox organized by book chapter.

   Note: At the time of writing, the ENVI 5.0 API documentation is very obscure in regard to programmatic access to training region data (shape files or ROIs). For this reason, the supervised classification algorithms of Chapters 6 and 7 are only offered for the ENVI Classic interface. Please check the website for eventual updates.

6. (This step is only necessary for the calculation of structure heights with `CALCHEIGHT_RUN` under ENVI Classic and can be skipped if desired.) Copy the file `MJC_CURSOR_MOTION` (in the AUXIL directory) into the `SAVE_ADD` directory. In

   `File/Preferences/User Defined Motion Routine`

   in the ENVI Classic main menu, enter the string: `MJC_CURSOR_MOTION`.

7. Some of the extensions can take advantage of NVIDIA's *Compute Unified Device Architecture* (CUDA), and require the GPULib API available from

   `http:/http://www.txcorp.com/home/gpulib`

   in order to run on the GPU. See the above URL for installation instructions. If CUDA/GPULib is not installed, the scripts will run in the normal way on the CPU.

C.2 Documentation

The ENVI extensions described below can all be run directly from the IDL console, provided that an ENVI interactive session is running. They can also be started from the ENVI Classic Menu or the ENVI 5 Toolbox using the menu items indicated in parentheses.

C.2.1 ENVI extensions for Chapter 4

C.2.1.1 Kernel principal components analysis (Section 4.4.2)

```
KPCA_RUN (Transform/Principal Components/Kernel PCA (CUDA))
KPCA_RUN5 (Toolbox/Extensions/Chapter4/Kernel PCA (CUDA))
```

The user is first queried for an input image filename and a training sample size, maximum 2000. If the latter is 0, then 100 representative training pixel vectors are chosen by the k-means algorithm. Otherwise, a random sample of the desired size is used. Next, the number of kernel principal components to retain (maximum 10) and the kernel parameter `NSCALE` are entered. The Gaussian kernel parameter γ is calculated from the latter as

$$\gamma = \frac{1}{2(\text{NSCALE}\ \sigma)^2}, \tag{C.1}$$

where $\sigma = \langle \|\boldsymbol{g}(\nu) - \boldsymbol{g}(\nu')\| \rangle$ is the average Euclidean distance between the training observations. The value of γ is printed to the IDL console. Next, the user can choose whether to center the kernel matrix on the training data alone (faster, see Canty and Nieslsen (2012)) or on all of the image pixels. Finally, the output destination (memory or file) is selected.

After centering and diagonalizing the kernel matrix, a plot of the eigenvalues is displayed. Then, if centering on all data was chosen, the projection is carried out in two passes through the image. On the first pass, the matrix column, row and overall sums required for centering are accumulated. These are applied on the second pass as each image pixel is projected. For centering on the training data, only one pass is needed. If CUDA/GPULib are available (this is detected automatically), then both centering and projection are performed on the GPU.

C.2.2 ENVI extensions for Chapter 5

C.2.2.1 Panchromatic sharpening of multispectral images using the discrete or à trous wavelet transform (Sections 5.3.4 and 5.3.5)

```
DWT_RUN (Transform/Image Sharpening/Discrete Wavelet Transform)
DWT_RUN5 (Toolbox/Extensions/Chapter5/DWT Pansharpening)
ATWT_RUN (Transform/Image Sharpening/A Trous Wavelet Transform)
ATWT_RUN5 (Toolbox/Extensions/Chapter5/ATWT Pansharpening)
```

The low- and high-resolution images must be geo-referenced (often they are acquired simultaneously from the same platform) and their spatial resolutions must differ by a factor of 2 or 4. This can be achieved by a prior re-sampling, if necessary. The user is queried for the (spatial/spectral subset of the) image to be sharpened and then for the corresponding panchromatic or high-resolution image. The latter image should overlap the low-resolution image completely.

Then the output destination (memory or file) must be entered. During the calculation, scatterplots are shown of the wavelet coefficients for the low- vs. high-resolution bands.

C.2.2.2 Calculation of the Wang–Bovik index for radiometric fidelity of a pan-sharpened image (Section 5.3.6)

```
QUALITY_INDEX_RUN (Transform/Image Sharpening/Quality Index)
QUALITY_INDEX_RUN5 (Toolbox/Extensions/Chapter5/Quality Index)
```

The user is first prompted for the reference multispectral image to which the sharpened image is to be compared and then for the image whose quality is to be determined. The reference image must completely overlap the pan-sharpened image. The latter is down-sampled to the same resolution as the reference image and then the image bands are matched to the nearest pixel using phase correlation, plots of which are shown during the calculation. The calculated indices are displayed in an IDL text widget.

C.2.2.3 Conversion of SAR imagery to ENVI standard files

```
POLSARINGEST_RUN (Radar/Polarimetric Tools/polSAR Covariance Matrix)
```

Geocoded, multi-look polarimetric SAR imagery suitable for processing with the filtering, classification and change detection algorithms described in the text may be obtained directly from the provider or generated from single-look complex (SLC) data. In the latter case, the open source software packages PolSARpro (European Space agency),

`http://earth.eo.esa.int/polsarpro/`

together with MapReady (Alaska Satellite Facility),

`http://www.asf.alaska.edu/downloads/software_tools`

are a good choice; see Gens et al. (2013). PolSARpro is first used to create multi-look images in covariance matrix format, which can then be exported to MapReady for georeferencing with or without a DEM. Alternative commercial solutions are the Gamma Software (Gamma Remote Sensing)

`http://www.gamma-rs.ch/`

and the SARscape add-on module for ENVI

http://www.exelisvis.com/ProductsServices/ENVI/ENVISARscape.aspx

The ENVI/IDL extension `POLSARINGEST_RUN` combines the outputs from the above preprocessing systems to a single, multi-band ENVI standard file in 32-bit floating point format. For full quad polarimetric data, it generates 9 bands, which are ordered as follows:

$$\begin{aligned}
&\mathrm{C11} = \langle |s_{hh}|^2 \rangle \\
&\mathrm{C12re} = \langle \sqrt{2} s_{hh} s_{hv}^* \rangle (\text{real part}) \\
&\mathrm{C12im} = \langle \sqrt{2} s_{hh} s_{hv}^* \rangle (\text{imaginary part}) \\
&\mathrm{C13re} = \langle s_{hh} s_{vv}^* \rangle (\text{real part}) \\
&\mathrm{C13im} = \langle s_{hh} s_{vv}^* \rangle (\text{imaginary part}) \\
&\mathrm{C22} = \langle 2|s_{hv}|^2 \rangle \\
&\mathrm{C23re)} = \langle \sqrt{2} s_{hv} s_{vv}^* \rangle (\text{real part}) \\
&\mathrm{C23im)} = \langle \sqrt{2} s_{hv} s_{vv}^* \rangle (\text{imaginary part}) \\
&\mathrm{C33} = \langle |s_{vv}|^2 \rangle .
\end{aligned}$$

For dual or single polarimetry, the bands corresponding to the missing matrix elements are not present. Thus, a dual polarimetric image will consist of four bands, e.g., C11, C12re, C12im, and C22; a single polarimetric image will consist of just one band, usually C11 or C33. The ENVI files generated by the script can be read by all of the SAR processing routines described below, and also by their Python counterparts described in Appendix D.

The user is prompted for the directory containing the georeferenced covariance matrix files (consisting of one file for each of the real and imaginary components), and then requested to choose (a spatial subset of) one of them. The other files are then read in automatically with the same spatial subset. Finally, an output filename is requested.

C.2.2.4 Multivariate estimation of equivalent number of looks for polarimetric SAR in covariance matrix format

```
ENLML_RUN (Radar/Polarimetric Tools/ENL Estimation)
LOOKUP.TXT
```

This method is not discussed in the text. It is a multivariate technique based on the maximum likelihood estimator explained in Anfinsen et al. (2009a) and is reported to be superior to the standard univariate method discussed in Section 5.4.2. The polarimetric SAR image should be in the ENVI standard format generated by the routine `POLSARINGEST_RUN` described in Section C.2.2.3 above.

The user is asked to select (a spatial subset of) the ENVI covariance matrix file and the desired window size (default 7×7). The local ENL values are estimated in a moving window and stored as a single-band, floating point raster image in ENVI memory. A histogram of the ENL values is plotted. In the ENVI Classic environment, the zoom window can be used to isolate a region in the ENL image with a homogeneous signal background (e.g., flat agricultural field) where the speckle statistics are well developed. ENVI's interactive histogram stretching facility will display a histogram of the ENL

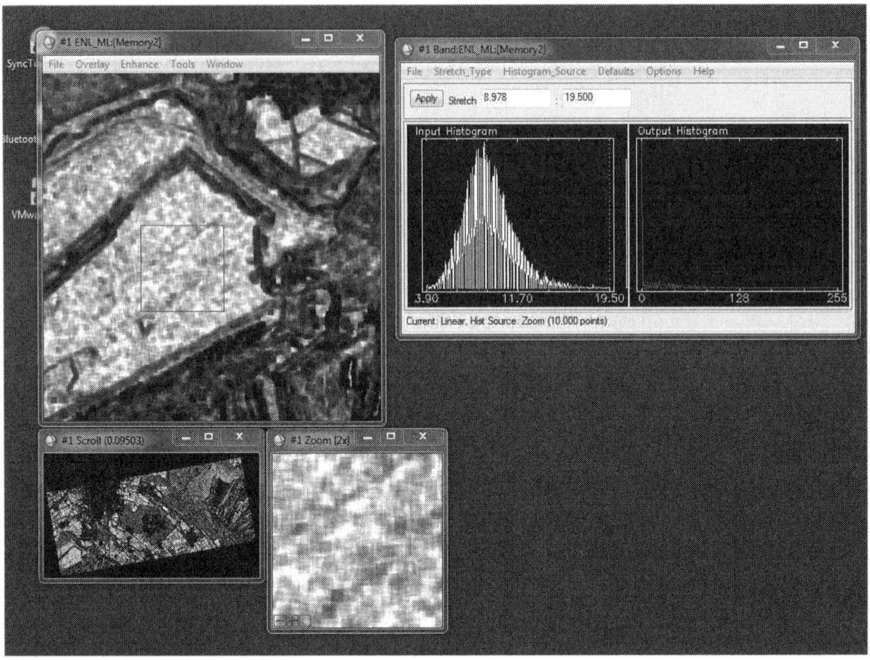

FIGURE C.1
Illustrating ENL determination in ENVI Classic.

values in the zoom window, the mid-point of which will give an ENL estimate for the full image. An example is shown in Figure C.1.

C.2.2.5 Minimum mean square error filtering of polarimetric SAR imagery (Section 5.4.3.1)

```
MMSE_FILTER_RUN (Radar/Polarimetric Tools/MMSE polSAR filter)
MMSE_FILTER_RUN5 (Toolbox/Extensions/Chapter5/MMSE polSAR filter)
```

The polarimetric SAR image is assumed to have been generated by the POLSARINGEST_RUN extension of Section C.2.2.3 above. The user is asked to enter the ENVI filename and the equivalent number of looks. The filtering operation occurs in two steps: first the span image is processed to determine the filter weights, then all components of the covariance matrix are filtered separately. The result may be stored to disk or to ENVI memory.

C.2.2.6 Gamma MAP filtering of polarimetric SAR imagery (Section 5.4.3.2)

```
GAMMA_FILTER_RUN (Radar/Polarimetric Tools/Gamma MAP polSAR filter)
GAMMA_FILTER_RUN (Toolbox/Extensions/Chapter5/Gamma MAP polSAR filter)
```

The polarimetric SAR image is assumed to have been generated by the `POLSARINGEST_RUN` extension of Section C.2.2.3 above. The user is asked to enter the ENVI filename, the equivalent number of looks and, finally, the number of iterations (default 1). Only the diagonal components of the covariance matrix are filtered. The result may be stored to disk or to ENVI memory.

C.2.2.7 Determine heights of vertical structures in high-resolution images (Section 5.5.3)

`CALCHEIGHT_RUN (Tools/Height Calculator)`

This routine makes use of rational function models (RFMs) accompanying ortho-ready imagery distributed by the image providers. It is implemented as a small graphical user interface (GUI) using IDL widgets. After displaying a high-resolution image and starting the GUI from the image window's `Tools` menu, an RFM file must be loaded from the GUI menu

`File/Load RPC File`

(extension RPC or RPB). If a digital elevation model (DEM) is available for the scene, this can also be entered with

`File/Load DEM File`

A DEM is not required, however. Click on the bottom of a vertical structure to set the base height and then shift-click on the top of the structure. Press the `CALC` button to display the structure's height, latitude, longitude and base elevation. The number in brackets next to the height is the minimum distance (in pixels) between the top pixel and a (projected) vertical line through the bottom pixel. It should be of the order of 1 or less. If no DEM is supplied, the base elevation is the average value for the whole scene. If a DEM is used, the base elevation is taken from it. The latitude and longitude are then ortho-rectified values.

C.2.2.8 Local solar incidence angle correction for multispectral images over rough terrain (Section 5.5.6)

`C_CORRECTION_RUN (Topographic/C-Correction)`
`C_CORRECTION_RUN5 (Toolbox/Extensions/Chapter5/C-Correction)`

Select the (spectral/spatial subset of the) image to be corrected and (optionally) a mask. Then, in the edit box, enter the solar elevation and azimuth in degrees and a size for the kernel used for slope/aspect determination (default 9×9). Then select the corresponding DEM file and the output destination (file or memory).

C.2.2.9 Tie-point extraction for image–image registration (Section 5.6.2)

```
CONTOUR_MATCH_RUN (Map/Registration/Contour Matching)
CONTOUR_MATCH_RUN5 (Toolbox/Extensions/Chapter5/Contour Matching)
```

This extension is primarily intended for fine adjustment, assuming that the images are already geo-referenced and, if necessary, re-sampled to the same ground sample distance. From the ensuing dialog windows, choose the base and warp image bands. Then enter the σ parameter for the LoG filter (default value is 2.5 pixels). Finally, give a filename (extension `.PTS`) for the tie-points (ground control points). If tie-points are found, they can be read from this file into ENVI's `Ground Control Points Selection` dialog. It may be necessary to edit them further to eliminate outliers.

C.2.3 ENVI extensions for Chapter 6

C.2.3.1 Bayes maximum likelihood classifier (Section 6.3.2)

```
MAXLIKE_RUN (Classification/Supervised/MaxLike (test output))
```

This routine is a wrapper for ENVI's maximum likelihood classifier, which reserves test data from the training regions of interest (ROIs) in the ratio 2:1 for training:test. The user is prompted for an input file, the training ROIs (which must be present beforehand) and output destinations for the classified image, the rule image (optional) and the test results. The test result file can be processed with the procedures `CT_RUN` and and `McNEMAR_RUN` discussed in Section C.2.4.2 below. The routine `RULE_CONVERT`, callable only from the ENVI command prompt, may be used to convert the ENVI rule image to probabilities; see Section 6.3.1.

C.2.3.2 Gaussian kernel classification (Section 6.4)

```
KERNEL_RUN (Classification/Supervised/Gaussian Kernel (CUDA))
```

This extension performs nonparametric classification of a multispectral image using the Gaussian kernel algorithm. The program will make use of CUDA, if it is installed, to calculate the Gauss kernel matrices. The user is prompted for an input file, the training ROIs (which must be present beforehand) and output destinations for the classified image, the probability image (optional), and the test results (optional). During calculation, a plot window shows the minimization of the misclassification rate with respect to the parameter σ as shown in Figure 6.5. The program uses spectral tiling and may be used for large images. The test result file can be processed with the procedures `CT_RUN` and `McNEMAR_RUN` discussed in Section C.2.4.2 below.

C.2.3.3 Neural network classifiers (Section 6.5 and Appendix B)

`FFN_RUN (Classification/Supervised/Neural Net (Kalman/Conj Gradient))`

The extension `FFN_RUN` classifies a multispectral image with a two-layer, feed-forward neural network trained with a combination of the Kalman filter algorithm and the scaled conjugate gradient algorithm. The user is prompted for an input file, the training ROIs (which must be present beforehand), output destinations for the classified image, the probability image (optional) and the test results (optional), the number of hidden neurons and, finally, whether or not to use validation. During the calculation, the cross-entropy cost function is displayed. Training can be interrupted at any time, whereupon processing continues to the next phase. After training, the Hessian matrix may be calculated, in which case its eigenvalues are written to a text window. The program uses spectral tiling and may be used for large images. The test result file can be processed with the procedures `CT_RUN` and and `McNEMAR_RUN` discussed in Section C.2.4.2 below.

C.2.3.4 Support vector machine classifier (Section 6.6.6)

`SVM_RUN (Classification/Supervised/SVM (test output))`

This routine is a wrapper for ENVI's support vector machine classifier, which reserves test data from the training regions of interest (ROIs) in the ratio 2:1 for training:test. The user is prompted for an input file and the training ROIs (which must be present beforehand) and output destination for the test results. Classification and rule images are written to memory only. SVM-specific parameters (kernel type, kernel parameters, pyramid depth) must be set programmatically. The test result file can be processed with the procedures `CT_RUN` and `McNEMAR_RUN` discussed in Section C.2.4.2 below.

C.2.4 ENVI extensions for Chapter 7

C.2.4.1 Probabilistic label relaxation postprocessing (Section 7.1.2)

```
PLR_RUN (Classification/Post Classification/PLR)
PLR_RECLASS_RUN (Classification/Post Classification/PLR Reclassify)
PLR_RUN5 (Toolbox/Extensions/Chapter7/PLR)
PLR_RECLASS_RUN5 (Toolbox/Extensions/Chapter7/PLR Reclassify)
```

The `PLR_RUN` or `PLR_RUN5` extension takes as input a class probability vector image (rule image) generated by any of the supervised classification extensions described in this appendix, as well as from the clustering routine `EM_RUN` described in Section C.2.5.4 below, and generates a modified class probability vector image. At the prompt, choose a class membership probabilities image and the number of iterations (default = 3). Then choose a destination either in memory or as a named file.

The generated rule image can then be processed with PLR_RECLASS_RUN or PLR_RECLASS_RUN5 to generate an improved classification image with better spatial coherence. At the prompt, choose a previous classification image if available (to synchronize the class colors) and then a class membership probabilities image. At the next prompt for a dummy unclassified class reply "Yes." Then choose a destination either in memory or as a named file.

C.2.4.2 Contingency tables and McNemar test (Section 7.2)

```
CT_RUN (Classification/Post Classification/Contingency Table)
MCNEMAR_RUN (Classification/Post Classification/McNemar Test)
CT_RUN5 (Toolbox/Extensions/Chapter7/Contingency Table)
MCNEMAR_RUN5 (Toolbox/Extensions/Chapter7/McNemar Test)
```

The procedures CT_RUN(5) and McNEMAR_RUN(5) are used to evaluate and compare test results generated by the supervised classifiers MAXLIKE_RUN, KERNEL_RUN, FFN_RUN, FFN3AB_RUN (see below) and SVM_RUN. They request input files with extension TST and write their outputs (contingency tables, accuracies, test statistics etc.) to the IDL console.

C.2.4.3 AdaBoost.M1 algorithm with three-layer neural network classifiers (Section 7.3)

```
FFN3AB_RUN (Classification/Supervised/Neural Net (AdaBoost))
```

The user is asked to choose an input file, the training ROIs (which must be present beforehand), output destinations for the classified image, the probability image (optional) and the test results (required). Then the user is prompted for the number of training epochs for each network in the sequence (default 5) and the numbers of neurons in each of the two hidden layers (default 10). During the calculation, the cross-entropy cost function for the current network is displayed as well as the overall training and generalization errors of the sequence; see Figure 7.3. Training can be interrupted at any time, whereupon the boosted sequence is terminated and the image is classified. The program uses spectral tiling and may be applied to large images. The test result file can be processed with the procedures CT_RUN(5) and McNEMAR_RUN(5) discussed above.

C.2.4.4 Maximum likelihood classification of polarimetric SAR imagery (Section 7.4)

```
MAXLIKESAR_RUN (Classification/Supervised/MaxLike polSAR)
```

The quad or dual polarimetric SAR image should have the been generated by the POLSARINGEST_RUN extension of Section C.2.2.3. The user is asked to enter the ENVI filename, then prompted for the training ROIs (which must be present beforehand), and finally for a filename for the test statistics. The classification result is written to ENVI memory only, but can be saved to disk manually. The test statistics file can be processed with the procedure `CT_RUN` or `CT_RUN5` discussed in Section C.2.4.2 above.

C.2.4.5 Kernel RX anomaly detection (Section 7.5.5)

```
KRX_RUN (Spectral/Kernel RX (CUDA))
KRX_RUN5 (Toolbox/Extensions/Chapter7/Kernel RX (CUDA))
```

Choose (a spatial/spectral subset of) a multi- or hyperspectral image in the first dialog. Then select the number of training samples for the kernel matrix (select 0 to determine them with the K-means algorithm). Next, enter the `NSCALE` parameter for the Gaussian kernel (the Gaussian kernel parameter γ is calculated as in Equation (C.1)), and finally the output destination for the anomaly image. If CUDA/GPULib is available, the anomaly values are calculated on the GPU.

C.2.5 ENVI extensions for Chapter 8

C.2.5.1 Kernel K-means clustering of multispectral imagery (Section 8.2.2)

```
KKMEANS_RUN (Classification/Unsupervised/Kernel K-Means (CUDA))
KKMEANS_RUN5 (Toolbox/Extensions/Chapter8/Kernel K-Means (CUDA))
```

Choose (a spatial/spectral subset of) a multispectral image in the first dialog. Then enter the number of random training samples and number of clusters in the corresponding prompts. The next prompt is for the kernel parameter `NSCALE` (the Gaussian kernel parameter γ is calculated as in Equation (C.1)). Finally, decide whether or not to save the clustered training data as ROIs and choose the output destination (memory or file). If CUDA/GPULib is available, then both training and generalization (classification) are performed on the GPU, resulting in a considerable speedup. This extension does not make use of the ENVI tiling facility and is not intended to be used with large images.

C.2.5.2 Agglomerative hierarchical clustering of multispectral imagery (Section 8.2.4)

```
HCL_RUN (Classification/Unsupervised/Agglomerative Hierarchical)
HCL_RUN5 (Toolbox/Extensions/Chapter8/Agglomerative Clustering)
```

Clustering is performed on a small random sample of image pixels (≤ 2000), and generalized to the entire image by using ENVI's built-in maximum likelihood supervised classification algorithm. Choose (a spatial/spectral subset of) a multispectral image in the first dialog. Then enter the number of random samples and number of clusters in the corresponding prompts. Finally, indicate whether the classified image is to be saved to disk or memory. During computation, the current number of clusters is displayed in the progress bar. The result can be used to cluster a larger spatial subset if so desired (this is prompted for). This extension does not make use of the ENVI tiling facility and is not intended to be used with large images.

C.2.5.3 Fuzzy K-means clustering of multispectral imagery (Section 8.2.5)

```
FKM_RUN (Classification/Unsupervised/Fuzzy K-Means)
FKM_RUN5 (Toolbox/Extensions/Chapter8/Fuzzy K-Means Clustering)
```

Clustering is performed on a random sample of image pixels ($\leq 10^5$), and generalized to the entire image by using a modification of the IDL distance classifier `CLUSTER_FKM.PRO` (in the AUXIL directory). Choose (a spatial/spectral subset of) a multispectral image at the prompt. Then enter the desired number of clusters and whether or not they should be saved as ROIs at the corresponding prompts. Finally, indicate whether the class image is to be saved to disk or memory. This extension does not make use of the ENVI tiling facility and is not intended to be used with large images.

C.2.5.4 Gaussian mixture clustering using the expectation maximization algorithm (Section 8.3)

```
EM_RUN (Classification/Unsupervised/Gaussian Mixture (CUDA))
EM_RUN5 (Toolbox/Extensions/Chapter8/Gaussian Mixture Clustering (CUDA))
```

Clustering occurs (optionally) at different scales, and both simulated annealing and spatial memberships can be included if desired. The program will also optionally generate class membership probability images, which can be postprocessed with probabilistic label relaxation (see Section C.2.4.1). Choose (a spatial/spectral subset of) a multispectral image at the prompt. Then enter the desired number of clusters and compressions (pyramid depth) at the corresponding prompts. In the `Clustering parameters` dialog, enter the initial annealing temperature (zero for no annealing) and spatial membership parameter `Beta` (zero for no spatial memberships). At the next prompt, indicate whether or not to save the clusters as ROIs. Finally, say whether the class image is to be saved to disk or memory and similarly for the membership probability image (optional). This extension does not make use of the ENVI tiling facility and is not intended to be used with large images.

C.2.5.5 Image cluster visualization using the Kohonen self-organizing map (Section 8.6)

```
SOM_RUN (Classification/Unsupervised/Self Organizing Map)
SOM_RUN5 (Toolbox/Extensions/Chapter8/Self Organizing Map)
```

At the prompts, enter the (spectral/spatial subset of the) image to be visualized, the dimension of the neural network cube (default $6 \times 6 \times 6$), and the output destination for the map. The network is trained with a random sample of 10^4 pixel vectors. A three-dimensional plot of the neuron cube projected onto the space of the first three image bands is generated; see Figure 8.7. The routine does not use the ENVI tiling facility for the final classification and is not intended to be used with large images.

C.2.5.6 Segmentation of a class image by finding connected components for each separate class (Section 8.7.1)

`SEGMENT_CLASS_RUN (Classification/Unsupervised/Cluster Segments)`

This routine is an alternative to the ENVI `.../Segmentation Image` command and, unlike that command, honors class labels. At the prompts, enter the (spatial subset of the) classification image to be segmented and select the classes to include in segmentation. Then choose the minimum segment size to accept, the connectivity (four or eight) for blobbing and, finally, the output destination (memory or file).

C.2.5.7 Mean shift segmentation of multispectral images (Section 8.7.3)

```
MEAN_SHIFT_RUN (Classification/Unsupervised/Mean Shift Segmentation)
MEAN_SHIFT_RUN5 (Toolbox/Extensions/Chapter8/Mean Shift Segmentation)
```

The user is prompted for the (spectral/spatial subset of the) multispectral (or gray-scale) image to be segmented, and then, in an edit widget, for the spatial and spectral bandwidths (defaults 15) and minimum segment size (default 30 pixels). Finally, the user must choose a destination (memory or file) for the output. Three images are generated:

1. The segmented image (ENVI standard)
2. The segment boundaries (ENVI classification file)
3. The segment labels (ENVI standard)

The boundaries can be overlaid onto the segmented image (or segment labels) with the `Overlay/Classification` command in the ENVI Classic image menu. The routine associates pixels along the current mean shift path with the final mode reached on that path, leading to an unavoidable fragmentation of some of the segments. To counter this effect to some extent, the segmented image is postprocessed with a 3×3 median filter.

C.2.6 ENVI extensions for Chapter 9

C.2.6.1 Iteratively re-weighted multivariate alteration detection (Section 9.4)

```
MAD_RUN (Basic Tools/Change Detection/iMAD/Compute new stats...)
MAD_RUN5 (Toolbox/Extensions/Chapter9/Compute new stats AND run iMAD)
MAD_VIEW_RUN (Basic Tools/Change Detection/MAD View)
```

As with other linear spectral transformations in the ENVI environment such as PCA and MNF, there are two ways to run iMAD:

1. by computing image statistics from the current data, i.e., the bitemporal image itself, or

2. by reading an existing statistics file and applying it to the current data.

Method 1

The first of these is the more common. Begin by choosing (spatial/spectral subsets of) two multispectral images at the prompts. When entering the first image, a mask band can be optionally specified. Then enter the maximum number of iterations (default 50), a penalization regularization parameter (default 0.0) and the output destinations for the MAD and canonical variates (CVs). Press `cancel` for whichever output is not desired. If the CVs are to be output to a file, the program will append _1 and _2 to the filenames in order to distinguish the two CV files which are created. If the two images are not of the same spectral/spatial dimension, the program will abort. If an image is not in band interleaved by pixel (BIP) or band interleaved by line (BIL) format, you will be prompted to allow it to be converted to BIP in place (the image must reside on disk in this case). Finally, you are prompted for a statistics filename (ASCII file with extension `MAD`) where you can (optionally) save the transformation coefficients calculated with the current data. The program also appends an additional band to the MAD variates image consisting of the chi-square value (sum of squares of standardized MAD values) of each pixel.

There are two primary reasons for masking the images:

1. "Black edge" pixels. Often, when processing full scenes, the image margins contain no data. The MAD algorithm may then misinterpret these pixels as the no-change background and converge to them. Since the edge pixels are constants (usually zeroes), the weighted covariance matrices will quickly become degenerate and the program will abort, usually with the message `CHOLDC FAILED`.

2. Large water bodies. Generally, water bodies provide a good no-change background against which to measure change. However, if they are large and if illumination effects (e.g., due to waves, solar glare) lead to a uniform difference in reflectance between the two acquisitions, then

they can constitute a false no-change background to which the MAD algorithm may converge.

Both effects can be countered by appropriate masking. Cloud cover, on the other hand, is usually never a problem, since it corresponds to genuine change.

Method 2

The second method is provided primarily for situations in which the MAD algorithm fails to converge because the amount of real change between the acquisitions is too large: a useful no-change background is not found. If this is the case, it may be possible, by experimentation, to find a spatial subset of the data for which convergence is possible. The generated statistics file (see above) may then be used to generalize to the entire scene. After responding to the prompt for an existing stats file, enter the (spatial/spectral) subsets of the two images (the spectral subsets must, of course, concur with those used to generate the stats file) and the destinations for the MAD and canonical variates. The MAD transformation is then performed immediately.

These extensions use ENVI spectral tiling and can be run on large datasets, e.g., Landsat TM full scenes.

Viewing

`MAD_VIEW_RUN` is a simple ENVI/IDL GUI for viewing MAD or MAD/MNF variates. It can be used to set marginal thresholds on any three MAD variates simultaneously and to stretch the data over different dynamic ranges. It is provided with a rudimentary help file in PDF format.

C.2.6.2 Polarimetric SAR Change Detection

```
WISHARTCHANGE_RUN (Radar/Polarimetric Tools/SAR Change Detection)
WISHARTCHANGE_RUN5 (Toolbox/Extensions/Chapter9/SAR Change Detection)
```

The two polarimetric SAR images should have been generated by the `POLSAR-INGEST_RUN` extension of Section C.2.2.3, and not necessarily have been co-registered. The user is asked to enter the ENVI filename and the equivalent number of looks for both images. The images must have the same polarimetry, which can be any of the following:

- full polarimetry
- dual polarimetry
- single polarization

Then, a desired significance level for change identification (default 0.01) and output destination (ENVI memory or disk) must be chosen. At the next prompt, the two images may be co-registered automatically if desired, in which

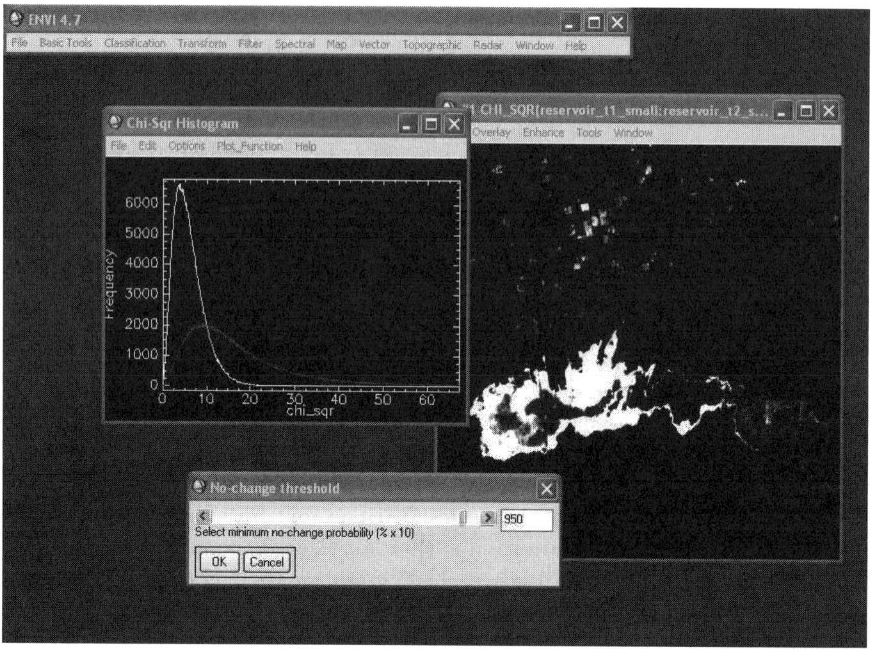

FIGURE C.2
Radiometric normalization. **(See color insert.)**

case a histogram of tie-point distance ratios (see Section 5.6.2.5) is displayed along with the number of detected tie-points (GCPs). The user can abort at this stage if an insufficient number of tie-points have been found. Two image files are created:

- a 4-band image consisting of the span of the first image chosen, the test statistic $-2\rho \ln Q$, the associated change probability $\Pr(-2\rho \ln Q) \leq z$ and the thresholded change probability at the desired significance;
- an ENVI classification image for change and no-change at the chosen significance level.

C.2.6.3 Scatterplot matching (Section 9.8.1)

```
BSLFCP_RUN (Basic Tools/Change Detection/Scatterplot Normalization)
BSLFCP_RUN5 (Toolbox/Extensions/Chapter9/Scatterplot Normalization)
```

This extension exploits the characteristic shape of NIR vs. red scatterplots in multispectral images for radiometric normalization. The method only applies for the two bands in question, e.g., bands 3 and 4 for Landsat TM imagery or bands 2 and 3 for ASTER VNIR data. The user is prompted for the red and near-infrared bands of the reference and target image and the output destination (memory or file). Scatterplots of the reference image and of the target image before and after normalization are displayed and the result written as a two-band ENVI standard image.

C.2.6.4 IR-MAD relative radiometric normalization (Section 9.8.2)

`RADCAL_RUN (Basic Tools/Change Detection/Radiometric Normalization)`
`RADCAL_RUN5 (Toolbox/Extensions/Chapter9/Radiometric Normalization)`

This extension takes advantage of the linear and affine invariance of the MAD transformation to perform a relative radiometric normalization of the images involved in the transformation.

At the prompts, choose the same spatial subset of the two images as were used in Method 1 in Section C.2.6.1 above to generate the IR-MAD image. The spectral subsets must be equal, but need not be the same as for the IR-MAD image. Then enter the chi-square band which was appended to the MAD image. You will then be prompted to set a no-change probability threshold for the pixels to be used in the normalization (default 95%); see Figure C.2. The plot on the left shows, in white, the theoretical chi-square distribution for no-change data and, in red, the actual distribution as determined for all of the pixels in the image. The threshold probability in the slider refers to the theoretical distribution. If the true distribution is severely shifted away from the theoretical one due to a predominance of change, it may be necessary to choose a lower threshold in order to have sufficiently many no-change pixels for a good normalization.

The next prompt is for a destination (memory or disk) for the normalized image, after which the normalization takes place. Orthogonal regression lines are plotted for each multispectral band and a table showing associated statistical parameters and some hypothesis test results is presented.

A final prompt asks if the current regression parameters are to be applied in order to normalize another file (e.g., a full scene). If so, that file must be chosen with the same spectral subset as for the reference and target images and a destination filename must be supplied.

D

Python Scripts

This appendix provides installation instructions and documentation for the Python scripts implementing the algorithms described in the text. The programs were tested with 32-bit Python 2.7 on MS Windows 7.0, and with Python 2.7 on 32 and 64-bit Linux (Ubuntu 12.04).

D.1 Installation

The Python interpreter is pre-installed on Mac OS and Linux. The latest Python 2.7x can be obtained for most operating systems from `http://www.python.org/`

D.1.1 Required packages

The following is a list of the Python extension packages which are imported in the various scripts, together with the URLs at which they can be obtained.*

`numpy`: `http://www.numpy.org/` (numerical Python)

`scipy`: `http://www.scipy.org/` (scientific Python)

`matplotlib`: `http://matplotlib.org/` (2D plotting library)

`gdal`: `https://pypi.python.org/pypi/GDAL/` (geo-spatial data abstraction library)

`mlpy`: `http://mlpy.sourceforge.net/` (machine learning Python)

`ctypes`: `http://python.net/crew/theller/ctypes/` (foreign function interface)

`shapely`: `https://pypi.python.org/pypi/Shapely` (manipulation and analysis of planar geometric objects)

*Windows users can obtain pre-compiled binaries for most of these at `http://www.lfd.uci.edu/~gohlke/pythonlibs/`

opencv: http://sourceforge.net/projects/opencvlibrary/ (open source computer vision library)

spy: http://spectralpython.sourceforge.net/ (spectral Python)

multyvac: https://www.multyvac.com/ (high-performance cloud computing platform)

auxil (additional auxiliary routines, included with the software on the author's website)

All Python scripts which accompany this book, including the auxil package, can be downloaded from

http://ms-image-analysis.appspot.com/static/homepage/software.html

or from

http://mcanty.homepage.t-online.de/software.html

The IR-MAD script imad.py requires the dynamic library prov_means.dll (Windows) or libprov_means.so (Linux, Mac OS). Compiled versions for 32-bit Python (Windows) and Python on 32-bit and 64-bit Linux are included with the software, together with the source code prov_means.c. The libraries should be placed in the OS path, e.g., C:\Windows or /usr/lib.

To install the auxil package, open a console in the unpacked directory and type the following:

```
python setup.py install
```

The scripts themselves are organized according to book chapter and can be run from the command line. Convenient environments are idle (included in most Python distributions) and ipython; see http://ipython.org/.

D.1.2 Eclipse and Pydev

For those who wish to program the examples given in the exercises, or modify/improve the scripts provided here, Eclipse (http://www.eclipse.org/) together with the plug-in Pydev (http://pydev.org/) provide an excellent, platform-independent Python programming development environment, very similar to that for IDL. The environment includes syntax highlighting, code completion and debugging.

D.2 Documentation

D.2.1 Utilities

The `auxil` package contains the following modules:

`auxil.auxil.py`

A collection of auxiliary routines for processing multispectral imagery.

`auxil.congrid.py`

Arbitrary re-sampling of an array to new dimension sizes. Mimics the `CONGRID()` function in IDL.

`auxil.header.py`

Defines an object class representing the text fields of an ENVI format header.

`auxil.png.py`

A pure Python PNG coder/decoder, see `http://pythonhosted.org/pypng/png.html`

`auxil.polsar.py`

Defines an object class to store fully polarimetric SAR data in multilook covariance matrix form.

`auxil.supervisedclass.py`

Defines object classes for supervised image classification: Bayes maximum likelihood, neural networks with backpropagation and scaled conjugate gradient training, and a support vector machine.

D.2.2 Scripts for Chapter 1

`dispms.py`

A command-line oriented routine to display any three spectral bands of a multispectral image as an RGB composite. Usage:

```
python dispms [-f filename] [-p pos] [-d dims] [- e enhancement]
```

The RGB band positions and spatial dimensions are quoted lists, e.g.,

```
-p "[0,1,3]" -d "[0,0,400,400]"
```

The dimensions list `dims` is of form `[X0,Y0,samples,lines]`. The enhancement modes are: 1 = linear byte stretch, 2 = linear stretch, 3 = linear 2% stretch, 4 = histogram equalization. Information not supplied on the command line is queried interactively.

D.2.3 Scripts for Chapter 4

D.2.3.1 Nonlinear principal components analysis (Section 4.4.2)

`kpca.py`

The user is first queried for a working directory, image filename and a training sample size. If the latter is 0, then 100 representative training pixel vectors are chosen by the k-means algorithm. Otherwise, a random sample of the desired size is used. Next, the number of kernel principal components to retain is entered and the destination output file selected. The user can then choose between a linear or Gaussian kernel. The Gaussian kernel parameter γ is calculated as $\gamma = 1/(2\sigma^2)$, where $\sigma = \langle \|\boldsymbol{g}(\nu) - \boldsymbol{g}(\nu')\| \rangle$ is the average Euclidean distance between the training observations. Finally, the output destination is selected. After centering on the training data and diagonalizing the kernel matrix, a plot of the eigenvalues is displayed and the projected image is stored to disk.

D.2.4 Scripts for Chapter 5

D.2.4.1 Panchromatic sharpening of multispectral images using the discrete or à trous wavelet transform (Sections 5.3.4 and 5.3.5)

`dwt.py`
`atwt.py`

The user is queried for the working directory, the (spatial/spectral subset of the) multispectral image to be sharpened, the corresponding panchromatic or high-resolution image and the output file. The multispectral image should overlap the panchromatic image completely, so that the panchromatic image defines the extent of the final pan-sharpened product. (The upper left-hand corner of the panchromatic image should be as close as possible to that of the MS image, since phase correlation is used to co-register them.) Then the MS to pan spatial resolution ratio (2 or 4), the MS band to be used for co-registration, and a fine adjustment parameter are queried. During the calculation regression coefficients and correlations of the wavelet coefficients for the low- vs. high-resolution bands are printed.

D.2.4.2 Conversion of SAR imagery to ENVI standard files

`polsaringest.py`

Geocoded, multi-look polarimetric SAR imagery suitable for processing with the filtering, classification and change detection algorithms described in the text may be obtained directly from the provider or generated from single-look complex (SLC) data. In the latter case, the open source software packages PolSARpro (European Space agency),

`http://earth.eo.esa.int/polsarpro/`

together with MapReady (Alaska Satellite Facility),

`http://www.asf.alaska.edu/downloads/software_tools`

are a good choice; see Gens et al. (2013). PolSARpro is first used to create multi-look images in covariance matrix format, which can then be exported to MapReady for georeferencing with or without a DEM. Alternative commercial solutions are the Gamma Software (Gamma Remote Sensing)

`http://www.gamma-rs.ch/`

and the SARscape add-on module for ENVI

http://www.exelisvis.com/ProductsServices/ENVI/ENVISARscape.aspx

The Python script `polsaringest.py` combines the outputs from the above preprocessing systems to a single, multi-band file in 32-bit floating point format. For full quad polarimetric data, it generates 9 bands, which are ordered as follows:

$$\begin{aligned}
&\text{C11} = \langle |s_{hh}|^2 \rangle \\
&\text{C12re} = \langle \sqrt{2} s_{hh} s_{hv}^* \rangle \text{(real part)} \\
&\text{C12im} = \langle \sqrt{2} s_{hh} s_{hv}^* \rangle \text{(imaginary part)} \\
&\text{C13re} = \langle s_{hh} s_{vv}^* \rangle \text{(real part)} \\
&\text{C13im} = \langle s_{hh} s_{vv}^* \rangle \text{(imaginary part)} \\
&\text{C22} = \langle 2|s_{hv}|^2 \rangle \\
&\text{C23re)} = \langle \sqrt{2} s_{hv} s_{vv}^* \rangle \text{(real part)} \\
&\text{C23im)} = \langle \sqrt{2} s_{hv} s_{vv}^* \rangle \text{(imaginary part)} \\
&\text{C33} = \langle |s_{vv}|^2 \rangle .
\end{aligned}$$

For dual or single polarimetry, the bands corresponding to the missing matrix elements are not present. Thus, a dual polarimetric image will consist of four bands, e.g., C11, C12re, C12im, and C22; a single polarimetric image will consist of just one band, usually C11 or C33. The files generated by the script can be read by all of the SAR processing routines described below, and also by their ENVI/IDL counterparts described in Appendix C.

The user is prompted for the directory containing the georeferenced covariance matrix files (consisting of one file for each of the real and imaginary components), and then requested to choose (a spatial subset of) one of them. The other files are then read in automatically with the same spatial subset. Finally, an output filename and desired format are requested.

D.2.4.3 Multivariate estimation of equivalent number of looks for polarimetric SAR in covariance matrix format

`enlml.py`
`lookup.txt`

This method is not discussed in the text. It is a multivariate technique based on the maximum likelihood estimator explained in Anfinsen et al. (2009a) and is reported to be superior to the standard univariate method discussed in Section 5.4.2. The polarimetric SAR image should be in the format generated by the script `polsaringest.py` described in Section D.2.4.2 above. The user is asked to select (a spatial subset of) the covariance matrix file, the desired window size (default 7×7) and an output filename and format. The local ENL values are estimated in a moving window and stored as a single-band, floating point raster image. A histogram of the ENL values is plotted.

D.2.4.4 Minimum mean square error filtering of polarimetric SAR imagery (Section 5.4.3.1)

`mmse_filter.py`

The polarimetric SAR image should be in the format generated by the script `polsaringest.py` described in Section D.2.4.2 above. The user is queried for the working directory, then for the input filename, the equivalent number of looks and, finally, the output filename. The filtering operation occurs in two steps: first the span image is processed to determine the filter weights, then all components of the covariance matrix are filtered separately.

D.2.4.5 Gamma maximum a posteriori filtering of polarimetric SAR imagery (Section 5.4.3.2)

`gamma_filter.py`

The polarimetric SAR image should have the same format as in Section D.2.4.2 above. The user is queried for the working directory, then for the input filename, the equivalent number of looks, the number of iterations and, finally, the output filename. Only the diagonal elements of the covariance matrix are filtered.

D.2.5 Scripts for Chapter 6

D.2.5.1 Supervised classification of multispectral images with maximum likelihood, neural network and support vector machine (Sections 6.3, 6.5, 6.6 and Appendix B)

`classify.py`

The user is first queried for a working directory and input image filename and (optionally) a spectral subset. The image can be in any format recognized

by GDAL, but must be geo-referenced. Then the user can choose between maximum likelihood, neural network (with backpropagation or scaled conjugate gradient training), or support vector machine algorithms. Next an ESRI shape file, which defines the regions of interest in the input image, and an output filename for the test results must be entered. Following this, the user must enter the filename for the classification (thematic map) image and (optionally) a filename for the class membership probability image. The latter is not available for the maximum likelihood classifier. Finally, if the neural network classifier was selected, the number of hidden neurons must be entered. (A cross-entropy plot will be displayed after training has completed.) If the output format chosen is ENVI, then an ENVI header corresponding to the ENVI filetype "classification" will be written. The test result file can be processed with the scripts `ct.py` and `mcnemar.py` discussed in Section D.2.6.3 below.

D.2.6 Scripts for Chapter 7

D.2.6.1 Probabilistic label relaxation postprocessing (Section 7.1.2)

`plr.py`
`plr_reclass.py`

The script takes as input a class probability vector image generated by most of the supervised classification extensions described in this appendix, as well as from the clustering routine `em.py` described in Section D.2.7.2 below, and generates a modified class probability vector image (rule image). At the prompt, choose a class membership probabilities image and the number of iterations (default = 3). Then choose a destination filename and format. The generated rule image can then be processed with the script `plr_reclass.py` to produce an improved classification image with better spatial coherence. At the prompt, choose a class membership probabilities image. Then choose an output filename and format. If the format is ENVI, then an ENVI classification file will be written.

D.2.6.2 Neural network supervised classification with cross-validation (Section 7.2.2)

`ffncg.py`

This script requires registration on the Multyvac website

`http://www.multyvac.com/`

The user is first queried for a working directory, an input image filename and the training data shapefile. The image can be in any format recognized by GDAL, but must be georeferenced. Next, enter the number of hidden neurons and the filename for the classification image (thematic map). Training and classification take place on the host computer using 9/10th of the training data. The classification method used is a two-layer feed-forward network with

scaled conjugate gradient training. Upon completion, a cross-entropy plot characterizing the training phase is displayed. When the plot is closed, the training data are uploaded to the cloud service and a 10-fold cross-validation is carried through. The results (misclassification rate and standard deviation) are printed to the standard output.

D.2.6.3 Contingency tables and McNemar test (Section 7.2)

`ct.py`
`mcnemar.py`

The scripts `ct.py` and `mcnemar.py` are used to evaluate and compare test results generated by the supervised classifiers discussed in this appendix. They request input files with extension `tst` and generate their outputs (contingency tables, accuracies, test statistics, etc.) on the standard output.

D.2.6.4 Anomaly detection with the RX-algorithm (Section 7.5.4)

`rx.py`

The user is first queried for a working directory and input image filename. The image can be in any format recognized by GDAL. Then the output filename and desired format must be entered. The anomaly image (the Mahalanobis distance of each pixel vector to the mean image background) is written to disk.

D.2.7 Scripts for Chapter 8

D.2.7.1 Kernel K-means clustering of multispectral imagery (Section 8.2. 2)

`kkmeans.py`

The user is first queried for a working directory and input image filename. The image can be in any format recognized by GDAL. Then spatial and/or spectral subsets can be entered along with a training data sample size to estimate the kernel and the desired number of clusters. Finally, the output filename and format and the kernel type (Gaussian or linear) must be chosen. If the output format is ENVI, then an ENVI classification file header is generated.

D.2.7.2 Gaussian mixture clustering using the expectation maximization algorithm (Section 8.3)

`em.py`

Clustering occurs optionally at different scales, and both simulated annealing and spatial memberships can be included if desired. The user is queried

for a working directory and input image filename. The image may be in any format recognized by GDAL. Then, at the corresponding prompts, spectral and/or spatial subsets may be chosen, followed by the number of clusters, the number of compressions (initial and final pyramid depths), initial annealing temperature (zero for no annealing), and spatial membership parameter `beta` (zero for no spatial memberships). Finally, the user is prompted for the desired output format and filename for the classification image and (optionally) for a probability image. If ENVI format is chosen, an ENVI classification file will be written.

D.2.8 Scripts for Chapter 9

D.2.8.1 Iteratively re-weighted multivariate alteration detection (Section 9.4)

`iMad.py`

When running the `iMad.py` script, you are first prompted for an input directory and then for the (spectral and spatial subsets of) the first and second input images. The input subsets must be co-registered and have the same spatial/spectral dimensions. They can be in any format recognized by GDAL. The following prompts are for a regularization or penalization factor (default 0) and for an output filename, which again can be in any GDAL multispectral format. The script prints the convergence criterion `Delta` and the canonical correlations for each iteration. Computation terminates when `Delta` < 0.001 or after 100 iterations. The output consists of the stacked MAD variates in order of decreasing variance followed by the chi-square image.

D.2.8.2 Polarimetric SAR change detection

`wishartchange.py`

The two polarimetric SAR images should have the same format as in Section D.2.4.2 above and be co-registered. The user is asked to choose a working directory, and then to enter the input filename and the equivalent number of looks for each image successively. The images must have the same polarimetry, which can be any of the following:

- full polarimetry
- dual polarimetry
- single polarization

and have the same spatial dimension. Finally, an output filename and desired format is queried. After completion, a 3-band image consisting of the span of the first image chosen, the test statistic $-2\rho\ln Q$ (Conradsen et al., 2003), and the associated change probability $\Pr(-2\rho\ln Q) \le z)$ is written. The latter can be thresholded to obtain a change map at any desired significance level (e.g., at 0.99 for changes at the 1% significance level.)

D.2.8.3 IR-MAD relative radiometric normalization (Section 9.8)

`radcal.py`

This script takes advantage of the linear and affine invariance of the MAD transformation to perform a relative radiometric normalization of the images involved in the transformation.

The routine first prompts for an input directory, (spectral/spatial subsets of) the reference and target images, as well as for the spatial subset of the iMad image generated according to Section D.2.8.1 above (the chi-square band is found automatically). The spatial subsets of all three input images must have the same size and the spectral subsets of the reference and target images must match. After the prompt for an output filename, which can have any of the formats recognized by GDAL, the filename of a larger (e.g., full) scene can be (optionally) entered. This file must have the same spectral dimensions as the reference and target files and will be normalized with the regression coefficients determined by them. The result will be stored in the same format and with the same root name as the specified output filename, but with `_norm` appended. Finally the user is prompted for a minimum no-change probability threshold (default 0.95). The script prints, band-wise, the regression coefficients, and the results of statistical tests for equal means and variances of the reference and normalized target image bands. The tests are evaluated on the basis of the hold-out test pixels (train:test = 2:1). The script also plots the orthogonal regression lines for up to 10 spectral bands.

Mathematical Notation

X	random variable
x	realization of X, observation
$\boldsymbol{X}$	random (column) vector
$\boldsymbol{x}$	realization of $\boldsymbol{X}$ (vector observation)
$\boldsymbol{\mathcal{X}}$	data design matrix for $\boldsymbol{X}$
$G, g, \boldsymbol{G}, \boldsymbol{g}, \boldsymbol{\mathcal{G}}$	as above, designating pixel gray-values
$\boldsymbol{x}^\top$	transposed form of the vector $\boldsymbol{x}$ (row vector)
$\lVert\boldsymbol{x}\rVert$	length or (2-)norm of $\boldsymbol{x}$
$\boldsymbol{x}^\top\boldsymbol{y}$	inner product of two vectors (scalar)
$\boldsymbol{x}\cdot\boldsymbol{y}$	Hadamard (component-by-component) product
$\boldsymbol{x}\boldsymbol{y}^\top$	outer product of two vectors (matrix)
$\boldsymbol{C}$	matrix
$\boldsymbol{x}^\top\boldsymbol{C}\boldsymbol{y}$	quadratic form (scalar)
$\lvert\boldsymbol{C}\rvert$	determinant of $\boldsymbol{C}$
$\mathrm{tr}(\boldsymbol{C})$	trace of $\boldsymbol{C}$
$\boldsymbol{I}$	identity matrix
$\mathbf{0}$	column vector of zeroes
$\mathbf{1}$	column vector of ones
$\boldsymbol{\Lambda}$	$\mathrm{Diag}(\lambda_1 \dots \lambda_N)$ (diagonal matrix of eigenvalues)
$\frac{\partial f(x)}{\partial \boldsymbol{x}}$	partial derivative of $f(x)$ with respect to vector $\boldsymbol{x}$
$f(x)\rvert_{x=x^*}$	$f(x)$ evaluated at $x = x^*$
$\mathbf{i}$	$\sqrt{-1}$
$\lvert z\rvert$	absolute value of real or complex number z
z^*	complex conjugate of complex number z
Ω	sample space in probability theory
$\Pr(A \mid B)$	probability of event A conditional on event B
$P(x)$	distribution function for random variable X
$P(\boldsymbol{x})$	joint distribution function for random vector $\boldsymbol{X}$
$p(x)$	probability density function for X
$p(\boldsymbol{x})$	joint probability density function for $\boldsymbol{X}$
$\langle X\rangle,\ \mu$	mean (or expected) value
$\mathrm{var}(X),\ \sigma^2$	variance
$\langle\boldsymbol{X}\rangle,\ \boldsymbol{\mu}$	mean vector
$\boldsymbol{\Sigma}$	variance–covariance matrix
$\boldsymbol{S}$	estimator for $\boldsymbol{\Sigma}$ (random matrix)
$\boldsymbol{s}$	estimate of $\boldsymbol{\Sigma}$ (realization of $\boldsymbol{S}$)

$\hat{\boldsymbol{\mu}},\ \hat{\boldsymbol{\Sigma}}$	maximum likelihood estimates of $\boldsymbol{\mu},\ \boldsymbol{\Sigma}$
$\boldsymbol{\mathcal{K}}$	kernel matrix with elements $k(\boldsymbol{x}, \boldsymbol{x}')$
$\Phi(x)$	standard normal probability distribution function
$\phi(x)$	standard normal probability density function
$p_{\chi^2;n}(x)$	chi-square density function with n degrees of freedom
$p_{t;n}(x)$	Student-t density function with n degrees of freedom
$p_{f;m,n}(x)$	F-density function with m and n degrees of freedom
$p_{\mathcal{W}}(\boldsymbol{x}, m)$	Wishart distribution with m degrees of freedom
$p_{\mathcal{W}_C}(\boldsymbol{x}, m)$	complex Wishart distribution with m degrees of freedom
$(\hat{x}(0), \hat{x}(1)\dots)$	discrete Fourier transform of array $(x(0), x(1)\dots)$
$\boldsymbol{x} * \boldsymbol{y}$	discrete convolution of vectors $\boldsymbol{x}$ and $\boldsymbol{y}$
$\langle\phi, \psi\rangle$	inner product $\int \phi(x)\psi(x)dx$ of functions ϕ and ψ
$\mathcal{N}_i$	neighborhood of ith pixel
$\{x \mid c(x)\}$	set of elements x that satisfy condition $c(x)$
$\mathcal{K}$	set of class labels $\{1\dots K\}$
$\boldsymbol{\ell}$	vector representation of a class label
$u \in U$	u is an element of the set U
$U \otimes V$	Cartesian product set
$V \subset U$	V is a (proper) subset of the set U
$\arg\max_x f(x)$	the set of x which maximizes $f(x)$
$f : A \mapsto B$	the function f which maps the set A to the set B
$\mathbb{R}$	set of real numbers
$\mathbb{Z}$	set of integers
$=:$	equal by definition
$\square$	end of a proof

References

Aanaes, H., Sveinsson, J. R., Nielsen, A. A., Bovith, T., and Benediktsson, A. (2008). Model-based satellite image fusion. *IEEE Transactions on Geoscience and Remote Sensing*, 46(5):1336–1346.

Aboufadel, E. and Schlicker, S. (1999). *Discovering Wavelets*. J. Wiley and Sons.

Abrams, M., Hook, S., and Ramachandran, B. (1999). *ASTER User Handbook*. Technical report, Jet Propulsion Laboratory and California Institute of Technology.

Aiazzi, B., Alparone, L., Baronti, S., and Garzelli, A. (2002). Context-driven fusion of high spatial and spectral resolution images based on oversampled multiresolution analysis. *IEEE Transactions on Geoscience and Remote Sensing*, 40(10):2300–2312.

Albanese, D., Visintainer, R., Merler, S., Riccadonna, S., Jurman, G., and Furlanello, C. (2012). mlpy: Machine learning python. Cornell University Library. Internet http://arxiv.org/abs/1202.6548arxiv.org/abs/1202.6548.

Anderson, T. W. (2003). *An Introduction to Multivariate Statistical Analysis*. Wiley Series in Probability and Statistics, third edition.

Anfinsen, S., Doulgeris, A., and Eltoft, T. (2009a). Estimation of the equivalent number of looks in polarimetric synthetic aperture radar imagery. *IEEE Transactions on Geoscience and Remote Sensing*, 47(11):3795–3809.

Anfinsen, S., Eltoft, T., and Doulgeris, A. (2009b). A relaxed Wishart model for polarimetric SAR data. In *Proc. PolinSAR, 4th International Workshop on Science and Applications of SAR Polarimetry and Polarimetric Interferometry*, Frascati, Italy.

Beisl, U. (2001). *Correction of Bidirectional Effects in Imaging Spectrometer Data*. Remote Sensing Laboratories, Remote Sensing Series 37, Department of Geography, University of Zurich.

Bellman, R. (1961). *Adaptive Control Processes, A Guided Tour*. Princeton University Press.

Belousov, A. I., Verzakov, S. A., and von Frese, J. (2002). A flexible classification approach with optimal generalization performance: support vector machines. *Chemometrics and Intelligent Laboratory Systems*, 64:15–25.

Benz, U. C., Hoffmann, P., Willhauck, G., Lingfelder, I., and Heynen, M. (2004). Multi-resolution, object-oriented fuzzy analysis of remote sensing data for GIS-ready information. *ISPRS Journal of Photogrammetry & Remote Sensing*, 258:239–258.

Bilbo, C. M. (1989). Statistisk analyse af relationer mellem alternative antistoftracere. Master's thesis, Informatics and Mathematical Modeling, Technical University of Denmark, Lyngby. In Danish.

Bishop, C. M. (1995). *Neural Networks for Pattern Recognition.* Oxford University Press.

Bishop, C. M. (2006). *Pattern Recognition and Machine Learning.* Springer.

Bishop, Y. M. M., Feinberg, E. E., and Holland, P. W. (1975). *Discrete Multivariate Analysis, Theory and Practice.* Cambridge Press.

Bradley, A. P. (2003). Shift-invariance in the discrete wavelet transform. In Sun, C., Talbot, H., Ourselin, S., and Adriaansen, T., editors, *Proc. VIIth Digital Image Computing: Techniques and Applications*, pages 29–38.

Breiman, L. (1996). Bagging predictors. *Machine Learning*, 24(2):123–140.

Bridle, J. S. (1990). Probabilistic interpretation of feed-forward classification outputs, with relationships to statistical pattern recognition. In Soulié, F. F. and Hérault, J., editors, *Neurocomputing: Algorithms, Architectures and Applications*, pages 227–236. Springer.

Bruzzone, L. and Prieto, D. F. (2000). Automatic analysis of the difference image for unsupervised change detection. *IEEE Transactions on Pattern Analysis and Machine Intelligence*, 11(4):1171–1182.

Canny, J. (1986). A computational approach to edge detection. *IEEE Transactions on Pattern Analysis and Machine Intelligence*, 8(6):679–699.

Canty, M. J. (2009). Boosting a fast neural network for supervised land cover classification. *Computers and Geosciences*, 35:1280–1295.

Canty, M. J. and Nielsen, A. A. (2006). Visualization and unsupervised classification of changes in multispectral satellite imagery. *International Journal of Remote Sensing*, 27(18):3961–3975. Internet http://www.imm.dtu.dk/pubdb/p.php?3389.

Canty, M. J. and Nielsen, A. A. (2008). Automatic radiometric normalization of multitemporal satellite imagery with the iteratively re-weighted MAD transformation. *Remote Sensing of Environment*, 112(3):1025–1036. Internet http://www.imm.dtu.dk/pubdb/p.php?5362.

Canty, M. J., Nielsen, A. A., and Schmidt, M. (2004). Automatic radiometric normalization of multitemporal satellite imagery. *Remote Sensing of Environment*, 91(3-4):441–451. Internet http://www.imm.dtu.dk/pubdb/p.php?2815.

Canty, M. J. and Nieslsen, A. A. (2012). Linear and kernel methods for multivariate change detection. *Computers and Geosciences*, 38:107–114.

Chang, C.-C. and Lin, C.-J. (2011). A library for support vector machines. *ACM Transactions on Intelligent Systems and Technology,* ACM Digital Lib. Internet http://www.csie.ntu.edu.tw/ cjlin/libsvm/.

Cohen, J. (1960). A coefficient of agreement for nominal scales. *Educational and Psychological Measurement*, 20:37–46.

Comaniciu, D. and Meer, P. (2002). Mean shift: a robust approach toward feature space analysis. *IEEE Transactions on Pattern Analysis and Machine Intelligence*, 24(5):603–619.

Congalton, R. G. and Green, K. (1999). *Assessing the Accuracy of Remotely Sensed Data: Principles and Practices.* Lewis Publishers.

Conradsen, K., Nielsen, A. A., Schou, J., and Skriver, H. (2003). A test statistic in the complex Wishart distribution and its application to change detection in polarimetric SAR data. *IEEE Transactions on Geoscience and Remote Sensing*, 41(1):3–19. Internet http://www2.imm.dtu.dk/pubdb/views/publication_details.php?id=1219.

Coppin, P., Jonckheere, I., Nackaerts, K., and Muys, B. (2004). Digital change detection methods in ecosystem monitoring: A review. *International Journal of Remote Sensing*, 25(9):1565–1596.

Cristianini, N. and Shawe–Taylor, J. (2000). *Support Vector Machines and Other Kernel-based Learning Methods.* Cambridge University Press.

Daubechies, I. (1988). Orthonormal bases of compactly supported wavelets. *Communications on Pure and Applied Mathematics*, 41:909–996.

Dhillon, I., Guan, Y., and Kulis, B. (2005). *A Unified View of Kernel K-Means, Spectral Clustering and Graph Partitioning.* Technical Report UTCS TR-04-25, University of Texas at Austin.

Dietterich, T. G. (1998). Approximate statistical tests for comparing supervised classification learning algorithms. *Neural Computation*, 10(7):1895–1923.

Du, Y., Teillet, P. M., and Cihlar, J. (2002). Radiometric normalization of multitemporal high-resolution images with quality control for land cover change detection. *Remote Sensing of Environment*, 82:123–134.

Duda, R. O. and Hart, P. E. (1973). *Pattern Classification and Scene Analysis.* J. Wiley and Sons.

Duda, R. O., Hart, P. E., and Stork, D. G. (2001). *Pattern Classification.* Wiley Interscience, second edition.

Duda, T. and Canty, M. J. (2002). Unsupervised classification of satellite imagery: choosing a good algorithm. *International Journal of Remote Sensing*, 23(11):2193–2212.

Dunn, J. C. (1973). A fuzzy relative of the isodata process and its use in detecting compact well-separated clusters. *Journal of Cybernetics*, PAM1-1:32–57.

Exelis/DLR (2012). *TerraSAR-X/TanDEM-X SSC/CoSSC Reader for IDL and ENVI.* Technical Report 2011-ITT-DLR-0001, Exelis/DLR.

Fahlman, S. E. and LeBiere, C. (1990). The cascade correlation learning architecture. In Touertzky, D. S., editor, *Advances in Neural Information Processing Systems 2*, pages 524–532. Morgan Kaufmann.

Fanning, D. W. (2000). *IDL Programming Techniques.* Fanning Software Consulting.

Fraley, C. (1996). *Algorithms for Model-Based Gaussian Hierarchical Clustering.* Technical report 311, Department of Statistics, University of Washington, Seattle.

Freund, J. E. (1992). *Mathematical Statistics.* Prentice-Hall, fifth edition.

Freund, Y. and Shapire, R. E. (1996). Experiments with a new boosting algorithm. In *Proceedings of the Thirteenth International Conference on Machine Learning*, pages 148–156. Morgan Kaufmann.

Freund, Y. and Shapire, R. E. (1997). A decision-theoretic generalization of on-line learning and an application to boosting. *Journal of Computer and System Sciences*, 55:119–139.

Fukunaga, K. (1990). *Introduction to Statistical Pattern Recognition.* Academic Press, second edition.

Fukunaga, K. and Hostetler, L. D. (1975). The estimation of the gradient of a density function, with applications to pattern recognition. *IEEE Transactions on Information Theory*, IT-21:32–40.

Galloy, M. (2011). *Modern IDL: A Guide to IDL Programming.* M. Galloy. Internet http://michaelgalloy.com/.

Gath, I. and Geva, A. B. (1989). Unsupervised optimal fuzzy clustering. *IEEE Transactions on Pattern Analysis and Machine Intelligence*, 3(3):773–781.

Gens, R., Atwood, D. K., and Pottier, E. (2013). Geocoding of polarimetric processing results: Alternative processing strategies. *Remote Sensing Letters*, 4(1):38–44.

Gonzalez, R. C. and Woods, R. E. (2002). *Digital Image Processing*. Addison-Wesley.

Goodman, J. W. (1984). Statistical properties of laser speckle patterns. In Dainty, J. C., editor, *Laser Speckle and Related Phenomena*, pages 9–75. Springer.

Goodman, N. R. (1963). Statistical analysis based on a certain multivariate complex Gaussian distribution (An Introduction). *Annals of Mathematical Statistics*, 34:152–177.

Green, A. A., Berman, M., Switzer, P., and Craig, M. D. (1988). A transformation for ordering multispectral data in terms of image quality with implications for noise removal. *IEEE Transactions on Geoscience and Remote Sensing*, 26(1):65–74.

Groß, M. H. and Seibert, F. (1993). Visualization of multidimensional image data sets using a neural network. *The Visual Computer*, 10:145–159.

Gumley, L. E. (2002). *Practical IDL Programming*. Morgan Kaufmann.

Haberächer, P. (1995). *Praxis der Digitalen Bildverarbeitungen und Mustererkennung*. Carl Hanser Verlag.

Harris, C. and Stephens, M. (1988). A combined corner and edge detector. In *Proceedings of the Fourth Alvey Vision Conference*, pages 147–151.

Harsanyi, J. C. (1993). *Detection and Classification of Subpixel Spectral Signatures in Hyperspectral Image Sequences*. Ph.D. Thesis, University of Maryland, 116 pp.

Harsanyi, J. C. and Chang, C.-I. (1994). Hyperspectral image classification and dimensionality reduction: an orthogonal subspace projection approach. *IEEE Transactions on Geoscience and Remote Sensing*, 32(4):779–785.

Hayken, S. (1994). *Neural Networks: A Comprehensive Foundation*. Macmillan.

Hertz, J., Krogh, A., and Palmer, R. G. (1991). *Introduction to the Theory of Neural Computation*. Addison-Wesley.

Hilger, K. B. (2001). Exploratory analysis of multivariate data. Ph.D. Thesis, IMM-PHD-2001-89, Technical University of Denmark.

Hilger, K. B. and Nielsen, A. A. (2000). Targeting input data for change detection studies by suppression of undesired spectra. In *Proceedings of a Seminar on Remote Sensing and Image Analysis Techniques for Revision of Topographic Databases, KMS,* The National Survey and Cadastre, Copenhagen, Denmark, February 2000.

Hotelling, H. (1936). Relations between two sets of variates. *Biometrika*,

28:321–377.

Hu, M. K. (1962). Visual pattern recognition by moment invariants. *IRE Transactions on Information Theory*, IT-8:179–187.

Jensen, J. R. (2005). *Introductory Digital Image Analysis: A Remote Sensing Perspective*. Prentice Hall.

Kendall, M. and Stuart, A. (1979). *The Advanced Theory of Statistics*, volume 2. Charles Griffen & Company Limited, fourth edition.

Kohonen, T. (1989). *Self-Organization and Associative Memory*. Springer.

Kruse, F. A., Lefkoff, A. B., Boardman, J. B., Heidebrecht, K. B., Shapiro, A. T., Barloon, P. J., and Goetz, A. F. H. (1993). The spectral image processing system (SIPS), interactive visualization and analysis of imaging spectrometer data. *Remote Sensing of Environment*, 44:145–163.

Kurz, F., Charmette, B., Suri, S., Rosenbaum, D., Spangler, M., Leonhardt, A., Bachleitner, M., Stätter, R., and Reinartz, P. (2007). Automatic traffic monitoring with an airborne wide-angle digital camera system for estimation of travel times. In Stilla, U., Mayer, H., Rottensteiner, F., Heipke, C., and Hinz, S., editors, *Photogrammetric Image Analysis*. International Archives of the Photogrammetry, Remote Sensing and Spatial Information Service PIA07, Munich, Germany.

Kwon, H. and Nasrabadi, N. M. (2005). Kernel RX-algorithm: a nonlinear anomaly detector for hyperspectral imagery. *IEEE Transactions on Geoscience and Remote Sensing*, 43(2):388–397.

Lang, H. R. and Welch, R. (1999). *Algorithm Theoretical Basis Document for ASTER Digital Elevation Models*. Technical report, Jet Propulsion Laboratory and University of Georgia.

Langtangen, H. P. (2009). *Python Scripting for Computational Science*. Springer.

Lee, J.-S., Grunes, M. R., and de Grandi, G. (1999). Polarimetric SAR speckle filtering and its implication for classification. *IEEE Transactions on Geoscience and Remote Sensing*, 37(5):2363–2373.

Lee, J.-S., Grunes, M. R., and Kwok, R. (1994). Classification of multilook polarimetric SAR imagery based on complex Wishart distribution. *International Journal of Remote Sensing*, 15(11):2299–2311.

Li, H., Manjunath, B. S., and Mitra, S. K. (1995). A contour-based approach to multisensor image registration. *IEEE Transactions on Image Processing*, 4(3):320–334.

Li, S. Z. (2001). *Markov Random Field Modeling in Image Analysis*. Computer Science Workbench. Springer, second edition.

Liao, X. and Pawlak, M. (1996). On image analysis by moments. *IEEE Transactions on Pattern Analysis and Machine Intelligence*, 18(3):254–266.

Maas, S. J. and Rajan, N. (2010). Normalizing and converting image DC data using scatter plot matching. *Remote Sensing*, 2(7):1644–1661.

Mallat, S. G. (1989). A theory for multiresolution signal decomposition: the wavelet representation. *IEEE Transactions on Pattern Analysis and Machine Intelligence*, 11(7):674–693.

Mardia, K. V., Kent, J. T., and Bibby, J. M. (1979). *Multivariate Analysis*. Academic Press.

Masters, T. (1995). *Advanced Algorithms for Neural Networks, A C++ Sourcebook*. J. Wiley and Sons.

Mather, P. and Koch, M. (2010). *Computer Processing of Remotely-Sensed Images: An Introduction*. Wiley, fourth edition.

Milman, A. S. (1999). *Mathematical Principles of Remote Sensing*. Sleeping Bear Press.

Moeller, M. F. (1993). A scaled conjugate gradient algorithm for fast supervised learning. *Neural Networks*, 6:525–533.

Müller, K.-R., Mika, S., Rätsch, G., Tsuda, K., and Schölkopf, B. (2001). An introduction to kernel-based learning algorithms. *IEEE Transactions on Neural Networks*, 12(2):181–202.

Murphey, Y. L., Chen, Z., and Guo, H. (2001). Neural learning using AdaBoost. In *Proceedings. IJCNN apos;01. International Joint Conference on Neural Networks*, volume 2, pages 1037–1042.

Mustard, J. F. and Sunshine, J. M. (1999). Spectral analysis for Earth science: investigations using remote sensing data. In Rencz, A., editor, *Manual of Remote Sensing*, pages 251–307. J. Wiley and Sons, second edition.

Nielsen, A. A. (2001). Spectral mixture analysis: linear and semi-parametric full and iterated partial unmixing in multi- and hyperspectral data. *Journal of Mathematical Imaging and Vision*, 15:17–37.

Nielsen, A. A. (2007). The regularized iteratively reweighted MAD method for change detection in multi- and hyperspectral data. *IEEE Transactions on Image Processing*, 16(2):463–478. Internet http://www.imm.dtu.dk/pubdb/p.php?4695.

Nielsen, A. A. and Canty, M. J. (2008). Kernel principal component analysis for change detections. In *SPIE Europe Remote Sensing Conference, Cardiff, Great Britain, 15-18 September*, volume 7109.

Nielsen, A. A., Conradsen, K., and Simpson, J. J. (1998). Multivariate alteration detection (MAD) and MAF post-processing in multispectral, bitemporal image data: new approaches to change detection studies. *Remote Sensing of Environment*, 64:1–19. Internet http://www.imm.dtu.dk/pubdb/p.php?1220.

Núnez, J., Otazu, X., Fors, O., Prades, A., Palà, V., and Arbiol, R. (1999). Multiresolution-based image fusion with additive wavelet decomposition. *IEEE Transactions on Geoscience and Remote Sensing*, 37(3):1204–1211.

Oliver, C. and Quegan, S. (2004). *Understanding Synthetic Aperture Radar Images.* SciTech.

Palubinskas, G. (1998). K-means clustering algorithm using the entropy. In *SPIE European Symposium on Remote Sensing, Conference on Image and Signal Processing for Remote Sensing,* September, Barcelona, volume 3500, pages 63–71.

Patefield, W. M. (1977). On the information matrix in the linear functional problem. *Journal of the Royal Statistical Society, Series C*, 26:69–70.

Philpot, W. and Ansty, T. (2013). Analytical description of pseudoinvariant features. *IEEE Transactions on Geoscience and Remote Sensing*, 51(4):2016–2021.

Pitz, W. and Miller, D. (2010). The TerraSAR-X satellite. *IEEE Transactions on Geoscience and Remote Sensing*, 48(2):615–622.

Polikar, R. (2006). Ensemble-based systems in decision making. *IEEE Circuits and Systems Magazine*, Third Quarter 2006:21–45.

Press, W. H., Teukolsky, S. A., Vetterling, W. T., and Flannery, B. P. (2002). *Numerical Recipes in C++.* Cambridge University Press, second edition.

Price, D., Knerr, S., Personnaz, L., and Dreyfus, G. (1995). Pairwise neural network classifiers with probabilistic outputs. In Fischler, M. A. and Firschein, O., editors, *Neural Information Processing Systems*, pages 1109–1116. MIT Press.

Prokop, R. J. and Reeves, A. P. (1992). A survey of moment-based techniques for unoccluded object representation and recognition. *Graphical Models and Image Processing*, 54(5):438–460.

Quam, L. H. (1987). Hierarchical warp stereo. In Fischler, M. A. and Firschein, O., editors, *Readings in Computer Vision*, pages 80–86. Morgan Kaufmann.

Radke, R. J., Andra, S., Al-Kofahi, O., and Roysam, B. (2005). Image change detection algorithms: a systematic survey. *IEEE Transactions on Image Processing*, 14(4):294–307.

Ranchin, T. and Wald, L. (2000). Fusion of high spatial and spectral resolution images: the ARSIS concept and its implementation. *Photogrammetric Engineering and Remote Sensing*, 66(1):49–61.

Reddy, B. S. and Chatterji, B. N. (1996). An FFT-based technique for translation, rotation and scale-invariant image registration. *IEEE Transactions on Image Processing*, 5(8):1266–1271.

Redner, R. A. and Walker, H. F. (1984). Mixture densities, maximum likelihood and the EM algorithm. *SIAM Review*, 26(2):195–239.

Reed, I. S. and Yu, X. (1990). Adaptive multiple band CFAR detection of an optical pattern with unknown spectral distribution. *IEEE Transactions on Acoustics, Speech, and Signal Processing*, 38(10):1760–1770.

Riano, D., Chuvieco, E., Salas, J., and Aguado, I. (2003). Assessment of different topographic corrections in LANDSAT-TM data for mapping vegetation types. *IEEE Transactions on Geoscience and Remote Sensing*, 41(5):1056–1061.

Richards, J. A. (2009). *Remote Sensing with Imaging Radar*. Springer.

Richards, J. A. and Jia, X. (2006). *Remote Sensing Digital Image Analysis (4th Ed.)*. Springer.

Ripley, B. D. (1996). *Pattern Recognition and Neural Networks*. Cambridge University Press.

Schaum, A. and Stocker, A. (1997). Spectrally selective target detection. In *Proceedings of the International Symposium on Spectral Sensing Research*.

Schölkopf, B., Smola, A., and Müller, K.-R. (1998). Nonlinear component analysis as a kernel eigenvalue problem. *Neural Computation*, 10(5):1299–1319.

Schott, J. R., Salvaggio, C., and Volchok, W. J. (1988). Radiometric scene normalization using pseudo-invariant features. *Remote Sensing of Environment*, 26:1–16.

Schowengerdt, R. A. (1997). *Remote Sensing, Models and Methods for Image Processing*. Academic Press, second edition.

Schroeder, T. A., Cohen, W. B., Song, C., Canty, M. J., and Zhiqiang, Y. (2006). Radiometric calibration of Landsat data for characterization of early successional forest patterns in western Oregon. *Remote Sensing of Environment*, 103(1):16–26.

Schwenk, H. and Bengio, Y. (2000). Boosting neural networks. *Neural Computation*, 12(8):1869–1887.

Settle, J. J. (1996). On the relation between spectral unmixing and subspace projection. *IEEE Transactions on Geoscience and Remote Sensing*,

34(4):1045–1046.

Shah, S. and Palmieri, F. (1990). Meka: a fast, local algorithm for training feed-forward neural networks. *Proceedings of the International Joint Conference on Neural Networks, San Diego*, I(3):41–46.

Shawe–Taylor, J. and Cristianini, N. (2004). *Kernel Methods for Pattern Analysis*. Cambridge University Press.

Shekarforoush, H., Berthod, M., and Zerubia, J. (1995). *Subpixel Image Registration by Estimating the Polyphase Decomposition of the Cross-power Spectrum*. Technical report 2707, Institut National de Recherche en Informatique et en Automatique (INRIA).

Siegel, S. S. (1965). *Nonparametric Statistics for the Behavioral Sciences*. McGraw-Hill.

Singh, A. (1989). Digital change detection techniques using remotely-sensed data. *International Journal of Remote Sensing*, 10(6):989–1002.

Smith, S. M. and Brady, J. M. (1997). SUSAN: a new approach to low-level image processing. *International Journal of Computer Vision*, 23(1):45–78.

Solem, J. E. (2012). *Programming Computer Vision with Python*. O'Reilly.

Strang, G. (1989). Wavelets and dilation equations: a brief introduction. *SIAM Review*, 31(4):614–627.

Strang, G. and Nguyen, T. (1997). *Wavelets and Filter Banks*. Wellesley-Cambridge Press, second edition.

Stuckens, J., Coppin, P. R., and Bauer, M. E. (2000). Integrating contextual information with per-pixel classification for improved land cover classification. *Remote Sensing of Environment*, 71(2):82–96.

Sulsoft (2003). *AsterDTM 2.0 Installation and User's Guide*. Technical report, SulSoft Ltd, Porto Alegre, Brazil.

Tao, C. V. and Hu, Y. (2001). A comprehensive study of the rational function model for photogrammetric processing. *Photogrammetric Engineering and Remote Sensing*, 67(12):1347–1357.

Teague, M. (1980). Image analysis by the general theory of moments. *Journal of the Optical Society of America*, 70(8):920–930.

Teillet, P. M., Guindon, B., and Goodenough, D. G. (1982). On the slope-aspect correction of multispectral scanner data. *Canadian Journal of Remote Sensing*, 8(2):84–106.

Theiler, U. and Matsekh, A. M. (2009). *Total Least Squares for Anomalous Change Detection*. Technical Report LA-UR-10-01285, Los Alamos National Laboratory.

Tran, T. N., Wehrens, R., and Buydens, L. M. C. (2005). Clustering multispectral images: a tutorial. *Chemometrics and Intelligent Laboratory Systems*, 77:3–17.

Tsai, V. J. D. (1982). Evaluation of multiresolution image fusion algorithms. In *Proceedings of the International Symposium on Remote Sensing of Arid and Semi-Arid Lands,* Cairo, Egypt, pages 599–616.

van Niel, T. G., McVicar, T. R., and Datt, B. (2005). On the relationship between training sample size and data dimensionality: Monte Carlo analysis of broadband multi-temporal classification. *Remote Sensing of Environment*, 98(4):468–480.

von Luxburg, U. (2006). *A Tutorial on Spectral Clustering.* Technical Report TR-149, Max-Planck-Institut für biologische Kybernetik.

Vrabel, J. (1996). Multispectral imagery band sharpening study. *Photogrammetric Engineering and Remote Sensing*, 62(9):1075–1083.

Wang, Z. and Bovik, A. C. (2002). A universal image quality index. *IEEE Signal Processing Letters*, 9(3):81–84.

Weiss, S. M. and Kulikowski, C. A. (1991). *Computer Systems That Learn.* Morgan Kaufmann.

Welch, R. and Ahlers, W. (1987). Merging multiresolution SPOT HRV and LANDSAT TM data. *Photogrammetric Engineering and Remote Sensing*, 53(3):301–303.

Westra, E. (2013). *Python Geospatial Development (2nd edition).* Packt Publishing.

Wiemker, R. (1997). An iterative spectral-spatial Bayesian labeling approach for unsupervised robust change detection on remotely sensed multispectral imagery. In *Proceedings of the 7th International Conference on Computer Analysis of Images and Patterns*, volume LCNS 1296, pages 263–370.

Winkler, G. (1995). *Image Analysis, Random Fields and Dynamic Monte Carlo Methods.* Applications of Mathematics. Springer.

Wu, T.-F., Lin, C.-J., and Weng, R. C. (2004). Probability estimates for multi-class classification by pairwise coupling. *Journal of Machine Learning Research*, 5:975–1005.

Xie, H., Hicks, N., Keller, G. R., Huang, H., and Kreinovich, V. (2003). An ENVI/IDL implementation of the FFT-based algorithm for automatic image registration. *Computers and Geosciences*, 29:1045–1055.

Yang, X. and Lo, C. P. (2000). Relative radiometric normalization performance for change detection from multi-date satellite images. *Photogrammetric Engineering and Remote Sensing*, 66:967–980.

Yocky, D. A. (1996). Artifacts in wavelet image merging. *Optical Engineering*, 35(7):2094–2101.

Index

D